24 Update in Intensive Care and Emergency Medicine

Edited by J. L. Vincent

Springer
Berlin
Heidelberg
New York
Barcelona
Budapest
Hong Kong
London
Milan
Paris
Santa Clara
Singapore
Tokyo

M. P. Fink D. Payen (Eds.)

Role of Nitric Oxide in Sepsis and ARDS

With 89 Figures and 32 Tables

Springer

Series Editor

Prof. Dr. Jean-Louis Vincent
Clinical Director, Department of Intensive Care
Erasme University Hospital
Route de Lennik 808, 1070 Brussels, Belgium

Volume Editors

Prof. Dr. M. P. Fink
Department of Surgery, Beth Israel Hospital
330 Brookline Avenue, Boston, MA 02215, USA

Prof. Dr. Didier Payen
Department of Anesthesiology and Intensive Care,
Lariboisière University Hospital, 2 rue Ambroise Paré,
75475 Paris Cedex 10, France

ISBN-13: 978-3-642-79922-8 e-ISBN-13: 978-3-642-79920-4
DOI: 10.1007/978-3-642-79920-4

Die Deutsche Bibliothek – CIP-Einheitsaufnahme
Role of nitric oxide in sepsis and ARDS: with 32 tables / M. P. Fink ; D. Payen (ed.). – Berlin ; Heidelberg ; New York ; Barcelona ; Budapest ; Hong Kong ; London ; Milan ; Paris ; Tokyo : Springer, 1995
(Update in intensive care and emergency medicine ; 24)

NE: Fink, Mitchell P. [Hrsg.]; GT

Softcover reprint of the hardcover 1st edition 1995

SPIN: 10507062 19/3130-5 4 3 2 1 0 – Printed on acid-free paper

Table of Contents

Nitric Oxide Inhalation in RDS and ARDS

List of Contributors

Abman S. H.
Dept. of Pediatrics, Children's Hospital, 1056 E. 19th Street, Denver CO 80218-1088, USA

Archer S. L.
Dept. of Medicine, Minneapolis VA Medical Center, 1 Veterans Drive, Minneapolis MN 55417, USA

Bennett E. D.
Dept. of Intensive Care, St George's Hospital, Blackshaw Road, London SW17 0RE, United Kingdom

Billiar T. R.
Dept. of Surgery, University of Pittsburgh, 200 Lothrop Street, Pittsburgh PA 15213-2582, USA

Booke M.
Dept. of Anesthesiology, The University of Texas Medical Branch, Galveston TX 77555-0833, USA

Brunet F.
Dept. of Intensive Care, Cochin Hospital, 27 rue du Faubourg Saint-Jacques, 75679 Paris cedex 14, France

Cobb J. P.
Dept. of Critical Care, National Institutes of Health, Building 10 room 7D43, Bethesda MD 20892, USA

Cohen J.
Dept. of Infectious Diseases and Bacteriology, Royal Postgraduate Medical School, Du Cane Road, London W12 0NN, United Kingdom

Danner R. L.
Dept. of Critical Care, National Institutes of Health, Building 10 room 7D43, Bethesda MD 20892, USA

Dhainaut J. F.
Dept. of Intensive Care, Cochin Hospital, 27 rue du Faubourg Saint-Jacques, 75679 Paris cedex 14, France

Dinh-Xuan A. T.
Dept. of Physiology, Cochin Hospital, 27 rue du Faubourg Saint-Jacques, 75679 Paris cedex 14, France

Evans T. W.
Dept. of Thoracic Medicine, National Heart and Lung Institute, Dovehouse Street, London SW3 6NP, United Kingdom

Falke K. J.
Dept. of Anesthesiology, Rudolf Virchow University Hospital, Augustenburger Platz 1, 13344 Berlin, Germany

Fink M. P.
Dept. of Surgery, Beth Israel Hospital, 330 Brookline Avenue, Boston MA 02215, USA

Fonseca G. A.
Dept. of Genito-Urinary Oncology, The University of Texas MD Anderson Cancer Center, 1515 Holcombe Boulevard, Houston, TX 77030, USA

Frostell C. G.
Dept. of Anesthesiology and Intensive Care, Danderyd Hospital, 182 88 Danderyd, Sweden

Gatecel C.
Dept. of Intensive Care, Lariboisière Hospital, 2 rue Ambroise Paré, 75010 Paris, France

Gerlach H.
Dept. of Anesthesiology, Rudolf Virchow University Hospital, Augustenburger Platz 1, 13344 Berlin, Germany

Griffiths M. J. D.
Dept. of Thoracic Medicine, National Heart and Lung Institute, Dovehouse Street, London SW3 6NP, United Kingdom

Gustafsson L. E.
Dept. of Physiology and Pharmacology, Karolinska Institute, 171 77 Stockholm, Sweden

Hurford W. E.
Dept. of Anesthesiology, Massachusetts General Hospital, 32 Fruit Street, Boston MA 02114, USA

Kilbourn R. G.
Dept. of Genito-Urinary Oncology, The University of Texas MD Anderson Cancer Center, 1515 Holcombe Boulevard, Houston, TX 77030, USA

Kirsch J. R.
Dept. of Anesthesiology, The Johns Hopkins Medical Institutions, 600 North Wolfe Street, Baltimore MD 21287-4963, USA

Kong R.
Dept. of Intensive Care, Lariboisière Hospital, 2 rue Ambroise Paré, 75010 Paris, France

Kumar A.
Dept. of Medicine, University of Wisconsin Hospital and Clinics, Madison WI 53792-3240, USA

Leeman M.
Dept. of Intensive Care, Erasme University Hospital, route de Lennik 808, 1070 Brussels, Belgium

Lefer A. M.
Dept. of Physiology, Jefferson Medical College, 1020 Locust Street, Philadelphia PA 19107-6799, USA

Lönnqvist P. A.
Dept. of Physiology and Pharmacology, Karolinska Institute, 171 77 Stockholm, Sweden

Mayer B.
Dept. of Pharmacology, Karl-Franzens University, Universitätsplatz 2, 8010 Graz, Austria

Mercier J. C.
Dept. of Intensive Care, Robert Debré Hospital, 48 Boulevard Sérurier, 75019 Paris, France

Michelakis E.
Dept. of Medicine, Minneapolis VA Medical Center, 1 Veterans Drive, Minneapolis MN 55417, USA

Parratt J. R.
Dept. of Physiology, University of Strathclyde, Royal College, Glasgow G1 1XW, United Kingdom

Parrillo J. E.
Dept. of Medicine, Rush-Presbyterian-St Luke's Medical Center, 1653 West Congress Parkway, Chicago, IL 60612, USA

Payen D. M.
Dept. of Intensive Care, Lariboisière Hospital, 2 rue Ambroise Paré, 75010 Paris, France

Persson M. G.
Dept. of Physiology and Pharmacology, Karolinska Institute, 171 77 Stockholm, Sweden

Preiser J. C.
Dept. of Intensive Care, Erasme University Hospital, route de Lennik 808, 1070 Brussels, Belgium

Radomski M. W.
Dept. of Pharmacology, Heritage Medical Research Center, University of Alberta, Edmonton Alberta T6G 2S2, Canada

Rossaint R.
Dept. of Anesthesiology, Rudolf Virchow University Hospital, Augustenburger Platz 1, 13344 Berlin, Germany

Salas E.
Dept. of Pharmacology, Heritage Medical Research Center, University of Alberta, Edmonton Alberta T6G 2S2, Canada

Salzman A. L.
Dept. of Surgery, Beth Israel Hospital, 330 Brookline Avenue, Boston MA 02215, USA

Shah A. M.
Dept. of Cardiology, University of Wales College of Medicine, Heath Park, Cardiff CF4 4XN, United Kingdom

Shears L. L.
Dept. of Surgery, University of Pittsburgh, 200 Lothrop Street, Pittsburgh PA 15213-2582, USA

Shultz P. J.
Dept. of Medicine, Minneapolis VA Medical Center, 1 Veterans Drive, Minneapolis MN 55417, USA

Sriskandan S.
Dept. of Infectious Diseases and Bacteriology, Royal Postgraduate Medical School, Du Cane Road, London W12 0NN, United Kingdom

Stoclet J. C.
Dept. of Pharmacology, Université Louis Pasteur, BP 10F, 67048 Strasbourg, France

Thiemermann C.
Dept. of Pharmacology, St Bartholomew's Hospital Medical College, Chartherhouse Square, London EC1M 6BQ, United Kingdom

Traber D. L.
Dept. of Anesthesiology, The University of Texas Medical Branch, Galveston TX 77555-0833, USA

Traber L. D.
Dept. of Anesthesiology, The University of Texas Medical Branch, Galveston TX 77555-0833, USA

Traystman R. J.
Dept. of Anesthesiology, The Johns Hopkins Medical Institutions, 600 North Wolfe Street, Baltimore MD 21287-4963, USA

Unno N.
Dept. of Surgery, Beth Israel Hospital, 330 Brookline Avenue, Boston MA 02215, USA

Van Dervort A. L.
Dept. of Critical Care, National Institutes of Health, Building 10 room 7D43, Bethesda MD 20892, USA

Vincent J. L.
Dept. of Intensive Care, Erasme University Hospital, route de Lennik 808, 1070 Brussels, Belgium

Wilkinson J.
Dept. of Infectious Diseases and Bacteriology, Royal Postgraduate Medical School, Du Cane Road, London W12 0NN, United Kingdom

Zapol W. M.
Dept. of Anesthesiology, Massachusetts General Hospital, 32 Fruit Street, Boston MA 02114, USA

Zhang H.
Dept. of Intensive Care, Erasme University Hospital, route de Lennik 808, 1070 Brussels, Belgium

Common Abbreviations

ADP	Adenosine diphosphate
AGP	Alpha-acid glycoprotein
ALI	Acute lung injury
ARDS	Acute respiratory distress syndrome
ATP	Adenosine triphosphate
cAMP	Cyclic adenosine monophosphate
CBF	Cerebral blood flow
cDNA	Chromo-deoxyribonucleic acid
cGMP	3′,5′-Cyclic guanosine monophosphate
cNOS	Constitutive nitric oxide synthase
COPD	Chronic obstructive pulmonary disease
D-NMMA	N^G-monomethyl-D-arginine
DO_2	Oxygen delivery
EDRF	Endothelium-derived relaxing factor
EKG	Electrocardiogram
eNOS	Endothelial cells nitric oxide synthase
FAD	Flavin adenine dinucleotide
FiO_2	Inspired oxygen fraction
FMN	Flavin mononucleotide
GI	Gastrointestinal
GTP	Guanosine triphosphate
H_4 biopterin	Tetrahydrobiopterin
ICAM-1	Intercellular adhesion molecule
ICU	Intensive care unit
IFN	Interferon gamma
IL	Interleukin
iNOS	Inducible nitric oxide synthase
L-NAME	N^G-nitro-L-arginine methyl esther
L-NMMA	N^G-monomethyl-L-arginine
L-NAA	N^W-amino-L-arginine

L-NNA	N^W-nitro-L-arginine
LPS	Lipopolysaccharide
LVEF	Left ventricular ejection fraction
MCA	Middle cerebral artery
MCAT	Middle cationic amino acid transporter
MODS	Multiple organ dysfunction syndrome
MOF	Multiple organ failure
mRNA	Messenger ribonucleic acid
NAC	N-acetylcysteine
NADPH	Nicotinamide adenine dinucleotide phosphate
nNOS	Neurones nitric oxide synthase
NO	Nitric oxide
NOS	Nitric oxide synthase
PG	Prostaglandin
PMN	Polymorphonuclear leukocytes
rmIFN	Recombinant murine interferon gamma
SIN-1	3-morpholinosydnonimine
SIRS	Systemic inflammatory response syndrome
SNAP	S-nitroso-N-acetylpenicillamine
TGF	Transforming growth factor
TNF	Tumor necrosis factor
TX	Thromboxane
VO_2	Oxygen consumption/uptake

Biochemistry of Nitric Oxide and Regulation of Gene Expression

Biosynthesis of Nitric Oxide: An Overview

B. Mayer

Introduction

Soon after the first reports on mammalian nitric oxide (NO) synthesis in 1987 [1, 2], cytosols of cytokine-activated macrophages were shown to contain an enzymatic activity catalyzing an NADPH-dependent conversion of the amino acid L-arginine to NO [3]. Subsequently, enzymatic NO formation was described for a variety of tissues. Consistent with the apparent involvement of Ca^{2+} in NO-mediated signal transduction in blood vessels [4] and neurons [5], NO synthesis turned out to require micromolar concentrations of free Ca^{2+} in endothelial cells [6] and brain [7, 8], whereas enzymatic NO formation occurred in a Ca^{2+}-independent manner in cytokine-activated macrophages [3]. In the meantime our knowledge about this novel biochemical pathway has rapidly grown. The present chapter briefly reviews molecular mechanisms and regulation of NO biosynthesis with special focus on the enzymology of NO-synthesizing enzymes, the NO synthases (NOS).

NO Synthase Isoforms

NO is synthesized enzymatically from the amino acid L-arginine by different NOS (EC 1.14.13.39) isozymes which have been purified and cloned from a number of species including man (for recent reviews see [9–11]). Chromosomal mapping of the genes encoding NOS hints at the existence of three distinct gene products [12]. Two of these NOS isozymes are constitutively expressed in neurones (nNOS) and vascular endothelial cells (eNOS), while expression of a third isoform (iNOS) is induced by various cytokines in macrophages and a number of other nucleated mammalian cells, including hepatocytes, vascular smooth muscle, and glial cells. Since expression of nNOS and eNOS is not confined to neurons and endothelial cells, respectively, it has been suggested to use an alternative nomenclature (type I–III) (Table 1) for unambiguous identification of the three isozymes [13].

Constitutive NOS are Ca^{2+}/calmodulin-dependent enzymes and physiologically activated by hormones or neurotransmitters enhancing the intracellular concentrations of free Ca^{2+}. The Ca^{2+}-independent iNOS contains cal-

Table 1. NO synthase isoforms

Isoform	Abbrev.	Type	MW (kDa)	Ca^{2+}-dep.	Con-stitutive	Cells and tissues
neuronal	nNOS	I	160	yes	yes	– central and peripheral neurons – skeletal muscle
endothelial	eNOS	III	130	yes	yes	– endothelial cells – hippocampal neurons – kidney and bronchiolar epithelium
inducible	iNOS	II	130	no	no (?)	– macrophages, glia – hepatocytes, myocytes – smooth muscle

For a comprehensive review of iNOS expression in different tissues see [13]

modulin as a tightly bound subunit even under nominally Ca^{2+}-free conditions [14], suggesting that all NOS are dependent on calmodulin, but that only the constitutive isoforms require micromolar Ca^{2+} for calmodulin binding. Conceivably, the post-translational assembly of the iNOS-calmodulin complex could be regulated by specific mechanisms and provide a means for cells to control the activity of iNOS.

Mechanism of NO Biosynthesis

The different iNOS catalyze the same reaction and exhibit similar biochemical properties. As shown in Figure 1, the reaction is a two step redox-process. L-arginine is hydroxylated to the intermediate N^G-hydroxy-L-arginine first, which is not released in significant quantities from the enzyme but immedia-

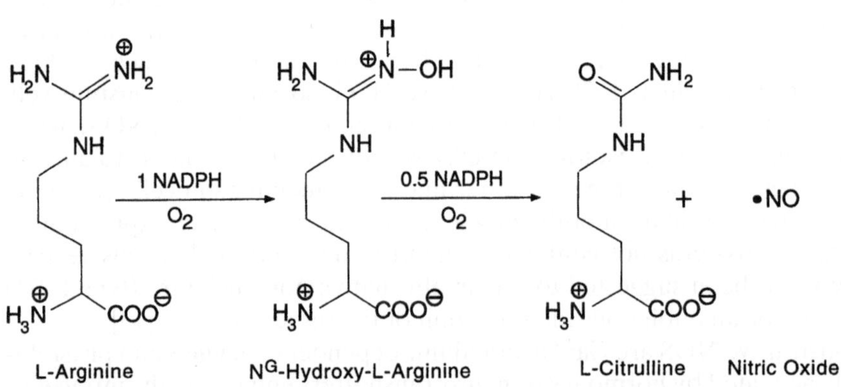

Fig. 1. NO synthase-catalyzed oxidation of L-arginine to L-citrulline and NO

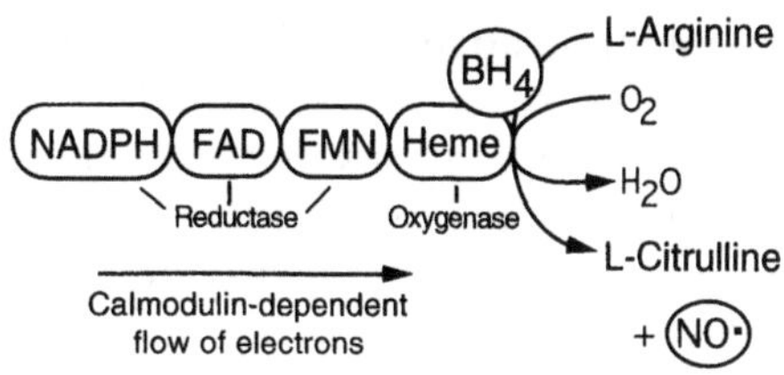

Fig. 2. Prosthetic groups and catalytic domains of NO synthase (see common abbreviations pp. XI, XII)

tely undergoes oxidative cleavage to yield NO and L-citrulline [15, 16]. The oxygen atoms incorporated during each of the two reaction steps are derived from molecular oxygen [17, 18]. The possible reaction mechanisms of enzymatic NO formation were reviewed in detail recently [19–21].

Using different enzyme preparations, several laboratories have shown that NOS contain ferroprotoporphyrin IX (heme) and the reduced flavins FAD and FMN as prosthetic groups [22–29]. The catalytically active heme iron is involved in reductive activation of molecular oxygen and substrate binding [16, 30, 31]. The required reducing equivalents are derived from NADPH and shuttled in a calmodulin-dependent manner through the flavins to the heme [32, 33]. Thus, as schematically shown in Figure 2, NOS resemble the two-component microsomal cytochrome P-450 hydroxylating systems, which consist of a FAD- and FMN-containing cytochrome P-450 reductase shuttling NADPH-derived electrons to the prosthetic heme group of tightly associated cytochromes P-450 [34, 35]. Indeed, the C-terminal half of NOS shows pronounced sequence similarities to cytochrome P-450 reductase [36], and the bidomain structure of NOS was recently confirmed by separation of the reductase and oxygenase domains subsequent to tryptic cleavage of the rat brain enzyme [37]. As such a "self-sufficient" cytochrome P-450, NOS is unique among mammalian enzymes, but similar proteins are found in lower organisms, such as *bacillus megaterium* [38, 39] and *fusarius oxysporum* [40].

Tetrahydrobiopterin as a Cofactor in NO Biosynthesis

Tetrahydrobiopterin (H_4 biopterin) is known since the early 1960s as cofactor of monoxygenases hydroxylating the aromatic amino acids phenylalanine, tyrosine, and tryptophan [41, 42]. In 1989, two groups have identified H_4 biopterin as cofactor of macrophage iNOS [43, 44]. Several analytical methods, including mass spectrometry, revealed presence of micromolar concentrations of H_4 biopterin in soluble fractions of the cytokine-activated cells. Subsequent to separation from low molecular mass components, macrophage iNOS was stimulated by added H_4 biopterin but showed no absolute requirement for the exogenous pteridine. In a later study [45], we found that nNOS purified from porcine brain also exhibited an apparently pteridine-independent basal activity which was increased about 4-fold upon addition of H_4 biopterin. These initial studies have suggested that H_4 biopterin might positively modulate NO synthesis without being an essential cofactor in the reaction. Nevertheless, determination of reduced biopterins in purified preparations of nNOS revealed that H_4 biopterin co-purified with the enzyme, indicating that the endogenously present pteridine may have accounted for the basal enzyme activity observed in the absence of the added cofactor [22]. Conceivably, the amount of H_4 biopterin remaining tightly bound to isolated NOS could depend on the techniques applied for tissue homogenization and/or protein purification, possibly explaining why other laboratories have reported on preparation of apparently H_4 biopterin-independent NOS [46–48]. In the meantime, the essential role of H_4 biopterin in NO biosynthesis is well established, and positive correlation of basal NOS activity with the amount of endogenous H_4 biopterin has been confirmed with iNOS purified from rat [49] and human [50] cerebellum, activated macrophages [51, 52], cultured endothelial cells [29], Sf9 cells infected with rat brain NOS-recombinant baculovirus [53], and in the slime mold *physarum polycephalum* [54].

NOS appear to have fairly high affinities for H_4 biopterin, since nanomolar (nM) concentrations were found to be sufficient for half-maximal stimulation of the different isoforms. Binding studies using tritium-labelled H_4 biopterin as radioligand revealed a KD of 40 nM and an unusually high selectivity of the pteridine site of nNOS, as the 6S-stereoisomer of H_4 biopterin as well as several synthetic pteridine derivatives, which are active cofactors of aromatic amino acid hydroxylases, were only poor antagonists of H_4 biopterin binding [55]. This high affinity of the pteridine site could be the reason why H_4 biopterin copurifies with NOS, but the kinetic binding data do not support this view. Based on the calculated half-life of ≈ 2 min for dissociation of the radioligand, NOS should have been purified as virtually H_4 biopterin-free apoenzyme. Moreover, H_2 biopterin and 7-nitroindazole potently antagonized functional effects as well as binding of exogenous H_4 biopterin but failed to displace the endogenously bound cofactor [55, 56].

These results suggest that purification of NOS may yield two functionally intact and otherwise indistinguishable species of nNOS: the holoenzyme containing irreversibly bound H_4 biopterin, and a pteridine-deficient form which can be reconstituted in a reversible manner by addition of H_4 biopterin. The reason for this apparent heterogeneity of purified NOS is unclear. Neither it is known which of the two postulated species occurs physiologically in cells and which one represents an artifact produced during work-up of the tissue. These questions may be of considerable practical relevance with respect to the pharmacological effects of pteridine derivatives being currently developed to serve as isoform-selective NOS inhibitors. Since the different iNOS exhibit similar affinities for H_4 biopterin, one would not expect a pronounced isoform selectivity of those drugs. However, if the H_4 biopterin-containing holoenzyme should turn out as the predominant species of NOS occurring physiologically, H_4 biopterin antagonists should not have any acute effects on NO formation by constitutive NOS, but could prevent expression of functionally intact iNOS during inflammatory processes.

Enzymatic Formation of Reactive Oxygen Species by NOS

Binding of Ca^{2+}/calmodulin to nNOS activates the enzyme to oxidize NADPH and transfer the reducing equivalents to molecular oxygen [22]. The electron shuttle does not require the presence of L-arginine and leads to generation of superoxide anion and hydrogen peroxide when the concentrations of the substrate are low [57, 58]. This NADPH oxidase activity of nNOS is sensitive to miconazole and cyanide, which do not block the cytochrome c reductase activity of the enzyme, suggesting that oxygen activation is catalyzed by the heme and not the flavins [16, 30]. Exogenous H_4 biopterin does not affect rates of hydrogen peroxide formation catalyzed by pteridine-deficient nNOS in the absence of L-arginine [59], suggesting that NOS, in contrast to other pteridine-dependent monooxygenases [42], does not require H_4 biopterin as a cofactor for reductive oxygen activation. Accordingly, both the substrate L-arginine and the cofactor H_4 biopterin have to be present in saturating concentrations to efficiently prevent formation of reactive oxygen species by nNOS and probably also by eNOS (see below). Macrophage iNOS, however, exhibits only poor NADPH oxidase activity in the absence of L-arginine, indicating that uncoupled oxygen reduction of iNOS is down-regulated to prevent continuous generation of oxygen radicals by the Ca^{2+}-independent isozyme [33].

Two reports [60, 65] indicate that the enzymatic formation of reactive oxygen species we and others have observed with purified NOS does also occur in intact cells under conditions of reduced availability of L-arginine or H_4 biopterin. Culcasi et al. [60] have shown that N-methyl-D-aspartate-stimulated cultured neurones released superoxide anions in a NOS-dependent manner when the L-arginine concentrations in the culture medium were decreased by treatment with arginase. In the brain, free arginine is located predominantly in glial cells [61], suggesting that neurotransmitter-induced activation of nNOS may require transfer of the NOS substrate from astrocytes to neurones [62]. Conceivably, inefficient supply of L-arginine could result in nNOS-mediated generation of superoxide, hydrogen peroxide, and peroxynitrite, reactive species thought to contribute to excitotoxicity in the brain [63, 64]. More recently, Consentino and Katusic [65] have provided evidence that H_4 biopterin-depletion of canine coronary arteries may result in eNOS-catalyzed formation of hydrogen peroxide, which apparently mediates endothelium-dependent relaxations in the pteridine-depleted tissues. As the authors have pointed out "this initially compensatory response, triggered by a dysfunctional (endothelial) nitric oxide synthase, may represent an important mechanism underlying oxidative vascular injury" [65].

Feedback Inhibition of NOS by NO

An important issue presumably pertinent to the physiology and clinical application of NO is feedback inhibition of NOS. It is known that NO exhibits considerably high affinity for ferrous heme, and one of the most pronounced biological effects of NO, the activation of soluble guanylyl cyclase, results from an interaction with the prosthetic heme group of the cyclase [66, 67]. On the other hand, NO blocks different iron-containing enzymes [68] and has been described as potent inhibitor of microsomal cytochromes P-450 [69–71]. In keeping with the identity of NOS as cytochrome P-450, all three isozymes are inhibited by NO [72–76], apparently due to binding of NO to the prosthetic heme group of NOS [77]. It has been suggested that H_4 biopterin may function as cofactor preventing feedback inhibition of brain NOS [78], but recent data from my laboratory indicate that the effect of H_4 biopterin is due to an autooxidation reaction resulting in generation of superoxide which rapidly reacts with NO to give peroxynitrite [79]. Our study showed that feedback inhibition of NOS does occur even in the presence of H_4 biopterin when superoxide is rapidly inactivated to hydrogen peroxide by including high amounts of superoxide dismutase in the enzyme assays. However, it is unclear at present whether NO concentrations may become sufficiently high in cells to feedback to NOS, and to what extend superoxide-mediated inactivation of NO occurs under physiological or pathological conditions.

Effects of Inhibitors on the Multiple Catalytic Functions of NOS

As briefly described in this chapter, NOS exhibits several catalytic functions which involve different prosthetic groups. Reduction of cytochrome c needs only NADPH and the reduced flavins as cofactors, whereas the heme moiety is additionally requisite for uncoupled oxygen activation and thus for enzymatic generation of hydrogen peroxide. Finally, H_4 biopterin is required for coupling of heme-catalyzed oxygen activation to L-arginine metabolism. Comparative measurements of these different catalytic activities has proven useful for investigating the molecular mechanisms underlying the effects of various NOS inhibitors. The compounds listed in Table 2 bind in an L-arginine-competitive manner either to the substrate site of NOS or to the heme, and do not affect flavin-mediated electron transfer as revealed by the lack of

Table 2. Effects of L-arginine-competitive inhibitors on the different catalytic functions of nNOS [16, 55, 56, 80–84]

Inhibitor	Full activity (K_i, μM)	NADPH oxidase (K_i or IC_{50}, μM)	Cytochrome c reductase
N^G-methyl-L-arginine	0.17	>1000	not inhibited
N^G-nitro-L-arginine	0.17	0.20	not inhibited
L-thiocitrulline	0.06–2.0	3.5	not inhibited
Imidazole	260	280	not inhibited
7-nitroindazole	2.8	0.28	not inhibited

inhibition of cytochrome c reduction. It may be of pharmacological and clinical relevance that some of the L-arginine-based inhibitors, e.g. N^G-nitro-L-arginine and its methyl ester, block the substrate-independent NADPH oxidase (superoxide forming) activity of NOS, whereas N^G-methyl-L-arginine and other L-arginine derivatives do not. Inhibition of uncoupled oxygen activation by N^G-nitro-L-arginine is apparently due to a rather slow, reversible inactivation of the constitutive isoforms of NOS [80, 81] and to interference of the inhibitor with heme iron reduction [82]. L-Thiocitrulline exerts similar effects, but enzyme inhibition is much faster in onset and more rapidly reversed than the effects of the nitro compounds [82, 83]. Imidazole, a classical inhibitor of cytochrome P-450, binds to the heme iron of NOS resulting in inhibition of oxygen reduction and NO synthesis. L-arginine competitive inhibition of NOS by this heme-site inhibitor demonstrates the tight interaction of the substrate binding domain with the heme [84]. This was further confirmed by experiments involving 7-nitroindazole, an antinociceptive drug with *in vivo* selectivity for inhibition of nNOS [85]. Albeit structurally unrelated to any of the NOS substrates or cofactors, 7-nitroindazole apparently competes for L-arginine and H_4 biopterin binding to nNOS [56]. Inhibition of uncoupled oxygen activation, which neither requires L-arginine nor H_4 biopterin suggests that the drug binds to the heme and thereby antagonizes substrate and cofactor binding. Since all NOS isoforms exhibit comparable sensitivity to inhibition by 7-nitroindazole, its *in vivo* selectivity may be due to tissue-specific mechanisms involved in uptake or metabolism of the drug.

Conclusion

Recent work on the molecular mechanisms of NO biosynthesis has involved biochemical and biophysical characterization of different NOS isoforms at the protein level as well as studies on the complex regulation of gene transcription and mRNA translation. These studies revealed that NOS are proteins of unique complexity whose activity and expression levels are tightly regulated by a variety of physiological and pathophysiological processes. Even though our knowledge about regulatory principles of NO synthesis has rapidly grown during the past few years, the field still rapidly develops. The astute combination of pharmacology, biochemistry, and molecular biology holds promise that we will soon understand more precisely how mammalian cells regulate their NO production and to what extent disorders in the NO system contribute to the pathogenesis of human disease.

Acknowledgement: The experimental work referred to from the author's laboratory was supported by the Fonds zur Förderung der Wissenschaftlichen Forschung in Austria.

References

1. Palmer RMJ, Ferrige AG, Moncada S (1987) Nitric oxide release accounts for the biological activity of endothelium-derived relaxing factor. Nature 327:524–526
2. Ignarro LJ, Buga GM, Wood KS, Byrns RE, Chaudhuri G (1987) Endothelium-derived relaxing factor produced and released from artery and vein is nitric oxide. Proc Natl Acad Sci USA 84:9265–9269
3. Marletta MA, Yoon PS, Iyengar R, Leaf CD, Wishnok JS (1988) Macrophage oxidation of L-arginine to nitrite and nitrate: Nitric oxide is an intermediate. Biochemistry 27:8706–8711
4. Moncada S, Palmer RM, Higgs EA (1989) Biosynthesis of nitric oxide from L-arginine: A pathway for the regulation of cell function and communication. Biochem Pharmacol 38:1709–1715
5. Garthwaite J, Charles SL, Chess-Williams R (1988) Endothelium-derived relaxing factor release on activation of NMDA receptors suggests role as intercellular messenger in the brain. Nature 336:385–388
6. Mayer B, Böhme E (1989) Ca^{2+}-dependent formation of an L-arginine-derived activator of soluble guanylyl cyclase in bovine lung. FEBS Lett 256:211–214
7. Bredt DS, Snyder SH (1989) Nitric oxide mediates glutamate-linked enhancement of cGMP levels in the cerebellum. Proc Natl Acad Sci USA 86:9030–9033
8. Knowles RG, Palacios M, Palmer RMJ, Moncada S (1989) Formation of nitric oxide from L-arginine in the central nervous system: A transduction mechanism for stimulation of the soluble guanylate cyclase. Proc Natl Acad Sci USA 86:5159–5162
9. Marletta MA (1993) Nitric oxide synthase structure and mechanism. J Biol Chem 268: 12231–12234
10. Mayer B (1993) Molecular characteristics and enzymology of nitric oxide synthase and soluble guanylyl cyclase in the CNS. Semin Neurosci 5:197–205
11. Knowles RG, Moncada S (1994) Nitric oxide synthases in mammals. Biochem J 298: 249–258
12. Nathan C, Xie QW (1994) Regulation of biosynthesis of nitric oxide. J Biol Chem 269:13725–13728
13. Förstermann U, Closs EI, Pollock JS, et al (1994) Nitric oxide synthase isozymes: Characterization, purification, molecular cloning, and functions. Hypertension 23: 1121–1131
14. Cho HJ, Xie QW, Calaycay J, et al (1992) Calmodulin is a subunit of nitric oxide synthase from macrophages. J Exp Med 176:599–604
15. Stuehr DJ, Kwon NS, Nathan CF, Griffith OW, Feldman PL, Wiseman J (1991) N omega-hydroxy-L-arginine is an intermediate in the biosynthesis of nitric oxide from L-arginine. J Biol Chem 266:6259–6263
16. Klatt P, Schmidt K, Uray G, Mayer B (1993) Multiple catalytic functions of brain nitric oxide synthase. Biochemical characterization, cofactor requirement and role of N^G-hydroxy-L-arginine as an intermediate. J Biol Chem 268:14781–14787
17. Kwon NS, Nathan CF, Gilker C, Griffith OW, Matthews DE, Stuehr DJ (1990) L-citrulline production from L-arginine by macrophage nitric oxide synthase. The ureido oxygen derives from dioxygen. J Biol Chem 265:13442–13445
18. Leone AM, Palmer RMJ, Knowles RG, Francis PL, Ashton DS, Moncada S (1991) Constitutive and inducible nitric oxide synthases incorporate molecular oxygen into both nitric oxide and citrulline. J Biol Chem 266:23790–23795
19. Feldman PL, Griffith OW, Stuehr DJ (1993) The surprising life of nitric oxide. Chem Eng News 71:26–38
20. Korth HG, Sustmann R, Thater C, Butler AR, Ingold KU (1994) On the mechanism of the nitric oxide synthase-catalyzed conversion of N-omega-hydroxy-L-arginine to citrulline and nitric oxide. J Biol Chem 269:17776–17779
21. Marletta MA (1994) Nitric oxide synthase: Aspects concerning structure and catalysis. Cell 78:927–930

22. Mayer B, John M, Heinzel B, et al (1991) Brain nitric oxide synthase is a biopterin- and flavin-containing multifunctional oxido-reductase. FEBS Lett 288:187–191
23. Stuehr DJ, Cho HJ, Kwon NS, Weise MF, Nathan CF (1991) Purification and characterization of the cytokine-induced macrophage nitric oxide synthase: An FAD- and FMN-containing flavoprotein. Proc Natl Acad Sci USA 88:7773–7777
24. Hevel JM, White KA, Marletta MA (1991) Purification of the inducible murine macrophage nitric oxide synthase. Identification as a flavoprotein. J Biol Chem 266: 22789–22791
25. White KA, Marletta MA (1992) Nitric oxide synthase is a cytochrome-P-450 type hemoprotein. Biochemistry 31:6627–6631
26. Stuehr DJ, Ikeda-Saito M (1992) Spectral characterization of brain and macrophage nitric oxide synthases: Cytochrome-P-450-like heme proteins that contain a flavin semiquinone radical. J Biol Chem 267:20547–20550
27. McMillan K, Bredt DS, Hirsch DJ, Snyder SH, Clark, JE, Masters BSS (1992) Cloned, expressed rat cerebellar nitric oxide synthase contains stoichiometric amounts of heme, which binds carbon monoxide. Proc Natl Acad Sci USA 89:11141–11145
28. Klatt P, Schmidt K, Mayer B (1992) Brain nitric oxide synthase is a haemoprotein. Biochem J 288:15–17
29. Pollock JS, Werner ER, Mitchell JA, Förstermann U (1993) Particulate endothelial nitric oxide synthase: Requirement and content of tetrahydrobiopterin, FAD, and FMN. Endothelium 1:147–152
30. Wolff DJ, Datto GA, Samatovicz RA, Tempsick RA (1993) Calmodulin-dependent nitric oxide synthase: Mechanism of inhibition by imidazole and phenylimidazoles. J Biol Chem 268:9425–9429
31. McMillan K, Masters BSS (1993) Optical difference spectrophotometry as a probe of rat brain nitric oxide synthase heme-substrate interaction. Biochemistry 32:9875–9880
32. Klatt P, Heinzel B, John M, Kastner M, Böhme E, Mayer B (1992) Ca^{2+}/calmodulin-dependent cytochrome c reductase activity of brain nitric oxide synthase. J Biol Chem 267:11374–11378
33. Abu-Soud HM, Stuehr DJ (1993) Nitric oxide synthases reveal a novel role for calmodulin in controlling electron transfer. Proc Natl Acad Sci USA 90:10769–10772
34. Guengerich FP (1991) Reactions and significance of cytochrome P-450 enzymes. J Biol Chem 266:10019–10022
35. Degtyarenko KN, Archakov AI (1993) Molecular evolution of P-450 superfamily and P-450-containing monooxygenase systems. FEBS Lett 332:1–8
36. Bredt DS, Hwang PM, Glatt CE, Lowenstein C, Reed RR, Snyder SH (1991) Cloned and expressed nitric oxide synthase structurally resembles cytochrome P-450 reductase. Nature 351:714–718
37. Sheta EA, McMillan K, Masters BSS (1994) Evidence for a bidomain structure of constitutive cerebellar nitric oxide synthase. J Biol Chem 269:15147–15153
38. Narhi LO, Fulco AJ (1986) Characterization of a catalytically self-sufficient 119,000-dalton cytochrome P-450 monoxygenase induced by barbiturates in *bacillus megaterium*. J Biol Chem 261:7160–7169
39. Fulco AJ (1991) P-450-BM-3 and other inducible bacterial P-450 cytochromes: Biochemistry and regulation. Annu Rev Pharmacol Toxicol 31:177–203
40. Nakayama N, Shoun H (1994) Fatty acid hydroxylase of the fungus Fusarium oxysporum is possibly a fused protein of cytochrome P-450 and its reductase. Biochem Biophys Res Commun 202:586–590
41. Nichol CA, Smith GK, Duch DS (1985) Biosynthesis and metabolism of tetrahydrobiopterin and molybdopterin. Annu Rev Biochem 54:729–764
42. Kaufman S (1993) New tetrahydrobiopterin-dependent systems. Annu Rev Nutr 13: 261–286
43. Tayeh MA, Marletta MA (1989) Macrophage oxidation of L-arginine to nitric oxide, nitrite, and nitrate. Tetrahydrobiopterin is required as a cofactor. J Biol Chem 264: 19654–19658

44. Kwon NS, Nathan CF, Stuehr DJ (1989) Reduced biopterin as a cofactor in the generation of nitrogen oxides by murine macrophages. J Biol Chem 264:20496–20501
45. Mayer B, John M, Böhme E (1990) Purification of a Ca^{2+}/calmodulin-dependent nitric oxide synthase from porcine cerebellum. Cofactor-role of tetrahydrobiopterin. FEBS Lett 277:215–219
46. Bredt DS, Snyder SH (1990) Isolation of nitric oxide synthetase, a calmodulin-requiring enzyme. Proc Natl Acad Sci USA 87:682–685
47. Knowles RG, Palacios M, Palmer RM, Moncada S (1990) Kinetic characteristics of nitric oxide synthase from rat brain. Biochem J 269:207–210
48. Dwyer MA, Bredt DS, Snyder SH (1991) Nitric oxide synthase. Irreversible inhibition by L-N^G-nitroarginine in brain *in vitro* and *in vivo*. Biochem Biophys Res Commun 176:1136–1141
49. Schmidt HHHW, Smith RM, Nakane M, Murad F (1992) Ca^{2+}/calmodulin-dependent NO synthase type-I: A biopteroflavoprotein with Ca^{2+}/calmodulin-independent diaphorase and reductase activities. Biochemistry 31:3243–3249
50. Klatt P, Heinzel B, Mayer B, et al (1992) Stimulation of human nitric oxide synthase by tetrahydrobiopterin and selective binding of the cofactor. FEBS Lett 305:160–162
51. Hevel JM, Marletta MA (1992) Macrophage nitric oxide synthase: Relationship between enzyme-bound tetrahydrobiopterin and synthase activity. Biochemistry 31: 7160–7165
52. Baek KJ, Thiel BA, Lucas S, Stuehr DJ (1993) Macrophage nitric oxide synthase subunits: Purification, characterization, and role of prosthetic groups and substrate in regulating their association into a dimeric enzyme. J Biol Chem 268:21120–21129
53. Harteneck C, Klatt P, Schmidt K, Mayer B (1994) Baculovirus-mediated expression of rat brain nitric oxide synthase and characterization of the purified enzyme. Biochem J 304:683–686
54. Werner-Felmayer G, Golderer G, Werner ER, Gröbner P, Wachter H (1994) Pteridine biosynthesis and nitric oxide synthase in *Physarum Polycephalum*. Biochem J 304: 105–111
55. Klatt P, Schmid M, Leopold E, Schmidt K, Werner ER, Mayer B (1994) The pteridine binding site of brain nitric oxide synthase: Tetrahydrobiopterin binding kinetics, specificity, and allosteric interaction with the substrate domain. J Biol Chem 269: 13861–13866
56. Mayer B, Klatt P, Werner ER, Schmidt K (1994) Molecular mechanism of inhibition of porcine brain nitric oxide synthase by the antinociceptive drug 7-nitroindazole. Neuropharmacology 33:1253–1259
57. Heinzel B, John M, Klatt P, Böhme E, Mayer B (1992) Ca^{2+}/calmodulin-dependent formation of hydrogen peroxide by brain nitric oxide synthase. Biochem J 281:627–630
58. Pou S, Pou WS, Bredt DS, Snyder SH, Rosen GM (1992) Generation of superoxide by purified brain nitric oxide synthase. J Biol Chem 267:24173–24176
59. Mayer B, Heinzel B, Klatt P, John M, Schmidt K, Böhme E (1992) Nitric oxide synthase-catalyzed activation of oxygen and reduction of cytochromes: Reaction mechanisms and possible physiological implications. J Cardiovasc Pharmacol 20:S54–S56
60. Culcasi M, Lafoncazal M, Pietri S, Bockaert J (1994) Glutamate receptors induce a burst of superoxide via activation of nitric oxide synthase in arginine-depleted neurons. J Biol Chem 269:12589–12593
61. Aoki E, Sembar R, Mikoshiba K, Kashiwamata S (1991) Predominant localization in glial cells of free L-arginine. Immunocytochemical evidence. Brain Res 547:190–192
62. Pow DV (1994) Immunocytochemical evidence for a glial localisation of arginine, and a neuronal localisation of citrulline in the rat neurohypophysis: Implications for nitrergic transmission. Neurosci Lett 181:141–144
63. Coyle JT, Puttfarcken P (1993) Oxidative stress, glutamate, and neurodegenerative disorders. Science 262:689–695
64. Lipton SA, Choi YB, Pan ZH, et al (1993) A redox-based mechanism for the neuroprotective and neurodestructive effects of nitric oxide and related nitroso-compounds. Nature 364:626–632

65. Cosentino F, Katusic ZS (1995) Tetrahydrobiopterin and dysfunction of endothelial nitric oxide synthase in coronary arteries. Circulation 91:139–144
66. Ignarro LJ (1991) Heme-dependent activation of guanylate cyclase by nitric oxide: A novel signal transduction mechanism. Blood Vessels 28:67–73
67. Wedel B, Humbert P, Harteneck C, et al (1994) Mutation of His-105 of the β1-subunit yields a nitric oxide-insensitive form of soluble guanylyl cyclase. Proc Natl Acad Sci USA 91:2592–2596
68. Henry Y, Lepoivre M, Drapier J-C, Ducrocq C, Boucher J-L, Guissani A (1993) EPR characterization of molecular targets for NO in mammalian cells and organelles. FASEB J 7:1124–1134
69. Wink DA, OsawaY, Darbyshire JF, Jones CR, Eshenaur SC, Nims RW (1993) Inhibition of cytochromes-P450 by nitric oxide and a nitric oxide-releasing agent. Arch Biochem Biophys 300:115–123
70. Khatsenko OG, Gross SS, Rifkind AB, Vane JR (1993) Nitric oxide is a mediator of the decrease in cytochrome-P450-dependent metabolism caused by immunostimulants. Proc Natl Acad Sci USA 90:11147–11151
71. Stadler J, Trockfeld J, Schmalix WA, et al (1994) Inhibition of cytochromes P-450-1A by nitric oxide. Proc Natl Acad Sci USA 91:3559–3563
72. Rogers NE, Ignarro LJ (1992) Constitutive nitric oxide synthase from cerebellum is reversibly inhibited by nitric oxide formed from L-arginine. Biochem Biophys Res Commun 189:242–249
73. Buga GM, Griscavage JM, Rogers NE, Ignarro LJ (1993) Negative feedback regulation of endothelial cell function by nitric oxide. Circ Res 73:808–812
74. Griscavage JM, Rogers NE, Sherman MP, Ignarro LJ (1993) Inducible nitric oxide synthase from a rat alveolar macrophage cell line is inhibited by nitric oxide. J Immunol 151:6329–6337
75. Assreuy J, Cunha FQ, Liew FY, Moncada S (1993) Feedback inhibition of nitric oxide synthase activity by nitric oxide. Br J Pharmacol 108:833–837
76. Rengasamy A, Johns RA (1993) Regulation of nitric oxide synthase by nitric oxide. Mol Pharmacol 44:124–128
77. Wang JL, Rousseau DL, Abu-Soud HM, Stuehr DJ (1994) Heme coordination of NO in NO synthase. Proc Natl Acad Sci USA 91:10512–10516
78. Griscavage JM, Fukuto JM, Komori Y, Ignarro LJ (1994) Nitric oxide inhibits neuronal nitric oxide synthase by interacting with the heme prosthetic group: Role of tetrahydrobiopterin in modulating the inhibitory action of nitric oxide. J Biol Chem 269:21644–21649
79. Mayer B, Klatt P, Werner ER, Schmidt K (1995) Kinetics and mechanism of tetrahydrobiopterin-induced oxidation of nitric oxide. J Biol Chem 270:655–659
80. Mayer B, Schmid M, Klatt P, Schmidt K (1993) Reversible inactivation of endothelial nitric oxide synthase by N^G-nitro-L-arginine. FEBS Lett 333:203–206
81. Klatt P, Schmidt K, Brunner F, Mayer B (1994) Inhibitors of brain nitric oxide synthase. Binding kinetics, metabolism, and enzyme inactivation. J Biol Chem 269:1674–1680
82. Abu-Soud HM, Feldman PL, Clark P, Stuehr DJ (1994) Electron transfer in the nitric-oxide synthases: Characterization of L-arginine analogs that block heme iron reduction. J Biol Chem 269:32318–32326
83. Frey C, Narayanan K, McMillan K, et al (1994) L-thiocitrulline: A stereospecific, heme-binding inhibitor of nitric-oxide synthases. J Biol Chem 269:26083–26091
84. Mayer B, Klatt P, Werner ER, Schmidt K (1994) Identification of imidazole as L-arginine-competitive inhibitor of porcine brain nitric oxide synthase. FEBS Lett 350:199–202
85. Moore PK, Wallace P, Gaffen Z, Hart SL, Babbedge RC (1993) Characterization of the novel nitric oxide synthase inhibitor 7-nitroindazole and related indazoles: Antinociceptive and cardiovascular effects. Br J Pharmacol 110:219–224

Biochemistry and Synthesis of NO in Sepsis

L. L. Shears and T. R. Billiar

Introduction

Nitric oxide (NO), although credited as being the smallest biologic product of mammalian cells, has generated a tidal wave of interest and investigation regarding its role as a fundamental intercellular messenger and mediator of such complex pathophysiologic conditions as sepsis. Going from relative obscurity, this molecule has become the focus of nearly 2000 publications per year, earning it the dubious title "molecule of the year" in 1992 by *Science* (Washington DC). Though nitrogen oxides were suggested to be synthetic products of mammalian cells as early as 1916 by Mitchell et al. [1], it was not until the 1980s that NO was considered to be more than a toxic atmospheric gas. The aim of this chapter will be to discuss the biochemistry and synthesis of NO, its regulation and the potential cytotoxic and cytoprotective roles in which it is involved during sepsis.

Biochemistry and Bidsynthesis of NO

NO is a free radical which can freely diffuse across cell membranes. NO is produced from one of the guanidino nitrogens of L-arginine [2, 3] by a family of NO synthases forming the free radical NO, with L-citrulline being a byproduct of this reaction (Fig. 1). NO is an unstable molecule having a half-life of only a few seconds (< 6 sec). The lability of NO is due to its rapid conversion to one of its stable metabolites, nitrate and nitrite, in the presence of oxygen or oxygen radicals such as superoxide. In aqueous solution, NO rapidly reacts with superoxide to form peroxynitrite, a powerful oxidant

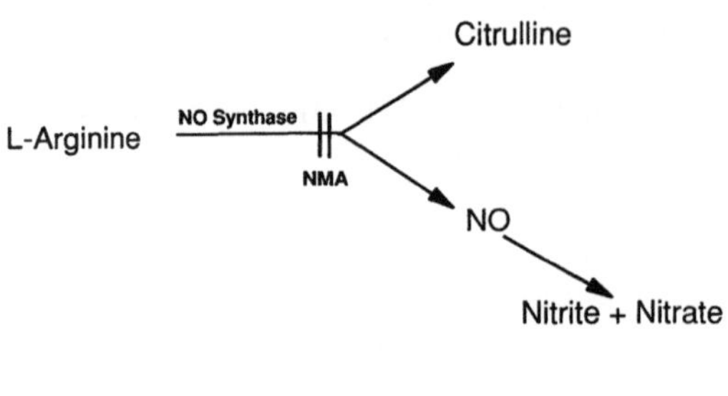

Fig. 1. Nitric oxide pathway. NO is synthesized from L-arginine by the enzyme NO synthase, forming the free-radical NO and citrulline as byproducts. NO is rapidly oxidized to the stable endproducts nitrite and nitrate. NMA competitively inhibits NO synthase

which is less diffusible than NO which can also cause tissue damage but alternatively may reduce the toxic potential of O_2-radicals by degrading it to non-toxic endproducts. NO has three redox forms which determine its reactivity and properties [4]. NO can exist as neutral nitric oxide ($^{\cdot}NO$), a nitrosonium cation (NO^+), or a nitrosyl anion (NO^-). The nitrosonium form of NO reacts with nucleophiles such as aromatic compounds and electron rich bases resulting in the nitrosylation of these compounds. Nitrosylation reactions produce metallo-nitrosyl compounds which can mediate physiological responses or pathologic changes such as carcinogenesis. The neutral form of NO avidly binds the metalloproteins such as hemoglobin and cytochromes involved in the electron transport system, resulting in inactivation of NO but also modulating the function of some of these proteins such as the inactivation of certain electron transport enzymes. A functional role for the nitrosyl anion has not yet been described.

NO is the product of the family of enzymes referred to as NO synthases (NOS). A major portion of NOS resembles cytochrome P-450 structurally with binding sites for flavin adenine mononucleotide, flavin adenine dinucleotide, and heme. Currently, three isoforms of this enzyme have been isolated, purified and the cDNAs and genes cloned. These three isoforms vary in subcellular location, amino acid sequences, regulation and functional roles. Two of these isoforms are continuously present and are termed constitutive NOS (cNOS). Since always present, the constitutive NOS enzymes can be immediately activated to produce small amounts of NO, which allow it to serve in the physiologic regulatory activities of various tissues. The cNOS group can be further subdivided into endothelial (eNOS) and neuronal isoforms (nNOS). The neuronal cNOS cDNA has been cloned from rat and human cerebellum [5, 6], while endothelial cNOS cDNA has been cloned from bovine [7–9] and human endothelial cells [10, 11]. The gene for neuronal cNOS has been mapped to chromosome 12 in humans while the endothelial cNOS has been mapped to chromosome 7. eNOS is mostly membrane bound via a myristylation site while nNOS is located in the cytosol of central and peripheral neurons. The activation of both cNOS isoforms is short lived allowing the NO generated to serve as a diffusible signaling molecule, thus mediating numerous physiological processes. The activity of cNOS isoforms is dependent on calmodulin binding to the enzymes after an elevation in resting intracellular calcium levels. Endothelial NOS possesses both paracrine and autocrine activity which is believed to be instrumental in keeping vessels in a chronic state of vasodilation which requires approximately 60% of the norepinephrine produced daily by the body to counteract its vasodilatory effects. The vasodilatory activity of NO is also counteracted by endothelin, the most potent vasoconstrictor known. NO also inhibits platelet and neutrophil adhesion to endothelial cells and possibly inhibits vascular smooth muscle proliferation as will be discussed later. Neuronal cNOS likely is involved in cell-to-cell communication, memory formation, neurotransmission and excitotoxicity.

The third isoform is not typically present in resting cells but instead the expression of the gene and the protein must be induced, hence the term indu-

Table 1. Cell types documented to express iNOS (non-human)

Macrophages	Pancreatic islet cells
Renal mesangial cells	EMT6 adenocarcinoma cell line
Kupffer cells	Chondrocytes
Hepatocytes	Fibroblasts
Vascular smooth muscle cells	Keratinocytes
Cardiac myocytes	Renal tubule cells
Inflammatory neutrophils	Splenocytes
Lymphocytes	Bone marrow cells
Endothelial cells	Astrocytes

Table 2. Human nitric oxide synthase isoforms

Isoform	Chromosome (human)	Tissue proposed	Functions
eNOS	7	endothelial cells	– vasomotor regulation – platelet homeostasis – inhibition of smooth muscle proliferation
nNOS	12	neuronal tissue skeletal muscle	– cell-to-cell communication – memory formation – neurotransmission – excitotoxicity
		bronchial cells cpithelial cells	
iNOS	17	monocytes macrophages hepatocytes bone marrow keratinocytes cardiac myocytes pancreatic beta islets	antitumor antimicrobial hepatic cytoprotection immune modulation autoimmune injury free radical scavenger

cible NOS (iNOS). Though initially found in macrophages, nearly all mammalian tissues can be induced to form iNOS (Table 1). Like nNOS, iNOS is also a cytosolic enzyme. The iNOS gene has recently been mapped to chromosome 17 in the human [12]. The iNOS gene consists of 26 exons and 25 introns. Translation of the mRNA produces a polypeptide of 1153 amino acids with a calculated molecular mass of 131 kD (Fig. 2). The active enzyme is a homodimer.

Table 2 compares the actions of the constitutive and inducible isoforms. NO production by iNOS provides the host with antitumor and antimicrobial activity. In striking contrast, NO produced by iNOS also has cytoprotective actions. Unlike the cNOS isoforms, iNOS expression results in a sustained

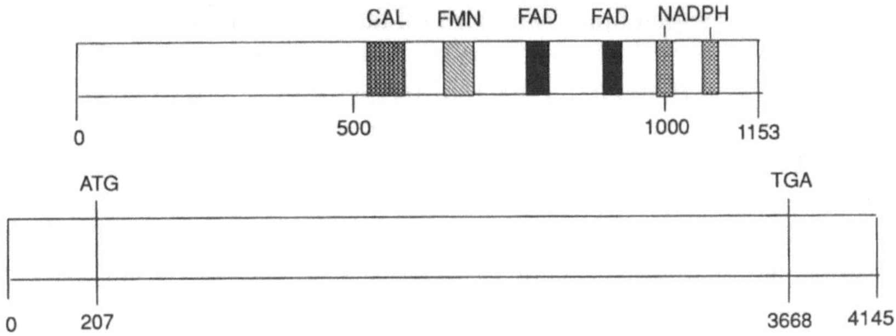

Fig. 2. Schematic diagram of human iNOS cDNA and iNOS protein. The top structure is the 1153-amino acid human hepatocyte iNOS protein. Putative cofactor binding sites are indicated as follows: CAL, calmodulin; FMN, flavin mononucleotide; FAD, flavin adenine dinucleotide; and NADPH, reduced nicotinamide-adenine dinucleotide phosphate. The bottom structure represents the cDNA for human hepatocyte iNOS. ATG = start codon; TGA = stop codon

production of NO for the life of the activated enzyme which may last up to 10 h. This sustained activation of iNOS expression is likely due to the presence of a calmodulin molecule tightly bound to the iNOS subunits. This results in a lack of need for elevations in intracellular calcium concentrations for enzyme activation. Thus, due to the large amount of NO which can be synthesized after activation of iNOS, it is the inducible isoform of NOS which is most likely the mediator of many of the responses attributed to NO during sepsis. For example, the sustained production of large amounts of NO by iNOS accounts for the severity of the refractory hypotension observed during septic insults which often times is unresponsive to even large amounts of vasopressors. Thus, we will confine most of this discussion to this particular isoform and how it is regulated during sepsis.

Regulation of iNOS Gene Expression during Sepsis

Without stimulation by endotoxin or cytokines, cells normally do not express iNOS. Stimulation of iNOS expression is usually in response to perturbations which generate inflammatory or immunologic reactions resulting in endotoxin and cytokine release. Certain agents including 5,6-dimethylxanthenone-4-acetic acid, picolinic acid, cyclic adenosine monophosphaste (cAMP) elevating agents, UV-light, ozone, and gram-positive microbes have also been shown to induce or at least synergize in the induction of iNOS by lipopolysaccharide (LPS) or cytokines. The diversity of stimuli which can induce NOS would suggest that the mechanism of iNOS gene activation is likely quite complex. A promoter/enhancer region for the iNOS gene has been identified in murine macrophages. This region includes a TATA box 30 base pairs upstream of the mRNA transcription site along with at least 24 oligonucleotide elements in the proximal 1.7 kilobases which are homologous to consensus sequences which bind transcription factors involved in the inducing other genes by cytokines. LPS induction of iNOS in murine macrophages is dependent on a NF-kB sequence located between nucleotides −85 to −76 with binding to this region by a cycloheximide-sensitive complex containing both the p50/c-rel and p50/RelA heterodimers of NF-kB, along with additional, unidentified nuclear proteins. Recent work in our laboratory has failed to demonstrate NF-kB dependence for the induction of the human form of the iNOS gene in hepatocytes and further work is currently underway to identify the promoter/enhancer regions in the human iNOS isoform.

NO production following cell stimulation is delayed for a few hours while the iNOS enzyme and possibly essential cofactors are synthesized. The regulation of gene expression following induction can occur via several mechanisms, one of which is upregulation of the transcriptional and translational processes of the gene of interest. Induction of NO production in murine macrophages has been shown to be at the transcriptional level by nuclear run-on analysis, indicating that a major part of iNOS regulation includes the upregulation of gene transcription. The stabilization of the newly transcribed messenger RNA may also contribute to increased mRNA. Detectable levels of iNOS mRNA following stimulation of cultured human hepatocytes with LPS does not occur until 3–4 h after stimulation, and maximal levels are not detected for 6–8 h, demonstrating the probable need for both transcription and translation prior to enzyme production and expression. LPS, the component of the cell wall of gram-negative bacteria responsible for the induction of the sepsis syndrome, has been demonstrated to be an important inducer of iNOS synthesis both *in vitro* and *in vivo*. LPS is known to trigger macrophage release of many cytokines *in vivo* including tumor necrosis factor (TNF), interleukin-1 (IL-1) , IL-6 and interferon-gamma (IFN-γ) which are involved in the synergistic activation of iNOS gene expression. *In vitro* studies by Geller et al. [13] demonstrated that INF-γ and TNF-α acted synergistically to increase iNOS expression. IL-6 when given alone was shown not to induce iNOS mRNA synthesis, but when added to the cytokine combination of INF-γ and TNF-α resulted in a further 25% increase in mRNA production. Induction of rodent macrophage iNOS mRNA by LPS plus IFN-γ also reflects an increase in iNOS gene transcription without changes in iNOS mRNA stability.

In response to LPS and cytokines such as IL-1, IL-6 and TNF, the liver produces large amounts of acute phase reactants. Geller et al. [13] were able to demonstrate that, although the expression of acute phase reactants and NO are induced by similar mediators, the expression of hepatocyte iNOS is not a part of the acute phase response induced by remote inflammation, and that iNOS is differentially regulated from the acute phase proteins. This determination was made by measuring the gene expression of iNOS as well as the acute phase reactant, alpha-acid glycoprotein (AGP), by Northern blot analysis after exposing rat hepatocytes *in vivo* to LPS or to the acute phase response produced by hindlimb turpentine injection. AGP and iNOS mRNA were upregulated in response to LPS administration. However, only AGP mRNA was upregulated by the administration of the hindlimb turpentine injection. This demonstrates that iNOS expression is not necessarily coinduced with the acute phase reactants.

The regulation of iNOS expression (reviewed in [14]) can also occur by altering the stability of the newly formed mRNA. INF-γ has no effect on basal iNOS transcription in murine macrophages but does result in an increase in iNOS mRNA due to increased iNOS mRNA stability. Transforming growth factor-β (TGF-β) suppresses murine macrophage expression of iNOS by decreasing the stability of newly formed iNOS mRNA and reducing translational efficiency. The cytokines IL-4, IL-8, IL-10, and macrophage deactivating factor also suppress iNOS activity in murine macrophages possibly by downregulating transcriptional processes and reducing mRNA stability. Dexamethasone, which is released during acute phase responses, nearly abolishes iNOS expression in a number of cell types including hepatocytes possibly acting after transcription has taken place.

As mentioned previously, iNOS activity is dependent upon the availability of the cofactors NADPH, flavin, and tetrahydrobiopterin. Thus, regulation of NO synthesis by iNOS may be dependent upon the availability of certain cofactors and substrates. Arginine for example is the only nitrogen donor which can be used to form NO by NOS. Thus, the availability of arginine could regulate the rate of NO production. Hepatocyte and macrophage uptake of arginine has been shown to be elevated during sepsis or in response to stimulators of iNOS expression i.e. LPS and cytokines. This increase in arginine uptake in macrophages is likely due to the upregulation of the newly described cationic amino acid transporter MCAT-2B [15]. Basal arginine uptake occurs via MCAT-1 which is found constitutively on the cell surface.

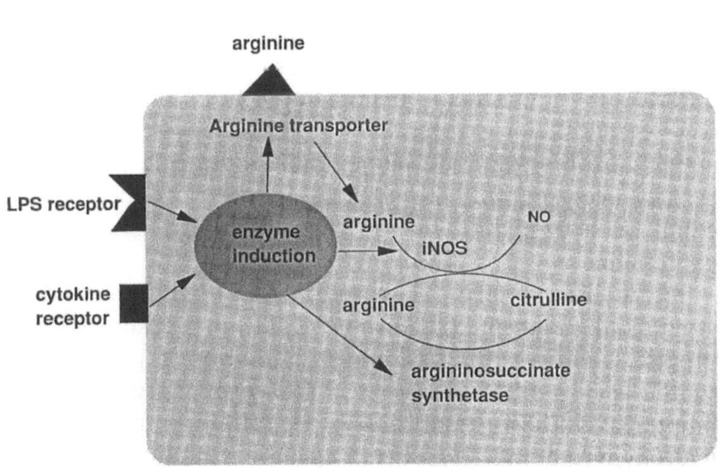

Fig. 3. LPS and cytokines may induce the expression of arginine membrane transporters which increase intracellular arginine uptake. Also induced is argininosuccinate synthetase which promotes the recycling of citrulline to arginine to provide sufficient substrate for NO synthesis

Endotoxin and various cytokines result in the upregulation of MCAT-2B expression. Although MCAT-2B expression is upregulated by the same inducers of iNOS in macrophages, the two are not necessarily coregulated since dexamethasone has been shown to inhibit the induction of iNOS but not MCAT-2B. This upregulation of the MCAT-2B amino acid transporter during sepsis would thus ensure adequate intracellular arginine concentrations for the large production of NO by iNOS during septic insults. A second means of providing adequate intracellular arginine levels to support NO synthesis would be through the recycling of citrulline, the coproduct of NOS-catalyzed metabolism of arginine, to form arginine. This conversion is carried out by the action of argininosuccinate synthetase and argininosuccinate lyase. Nussler et al. [16] demonstrated that LPS and INF-γ coinduce iNOS and argininosuccinate synthetase activity in murine macrophages. This induction of argininosuccinate synthetase was not blocked by NOS inhibitors, indicating that argininosuccinate synthetase induction is not the consequence of depleting cellular arginine levels by the coinduction of NOS. Since serum arginine levels can be limiting for NO synthesis, the upregulation of enzymes necessary for recycling citrulline to provide adequate arginine levels for NO synthesis would likely play a significant role in regulating NO production (Fig. 3).

NADPH is an essential cofactor during the synthesis of NO in that it donates the electrons necessary for NO synthesis. The requirement of NOS for NADPH would suggest that regulation of NADPH availability could play a role in the regulation of NO production by iNOS. One mechanism of NADPH synthesis is via the NADPH-generating pentose-phosphate pathway which is dependent on the rate limiting enzyme glucose-6-phosphate dehydrogenase (G-6-PDH). Addition of glucose, which would upregulate G-6-PDH expression, to cultured pig enterocytes stimulates the conversion of arginine to citrulline [17]. This is consistent with the notion that NADPH availability is increased by the pentose phosphate pathway. Additionally, G-6-PDH activity has been demonstrated to be induced in murine bone marrow-derived macrophages by LPS. LPS-dependent increases in G-6-PDH activity are blocked by cycloheximide indicating that regulation of enzyme activity occurred at a translational or pretranslational step. Thus NADPH could be upregulated by the same mediators shown to stimulate iNOS expression.

Tetrahydrobiopterin (H_4 biopterin), another cofactor for iNOS, is synthesized from guanosine triphoshate (GTP) in a pathway where GTP-cyclohydrolase 1 which is the first and rate-limiting step for H_4 biopterin *de novo* synthesis (Fig. 4). Biosynthesis of NO from L-arginine by NOS requires H_4 biopterin synthesis. Inhibition of H_4 biopterin synthesis by 2,4-diamino-6-hydroxypyrimidine (DAHP), a specific inhibitor for GTP cyclohydrolase, results in a dose-dependent reduction in NO production in smooth muscle cells. By administering authentic H_4 biopterin and sepiapterin, a substrate for the pterin salvage pathway of H_4 biopterin synthesis, restoration of NO synthesis by the macrophages was noted, suggesting that macrophages require

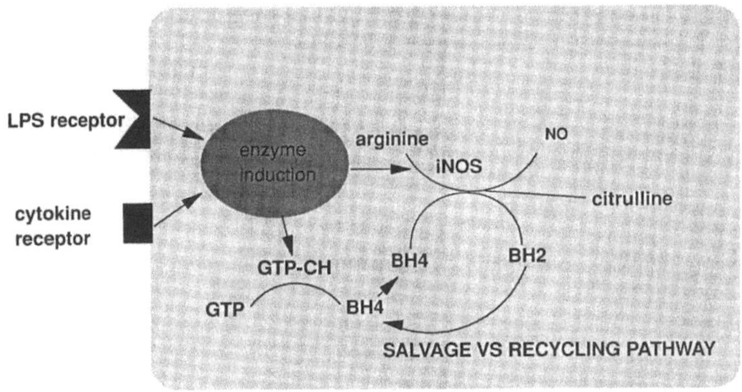

Fig. 4. LPS and cytokines induce production of GTP cyclohydrolase (GTP-CH), the rate limiting step for H_4 biopterin (BH4) synthesis. It remains unclear whether enzymes involved in the salvage pathway are also coinduced

active synthesis of H_4 biopterin to produce NO. H_4 biopterin can also be provided through a recycling pathway that converts oxidized quinoid-dihydrobiopterin to H_4 biopterin or a salvage pathway that converts dihydrobiopterin to H_4 biopterin. Most H_4 biopterins used in the production of NO comes from the *de novo* pathway in smooth muscle cells [18]. In unstimulated rat pulmonary artery smooth muscle cells, GTP-cyclohydrolase levels are very low. However, GTP-cyclohydrolase production is induced when cells are stimulated with LPS and cytokines. Whether the salvage pathway enzymes are upregulated is not yet known. These findings would suggest that the same mediators which induce NO synthesis during sepsis are also involved in increasing the expression of its substrates and cofactors.

Mechanism of NO Action during Sepsis

Several sequelae of the sepsis syndrome have been ascribed to the actions of NO. The cellular actions of NO can result in enzyme activation or enzyme inactivation (Table 3). The cellular responses to these enzyme perturbations are largely determined by which cells are producing the NO and how much NO is synthesized. The small amounts of NO produced by the cNOS isoforms can activate the NO-sensitive enzymes in the cell or surrounding cells. The short half-life of NO allows for only paracrine and autocrine activities to be imparted. Both the short half-life of NO in oxygenated solutions and the binding of NO by hemoglobin prevent any endocrinologic activity by NO. When cells are stimulated to express iNOS, the large amounts of NO produced not only activate NO-sensitive enzymes but can also bind to iron-containing enzymes resulting in their inactivation.

NO produced in endothelial cells can diffuse into the surrounding vascular smooth muscle cells where NO reacts with the ferrous ion in the heme

Table 3. Direct effects of NO on the activity of specific enzymes

Inhibition	Activation
Aconitase	Soluble guanylate cyclase
NADPH-ubiquinone oxidoreductase	Cyclooxygenase
Succinate-ubiquinone oxidoreductase	
Ribonucleotide reductase	
NADPH-oxidase	
Glyceraldehyde-3-phosphate-dehydrogenase	
Cyclooxygenase	
Catalase	

prosthetic group of the soluble guanylate cyclase resulting in enzyme activation, and increased formation of intracellular concentrations of cyclic guanosine monophosphate (cGMP). In vascular smooth muscle cells, increased cGMP impairs calcium entry into the cell which results in vascular smooth muscle relaxation and hence dilation of the vessel. Inhibition of NO with competitive inhibitors such as N^G-monomethyl-L-arginine abolishes the tonic vasodilatory action of eNOS resulting in a corresponding increase in resting blood pressure. In rodents, endothelial cells can also express iNOS in response to TNF and IFN-γ while vascular smooth muscle cells from humans and rodents express iNOS in response to IL-1, TNF, IFN-γ or LPS. Thus expression of this high output enzyme in the vessel wall is thought to account for the excessive vasodilation characteristic of septic shock. Infusion of NO inhibitors to septic patients can also inhibit the production of NO by iNOS resulting in restoration of systemic blood pressures. The induction of iNOS synthesis by cytokines and LPS leads to vascular relaxation which is resistant to vasoconstrictors, and that can be prevented by treating with glucocorticoids and inhibitors of NOS. The magnitude of the increase in NO generation during endotoxic shock is directly related to the degree of hypotension seen clinically. Thus, NO released by iNOS activation accounts for the vasodilation and resistance to vasopressors characteristic of septic shock, as well as the hypotension induced by cytokine therapy in patients with cancer.

Activation of soluble guanylate cyclase is also the mechanism by which NO inhibits platelet adherence and aggregation. NO acts synergistically with prostacyclin, another inhibitor of platelet aggregation which acts by increasing concentrations of cAMP in the platelet, to inhibit platelet aggregation. Unlike prostacyclin, NO also inhibits platelet adhesion to endothelial cells. Platelets themselves are capable of synthesizing NO and inhibit platelet activation through a negative feedback mechanism. These actions on platelets may be beneficial in preventing microvascular thrombosis in various organs during sepsis when blood pressures are low allowing NO to serve in a cytoprotective manner during low perfusion states.

NO can also inhibit leukocyte adherence to endothelial cells. NO and superoxide are believed to be capable of regulating mast cell activation and degranulation, events known to stimulate leukocyte adhesion. NO inhibition of mast cell activation and reduction in superoxide concentrations would reduce the release of pro-inflammatory agents and prevent leukocyte adhesion to endothelial cells.

Studies of murine tumor cell lines have shown that NO inhibits three iron-containing mitochondrial enzymes: aconitase of the Krebs cycle and NADPH-ubiquinone oxidoreductase and succinate-ubiquinone oxidoreductase of the electron transport chain. The inhibition of these enzymes results in reduced cellular respiration. NO also causes the inactivation of glyceraldehyde-3-phosphate dehydrogenase, an enzyme involved in both glycolysis and gluconeogenesis. The consequences of NO-induced inhibition of glyceraldehyde-3-phosphate dehydrogenase are not well established. NO has been shown to profoundly suppress total cellular protein synthesis in hepatocytes by an unknown mechanism. These inhibitory properties of iNOS may promote hepatic dysfunction during sepsis.

Role of NO and NO Inhibitors during Sepsis

Severe sepsis remains a leading cause of mortality in the ICU, contributing to or directly causing >40000 deaths per year in the United States. As our ability to keep patients alive that would previously have died following severe traumatic insults, more and more of these patients will be susceptible to septic events and succumb as a result of these insults. The proposal that NO is the final pathway to the development of hypotension following septic insults has been supported by many *in vitro* and *in vivo* observations in various species including man [19–22]. A review of 39 critically ill trauma and septic patients by Ochoa et al. [23] demonstrated convincingly that systemic nitrite and nitrate levels were significantly elevated and associated with low systemic vascular resistance and increased serum endotoxin levels in septic patients when compared to non-septic trauma patients. Similar findings were observed in patients undergoing oncologic immunotherapy with IL-2. The nitrogenous endproduct levels correlated well with the degree of hypotension in these patients. Further support that increased NO synthesis mediates septic shock [24–27] came from models demonstrating that the use of the NO inhibitor N^G-nitro-L-arginine methyl esther (L-NAME) resulted in dose-dependent increases in blood pressure when given in bolus or continuous infusion forms. Petros et al. [28] demonstrated in a randomized, double blind, placebo-controlled study that the use of the NO inhibitor, N^G monomethyl-L-arginine (L-NMMA), produced a dose-dependent increase in mean arterial pressure, systemic vascular resistance, pulmonary vascular resistance, central venous pressure, and pulmonary arterial occlusion pressure, and a decrease in cardiac output and heart rate. Although reductions in cardiac output could occur as a result of coronary vasoconstriction, no EKG changes were noted in these patients and the reductions in cardiac output were likely a result of the restoration of systemic vascular resistance. Similar findings have been reported in animal models [29–33]. High doses of L-NMMA resulted in an increase in mean arterial pressure of approximately 20 mm Hg lending support to the role of NO in causing the hypotension seen during septic shock. Palmer et al. [34], in an endotoxin-mediated shock model in rats, found that the administration of endotoxin caused an initial transient fall in blood pressure followed by a continuous fall in blood pressure over the next 5 h. Pretreatment of these animals with dexamethasone, which has been demonstrated to inhibit NOS induction, prevented the development of hypotension and shock in these animals. Bolus therapy with the NO inhibitor L-NMMA 90 min after the administration of endotoxin in low doses (10 to 30 mg/kg) also prevented the fall in blood pressure and reduced the levels of plasma nitrates.

Stimulation of iNOS expression in the heart may be associated with impaired contractility. This idea is based on the observation of models of isolated papillary muscles and isolated myocytes where NO has been demonstrated to cause a reduction in cardiac contractility. Whether NO represents a cardiac depressant factor in sepsis remains to be proven.

Although NO production during sepsis is believed to result in hypotension and the deleterious problems associated with hypoperfusion, the use of NO inhibitors in both animal studies and human clinical trials has led to the discovery that NO may also play a role in cytoprotection during sepsis. Killed *Corynebacterium parvum* (C. parvum) injection into rats results in moderate hepatic damage and increases in circulating nitrite and nitrate levels due to chronic hepatocellular inflammation [35–38]. C. parvum injection followed by LPS injection results in massive hepatic necrosis as determined histologically and by measurements of hepatocellular enzymes including aspartate aminotransferase and ornithine carbamoyltransferase. Injection of C. parvum alone followed by L-NMMA administration has no effect on the degree of hepatocellular injury. When C. parvum injection is followed by an LPS bolus, the administration of L-NMMA resulted in a dose-dependent suppression in plasma nitrite and nitrate levels and an increase in the degree of hepatocellular injury. L-arginine partially reversed the changes induced by L-NMMA. This paradox would suggest that NO plays some cytoprotective role during endotoxemia. The use of NO inhibitors in humans mentioned earlier also resulted in increased levels of hepatocellular enzymes although the increases were not statistically significant.

The mechanism behind this cytoprotective role of NO is partially elucidated. It is likely that NO prevents platelet adherence and aggregation preventing microvascular thrombosis during episodes of organ hypoperfusion seen with sepsis. Both NO and prostaglandins, each increased by LPS infusion, are known to act as vasodilators and inhibitors of platelet aggregation. Harbrecht et al. [39] demonstrated that inhibition of NO synthesis with L-NMMA and prostaglandin I_2 (PGI_2) with aspirin acted synergistically to increase hepatic injury. Aspirin alone had no effect. Infusion of PGI_2 decreased the liver damage. It is likely that prostaglandins act in concert with NO to limit the degree of hepatic injury during sepsis. However, NO appears to be more important in maintaining perfusion by this mechanism. A second mechanism by which NO maintains perfusion in solid organs such as the liver is by preventing increases in vascular resistance. We have shown in isolated perfused livers from rats pretreated with LPS, that NO prevents the α-agonist-reduced increases in intrahepatic resistance. It is likely that NO combats the vasoconstrictive actions of circulating catecholamines and other vasoconstrictors elevated in sepsis.

It is also likely that NO acts as an important free radical scavenger and limits the toxicity associated with free radicals such as superoxide. When NO reacts with superoxide radicals formed during sepsis, the byproduct of the reaction is most likely peroxynitrite which can degrade either to less toxic (NO_2^- or NO_3^-) or more toxic ($OH^\cdot$ and $^\cdot NO_2$) products depending on the local environment. This generation of peroxynitrite may be a mode of detoxifying $O_2^{\cdot-}$. The subsequent route of peroxynitrite degradation determines toxicity versus protection. In alkaline conditions, peroxynitrite is a stable compound. However, once it is protonated to form peroxynitrous acid, peroxynitrous acid can rapidly decay to form the non-toxic breakdown products H^+ and NO_3^- or the toxic products NO_2^- and $OH^\cdot$. The route of decomposition is determined by the configuration of the peroxynitrous acid molecule. In the favored low energy cis-configuration, the non-toxic breakdown products are formed. In the higher energy transconfiguration, the formation of toxic byproducts is favored.

It is known that the reactive oxygen species derived from xanthine oxidase, $O_2^{\cdot-}$, H_2O_2, and $^\cdot OH$, as well as peroxynitrite can cause cellular injury through membrane lipid peroxidation. Rubbo et al. [40] have recently demonstrated that NO can mediate the membrane lipid peroxidation caused by $O_2^{\cdot-}$, H_2O_2, and $^\cdot OH$. In this series of experiments, Rubbo delivered either authentic NO or NO released from S-nitroso-N-acetylpenicillamine (SNAP) or S-nitrosoglutathione (GSNO) to lipid oxidation systems exposed to the xanthine oxidase products. NO alone does not induce lipid peroxidation, but at low concentrations NO was shown to stimulate $O_2^{\cdot-}$, H_2O_2, and $^\cdot OH$ lipid oxidation likely through the generation of peroxynitrite. However, at higher rates of NO delivery, NO acted as an antioxidant and in a dose-dependent fashion reduced lipid peroxidation by these reactive oxygen species. This antioxidant effect occurs by terminating the lipid radical chain propagation reactions through direct reaction with the lipid alkoxyl and peroxyl radical products formed when peroxynitrite and the xanthine oxidase products reacted with the lipids. Thus as can be seen, NO can either potentiate the peroxidation of cell membranes or prevent this injury depending upon the conditions within the local environment.

Another proposed mechanism for the cytoprotective role of NO is through the induction of heme-oxygenase synthesis by NO. Cytokine and LPS induction of NO synthesis would result in increased levels of hemeoxygenases which would then metabolize heme and hemoproteins to iron and biliverdin. Increased intracellular iron concentrations would then promote ferritin synthesis. In addition to acting as an iron chelator, ferritin is also able to act as a free radical scavenger which could prevent superoxide, NO, and hydroxyl ion mediated cell injury.

Conclusion

Much more information will be required to determine which reactions mediated by NO are clinically relevant. At present, NO production during sepsis has been demonstrated to play both cytotoxic and cytoprotective roles (Table 4). The processes by which NO can mediate both cytoprotective and cytotoxic actions are clearly complex. As more information is obtained regarding the various activities of NO, we will continue to come closer to finding ways to mediate the hypotension generated by NO during sepsis and possibly not remove the cytoprotection offered by NO. The end result will hopefully be an improved survival for patients suffering septic insults.

Table 4. Potential cytoprotective and cytotoxic mechanisms of NO

Cytoprotection	Cytotoxicity
Inhibition of platelet aggregation	Refractory hypotension
Prevention of microvascular thrombosis	Enzyme inactivation
Inhibition of leukocyte adhesion	Generation of toxic free radicals
Inactivation of oxygen free radicals	
Reduced organ vascular resistance	

References

1. Mitchell HH, Shonle H, Grindley H (1916) The origin of nitrate in the urine. J Biol Chem 24:461
2. Kwon N, Nathan C, Gilker C (1990) L-citrulline production from L-arginine by macrophage nitric oxide synthase: The ureido oxygen derives from dioxygen. J Biol Chem 265:442–445
3. Leone A, Palmer R, Knowles R (1991) Constitutive and inducible nitric oxide synthases incorporate molecular oxygen into both nitric oxide and citrulline. J Biol Chem 266:790–795
4. Kam P, Govender G (1994) Nitric oxide: Basic science and clinical application. Anaesthesia 49:5115–5121
5. Bredt DS, Hwang PM, Glatt CE, Lowenstein C, Reed RR, Snyder SH (1991) Cloned and expressed nitric oxide synthase structurally resembles cytochrome P-450 reductase. Nature 351:714–718
6. Nakane M, Schmidt HHHW, Pollock JS, Forstermann U, Murad F (1993) Cloned human brain nitric oxide synthase is highly expressed in skeletal muscle. FEBS Lett 316:175–180
7. Lamas S, Marsden PA, Li GI, Tempst P, Michel L (1992) Endothelial nitric oxide synthase: Molecular cloning and characterization of a distinct constitutive enzyme isoform. Proc Natl Acad Sci USA 89:6348–6352
8. Sessa WC, Harrison JK, Barber CM (1992) Molecular cloning and expression of cDNA encoding endothelial cell nitric oxide synthase. J Biol Chem 267:15274–15276
9. Nishida K, Harrison DG, Navas JP (1992) Molecular cloning and characterization of the constitutive bovine aortic endothelial cell nitric oxide synthase. J Clin Invest 90:2092–2096

10. Janssens SP, Shimouchi A, Quertermous T, Bloch DB, Block KD (1992) Cloning and expression of a cDNA encoding human endothelium-derived relaxing factor/nitric oxide synthase. J Biol Chem 267:14519–14522
11. Marsden PA, Schappert KT, Chen HS (1992) Molecular cloning and characterization of human endothelial nitric oxide synthase. FEBS Lett 307:287–293
12. Geller D, Lowenstein C, Shapiro R, et al (1993) Molecular cloning and expression of inducible nitric oxide synthase from human hepatocytes. Proc Natl Acad Sci USA 90:3491–3495
13. Geller D, Freeswick P, Nguyen D, Nussler A, et al (1994) Differential induction of nitric oxide synthase in hepatocytes during endotoxemia and the acute-phase response. Arch Surg 129:165–171
14. Morris S, Billiar T (1994) New insights into the regulation of inducible nitric oxide synthesis. Am J Physiol 266:E829–E839
15. Closs E, Lyons R, Mitchell M, Cunningham A (1993) Characterization of the third member of the MCAT family of cationic amino acid transporter. Identification of a domain that determines the transport properties of the MCAT proteins. J Biol Chem 268:20796–20800
16. Morris S, Nakayama D, Nussler A, et al (1995) Co-induction of NO synthase and argininosuccinate synthase gene expression. Implications for regulation of NO synthesis. Proc 3rd Int Meet Biol Nitric oxide (In press)
17. Blachier F, Touil H, Vrillon B, Posho L, Duee P (1991) Stimulation by D-glucose of the direct conversion of arginine to citrulline in enterocytes isolated from pig jejunum. Biochem Biophys Res Commun 177:1171–1177
18. Nakayama D, Geller D, Silvio M, et al (1994) Increased activity of de novo tetrahydrobiopterin synthesis in pulmonary artery smooth muscle cells stimulated to produce nitric oxide. Am J Physiol 266:L455–L460
19. Palmer R, Ashton D, Moncada S (1988) Vascular endothelial cells synthesize nitric oxide from L-arginine. Nature 333:664–666
20. Palmer R, Ferrige A, Moncada S (1987) Nitric oxide release accounts for the biological activity of endothelium-derived relaxing factor. Nature 327:524–525
21. Furchgott R, Zawadzki J (1980) The obligatory role of endothelial cells in the relaxation of arterial smooth muscle by acetylcholine. Nature 288:272–276
22. Furchgott R (1980) Studies on relaxation of rabbit aorta by sodium nitrate: Basis for the proposal that the acid activatable component of the inhibitory factor from retractor penis is inorganic nitrate and the endothelium-derived relaxing factor is nitric oxide. In: Vanhoutte PM (ed) Mechanisms of vasodilatation. Raven, New York, pp 401–414
23. Jacob T, Ochoa J, Udekwu A, et al (1993) Nitric oxide production is inhibited in trauma patients. J Trauma 35:590–597
24. Tannenbaum S, Fett D, Young V (1978) Nitrite and nitrate are formed by endogenous synthesis in the human intestine. Science 200:1487–1489
25. Green L, Tannenbaum S, Goldman P (1981) Nitrate synthesis in the germfree and conventional rat. Science 212:56–58
26. Wagner D, Young V, Tannenbaum S (1983) Mammalian nitrate biosynthesis: Incorporation of NH3 into nitrate is enhanced by endotoxin treatment. Proc Natl Acad Sci USA 80:4518–4521
27. Stuehr D, Marletta M (1985) Mammalian nitrate biosynthesis: Mouse macrophages produce nitrite and nitrate in response to Escherichia coli lipopolysaccharide. Proc Natl Acad Sci USA 82:7738–7742
28. Petros A, Bennett D, Vallance P (1991) Effect of nitric oxide synthase inhibitor on hypotension in patients with septic shock. Lancet 338:1557–1578
29. Billiar T, Curran R, Harbrecht B (1990) Modulation of nitrogen oxide synthesis in vivo: N^G-monomethyl-L-arginine inhibits endotoxin-induced nitrite/nitrate biosynthesis while promoting hepatic damage. J Leuk Biol 48:565–569
30. Kilbourn R, Jubran A, Gross S (1991) Reversal of endotoxin-mediated shock by N^G-methyl- L-arginine, an inhibitor of nitric oxide synthesis. Biochem Biophys Res Commun 172:1132–1138

31. Kilbourn R, Gross S, Jubran A (1990) N^G-methyl-L-arginine inhibits tumor necrosis factor-induced hypotension: Implications for the involvement of nitric oxide. Proc Natl Acad Sci USA 84:3629–3632
32. Robertson F, Offner P, Ciceri D, Becker W, Pruitt B (1994) Detrimental hemodynamic effects of nitric oxide synthase inhibition in septic shock. Arch Surg 129:149–156
33. Kubes P, Kanwar S, Niu X, Gaboury J (1993) Nitric oxide synthesis inhibition induces leukocyte adhesion via superoxide and mast cells. FASEB J 7:1293–1299
34. Palmer R (1993) The discovery of nitric oxide in the vessel wall: A unifying concept in the pathogenesis of sepsis. Arch Surg 128:396–401
35. Billiar T, Hoffman R, Curran R, Langrehr J, Simmons R (1992) A role for inducible nitric oxide biosynthesis in the liver in inflammation and in the allogeneic immune response. J Lab Clin Med 120:192–197
36. Nussler A, Billiar T (1993) Inflammation, immunoregulation, and inducible nitric oxide synthase. J Leuk Biol 54:171–178
37. Billiar T, Curran R, Harbrecht B, Stuehr D, Demetris A, Simmons R (1990) Modulation of nitrogen oxide synthesis in vivo: N^G-monomethyl-L-arginine inhibits endotoxin-induced nitrite/nitrate biosynthesis while promoting hepatic damage. J Leuk Biol 48: 565–569
38. Harbrecht B, Billiar T, Stadler J, et al (1992) Inhibition of nitric oxide synthesis during endotoxemia promotes intrahepatic thrombosis and an oxygen radical-mediated hepatic injury. J Leuk Biol 52:390–394
39. Harbrecht B, Stadler J, Demetris A, Simmons R, Billiar T (1994) Nitric oxide and prostaglandins interact to prevent hepatic damage during murine endotoxemia. Am J Physiol 266:G1004–G1010
40. Rubbo H, Radi R, Trujillo M, et al (1995) Nitric oxide regulation of superoxide and peroxynitrite-dependent lipid peroxidation: Formation of novel nitrogen-containing oxidized lipid derivatives. J Biol Chem (In press)

Effects of Hypoxia on Inducible Nitric Oxide Synthase (iNOS) in Renal Mesangial Cells

E. Michelakis, P. J. Shultz, and S. L. Archer

"The great tragedy of Science:
The staying of a beautiful hypothesis by an ugly fact."
Thomas Huxley, 1825–1895 (*Collected Essays*, Ch VIII, "Biogenesis and Abiogenesis")

Introduction

The effects of hypoxia on the molecular biology and enzyme activity of nitric oxide synthase (NOS) have, until recently, been unclear due to problems measuring nitric oxide (NO), failure to examine the entire NOS pathway, and variable definitions of hypoxia. This chapter describes recent advances in our understanding of the effects of hypoxia on the inducible isoform of nitric oxide synthase (iNOS). The renal mesangial cell was chosen as a model in which to study the effects of hypoxia on iNOS because these cells make NO in sepsis, lack the constitutive form of NOS (cNOS) and, when primed with lipopolysaccharide (LPS), are prolific generators of NO. The importance of examining the iNOS pathway at multiple levels, from induction of messenger ribonucleic acid (mRNA) and iNOS activity to the postsynthetic fate of NO and activation of molecular targets, is explained.

The renal mesangial cell is a smooth muscle-like cell found in the glomerulus which is both a producer of and target for NO. Although first described in 1927 [1], it was not until early in 1960s, with the advent of electron microscopy, that scientists were convinced of the existence of mesangial cells [2]. Recent advances in glomerular tissue culture have revealed the complexity of the biology of the mesangial cell. They are pluripotential, and like other connective tissue cells, are able to vary their functions in response to altered environments.

Besides its structural role as a support structure for the glomerular capillary network, the mesangial cell is responsible for the formation and degradation of the mesangial matrix, the processing of circulating macromolecules and immune complexes and, through its contractile characteristics, the regulation of glomerular capillary flow and filtration rate. By its ability to secrete a wide variety of growth factors and mediators, the mesangial cell can influence the proliferation and function of cells not only in its vicinity but distally as well. The many functions of mesangial cells are outlined in Table 1 and have been described elsewhere [3–5].

Table 1. Role of NO in renal physiology and pathophysiology

	Ref.
1. Maintenance of GFR	[10]
– afferent and efferent arteriolar tone	
– filtration coefficient (through mesangial cell relaxation)	
2. Regulation of cortical and medullary blood flow	[53]
3. Role in tubular hypoxia/reoxygenation injury	[54]
4. Role in tubuloglomerular feedback and natriuresis	[55, 56]
5. Protective role against glomerular thrombosis	[57]
6. Inflammation	
– role in experimental nephrotoxic nephritis	[58]
– role in Heymann nephritis	[59]
7. Inhibition of mesangial cell proliferation	[10]
– Possible importance in the progression of glomerulosclerosis	

GFR=Glomerular filtration rate

Table 2. The renal mesangial cell has many functions [10]

	Mesangial cell functions
Smooth muscle cell properties – contain actin-myosin filaments – contract in response to AII and other vasoconstrictors	– Contraction alters Kf, GFR and movement of macromolecules through the mesangium
Inflammatory properties – production of free radicals – expression of Fc receptors	– Uptake and degradation of immune complexes
Synthetic properties – production and degradation of the mesangial matrix	– Structural support of the glomerulus
Production of growth factors – IL-3, IL-6, TNF, PDGF	– Control of cellular proliferation
Production of NO by iNOS	– Modulation of renal hemodynamics locally (arteriolar tone) and distally (tubuloglomerular feedback)

Kf: glomerular filtration coefficient; GFR: glomerular filtration rate; PDGF: platelet-derived growth factor; IL: interleukin; AII: angiotensin II; TNF: tumor necrosis factor

The mesangial cell has a second messengers system similar to that of smooth muscle cells [6] and can respond to vasoactive substances, including NO, with changes in shape [7]. As discussed later, the mesangial cell can also produce NO in response to mediators of sepsis, such as LPS and cytokines [8, 9]. Evidence that NO has a role in regulating renal blood flow and glomerular function *in vivo*, in health and disease, is accumulating . The role of NO in renal physiology and pathophysiology is summarized in Table 2 and reviewed elsewhere [10, 11].

The NOS Pathway and Hypoxia

NO is synthesized from the guanidino nitrogen of L-arginine and molecular oxygen, by dioxygenases known as NOS. At least three isoforms have been identified; two constitutive isoforms (neuronal (nNOS) and endothelial (eNOS)), and an inducible form (iNOS), first isolated in macrophages and subsequently identified in many cell types including vascular smooth muscle (see reviews [12, 13]).

cNOS is activated by Ca^{2+} transients and in response to endothelium-dependent vasodilators produces NO, accounting for the activity of endothelial derived relaxing factor (EDRF). Increases in cytosolic Ca^{2+} cause cNOS to bind calmodulin closely and this activates the enzyme which immediately produces NO [12]. iNOS is induced by endotoxins (e.g. LPS) and cytokines (interferons (IFN), interleukin (IL) 1, tumor necrosis factor (TNF)) by a process that requires several hours for deoxyribonucleic acid (DNA) transcription and protein synthesis to occur before NO synthesis is detectable. Calmodulin is always coupled with iNOS, so that the enzyme's activity is not modulated by Ca^{2+} transients. Nicotinamide adenine dinucleotide phosphate (NADPH) is a cosubstrate and flavin adenine mononucleotide (FMN), flavin adenine dinucleotide (FAD) and tetrahydrobiopterin (H_4 biopterin) are cofactors [12]. When activated, iNOS produces NO for prolonged periods of time. Increased levels of NO have been found in animals who have received cytokines [8] or who are septic [9]. Corticosteroids and some interleukins inhibit the induction of iNOS [14]. Both cNOS and iNOS have been purified and their cDNAs cloned. Endothelial cNOS and iNOS not only serve different physiologic functions, but their genes, which bear only a 40–50% homology, are located on different chromosomes (7 and 17, respectively) [13].

Control of transcription and translation is the major regulatory point in the synthesis of NO by iNOS, a major distinction from the constitutively expressed cNOS. Recently, McQuillan et al. [15] used Northern analysis to measure the effects of hypoxia on cNOS mRNA levels in cultured endothelial cells. They noted a 40–60% decrease in the cNOS mRNA levels after exposure of the cells to severe hypoxia (PO_2: 20 Torr) for 48 h (when compared with the normoxic levels). In contrast, endothelin-1 mRNA levels were increased by hypoxia. Not only was cNOS mRNA less abundant in severe hypoxia, it was also less stable. An indirect measure of NOS function, diaphorase activity, was also suppressed by hypoxia. Unlike our studies of hypoxic modulation of iNOS in the renal mesangium (based on reference [16]), McQuillan et al. [15] used more severe hypoxia, did not directly measure NO synthesis or examine intermediate grades of hypoxia, and dealt exclusively with cNOS. Studies of hypoxia's effects on the cNOS pathway cannot necessarily be extrapolated to iNOS given the differences in cofactor requirements, cellular location, and the disparate physiologic roles of these NOS isoforms.

Problems in Assessing the Effects of Hypoxia on NOS

The synthesis and postsynthetic fate of NO are susceptible to modulation by PO_2 at many levels (Fig. 1). When studying a non-specific stimulus, like hypoxia, one must examine its effects on each step in the NOS pathway before making a general conclusion regarding whether hypoxia enhances, inhibits or has no effect.

A second problem in assessing hypoxia's effect on NOS is that most investigators have relied on indirect measures of NO function (such as EDRF activity) to define the functional status of NOS. Bioassays are sensitive to, but not specific for, hypoxic alteration of NOS activity. Observing only the vascular response to hypoxia (e.g. increased renal flow in perfused kidneys) is potentially misleading as the effects of hypoxia on vascular tone and EDRF/NO activity are multifactorial. Hypoxia can alter receptor function, signal transduction and ion channel activity by NO-independent, hypoxia-sensitive mechanisms [17–19]. Hypoxic renal vasodilation for example is not mediated by changes in NO synthesis and is not inhibited by arginine analogs [20]. In the pulmonary circulation, severe hypoxia ($PO_2 < 30$ Torr, anoxia)

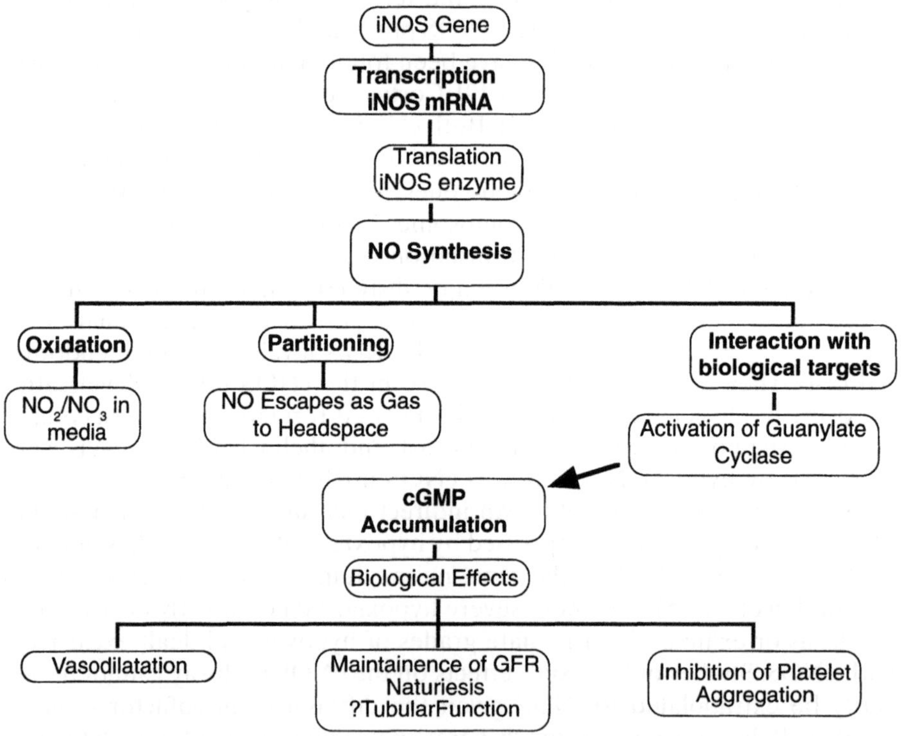

Fig. 1. Schematic of the iNOS Pathway. This diagram illustrates the many levels at which hypoxia could alter the iNOS pathway. The current study assessed the effects of graded hypoxia on steps presented in bold text. (From [16])

causes pulmonary vasodilatation in large part by opening adenosine triphosphate (ATP) sensitive potassium channels (K_{ATP}) [17, 18]. Furthermore, guanylate cyclase, a major target of NO, is redox modulated and can be activated or inhibited by oxygen radicals and peroxides other than NO [21]. The production of superoxide radical and hydrogen peroxide occurs as a byproduct of mitochondrial oxidative metabolism in direct proportion to the PO_2 [22–24]. Since hypoxia lowers production of radicals and peroxides, which promote survival of NO, increased NO levels are expected during hypoxia even if NOS function is unchanged or slightly depressed.

A third, related problem in assessing hypoxia's effects on NOS, is the reliance on L-arginine analogs (such as L-nitro-arginine methyl ester (L-NAME), or N^G-monomethyl-L-arginine (L-NMMA)), as "specific" NOS inhibitors [25]. The specificity of these inhibitors for NOS is imperfect [26, 27] and in the case of L-NMMA, the analog can actually be metabolized to NO, under certain circumstances [28]. Although the arginine analogs certainly inhibit NOS (including renal mesangial iNOS [29]) they can interfere with electron shuttling between metal groups in other enzymes thereby impairing NO-independent processes [27].

A fourth problem in reaching a consensus on the effects of "hypoxia" on the NOS pathway is the definition of hypoxia. Many authors refer to extreme O_2 deprivation as "hypoxia" when it should be considered anoxia (a stimulus which certainly will inhibit the function of an oxygenase such as iNOS). In this chapter, we define "moderate hypoxia" as PO_2 from 40–60 Torr, since the lowest PO_2 that man can tolerate for more than a brief period is around 40 Torr [30]. "Severe hypoxia" we will define as PO_2 of 20–40 Torr. To obtain a $PO_2 < 20$ Torr, even in cell culture experiments, one must exclude all oxygen from the system and flush the "headspace" (meaning the gas above the liquid in a closed container) with nitrogen; thus we feel such PO_2 represent anoxia. The effects of anoxia or severe hypoxia on iNOS function and induction are of questionable physiologic significance, but may have relevance in disease states in which blood supply is interrupted (such as infarction or ischemia).

Due to varying definitions of hypoxia, reliance on bioassays and indirect measures of NO synthesis, there are conflicting conclusions regarding the effects of hypoxia on NOS. Some investigators have found that hypoxia impairs NOS activity [31, 32]; others have demonstrated enhanced or preserved NO synthesis in hypoxia [33, 34]. All of these studies have dealt with constitutive forms of NOS.

Warren et al. [32] studying pulmonary artery rings perfused by the effluent of hypoxic, cultured endothelial cells, showed that, at a PO_2 of 42 Torr, endothelial-dependent relaxation was preserved, but at a PO_2 of 15 Torr it was diminished. EDRF activity rapidly recovered on return to normoxia.

Rengasamy et al. [35] used a co-incubation assay to study the effects of NO, generated from isolated cNOS, on vascular smooth muscle cells. Smooth muscle cell cGMP and radiolabelled citrulline were used to assess the activity of cNOS. Citrulline is an amino acid which is produced in a stochiometric relationship to NO by NOS. Severe hypoxia (PO_2 28 Torr) caused a substantial (50%), but reversible, decrease in the activity of isolated cNOS enzyme and these authors proposed that O_2 availability was the rate limiting factor in the synthesis of NO. These studies did not evaluate intermediate levels of hypoxia and thus failed to establish a threshold PO_2 at which this inhibition begins to occur.

In contrast, Isaacson et al. [34] in one of the few studies to measure NO production by an intact organ, showed increased NO production by chronically hypoxic lungs. Net production of NO and NO_2 was compared in lungs isolated from rats exposed to chronic hypobaric hypoxia (0.5 atmospheres for 1 month) and normoxic control rats. The NO and NO_2 were measured in the vascular compartment during a brief period of normoxic ventilation, not during ongoing hypoxia. Not only was basal NO synthesis increased but the relaxation evoked by endothelium-dependent vasodilators was preserved, a finding noted in similar studies of the pulmonary vasculature by others [33, 36]. In these studies of EDRF activity and cNOS-mediated NO synthesis, the severity of hypoxia was moderate ($PO_2 > 40$ Torr). In contrast, the studies in which hypoxia suppresses cNOS function [15, 35] or EDRF activity [32] employed severe hypoxia ($PO_2 < 30$ Torr). Thus, the severity of oxygen deprivation is an important determinant of hypoxia's effects on cNOS function and the threshold for hypoxic inhibition may be between 30–40 Torr.

Postsynthetic Metabolism of NO – Effects of Hypoxia

To determine whether iNOS function is altered by a given level of hypoxia, one can measure NO itself, its oxidation products or its effects on target molecules, such as guanylate cyclase. However, the biochemistry of NO is PO_2-dependent and one must consider the possibility that hypoxia might effect the parameters being measured through means which are not restricted to substrate deprivation and inhibition of NOS activity.

NO is a radical which exits preferentially as a gas. Its biologic half-life is short in the presence of oxygen, but it is quite stable in N_2, or in air, if NO is present in low concentrations. NO's postsynthetic behavior is characterized by its oxidation, partitioning, and interaction with molecular targets (Fig. 1).

Oxidation

In the absence of O_2, NO is quite stable but in oxygen its chemical half-life is measured in seconds. However, its biological half-life can last for minutes, because of the *in vivo* formation of complexes of NO with S-nitroso adducts of albumin [37] and activation of target molecules such as guanylate cyclase.

NO reacts readily with O_2 yielding a brown gas, nitrogen dioxide in the gas phase. In the aqueous phase, after leaving the cell and before entering the gas phase, NO can react with oxygen, yielding nitrite, NO_2. In the presence of strong oxidizing agents or hemoglobin, NO_2 can be converted to NO_3 [38]. The interaction of superoxide anion with NO generates peroxynitrite anion ($ONOO^-$) [39]. The effects of hypoxia on the cell's production of oxygen radicals and peroxides are well established [23, 24] and must be considered when using NO oxidation products, such as NO_2 and NO_3, to assess NOS activity in hypoxia.

Partitioning

NO has a high partition coefficient of more than 20/1 (molecules per unit volume gas/molecules per unit volume solution) [40]. When NO is injected into a test tube containing deoxygenated H_2O and a N_2 headspace, most of the NO which escapes oxidation leaves the liquid compartment and enters the headspace within seconds. Cell free *in vitro* experiments have shown that hypoxia favors the survival of NO. Thus, under hypoxic conditions, more NO survives and is able to leave the liquid and enter the headspace, from a fixed amount of authentic NO. If headspace NO were the only parameter measured, it might falsely appear that hypoxia increases NO synthesis. Figure 2 offers an illustration of this point. In this isolated, Krebs-albumin perfused

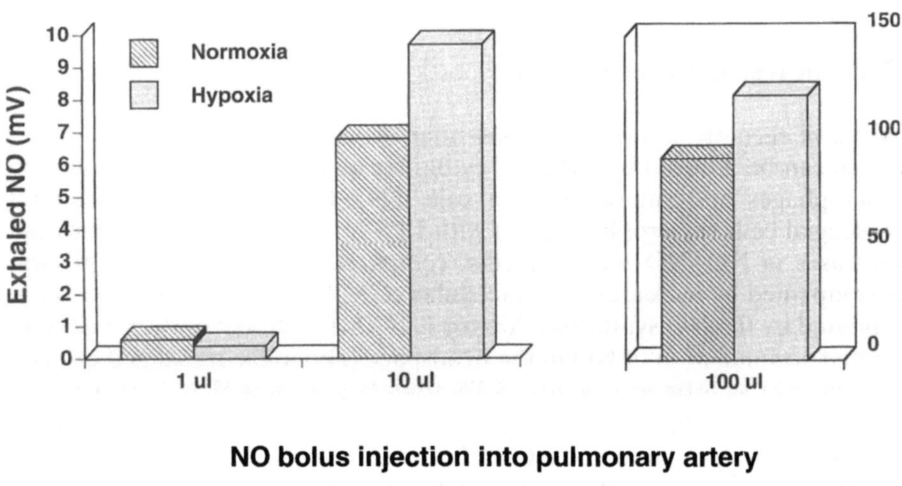

Fig. 2. Hypoxia increases the amount of NO which reaches the airway following injection of authentic NO into the pulmonary artery of the isolated rat lung. In this single rat lung, perfused with Krebs' solution containing 4% albumin, increasing doses of NO were injected into a pulmonary artery perfusion catheter. Expired NO was measured in breath as it left the endotracheal tube using chemiluminescence. (From [42] with permission.) The dose of NO is given as microliters of a saturated (~2 mM) solution. This illustrates how hypoxia enhances the survival of NO permitting it to partition to the gas phase. This illustrates how measuring only "headspace" or breath NO could mistakenly lead one to conclude that hypoxia increases NO synthesis when it is in fact simply facilitating partition of NO to the headspace

rat lung experiment, we compared the amount of NO which partitions to the breath following an injection of authentic NO into the pulmonary artery in normoxia versus hypoxia. It is evident that lowering the alveolar PO_2 increases observed NO levels in the breath, even though the dose of authentic NO administered to the pulmonary artery was constant.

Interaction with Molecular Targets

Apart from its vasodilatory properties, NO plays a role in neurotransmission, inflammation, immunity and thrombosis (for review see [41]). NO's involvement in such diverse pathophysiologic processes is the result of its ability to interact with many important moieties including thiols, heme groups and iron-sulfur centers. The reactivity of NO with some many molecular targets (hemoproteins, cytochromes, proteins) occurs because of its small molecular size, lipophilicity and an unpaired electron (for review see [42]). NO's primary molecular target is soluble guanylate cyclase (a heme protein) [43]. Activation of this enzyme results in accumulation of cGMP, which in turn promotes relaxation of vascular smooth muscle cells by activation of a calcium-dependent potassium channel [44] and decreases in cytoplasmic Ca^{2+} [45]. NO has both autocrine and paracrine functions, stimulating guanylate cyclase in the iNOS containing cell and nearby cells (Fig. 1).

iNOS in Renal Mesangial Cells

We have recently shown that glomerular mesangial cells possess an iNOS which can be induced by LPS and cytokines and is similar to that found in macrophages and smooth muscle cells [29]. Incubation of cultured rat mesangial cells (macrophage free) with LPS for 24 h resulted in significant increases in NO_2/NO_3 in the media, compared to controls [46]. This was accompanied by increases in intracellular cGMP. Both these increases were inhibited by the NO synthesis inhibitor L-NAME. Subsequently, we demonstrated accumulation of NO in the headspace (above the mesangial cell culture media) as little as 6 h after LPS stimulation using the chemiluminescence analyzer [29]. This assay has a detection threshold of approximately 1 part per billion for authentic NO and is much more sensitive than detecting NO by measuring its oxidation products using the Griess reaction [47]. This was also inhibited by L-NAME (Fig. 3). Although headspace NO is a sensitive marker of NO synthesis, it represents a very small portion of the total NO produced (Fig. 3); media NO_2 and NO_3 levels are much higher than headspace NO levels.

We used Northern hybridization, with a probe cloned from mouse macrophages, to demonstrate the presence of iNOS mRNA in cultured mesangial cells (Fig. 4). We could not detect any iNOS mRNA in unstimulated cells (i.e. cells not exposed to LPS or other cytokines) but iNOS mRNA was detect-

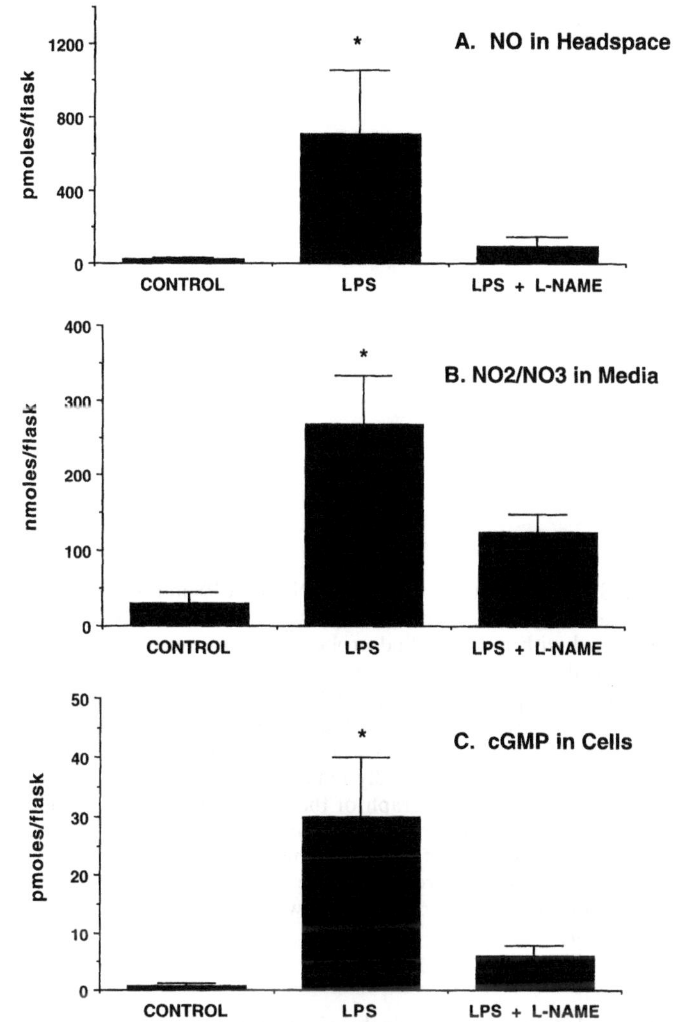

Fig. 3. Effects of lipopolysaccharide stimulation on iNOS function in renal mesangial cells. Comparison of NO in the headspace measured by chemiluminescence (*Panel A*), NO_2 and NO_3 in the media measured by the Greiss Reagent (*Panel B*) and intracellular cGMP measured by radioimmunoassay (*Panel C*) from the same rat mesangial cell cultures. Cells were incubated with MEM media containing 10% serum and 1 mM L-arginine only (control), LPS (10 μg/ml) or LPS + L-NAME (10^{-3} M) for 24 h in 75 cm^2 flasks. Data is the mean ± S.E.M. of 3 experiments, with each condition in duplicate flasks. * denotes $p<0.05$ vs control. (From [29] with permission)

able, without using the polymerase chain reaction for amplification, within 4 h of incubation with LPS. Using a cNOS mRNA probe, we found no basal or LPS-stimulated cNOS by Northern blot analysis in these cells [29]. We confirmed that glucocorticoids inhibit iNOS induction by showing that dexamethasone decreased media NO_2 and NO_3 levels of our cultured mesangial

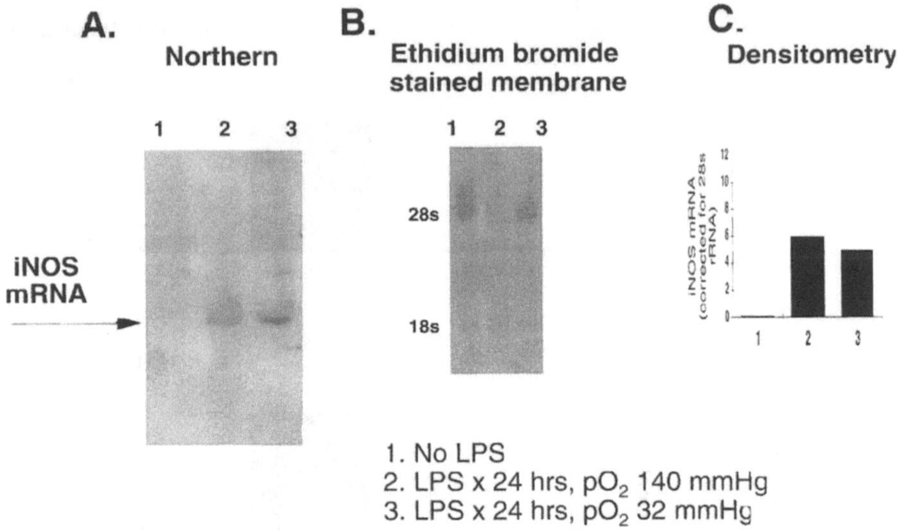

Fig. 4. Similar induction of iNOS mRNA by LPS in normoxia and severe hypoxia. *Panel A* shows an autoradiograph of a representative Northern blot of rat mesangial cell RNA, hybridized with radiolabelled probe for mouse macrophage iNOS. Lane 1 is RNA from control cells, incubated for 24 h in 95% air/5% CO_2. Lane 2 is RNA from cells incubated in the same atmosphere but with LPS (10 µg/ml) for 24 h, resulting in media PO_2 of 140 Torr. Lane 3 is RNA from cells incubated with LPS in 95% N_2/5% CO_2, resulting in media PO_2 of 32 Torr. 10 µg of total RNA were loaded in each lane. iNOS mRNA cannot be detected in control cells, but is markedly induced in cells incubated with LPS at both oxygen tensions. *Panel B* is a photograph of the ethidium bromide stained membrane used for the Northern hybridization in Panel A, demonstrating equivalent RNA loading and transfer. *Panel C* shows a graph of the densitometry units from each lane of Panel A, corrected for the densitometry units from the same lane in Panel B. A similar degree of iNOS mRNA induction by LPS is detected in cells with media PO_2s of 140 or 32 Torr. (From [29] with permission)

cells and prevented the induction of iNOS mRNA by LPS [29]. Ahn et al. [48] used *in situ* hybridization to localize iNOS mRNA in rat kidney. They found basal iNOS levels in tubular cells but only detected iNOS mRNA in the mesangial cells of LPS treated rats.

In the remainder of this chapter, we will discuss recent work [16] which uses the renal mesangial cell as a model in which to study the effects of graded hypoxia on the iNOS pathway. When compared with models commonly used to evaluate NOS modulation by hypoxia (arterial rings, perfused organs or intact animals), renal mesangial cells offer several advantages:

1) a one cell model which avoids the complexity of the endothelial cell-smooth muscle cell interaction.
2) absence of confounding effects of shear stress and blood flow.
3) absence of cNOS (as measured by Northern blot analysis).
4) large and sustained amounts of NO produced by the iNOS pathway.

Using confluent plates of second passage renal mesangial cells, we have studied the effects of hypoxia on the many steps in the iNOS pathway (Fig. 1):
a) Transcription/translation: by measuring the levels of iNOS mRNA;
b) NO synthesis by the translated enzyme: by using NOS inhibitors;
c) Oxidation of the produced NO: by measuring NO_2 and NO_3 in the media;
d) Partitioning of the produced NO: by measuring NO in the headspace; and
e) Interaction with biological targets: by measuring cGMP levels.

We assessed the effects of the following PO_2s on the iNOS pathway: 32 Torr (severe hypoxia), 45 Torr (moderate hypoxia) and 150 Torr (normoxia). We measured the effects of LPS on iNOS mRNA, NO in the headspace, NO_2 and NO_3 in the culture media and cGMP levels at each PO_2 [16]. As shown in Fig. 4, the induction of iNOS mRNA by LPS (10 μg/mL) was similar in severe hypoxia (30 Torr for 24 h) and normoxia. Thus, physiologically important hypoxia does not affect the transcription of iNOS gene, measured as the accumulation of iNOS mRNA. This is in contrast to the findings with endothelial cNOS mRNA by McQuillan et al. [15] who noted a 40–60% decrease in the levels of cNOS mRNA with more severe hypoxia (PO_2 20 Torr).

The increase in headspace NO levels after 24 h incubation with LPS was slightly less in severe hypoxia than in normoxia [29]. Interestingly, the greatest amount of headspace NO was found above cells incubated in moderate hypoxia (PO_2 42 Torr for 24 h) [16]. NO_2 and NO_3 levels in the media decreased to approximately 50% of the normoxic levels in moderate and severe hypoxia. The fact that headspace NO was increased in moderate hypoxia, while at the same time NO_2 and NO_3 levels were decreased is not surprising. It confirms *in vitro* cell-free experiments which showed that hypoxia facilitates the partitioning of NO to the headspace [47]. Only severe hypoxia decreased both the headspace NO and the media NO_2 and NO_3 levels. This suggests that iNOS activity is not reduced except in severe hypoxia; consistent with the findings of several investigators studying cerebellar and endothelial cNOS [32, 35, 49].

In support of the conclusion that the iNOS pathway is functionally intact at PO_2s as low as 45 Torr, we found cGMP levels were similar in normoxia and moderate hypoxia [16]. In the LPS-stimulated renal mesangial cell, cGMP is measured in the same cell that produces the NO, and NO synthesis occurs at a high rate for a prolonged period of time, minimizing the likelihood of measurement errors. Consequently, cGMP levels in our studies reflect the intracellular interaction of NO with guanylate cyclase. This may be less sensitive to postsynthetic oxidation and partitioning of NO than experiments in which NO is made by cNOS in the endothelium but cGMP is measured in a separate target cell (i.e. a vascular smooth muscle cell or fibroblast). In studies of cNOS using co-culture or bioassay techniques, moderate hypoxia has been reported to decrease cGMP levels [50, 51]. This may reflect a true difference in the sensitivity of cNOS versus iNOS to moderate hypoxia. Alternatively, hypoxia's effects on partitioning and oxidation of NO may be more evident in these cNOS studies where NO must exit the endothelium to reach the target vascular smooth muscle cell.

Most studies examining the effects of hypoxia on NOS have measured only one or two steps in the pathway and thus are unable to state the mechanism by which hypoxia acts. Rengasamy et al. [35] proposed that decreased O_2 supply was the rate limiting factor while Liu et al. [52] proposed that NO synthesis was depressed during hypoxia by impaired L-arginine uptake. Other possible explanations include hypoxic interference with the availability of NOS cofactors or hypoxic damage to NOS, although at least in the case of cNOS, the effects of hypoxia appear reversible [35]. Our finding that iNOS mRNA induction is resistant to severe hypoxia and that, above PO_2s of 40 Torr, NO synthesis and cGMP accumulation are preserved contrasts with those of McQuillan et al. [15]. This disparity emphasizes the need to study all the steps of the NOS pathway before concluding the nature of the effects of hypoxia on the NOS pathway and underlines the differences between the two isoforms of NOS.

Conclusion

The literature suggests cNOS mRNA production, mRNA stability and enzyme activity are inhibited by severe hypoxia ($PO_2 < 30$ Torr). There has been little evaluation of cNOS function at more physiologic PO_2s, other than measures of EDRF activity or bioassays, most of which suggest cNOS function is preserved at $PO_2\text{s} > 40$ Torr. We used cultured mesangial cells to study the effects of graded hypoxia on the induction and function of iNOS. Induction of iNOS by LPS is resistant to severe hypoxia (PO_2 32 Torr). iNOS function, measured as NO release to the headspace and media accumulation of NO_2 and NO_3, is preserved at $PO_2\text{s} > 45$ Torr but declines with more severe hypoxia. cGMP levels within LPS-stimulated renal mesangial cells are elevated to similar levels in cells studied at PO_2s of 150 and 45 Torr, suggesting molecular targeting of guanylate cyclase by NO is preserved even during moderate hypoxia.

Acknowledgments: The authors thank Kim Freude for her excellent work culturing renal mesangial cells and measuring nitric oxide and Martin Tristani-Firouzi MD for his help with the graphic presentation. The grant support for this research is provided by the Department of Veterans Affairs and NIH grant # R29-HL45735.

References

1. Zimmermann KW (1929) Über den Bau des Glomerulus der menschlichen Niere. Z Mikrosk Anat Forsch 18:520–552
2. Marinozzi V (1961) Struttura ed istofisiologia del glomerulo. Ati del II Aggiornamento professionale, Rome. Nefrologia Moderna 5:33–51

3. Shultz PJ, Raij L (1991) The glomerular mesangium: Role in initiation and progression of renal injury. Am J Kidn Dis (Suppl) 17:8–14
4. Johnson R, Floege J, Yoshimura A (1992) The activated mesangial cell: A glomerular "myofibroblast"? J Am Soc Nephrol 2:S190–S197
5. Latta H (1992) An approach to the stucture and function of the glomerular mesangium. J Am Soc Nephrol 2:S65–S73
6. Mene P, Cinotti GA, Pugliese F (1992) Signal transduction in mesangial cells. J Am Soc Nephrol 2:S100–S106
7. Shultz PJ, Schorer AE, Raij L (1990) Effects of endothelium-derived relaxing factor and nitric oxide on rat mesangial cells. Am J Physiol 258:F162–F167
8. Kilbourn RG, Gross SS, Jubran A, et al (1990) N^G-methyl-L-arginine inhibits tumor necrosis factor-induced hypotension: Implications for the involvement of nitric oxide. Proc Natl Acad Sci 87:3629–3632
9. Wang Q, Jacobs J, DeLeo J, et al (1991) Nitric oxide hemoglobin in mice and rats in endotoxic shock. Life Sciences 49:PL55–PL60
10. Raij L, Shultz PJ (1993) Endothelium-derived relaxing factor, nitric oxide: Effects on and production by mesangial cells and the glomerulus. J Am Soc Nephrol 3:1435–1441
11. Reyes AA, Karl IE, Klahr S (1994) Role of arginine in health and disease. Am J Physiol 36:F331–F346
12. Nathan C (1992) Nitric oxide as a secretory product of mammalian cells. FASEB J 6: 3051–3064
13. Knowles RG, Moncada S (1994) Nitric oxide synthases in mammals. Biochem J 298: 249–258
14. Geller DA, Nussler AK, Di Silvio M, et al (1993) Cytokines, endotoxin, and glucocorticoids regulate the expression of inducible nitric oxide synthase in hepatocytes. Proc Natl Acad Sci USA 90:522–526
15. McQuillan L, Leung GK, Marsden PA, et al (1994) Hypoxia inhibits expression of eNOS via transcriptional and posttranscriptional mechanisms. Am J Physiol 267: H1921–H1927
16. Archer SL, Freude KA, Schultz P (1995) The effect of graded hypoxia on the induction and function of inducible nitric oxide synthase (iNOS) in rat mesangial cells. Circ Res (In press)
17. Wiener CM, Dunn A, Sylvester JT (1991) ATP-dependent K^+ channels modulate vasoconstrictor responses to severe hypoxia in isolated ferret lungs. J Clin Invest 88: 500–504
18. Buescher P, Perse D, Pillai R, et al (1991) Energy state and vasomotor tone in hypoxic pig lungs. J Appl Physiol 70:1874–1881
19. Vallet B, Curtis SE, Winn MJ, et al (1994) Hypoxic vasodilation does not require nitric oxide (EDRF/NO) synthesis. J Appl Physiol 76:1256–1261
20. Hampl V, Weir EK, Archer SL (1994) Endothelium-derived nitric oxide is less important for basal tone regulation in the pulmonary than the renal circulation of the adult rat. J Vasc Med Biol 5:22–30
21. Braughler JM (1983) Soluble guanylate cyclase activation by nitric oxide and its reversal. Biochem Pharmacol 32:811–818
22. Freeman BA, Crapo JD (1981) Hyperoxia increases oxygen radical production in rat lungs and lung mitochondria. J Biol Chem 256:10986–10992
23. Archer SL, Nelson DP, Weir EK (1989) Simultaneous measurement of oxygen radicals and pulmonary vascular reactivity in the isolated rat lung. J Appl Physiol 67:1903–1911
24. Archer SL, Huang J, Henry T, et al (1993) A redox based oxygen sensor in rat pulmonary vasculature. Circ Res 73:1100–1112
25. Rees DD, Palmer RMJ, Hodson HF, et al (1989) A specific inhibitor of nitric oxide formation from L-arginine attenuates endothelium-dependent relaxation. Br J Pharmacol 96:418–424
26. Cocks TM, Angus JA (1991) Evidence that contraction of isolated arteries by L-NMMA and NOLA are not due to inhibition of basal EDRF release. J Cardiovasc Pharmacol 17:S159–S164

27. Peterson DA, Peterson DC, Archer S, et al (1992) The non-specificity of specific nitric oxide synthase inhibitors. Biochem Biophys Res Commun 187:797–801
28. Archer SL, Hampl V (1992) N^G-monomethyl-L-arginine causes nitric oxide synthesis in isolated arterial rings: Trouble in paradise. Biochem Biophys Res Commun 188: 590–596
29. Shultz PJ, Archer SL, Rosenberg ME (1994) Inducible nitric oxide synthase mRNA and activity in glomerular mesangial cells. Kidney Int 46:683–689
30. Groves BM, Reeves JT, Sutton JR, et al (1987) Operation Everest II: Elevated high-altitude pulmonary resistance unresponsive to oxygen. J Appl Physiol 63:521–530
31. Adnot S, Raffestin B, Eddahibi S, et al (1991) Loss of endothelium-dependent relaxant activity in the pulmonary circulation of rats exposed to chronic hypoxia. J Clin Invest 87:155–162
32. Warren JB, Maltby NH, McCormack D, et al (1989) Pulmonary endothelium-derived relaxing factor is impaired in hypoxia. Clin Sci 77:671–676
33. Emery CJ, Bee D, Barer GR (1981) Mechanical properties and reactivity of vessels in isolated perfused lungs of chronically hypoxic rats. Clin Sci 61:569–580
34. Isaacson TC, Hampl V, Weir EK, et al (1994) Increased endothelium-derived nitric oxide in hypertensive pulmonary circulation of chronically hypoxic rats. J Appl Physiol 76:933–940
35. Rengasamy A, Johns RA (1991) Characterization of endothelium-derived relaxing factor/nitric oxide synthase from bovine cerebellum and mechanism of modulation by high and low oxygen tensions. J Pharmacol Exp Ther 259:310–316
36. Russ RD, Walker BR (1993) Maintained endothelium-dependent pulmonary vasodilation following chronic hypoxia in the rat. J Appl Physiol 74:339–344
37. Stamler JS, Jaraki J, Osborne D, et al (1992) Nitric oxide circulates in mammalian plasma primarily as an S-nitroso adduct of serum albumin. Proc Natl Acad Sci 89: 7674–7677
38. Ignarro LJ, Fukuto JM, Griscavage JM, et al (1993) Oxidation of nitric oxide in aqueous solution to nitrite but not nitrate: Comparison with enzymatically formed nitric oxide from L-arginine. Proc Natl Acad Sci USA 90:8103–8107
39. Koppenol WH, Moreno JJ, Pryor WA, et al (1992) Peroxynitrite, a cloaked oxidant formed by nitric oxide and superoxide. Chem Res Toxicol 5:834–842
40. Zafiriou OC, McFarland M (1980) Determination of trace levels of nitric oxide in aqueous solution. Analytical Chemistry 52:1662–1667
41. Moncada S, Palmer RMJ, Higgs EA (1991) Nitric oxide: Physiology, pathophysiology and pharmacology. Pharmacol Rev 43:109–141
42. Archer S (1993) Measurement of nitric oxide in biological models. FASEB J 77: 349–360
43. Murad F, Arnold W, Mittal CK, et al (1979) Properties and regulation of guanylate cyclase and some proposed functions for cyclic GMP. Adv Cyclic Nucleotide Res 11:175–204
44. Archer SL, Huang JMC, Hampl V, et al (1994) Nitric oxide and cGMP cause vasorelaxation by activation of a charybdotoxin-sensitive K channel by cGMP-dependent protein kinase. Proc Natl Acad Sci USA 91:7583–7587
45. Twort CHC, van Breemen C (1988) Cyclic guanosine monophosphate-enhanced sequestration of Ca^{2+} by sarcoplasmic reticulum in vascular smooth muscle. Circ Res 62:961–964
46. Shultz PJ, Tayeh MA, Marletta MA, et al. (1991) Synthesis and action of nitric oxide in rat glomerular mesangial cells. Am J Physiol 261:F600– F606
47. Archer SL, Shultz PJ, Warren JB, et al (1995) Preparation of standards and measurement of nitric oxide, nitroxyl and related oxidation products. Methods 7 (In press)
48. Ahn KY, Mohaupt MG, Madsen KM, et al (1994) In situ hybridization localization of mRNA encoding inducible nitric oxide synthase in rat kidney. Am J Physiol 267: F748–F757
49. Kozniewska E, Oseka M, Stys T (1992) Effects of endothelium-derived nitric oxide on cerebral circulation during normoxia and hypoxia. J Cereb Blood Flow Metab 12: 311–317

50. Rodman DM, Yamaguchi T, Hasunuma K, et al (1990) Effects of hypoxia on endothelium-dependent relaxation of rat pulmonary artery. Am J Physiol 258:L207–L214
51. Johns RA, Linden JM, Peach MJ (1989) Endothelium-dependent relaxation and cyclic GMP accumulation in rabbit pulmonary artery are selectively impaired by moderate hypoxia. Circ Res 65:1508–1515
52. Liu S, Crawley DE, Barnes PJ, et al (1991) Endothelium-derived relaxing factor inhibits hypoxic pulmonary vasoconstriction in rats. Am Rev Respir Dis 143:32–37
53. Brezis M, Heyman S, Dinour D, et al (1991) Role of NO in renal medullary oxygenation. J Clin Invest 88:390–395
54. Yu L, Gengaro PE, Niederberger M, et al (1993) Nitric oxide: A mediator in rat tubular hypoxia/reoxygenation injury. Proc Natl Acad Sci 91:1691–1695
55. Thorup C, Persson EG (1994) Inhibition of locally produced NO resets tubuloglomerular feedback mechanism. Am J Physiol 267:F606–F611
57. Shultz PJ, Tolins JP (1993) Adaptation to increased dietary salt intake in the rat: Role of endogenous nitric oxide. J Clin Invest 91:642–650
56. Shultz PJ, Raij L (1992) Endogenously synthesized nitric oxide prevents endotoxin-induced glomerular thrombosis. J Clin Invest 90:1718–1725
58. Cattell V, Cook T, Moncada S (1990) Glomeruli synthesize nitrite in experimental nephrotoxic nephritis. Kidney Int 38:1056–1060
59. Cattel V, Largen P, de Heer E, et al (1991) Glomeruli synthesize nitrite in active Heymann nephritis: The source is infiltrating macrophages. Kidney Int 40:847–851

Vascular Smooth Muscle Function under Conditions of Sepsis and Endotoxemia

J. R. Parratt and J. C. Stoclet

Introduction

A pronounced and prolonged reduction in peripheral vascular resistance is one of the most marked and characteristic effects observed both in patients with the sepsis syndrome and in animals in which sepsis has been induced, or to which purified lipopolysaccharide (endotoxin, LPS) has been administered systemically. This circulatory failure is due to vasodilatation of most vascular beds resulting from a reduced ability to respond adequately to the transmitter released from sympathetic neurones supplying vascular smooth muscle (norepinephrine) [1] and to a variety of circulating vasoconstrictor substances such as epinephrine, angiotensin and vasopressin [1–3]. The vascular failure associated with sepsis and endotoxemia thus results from "vasoplegia". The importance of understanding the mechanisms involved in this vasoplegia and "unrelenting hypotension" [4] lies in the observation that this is a major contributory factor to mortality in both adult [5–8] and pediatric [9] patients with sepsis.

One early suggested mechanism for the loss of vascular responsiveness in endotoxemia was the presence in blood of a variety of circulating vasodilator substances including histamine, prostanoids and plasma kinins (reviewed in [10]). The fact that blood vessels isolated from animals previously given endotoxin also have a reduced responsiveness to endogenous vasoconstrictors [11–14] suggests that the defect is not due to circulating factors, since these would have been removed by the isolation and washing procedures inherent in investigating responses of isolated blood vessel preparations, but to either the *local* production of some powerful relaxant or to some fundamental defect in the ability of vascular smooth muscle cells to contract. For reasons that will be clear later, the first of these explanations is the most likely. The fact that alterations in vascular reactivity can be demonstrated in vessels isolated from endotoxemic or septic animals made an analysis of the mechanisms involved much easier, and such studies led directly to the finding that this is due, in the main, to the excessive production of nitric oxide (NO) by an induced NO synthase (iNOS). The evidence for this will be reviewed later.

There have been other explanations for the loss of vascular responsiveness in sepsis and endotoxemia, including downregulation of appropriate receptors, especially adrenoceptors, by high circulating levels of catecholamines. Levels of circulating catecholamines are known to be elevated under these conditions [15, 16]. Thus, the number of α-adrenoceptors is reduced in the aorta under conditions of endotoxemia and sepsis by as much as 50% [17]. Because there is no change in either agonist or antagonist affinity, but a reduction in the number of ligand recognition sites, it seems as though the presence of some circulating inhibitor which binds to the receptors under these conditions is unlikely. The fact that there is a general reduction in contractile, or vasoconstrictor responses to a variety of agonists, including calcium and depolarizing solutions of potassium chloride [14, 18, 19], make it unlikely that receptor downregulation is primarily responsible for vascular hyporeactivity either *in vivo* or *in vitro*, especially as there is no evidence for *generalized* receptor downregulation under these conditions [17].

The most likely explanation for the mechanism of the loss of vascular responsiveness in endotoxemia involves NO production by the vessel wall. Before this evidence is considered, some account of the normal role of endothelium-derived factors in regulating the activity of vascular smooth muscle is appropriate.

Regulation by Endothelial Cells of Vascular Smooth Muscle Function

The initial stimulus for studies on the relationship between endothelial and vascular smooth muscle cells were the experiments of Furchgott and Zawadzki [20]. They found that acetylcholine-induced relaxation of isolated artery preparations depended upon the presence of an intact endothelial layer. When this was removed mechanically, acetylcholine had no effect on such preparations and, in some cases, contraction resulted. The explanation is that acetylcholine acts on muscarinic receptors on the surface of endothelial cells resulting in the release of a substance, which Furchgott and Zawadzki called "endothelium-derived relaxing factor" (EDRF). This diffused across the short distance to the underlying vascular smooth muscle cells, interacted with the ion atom in the heme part of the soluble guanylyl cyclase molecule with the resultant increase in the production of cyclic GMP (cGMP). This both increases the uptake of calcium into intracellular storage sites within the smooth muscle cells and reduces calcium influx into the cell through calcium channels. The result is a lowering of calcium concentrations around the contractile elements, resulting in relaxation. We now know that EDRF is NO [21, 22] which is formed from one of the terminal guanido nitrogen atoms of L-arginine [23–25]. The reaction is stereospecific and a co-product of the reaction is L-citrulline [26]. Various analogs of L-arginine (and also of L-ornithine) inhibit the synthesis of NO from L-arginine in a dose-dependent and enantiomer-specific manner, and have been used to assess the role of NO production under a variety of physiological and pathophysiological conditions. These analogs include N^G-monomethyl-L-arginine

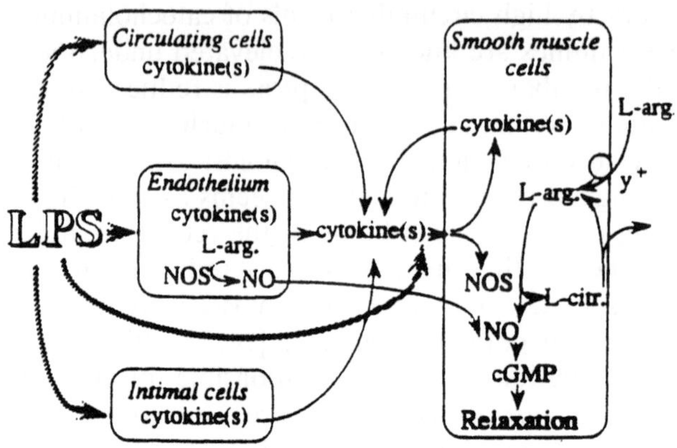

Fig. 1. The L-arginine pathway and the production of NO resulting in activation of guanylate cyclase and the elevation of cGMP both in endothelial cells and in smooth muscle cells following the administration of endotoxin (LPS). Although the predominant mechanism involves cytokine release, the possibility that endotoxin acts directly on a LPS-binding protein (CD 18 or CD 14) to directly stimulate the pathway is not excluded

(L-NMMA); [24], N^G-nitro-L-arginine methyl ester (L-NAME), [27,28]; and N-iminomethyl-L-ornithine (L-NIO). There seems little doubt that the L-arginine-NO pathway is active under physiological conditions and may even play a role in the normal regulation of arterial blood pressure [29]. An enzyme which stimulates the pathway, so called constitutive NO synthase (cNOS), is thus present under normal conditions in endothelial cells. There is no evidence that this particular enzyme is present under these conditions in vascular smooth muscle cells.

Stimulation of various receptors (e.g. muscarinic, bradykinin B_2 receptors, and receptors to various cytokines) results in an initial calcium transient [30] within endothelial cells, activation of the L-arginine-NO pathway and the release of NO. This results in an elevation of cGMP both within the endothelial cell and within the target cell, which in the vascular wall is the smooth muscle cell (Fig. 1).

Under conditions where endothelial function is impaired, with a resultant inhibition of NO production, responses to vasodilator agents released into the blood stream may also be modified.

Evidence for a Role for NO in Mediating Vascular Hyporesponsiveness

Since we have already reviewed this evidence on several occasions [31–34], only a brief review will be given here. The finding that NO, or a NO-containing molecule, is responsible for smooth muscle relaxation following activation of endothelial cells by a variety of agonists, and its identification with EDRF (reviewed by Moncada et al. [35]), together with the availability of inhibitors of the NO generating pathway, prompted a study to determine whether NO was responsible for the loss of responsiveness under conditions of endotoxemia. The title of the initial communication, in December 1989, to the Physiological Society in London is instructive. The title ("Impaired vascular reactivity in rat following endotoxin treatment can be endothelium-independent yet involves the L-arginine pathway") suggested first, that NO was responsible for the reduced vascular responsiveness, and secondly that the source of NO was not the endothelial cell. Since, as we have seen, the constitutive enzyme for the production of NO is not normally present in vascular smooth muscle cells, it suggested that an enzyme not normally present in these cells was responsible for the formation of NO. This was the first indication that endotoxin could induce, within hours, a NO forming enzyme [36].

For these studies [37–41], the aorta was removed from rats injected several hours earlier with *Escherichia coli* endotoxin. The vessels were then set up in an organ bath and revealed that in rings taken from endotoxin-treated animals, norepinephrine dose-response curves were shifted downwards and to

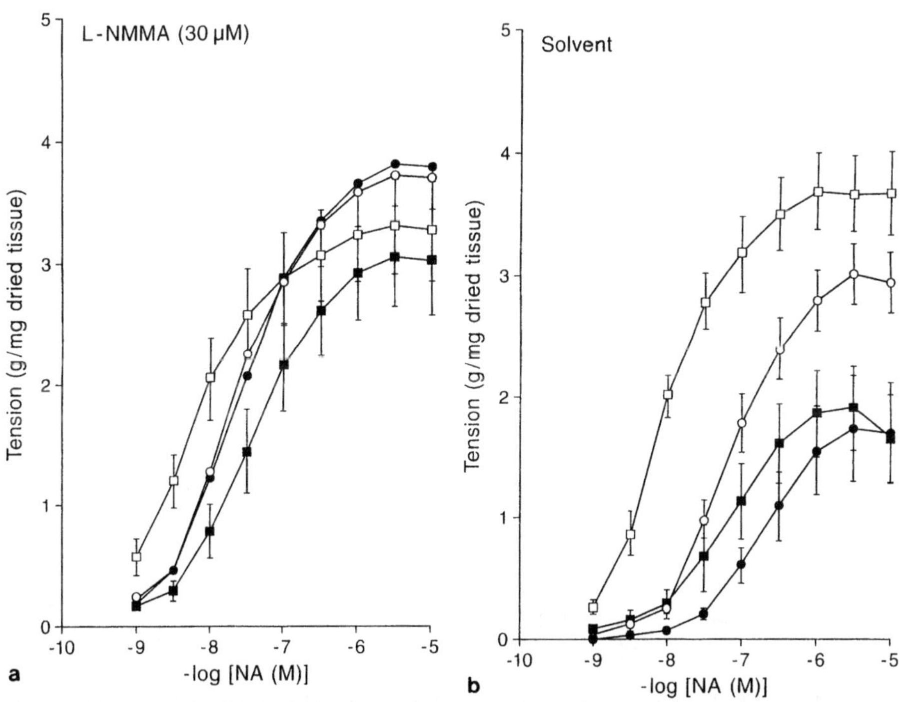

Fig. 2. Concentration response curves for norepinephrine (NA) in aortic rings with (open circles, closed circles) and without (open square, closed squares) endothelium, from control (open symbols) and LPS-treated rats (filled symbols) in the presence of solvent (**b**) or L-NMMA (**a**), 30 μM. Results are presented as the mean ± S.E.M. (n = 8). The inhibitor of the L-arginine-NO pathway restores responsiveness to norepinephrine in rats administered endotoxin

the right with a reduction in the maximum response (Fig. 2). This was true whether the vessels had an intact endothelium or whether the endothelium had been removed mechanically by gently rubbing the inside of the aorta. In most of the preparations, the responses of aortas taken from control vessels were less if the endothelium was present, implying that norepinephrine even under control conditions released a relaxant substance from these cells. The mediator responsible for the impaired contractions obtained in aortas removed from endotoxin-treated animals was NO because responses were restored to normal in the presence of various inhibitors of the L-arginine-NO pathway e.g. L-NMMA (Fig. 2) whereas D-NMMA was ineffective.

This conclusion was confirmed when cGMP measurements were made in aortic rings from control and LPS-treated rats when contracted with norepinephrine (Fig. 3). In vessels taken from LPS-treated rats, either with or without endothelium, large amounts of cGMP were formed compared to

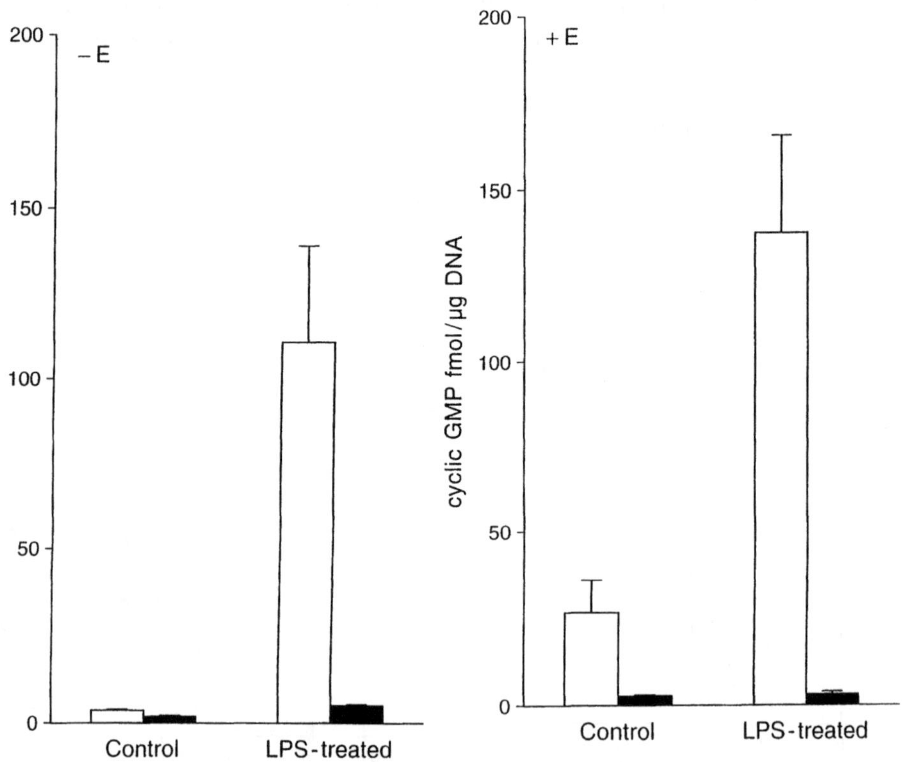

Fig. 3. The cGMP content of aortic rings with (+E) **b** and without (–E) **a** endothelium from control and endotoxin-treated rats when contracted with norepinephrine (10 μM) in the presence (filled columns) and absence (open columns) of L-NAME, 300 μM. Endotoxin results in the generation of large amounts of cGMP in preparations from endotoxin-treated rats whether the endothelium is present or not. That this is due to activation of the L-arginine-NO pathway is demonstrated by the abolition of these increases by an inhibitor of the L-arginine-NO pathway

controls. That this increase was due to the production of NO in these vessels was demonstrated by the fact that the elevated cGMP levels were not seen following treatment with L-NAME. Because increases were seen in vessels denuded of the endothelium, we can conclude that endotoxin induces an extra-endothelial activation of the NO pathway [37–41].

Incubation of Isolated Blood Vessels with Endotoxin Stimulates NO Formation

McKenna and his colleagues [42] showed that when blood vessels are incubated with a medium containing macrophages activated by endotoxin (or Interleukin-1, IL-1), subsequent responses of these vessels to norepinephrine are suppressed. We now know that this reduced responsiveness to norepinephrine is due to activation of the L-arginine-NO pathway [38] since the contraction-response curves, which are shifted downwards and to the right, are restored by L-NMMA (but not by D-NMMA) and also by methylene blue. Again, the major source of this NO production is not the endothelium since depressed responses are also seen in vessels denuded of endothelium and incubated with endotoxin [38].

The presence of the endothelium during *in vitro* incubation with LPS reduced the delay preceding the onset of hyporeactivity to norepinephrine in the rat aorta [43]. Further, the intraluminal addition of LPS-activated rabbit macrophages (which do not produce NO by themselves) rapidly diminished the responses of perfused carotid arteries to phenylephrine [44]. These findings suggest some form of co-operativity between cells which may modify NO-mediated vascular hyporeactivity and the rapidity of its onset after LPS administration. How soon this occurs *in vivo* may well be determined by the rapidity and extent of cytokine release and the possible role of other circulating cells.

Which Cell Types in the Vessel Wall generate NO in Response to LPS

The findings, outlined above, that LPS induces the intravascular production of a relaxing substance derived from L-arginine raises the question as to the cells involved in this production. It is now established that NO can be produced by several distinct forms of enzymes, each termed NO-synthase (NOS) [35, 45]. The enzyme present in endothelial cells (and responsible for the formation of Furchgott's EDRF) is expressed constitutively, is activated by the calcium-calmodulin complex and possesses phosphorylation sites allowing regulating by kininases. An enzyme can also be induced in certain cell types (initially characterized in activated murine macrophages), this enzyme is not sensitive to calcium and does not possess the regulatory phosphorylation sites of the endothelial form.

Soon after it was discovered that the L-arginine-NO pathway was involved in LPS-induced vascular hyporeactivity [36–38], it was reported that NOS activity could be induced by LPS in the rat aortic media-adventitia layer [41, 46]. This induction was prevented by dexamethasone or by cycloheximide. Since the aortic media-adventitia layer contains a significant proportion of various cell types, besides smooth muscle cells, and since smooth muscle cells in culture undergo phenotypic modulation [47] which may modify sensitivity to LPS as well as to cytokines (which are known to induce NOS in cultured smooth muscle cells [48–51]), it was decided to determine whether LPS was able to induce NO production in smooth muscle cells *in situ*. The media layers of rat thoracic aortas were dissected and exposed to LPS for 4 h; a cell suspension was then obtained by enzymatic digestion of the extracellular matrix and the cell suspension preplated for 2 h in Petri dishes in order to eliminate rapidly adhering cells e.g. residual endothelial

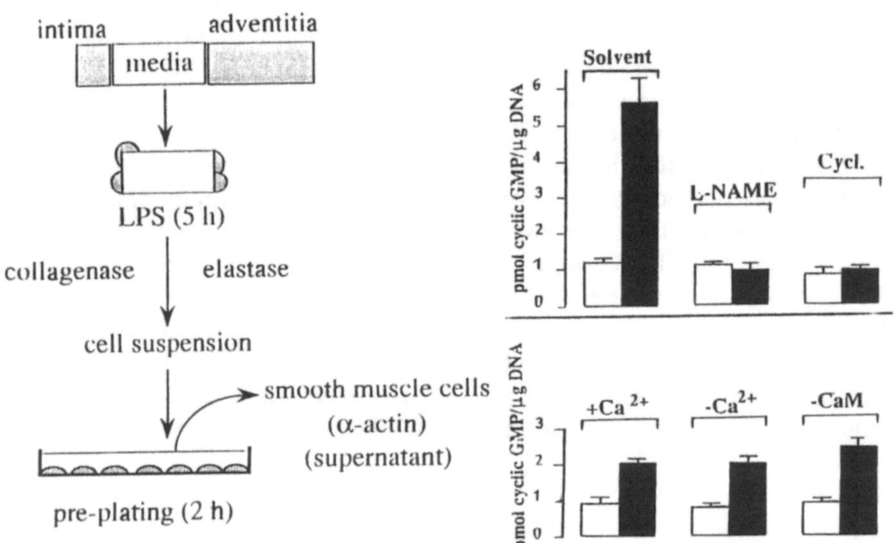

Fig. 4. Induction of NO production by LPS in rat aorta smooth muscle cells. As illustrated in the left part of the figure, the media layer was exposed to LPS (10 μg/mL) for 5 h. It was then submitted to enzymatic digestion and the cell suspension was preplated to eliminate non-vascular smooth muscle cells possibly contaminating the suspension. In those conditions, all cells contained α-smooth muscle actin. cGMP level in control □ and LPS ■ exposed cells was taken as an index of NO production. The right part of the figure shows the results obtained in 2 independent experiments (top and bottom, respectively), in the following experimental conditions: L-NAME 300 μM (L-NAME), cycloheximide (100 ng/mL) throughout the experiment (Cycl.), $CaCl_2$ (200 μM) in the bath ($+Ca^{2+}$), nominally calcium free medium with EGTA (1 mM) ($-Ca^{2+}$), the calmodulin antagonist calmidazinium, 3 μM (−CaM). None of those conditions had any effect on control cGMP levels, consistent with the absence of constitutive NO synthase. L-NAME, and cycloheximide abolished LPS-induced cGMP accumulation, which was unaffected by calcium and the calmodulin antagonist, consistent with the induction of NO-synthase by LPS in vascular smooth muscle cells

cells, macrophages (Fig. 4). This technique resulted in a suspension of pure smooth muscle cells, characterized by the presence of α-smooth muscle actin [51]. Exposure to LPS induced an increase in cGMP, which was abolished by L-NAME or by cycloheximide, but was unchanged in the absence of extracellular calcium or in the presence of the calmodulin antagonist calmidazolium [52]. If we assume that the L-NAME-inhibited elevation of cGMP content is an index of suppression of NO production, then these results indicate that LPS induces NOS in smooth muscle cells.

The Effect of LPS on Small Resistance Vessels

Most of the above studies have been carried out using large calibre vessels isolated from animals; these vessels play little part in resistance changes within individual vascular beds. Changes which might occur in resistance were investigated in small vessels (100–200 μm) removed either from the femoral or mesenteric vascular bed, mounted in organ chambers using a technique [53] which allows changes in developed tension to be measured using an isometric force transducer. In some of these small arteries, the vascular endothelium was gently removed with a tungstan wire. The vessels were removed from rats previously injected with *E. coli* LPS (or solvent) 4 h previously. These vessels had only a slightly decreased sensitivity to norepinephrine; however the addition of L-arginine (which had no effect on control vessels) caused a rapid and stereospecific relaxation of the small arteries from LPS-treated rats which was abolished both by methylene blue (an inhibitor of the activation of guanylyl cyclase by NO) or by the NOS inhibitor L-NAME [54]. Similar results were also obtained with small mesenteric arteries [55]. These findings demonstrate that LPS induces the production of a L-arginine-derived relaxing factor in small resistance arteries and that this production is mostly extra-endothelial and is almost entirely dependent upon extracellular L-arginine. This most likely contributes to the decline in peripheral vascular resistance, and ultimate circulatory failure, which is associated with high mortality in sepsis.

Changes in Vascular Responses induced by LPS *in vivo*

In experimental animals, the administration of LPS (or the induction of sepsis by a combination of cecal ligation and perforation) reduces vasoconstrictor responses to a variety of pressor agents [1–3, 18, 37] as well as to sympathetic nerve stimulation [56]. Although prostaglandin production also contributes, the main factor responsible for this loss of responsiveness now seems to be the generation of NO. This is illustrated in Fig. 5 which shows, in an anesthetized rat preparation, that the pressor responses to norepinephrine are reduced within 1 h of the commencement of an infusion of *E. coli* endotoxin in a dose (5 mg/kg/h) too small to significantly modify arterial pressure

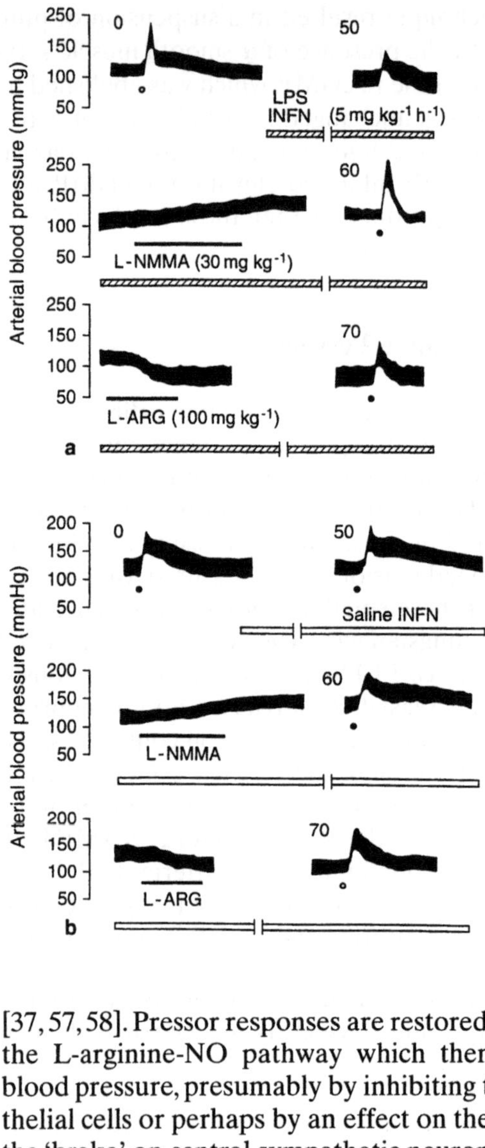

Fig. 5. Pressor responses to norepinephrine (at the closed circles) in anesthetized rats before and during an infusion of bacterial LPS derived from *E. coli* (5 mg/kg/h) or solvent (saline). Responses to norepinephrine (1 μg/kg at the closed circles) were depressed 50 min after commencing the infusion of LPS but these responses were restored to normal (middle panel) following the administration of L-NMMA. Administration of the substrate, L-arginine, resulted in a return of depressed responses to norepinephrine. Responses to norepinephrine in rats infused with saline were unaffected by either L-NMMA or by L-arginine. (From [37] with permission)

[37, 57, 58]. Pressor responses are restored following infusions of inhibitors of the L-arginine-NO pathway which themselves increase systemic arterial blood pressure, presumably by inhibiting the basal release of NO from endothelial cells or perhaps by an effect on the central nervous system to remove the 'brake' on central sympathetic neurones. This antagonism is competitive, since it can be reversed by the administration of a large dose of L-arginine (but not by D-arginine) and is stereospecific, since pressor responses are restored by L-NMMA but not by D-NMMA.

Inhibitors of the L-arginine-NO pathway such as L-NAME also restore normal pressor responses to sympathetic nerve stimulation as well as to exogenous norepinephrine. These studies were performed in preparations which allow electrical stimulation of the spinal sympathetic outflow [56] resulting in frequency-dependent pressor responses. These pressor responses are inhibited by endotoxin and restored by either L-NAME or by vasopressin [58].

We know little as yet concerning whether vasoconstrictor responses to sympathetic nerve stimulation are altered in all vascular beds innervated by such fibers or whether some vascular beds are affected more than others. This can be determined either by measuring blood flow in various vascular beds following sympathetic nerve stimulation under conditions of endotoxemia *in vivo*, or by isolating various vascular beds from rats administered endotoxin and studying their responses to nerve stimulation *ex vivo*. Using the latter approach, there is some evidence that endotoxin impairs norepinephrine release and that this effect is secondary to the increased production of NO and of prostanoids, possibly by the endothelium [59]. These studies were carried out in the isolated perfused mesenteric bed of the rat subjected to electrical field stimulation (which influences predominantly sympathetic vasoconstrictor fibers). In this particular vascular bed, although there is marked impairment of vasoconstrictor responses to electrical field stimulation, there is no alteration (unlike almost all other vascular beds) in the responses to norepinephrine itself. Because the impaired responsiveness to electrical field stimulation is not seen in preparations denuded of endothelium, the suggestion is that such stimulation releases NO from endothelial cells which feeds back to reduced norepinephrine release, presumably presynaptically, from sympathetic nerves. Results such as these [59] suggest that there is not only cross talk between endothelial cells and vascular smooth muscle cells (and also between endothelial cells and cardiac myocytes) but also between endothelial cells and sympathetic neurones. This effect on transmitter release may have pronounced consequences for the loss of vascular tone that occurs under conditions of clinical sepsis.

The other approach is to determine which vascular beds are particularly influenced by endotoxin and is best demonstrated in studies using conscious rats by Sheila Gardiner and her colleagues in Nottingham [60]. They used rats instrumented with pulsed Doppler probes (or electromagnetic flow probes) and examined regional changes in blood flow during the continuous administration of a dose of LPS which had minimal effects on systemic arterial blood pressure. These changes were assessed over a period of 24 h. In this preparation, there were initially transient increases in mesenteric blood flow and in renal blood flow but reductions in flow to the hindquarters vascular bed (predominantly skin and muscle). Later, there were reductions in mesenteric flow, as well as in hindquarters flow, although renal flow remained elevated. There are thus important regional changes in blood flow during the long-term infusion of LPS. Inhibition of NOS resulted in further impairment of mesenteric blood flow [61, 62] which may be due to the impaired release of norepinephrine described above [59].

It is not only vasoconstrictor responses that are impaired *in vivo* following the administration of LPS, and by inference, in patients with sepsis. There are also marked alterations in responses to certain endogenous vasodilator substances. In mesenteric arteries, vasodilator responses to bradykinin, histamine, substance P and acetylcholine are all inhibited by endotoxin, an effect that has been attributed to endothelial damage [63]. In a detailed study, Guc and his colleagues [58, 64] found, in pithed rats, that vasodilator responses to acetylcholine, 5-hydroxytryptamine. endothelin and bradykinin were all attenuated in rats given endotoxin, even when the blood pressure was restored by vasopressin. This reduced responsiveness could be due to endothelial dysfunction. We [65] have shown that responsiveness to acetylcholine is diminished in vessels incubated with LPS under *in vitro* conditions, whereas responses to NO generated by endothelium-independent vasodilators is unaffected. Again, this implies impaired endothelial function, for which there is other evidence early in endotoxemia.

These studies showing impaired vasodilator responses following sepsis and endotoxemia are difficult to reconcile with the evidence that it is the overproduction of NO that is responsible for the reduced vasoconstrictor responses. Working in Strasbourg, Fleming and her colleagues [43] showed that following an incubation of normal blood vessels, with or without endothelium, with endotoxin *in vitro,* there was an initial period of hyporesponsiveness to acetylcholine, at a time when responses to norepinephrine were unaffected (presumably due to enhanced NO production) but that later responsiveness to acetylcholine declined (presumably due to endothelial dysfunction).

There are also marked changes in hyperemic responses induced by blood vessel occlusion (post-occlusion hyperemia) following endotoxin administration. For example, in the mesenteric vascular bed, when the superior mesen-

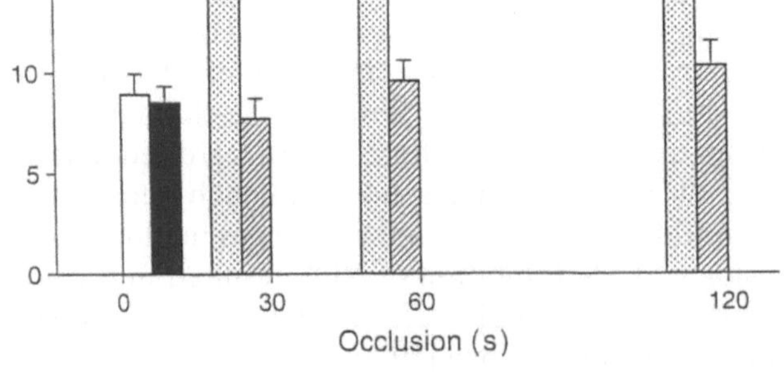

Fig. 6. The effect of endotoxin treatment on post-occlusion hyperemia in the mesenteric vascular bed of anesthetized rats. Endotoxin was administered (20 mg/kg i.p.) and 90 min later they were anesthetized with pentobarbitone. The magnitude of the post-occlusion hyperemia in the superior mesenteric artery was markedly reduced by endotoxin pretreatment although the resting flow (mL/min) was similar to the controls. (From [66] and submitted for publication)

teric artery is occluded for short periods of time, there is on release a marked hyperemia, the extent of which depends upon the foregoing occlusion period. This post-occlusion (or reactive) hyperemia is almost completely abolished in rats administered LPS in a dose which neither influences systemic arterial blood pressure or control flow to the mesenteric vascular bed (Fig. 6). This impairment of reactive hyperemia is also probably the result of endothelial dysfunction [66].

The general conclusion from all the above studies is that following endotoxin administration, and presumably during sepsis in patients, there is a general loss of vascular reactivity and that this applies to both vasoconstrictor and vasodilator responses. Further, there may be differences in responsiveness under these conditions in different vascular beds. Particularly important is the finding that responsiveness to norepinephrine is unimpaired in the mesenteric vascular bed, and we believe that this contributes to the failure of this organ in sepsis and supports the suggestion that the gut is a major target organ in shock associated with sepsis.

Inhibition of NOS as an Approach to the Treatment of Patients with Sepsis

Potentially, there are both beneficial and detrimental effects of inhibiting NOS under these conditions [67]. Among the detrimental effects of inhibiting NOS are damage to the small intestine, characterized by hyperemia, vasocongestion, an increase in vascular permeability and hemorrhage. At least in experimental studies, endotoxin-induced jejunal damage is enhanced rather than reduced by inhibitors of the L-arginine-NO pathway such as L-NMMA [68]. Another potentially detrimental effect is increased platelet adhesion to damaged endothelial surfaces and an increase in platelet accumulation, particularly in the lungs. Although changes in gas exchange have not been analyzed in patients following inhibition of the L-arginine-NO pathway, there are marked increases in pulmonary vascular resistance and pulmonary artery occlusion pressure [69].

The primary object of using this approach in patients would be the restoration of arterial blood pressure and responsiveness to sympathetic nerve stimulation and to exogenous catecholamines. Presumably, this would mean that doses of administered catecholamines (e.g. dopamine, norepinephrine) could be reduced. There is no doubt that the injection of these inhibitors both in normal human subjects and in patients with septic shock elevates systemic arterial pressure and increases vascular resistance in many (most?) vascular beds [70–73]. These changes in vascular resistance can be considerable [70, 73]. This of course could have markedly detrimental effects on oxygen delivery to cells in which aerobic metabolism is already impaired. This marked increase in resistance that occurs in many vascular beds could in part be due to a reduction in normal vasodilator responses to plasma kinins [74–76] which are released in considerable amounts during sepsis [10]. This marked increase in peripheral vascular resistance which occurs under conditions of sepsis when the L-arginine-NO pathway is inhibited is mainly responsible for the substantial reduction in cardiac output that can occur under these conditions [70, 72, 73]. This occurs despite an often increased myocardial contractility [77] which occurs because NO contributes to the reduction in myocardial contractility that occurs, at least under experimental conditions, following elevation of blood levels of LPS [77, 78]. This is discussed in detail elsewhere in this volume.

On balance, it seems rather premature to use classical inhibitors of the L-arginine-NO pathway as adjuncts to the therapy of patients with the sepsis syndrome, and there is no evidence, even in experimental preparations, for an improvement of mortality when this approach is used [67]. However, it could well be that the failure of this approach to date could be due, in part to the fact that the inhibitors were used too late in the sepsis process. Certainly, in some cases they have been used almost as a last resort. This is quite inappropriate and could result in the premature dismissal of drugs that inhibit NO production when, under carefully defined and controlled conditions, they might indeed prove helpful. Perhaps the most appropriate time to investigate whether decreasing NO production might be useful would be during the early, hyperdynamic phase of sepsis. This would be in order to interrupt the viscious cycle between marked and sustained peripheral vasodilatation and the increased cardiac output; this relationship between cardiac output and systemic vascular resistance is markedly altered in patients with the sepsis syndrome and accounts for the fundamental circulatory abnormality (reviewed in [79]).

Another possible approach would be to prevent the formation of NO only through the induced NOS. There is some evidence [80, 81] that using 'selective' inhibitors of the induced enzyme, one can restore responsiveness to norepinephrine (and to sympathetic nerve stimulation) without greatly modifying either systemic arterial pressure or peripheral vascular resistance [80, 82, 83]. This is an interesting approach which will be discussed in some detail by Christoph Thiemmermann later in this volume (pages 201–216).

Conclusion

The marked loss in vascular responsiveness which occurs during endotoxemia and in patients with sepsis, particularly during the hyperdynamic phase, is predominantly due to the induction of a NOS in various components of the blood vessel wall, including vascular smooth muscle cells. This induction is predominantly cytokine-mediated, although there is still the possibility that LPS can induce the enzyme through a direct interaction with its receptor on the surface of endothelial cells. The evidence that NO is responsible for many of the circulatory derangements (sustained peripheral vasodilatation, reduction in myocardial contractility) that occur in patients with sepsis is now overpowering. This has marked implications for the treatment of patients with sepsis but it is still unclear as to whether this approach is satisfactory because, as yet, there is no evidence at least using standard inhibitors of the pathway, for an improvement in mortality; rather the reverse. Possible future approaches would be to selectively inhibit the induced enzyme allowing the important protective functions of the constitutive enzyme to be maintained. Of course, the possibility exists that the reason for the induction of NO formation under these conditions is that, with endothelial dysfunction, these protective effects of the constitutively formed NO are lost and replaced by NO formed from the induced enzyme. In other words, this could be part of the body's response to infection to counteract a marked deterioration of endothelial function.

Acknowledgements. The author's work in this field was supported by the European Economic Community initially by grant (ST2J-0457C(EDB)) and currently by a BIOMED I grant (BMH 1-CT92-1993) together with R Busse, BL Furman, M Hecker, I Fleming, S Nagy and C Thiemmermann.

References

1. Parratt JR (1973) Myocardial and circulatory effects of E coli endotoxin: Modification of responses to catecholamines. Br J Pharmacol 47:12–25
2. Fink MP, Homer LD, Fletcher JR (1985) Diminished pressor response to exogenous norepinephrine and angiotensin II in septic, unanesthetized rats: Evidence for a prostaglandin-mediated effect. J Surg Res 38:335–342
3. Schaller MD, Waeber B, Nussberger J, Brunner HR (1985) Angiotensin II, vasopressin, and sympathetic activity in conscious rats with endotoxemia. Am J Physiol 249: H1086–H1092
4. Thijs LG, Groeneveld ABJ, Schneider AJ (1987) Changing hemodynamic concepts in septic shock. In: Kox W, Bihari D (eds) Septic shock and the adult respiratory distress syndrome. Springer, Berlin Heidelberg New York Tokyo, pp 79–94
5. Parker MM, Shelmaher JH, Bacharach SL, et al (1984) Profound but reversible myocardial depression in patients with septic shock. Ann Intern Med 100:483–490
6. Groenveld ABJ, Bronsveld W, Thijs LG (1986) Hemodynamic determinants of mortality in human septic shock. Surgery 99:265–274
7. Parrillo JE (1986) Cardiovascular dysfunction in humans with septic shock. In: Vincent JL (ed) Update in intensive care and emergency medicine, vol 1, Springer, Berlin Heidelberg New York Tokyo, pp 265–274
8. Parrillo JE (1985) Cardiovascular dysfunction in septic shock: New insights into a deadly disease. Internat J Cardiol 17:314—321
9. Pollack MM, Fields AI, Ruttimann UE (1984) Sequential cardiopulmonary variables of infants and children in septic shock. Crit Care Med 12:554–559
10. Parratt JR (1983) Neurohumoral agents and their release in shock. In: Altura BM, Lefer AM, Schumer W (eds) Handbook of shock and trauma, Vol 1 (Basic Science). Raven, New York, pp 311–336
11. Pomerantz K, Casey L, Fletcher JR, Ramwell PW (1982) Vascular reactivity in endotoxin shock: Effect of lidocaine or indomethacin pretreatment. Adv Shock Res 7: 191–198
12. McKenna TM, Martin FM, Chernow B, Briglia FA (1986) Vascular endothelium contributes to decreased aortic contractility in experimental sepsis. Circ Shock 19:267–273
13. Wakabayashi I, Hatake K, Kakishita E, Nagai K (1987) Diminution of contractile response of the aorta from endotoxin-injected rats. Eur J Pharmacol 141:117–122

14. Bigaud M, Julou-Schaeffer G, Parratt JR, Stoclet JC (1990) Endotoxin-induced impairment of vascular smooth muscle contractions elicited by different mechanisms. Eur J Pharmacol 190:185–192
15. Jones SB, Romano FD (1984) Plasma catecholamines in the conscious rat during endotoxicosis. Circ Shock 14:189–201
16. McKechnie K, Dean HG, Furman BL, Parratt JR (1985) Plasma catecholamines during endotoxin infusion in conscious unrestrained rats: Effects of adrenal demedullation and/or guanethidine treatment. Circ Shock 17:85–94
17. Carcillo JA, Litten RFZ, Suba EA, Roth BL (1988) Alterations in rat aortic $alpha_1$-adrenoceptors and $alpha_1$-adrenergic stimulated phosphoinositide hydrolysis in intraperitoneal sepsis. Circ Shock 26:331–339
18. McCaig DJ, Parratt JR (1980) Reduced myocardial response to calcium during endotoxin shock in the cat. Circ Shock 7:23–30
19. Gray GA, Julou-Schaeffer G, Oury K, Fleming I, Parratt JR, Stoclet JC (1990) An L-arginine-derived factor mediates endotoxin-induced vascular hyposensitivity to calcium. Eur J Pharmacol 191:89–92
20. Furchgott RF, Zawadzki JV (1980) The obligatory role of endothelial cells in the relaxation of arterial smooth muscle by acetylcholine. Nature (London) 288:373–376
21. Ignarro LJ, Buga GM, Wood KS, Byrns RE, Chaudhuri G (1987) Endothelium-derived relaxing factor produced and released from artery and vein is nitric oxide. Proc Natl Acad Sci USA 84:9265–9269
22. Palmer RMJ, Ferrige AG, Moncada S (1987) Nitric oxide release accounts for the biological activity of endothelium-derived relaxing factor. Nature 327:524–526
23. Palmer RMJ, Ashton DS, Moncada S (1988) Vascular endothelial cells synthesize nitric oxide from L-arginine. Nature 333:664–666
24. Palmer RMJ, Rees DD, Ashton DS, Moncada S (1988) L-arginine is the physiological precursor for the formation of nitric oxide. Biochem Biophys Res Commun 153: 1251–1256
25. Schmidt HH, Nau H, Wittfoht W, et al (1988) Arginine is a physiological precursor of endothelium-derived nitric oxide. Eur J Pharmacol 154:213–216
26. Palmer RMJ, Moncada S (1989) A novel citrulline-forming enzyme implicated in the formation of nitric oxide by vascular endothelial cells. Biochem Biophys Res Commun 158:348–352
27. Ishii K, Chang B, Kerwin JF Jr, Huang ZJ, Murad F (1990) N^W-Nitro-L-arginine: A potent inhibitor of endothelium-derived relaxing factor formation. Eur J Pharmacol 176:219–223
28. Moore PK, Al-Swayeh OA, Chong NWS, Evans RA, Gibson A (1990) L-N^G-nitro arginine (L-NOARG), a novel, L-arginine-reversible inhibitor of endothelium-dependent vasodilation in vitro. Br J Pharmacol 99:408–412
29. Rees DD, Palmer RMJ, Moncada S (1989) Role of endothelium-derived nitric oxide in the regulation of blood pressure. Proc Natl Acad Sci USA 86:3375–3378
30. Busse R, Mulsch A, Fleming I, Hecker M (1993) Mechanisms of nitric oxide release from the vascular endothelium. Circulation 87 (Suppl V):V18–V25
31. Julou-Schaeffer G, Gray GA, Fleming I, Schott C, Parratt JR, Stoclet JC (1990) Activation of the L-arginine-nitric oxide pathway is involved in vascular hyporeactivity induced by endotoxin. J Cardiovasc Pharmacol 17:S207–S212
32. Parratt JR, Stoclet JC (1992) Nitric oxide as a mediator of the vascular derangements of sepsis and endotoxemia. In: Lamy M, Thijs LG (eds) Mediators of Sepsis, Springer, Berlin, pp 174–189
33. Parratt JR, Stoclet JC, Fleming I (1993) The role of the L-arginine nitric oxide pathway in sepsis and endotoxemia with special reference to vascular impairment. In: Schlag G, Redl E (eds) Pathophysiology of shock, sepsis and organ failure. Springer, Berlin, pp 575–592
34. Parratt JR, Stoclet JC, Furman BL (1993) Substances mainly derived from vascular endothelium (endothelium-derived relaxing factor, or nitric oxide, and endothelin)

as chemical mediators in sepsis and endotoxemia. In: Naugebauer E, Holaday JW (eds) Handbook of mediators of septic shock. CRC Press, Boca Raton, pp 381–393
35. Moncada S, Palmer RMJ, Higgs EA (1990) Nitric oxide: Physiology, pathophysiology and pharmacology. Pharmacol Rev 43:109–142
36. Fleming I, Gray GA, Julou-Schaeffer G, Parratt JR, Schott C, Stoclet JC (1990) Impaired vascular reactivity in the rat following endotoxin treatment can be endothelium independent, yet involves the L-arginine pathway. J Physiol 423:18P (Abst)
37. Julou-Schaeffer G, Gray GA, Fleming I, Schott C, Parratt JR, Stoclet JC (1990) Loss of vascular responsiveness induced by endotoxin involves the L-arginine pathway. Am J Physiol 259:H1038–H1043
38. Fleming I, Gray GA, Julou-Schaeffer G, Parratt JR, Stoclet JC (1990) Incubation with endotoxin activates the L-arginine pathway in vascular tissue. Biochem Biophys Res Commun 171:562–568
39. Fleming I, Julou-Schaeffer GJ, Gray GA, Parratt JR, Stoclet JC (1991) Evidence that an L-arginine/nitric oxide dependent elevation of tissue cyclic GMP content is involved in depression of vascular reactivity by endotoxin. Br J Pharmacol 103:1047–1052
40. Fleming I, Gray GA, Parratt JR, Stoclet JC (1991) LPS-induced activation of the L-arginine NO pathway in endothelium denuded arterial tissue is time dependent and requires 'de nova' protein synthesis. Br J Pharmacol 102:123P (Abst)
41. Knowles RG, Salter M, Brooks SL, Moncada S (1990) Antiinflammatory glucocorticoids inhibit the induction by endotoxin of nitric oxide synthase in the lung, liver and aorta of the rat. Biochem Biophys Res Commun 172:1042–1148
42. McKenna TM, Reusch DW, Simpkins CO (1988) Macrophage-conditioned medium and interleukin 1 suppress vascular contractility. Circ Shock 25:187–196
43. Fleming I, Gray GA, Stoclet JC (1993) Influence of endothelium induction of the L-arginine nitric oxide pathway in rat aortas. Am J Physiol 264:H1200–H1207
44. Bernard C, Szekely B, Philip I, Wollman E, Payen D, Tedgui A (1992) Activated macrophages depress the contractility of rabbit carotids via a L-arginine nitric oxide dependent effector mechanism. Connection with amplified cytokine release. J Clin Invest 89:851–860
45. Nathan O (1992) Nitric oxide as a secreting product of mammalian cells. FASEB J 6: 3051–3064
46. Rees DD, Cellek S, Palmer RMJ, Moncada S (1990) Dexamethasone prevents the induction by endotoxin of a nitric oxide synthase and the associated effects on vascular tone: An insight into endotoxin shock. Biochem Biophys Res Commun 173:541–547
47. Chamley-Campbell JH, Campbell GR, Ross R (1979) The smooth muscle cell in culture. Physiol Rev 59:1–61
48. Beasley D, Schwartz JH, Brenner BM (1991) Interleukin 1 induced prolonged L-arginine-dependent cyclic guanosine monophosphate and nitrite production in rat vascular smooth muscle cells. J Clin Invest 87:602–608
49. Busse R, Mulsch A (1990) Induction of nitric oxide synthase by cytokines in vascular smooth muscle cells. FEBS Lett 275:87–90
50. Schini VB, Junquero DC, Scott-Burden R, Vanhoutte PM (1991) Interleukin-1β induces the production of a L-arginine derived relaxing factor from cultured smooth muscle cells from rat aorta. Biochem Biophys Res Commun 1:114–121
51. André P, Michel M, Schott C, Stoclet JC (1991) Endothelial and smooth muscle cells from the same rat aorta *in vitro*. Cell Dev Biol 27A:687–688
52. Fleming I, Gray GA, Schott C, Stoclet JC (1991) Inducible but not constitutive production of nitric oxide by vascular smooth muscle cells. Eur J Pharmacol 200:375–376
53. Mulvany MJ, Halpern W (1977) Contractile properties of small arterial resistance vessels in spontaneously hypertensive and normotensive rats. Circulation Res 41:19–26
54. Schneider F, Bucher B, Schott C, André A, Julou-Schaeffer G, Stoclet JC (1994) Effect of bacterial lipopolysaccharide on function of small femoral arteries from rat. Am J Physiol 266:H191–H198

55. Schneider F, Schott C, Stoclet JC, Julou-Schaeffer G (1992) L-arginine induces relaxation of small mesenteric arteries from endotoxin-treated rats. Eur J Pharmacol 211:269–272
56. Gray GA, Furman BL, Parratt JR (1990) Endotoxin-induced impairment of vascular reactivity in pithed rat: Role of arachidonic acid metabolites. Circ Shock 31:395–406
57. Gray GA, Schott C, Julou-Schaeffer G, Fleming I, Parratt JR, Stoclet JC (1991) An investigation of the effect of inhibitors of the L-arginine pathway on endotoxin-induced vascular hyporeactivity in vivo. Br J Pharmacol 103:1218–1224
58. Guc MO, Funnan BL, Parratt JR (1990) Endotoxin-induced impairment of vasopressor and vasodepressor responses in the pithed rat. Br J Pharmacol 101:913–919
59. Fatehi-Hassanabad Z, Furman BL, Parratt JR (1995) Endotoxin impairs sympathetic responses in the rat isolated perfused mesenteric bed. Role of nitric oxide in cyclooxygenase products. Br J Pharmacol (In press)
60. Gardiner SM, Kemp PA, Bennett T (1992) Regional haemodynamic changes during chronic infusion of lipopolysaccharide (LPS) in conscious rats. Br J Pharmacol 107: 290P (Abst)
61. Waller J, Gardiner SM, Bennett T (1994) The effect of aminoguanidine on the regional haemodynamic changes during a continuous low dose infusion of lipopolysaccharide in conscious rats. Br J Pharmacol 113:6P (Abst)
62. Gardiner SM, Kemp PA, Bennett T (1994) The effect of N-acetyl-5-5 hydroxytryptamine on regional haemodynamic responses to chronic endotoxaemia in conscious rats. Br J Pharmacol 112:525P (Abst)
63. Altura BM, Gebrewold A, Burton RW (1985) Failure of microscopic metarterioles to illicit vasodilator responses to acetylcholine, bradykinin, histamine and substance P after ischemic shock, endotoxemia and trauma: Possible role of endothelial cells. Microcirc Endoth Lymph 2:121–134
64. Guc MO, Gray GA, Furman BL, Parratt JR (1991) Endotoxin-induced impairment of vasodepressor responses in the pithed rat. Eur J Pharmacol 204:63–70
65. Gray GA, Julou-Schaeffer G, Fleming I, Parratt JR, Stoclet JC (1990) Endothelial function is impaired in aortae from lipopolysaccharide (LPS)-treated rats. J Mol Cell Cardiol 22:8
66. Fatehi-Hassanabad Z, Parratt JR, Furman BL (1994) Effects of endotoxin on the reactive hyperaemia in the rat mesenteric bed. Shock 2 (Suppl):46 (Abst)
67. Parratt JR (1995) Nitric oxide and cardiovascular dysfunction in sepsis and endotoxaemia: An introduction and an overview. In: Fourth Wiggers Bernard Conference. Berlin, Springer (In press)
68. Boughton-Smith NK, Hutcheson IR, Deacon AM, Whittle BJR, Moncada S (1990) Protective effect of S-nitroso-N-acetyl-penicillamine in endotoxin-induced acute intestinal damage in the rat. Eur J Pharmacol 191:485–488
69. Petros A, Lamb G, Leone A, Moncada S, Bennett D, Vallance P (1994) Effects of a nitric oxide synthase inhibitor in humans with septic shock. Cardiovasc Res 28:34–39
70. Petros A, Bennett D, Vallance P (1991) Effect of nitric oxide synthase inhibitors on hypotension in patients with septic shock. Lancet 338:1557–1558
71. Lorente JA, Landin L, de Pablo R, Renes E, Liste D (1993) L-arginine pathway in the sepsis syndrome. Crit Care Med 21:1287–1295
72. Lorente JA, Landin L, Renes E, Esteban A (1993) Regulation of vascular tone in sepsis. Intensive Care World 10:58–62
73. Lorente JA, Landin L, Esteban A (1994) Role of nitric oxide in the regulation of vascular tone in septic shock. In: Vincent JL (ed) Yearbook of intensive care and emergency medicine 1994. Springer, Berlin Heidelberg New York pp 75–89
74. Cockcroft JR, Chowienczyk PJ, Brett SE, Ritter JM (1994) Effect of N^G-monomethyl-L-arginine on kinin-induced vasodilation in the human forearm. Br J Clin Pharmacol 38:307–310
75. O'Kane KPJ, Webb DJ, Collier JG, Vallance PTJ (1994) Local N-N^G-monomethyl-arginine attenuates the vasodilator action of bradykinin in the human forearm. Br J Clin Pharmacol 38:311–316

76. Cockcroft JR, Chowienczyk PJ, Brett SE, Bender N, Ritter JM (1994) Inhibition of bradykinin-induced vasodilation in human forearm vasculature by icatibant, a potent B_2-receptor antagonist. Br J Clin Pharmacol 38:317–322
77. Nagy S, Kaszaki T, Wolford A, Parratt JR (1994) Effect of inhibition of nitric oxide synthase on myocardial contractility and cardiac output in hyperdynamic low-dose endotoxaemia. Shock 2:28P (Abst)
78. Brady AJB, Poole-Wilson PA, Harding SE, Warren JB (1992) Nitric oxide production within cardiac myocytes reduces their contractility in endotoxaemia. Am J Physiol 263:H1963–H1966
79. Parratt JR (1989) Alterations in vascular reactivity in sepsis and endotoxemia. In: Vincent JL (ed) Update in intensive care and emergency medicine, Vol 8. Springer, Berlin pp 299–308
80. Burns HO, Furman BL, Parratt JR (1993) Comparison of methylene blue and L-canavanine on endotoxin-induced impairment of vascular reactivity. Circ Shock (Suppl 1) P26 (Abst)
81. Paya D, Gray GA, Fleming I, Stoclet JC (1993) Effect of dexamethasone on the onset and persistence of vascular hyporeactivity induced by *E coli* lipopolysaccharide in rats. Circulatory Shock 41:103–112
82. Thiemermann C (1994) Role of the L-arginine-nitric oxide pathway in circulatory shock. Adv Pharmacol 28:45–49
83. Southan GJ, Szabo C, Thiemermann C (1995) Isothioureas: Potent inhibitors of nitric oxide synthases with variable isoform selectivity. Br J Pharmacol 114:510–516

Nitric Oxide as a Mediator in Gram-Positive Sepsis

J. Wilkinson, S. Sriskandan, and J. Cohen

Introduction

Sepsis and its sequelae are a cause of significant morbidity and mortality worldwide [1], due in part to a rise in the incidence of nosocomial infections as a consequence of the increased use of invasive procedures. Septic shock is now the commonest cause of death on intensive care units (ICU) in the US [2]. Traditionally, septic shock has been attributed to gram-negative infections, but it may be caused by gram-positive bacteria and other organisms. Gram-positive bacteria are increasingly being recognized as the cause of septic shock and sepsis syndromes [3]; in most series, they account for 40–50% of cases of bacteremia, and the mortality is at least as high as that seen complicating gram-negative infections.

The principal gram-positive bacteria which cause shock are *Staphylococcus aureus* and *Streptococcus pyogenes*. In addition, *Streptococcus pneumoniae* may be implicated, particularly in asplenic patients, and there is a curious but well recognized syndrome of profound shock due to viridans streptococci which occurs in neutropenic patients, often associated with the adult respiratory distress syndrome (ARDS) [4]. *Staphylococcus epidermidis* is commonly isolated from blood cultures but rarely causes shock.

Streptococcus pyogenes is a group A β-hemolytic streptococcus that causes a number of infections ranging from cellulitis to streptococcal toxic shock like syndrome (TSLS). Recently there has been an increase in the incidence of severe streptococcal infections, both in the US and UK. These severe infections are associated with an increased incidence of circulatory shock and death. They tend to affect otherwise healthy young people [3, 5, 6]. Profound hypotension is a clinical feature of TSLS; it is difficult to treat and associated with a high morbidity and mortality.

The cardiovascular manifestations of sepsis have been studied very extensively. Many mediators have been implicated, including histamine, products of arachidonic acid metabolism, endorphins, and endothelin. Lately, the role of nitric oxide (NO) has been appreciated. The potential role of NO in the pathogenesis of gram-negative septic shock is far more clearly understood than in gram-positive septic shock. In gram-negative sepsis, endotoxin (lipopolysaccharide, LPS) and cytokines cause induction of the inducible NO synthase (iNOS) in vascular smooth muscle cells leading to vasodilatation, circulatory collapse and shock. LPS [7], tumor necrosis factor-α (TNF-α), interleukin 1 (IL-1) and interferon gamma (IFNγ) have all been shown to be potent stimuli of NO production *in vitro* [8], and *in vivo*, patients with the sepsis syndrome and septic shock show increased levels of nitrite and nitrate in the blood [9]. Experimentally, the inhibition of the synthesis of NO by N^G-monomethyl-L-arginine (L-NMMA), a non-selective inhibitor of NOS, can reverse the hypotension seen in septic shock [10].

Less is known about the pathogenesis of gram-positive shock. Gram-positive organisms produce a number of exotoxins which are thought to play an important role in pathogenesis. These include the streptococcal pyrogenic exotoxins (SPE) of which there are three; A, B and C. SPE A is believed to act as a superantigen, inducing the production of large amounts of cytokines which can mediate tissue damage and shock [11]. SPE A and the products of sterile cell free group A streptococcal supernatants have been shown to induce TNF-α production by peripheral blood mononuclear cells [12].

Because of the similarity between some of the clinical features of gram-negative and gram-positive sepsis syndromes, it is of interest to ask whether NO is involved in the pathogenesis of shock due to gram-positive bacteria. To approach this we have carried out a number of experiments, both *in vitro* and *in vivo*, using as a model *Streptococcus pyogenes*.

Induction of NO *in vitro*

SPE A as an Inducer of NO

SPE A is a powerful mitogen and is able to induce a number of cytokines. The ability of recombinant SPE A to induce nitrite production was therefore of interest. For these studies, recombinant SPE A was prepared by using the polymerase chain reaction to amplify the gene from a wildtype strain of *S. pyogenes* (H250) known to produce SPE A, and then using a high efficiency vector to express it in *Escherichia coli*. The final concentration of pure SPE A was 340 μg/mL (LPS levels 3 ng/mL). This protein was used to immunize a rabbit, and we obtained an antibody to SPE A which reacted with both native and recombinant protein on a Western blot, and neutralized the mitogenic activity of recombinant SPE A.

The RAW 264 murine macrophage cell line was purchased from the European collection of animal cell cultures, Porton Down, UK, and maintained in DMEM containing 10% FCS and penicillin and streptomycin. 5×10^5 RAW cells were added to each well of a 24 well plate. The cells were then incubated in a 5% CO_2 rich atmosphere at 37°C for 6 h before stimulation. The Griess reaction was used to measure nitrite in the cell culture supernatant. SPE A alone had no effect, but in combination with recombinant murine rmIFNγ produced significant stimulation of NO production. However, some of this effect was apparently due to contaminating LPS since after the addi-

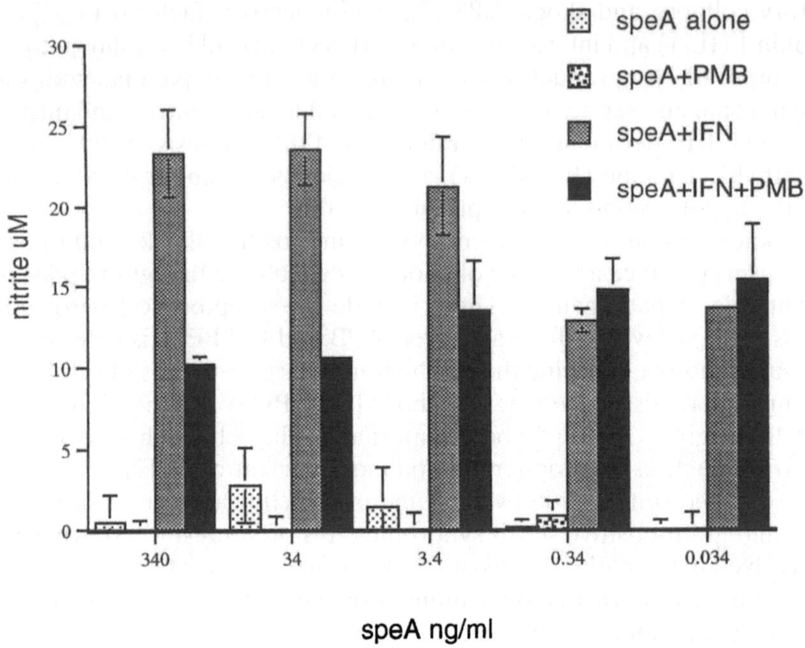

Fig. 1. Nitrite production by RAW cells stimulated with speA. speA alone has little effect, but there is activity when combined with rmIFNγ. In this experiment, rmIFNγ alone resulted in the release of 11 μM nitrite. speA = streptococcal pyrogenic exotoxin A; PMB = polymyxin B; IFN = rmIFNγ

tion of polymyxin B (PMB) only concentrations of SPE A >3.4 ng/mL produced a significant release of nitrite (Fig. 1).

We also performed an experiment with culture supernatant from *S. pyogenes* H250 (a strain known to produce SPE A) and rmIFNγ, in which pre- and post-immune anti-SPE A antiserum was added to the wells. The addition of the antiserum had no effect on nitrite production.

Effect of Streptococcal Culture Supernatants on NO Production

Both clinical isolates and type strains of *S. pyogenes* were used to prepare sterile, cell free supernatants. The supernatants were added at a 1 in 10 dilution in DMEM to the wells of the tissue culture plates, together with PMB at 5–10 μg/mL to neutralize any LPS contaminating the bacterial supernatants. rmIFNγ (Genzyme) was used at 200 u/mL to stimulate the cells. All the plates were incubated for 18 h.

Figure 2 shows the result of stimulation of RAW cells with streptococcal supernatant prepared in RPMI. Bacterial supernatant alone produced no significant increase in nitrite above background. rmIFNγ caused a small

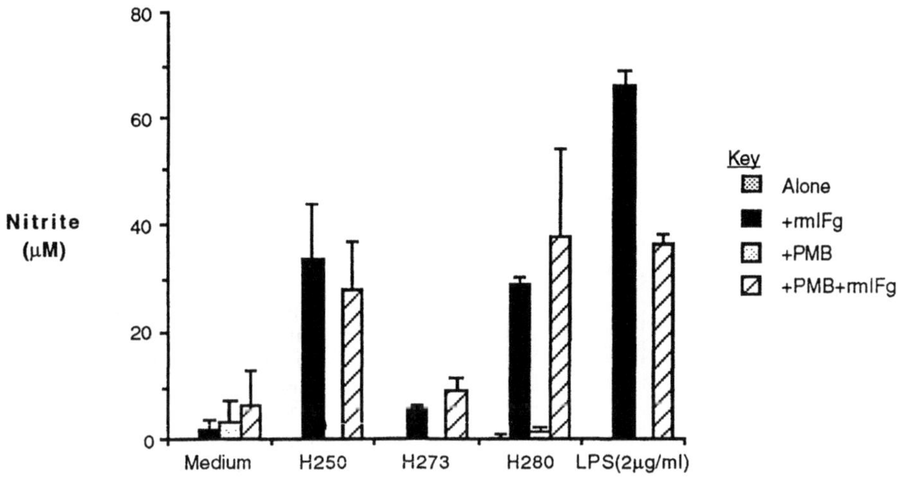

Fig. 2. Nitrite production by RAW cells stimulated with culture supernatant from 3 strains of *S. pyogenes*. rmIFg = rmIFNγ; PMB = polymyxin B

amount of nitrite production ($p<0.001$). Co-stimulation using bacterial supernatant and IFNγ caused a significant rise in nitrite production, seen with all supernatants ($p<0.001$). The addition of PMB alone had no effect on nitrite production, nor did it affect nitrite production by the combination of bacterial supernatant and IFNγ. There were significant differences between the supernatants in their ability to induce nitrite; H273 was the least potent ($p<0.001$). Similar results were obtained in 4 separate experiments.

Supernatants produced from H250 and H280 were heated for 10 min at 100° C and re-tested for their ability to induce NO from RAW cells in combination with rmIFNγ. Boiling did not impair NO induction, and this was not affected by PMB. Next, supernatants were fractionated by ultrafiltration. When a MW cut-off of 30000 kD was used, NO-inducing capacity was retained in the lower MW fractions. The addition of PMB had no effect. When the MW cut-off was reduced to 10000, both high and low MW fractions had equal effect.

The Effect of anti-TNF-α Antibody

To investigate the role of TNF-α, we added a hamster monoclonal antibody to rmTNF-α at 100 μg/mL to wells containing rmIFNγ or H250 supernatant. 1% hamster serum was used as a control, and PMB was included in all wells. Both H250 supernatant and rmTNF-α induced nitrite in synergy with rmIFNγ ($p<0.001$). In both cases this effect was significantly reduced by the addition of anti-TNF. Hamster serum had no effect (Fig. 3).

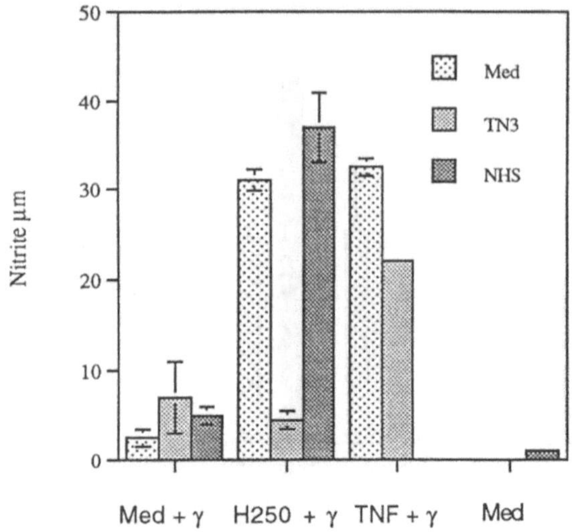

Fig. 3. The effect of monoclonal anti-TNF on nitrite production by RAW cells stimulated by supernatant from *S. pyogenes* H250. Med = medium alone; TN3 = hamster monoclonal antibody to rmTNF-α [TN3.19.12]; NHS = 1% normal hamster serum; γ = rmIFNγ

Induction of NO *in vivo*

We first determined whether recombinant SPE A was able to induce NO after intraperitoneal injection in BALB/c mice. Serum samples obtained at various times after injection were assayed for nitrate by the Griess reaction, but in none was there evidence of nitrate production. Next, we used a model of necrotizing fasciitis in mice, in which animals are given an intramuscular injection of 1×10^9 cfu of live *S. pyogenes*. Approximately 80% of mice become bacteremic, and most demonstrate elevated levels of IL-6. In contrast to the experiments with recombinant SPE A following infection with live organisms, we observed a 2–3 fold elevation of plasma nitrate levels.

Streptococci and the Induction of NO

Streptococcal supernatants, in synergy with rmIFNγ, are able to induce iNOS in a murine macrophage cell line. LPS is not the component responsible for this effect, as is demonstrated by the lack of effect of PMB. The component or components in supernatants which induce iNOS are heat resistant and have a molecular weight of less than 30000 kD. The autocrine production of TNF-α plays an important role in the induction of iNOS. Our data suggest that SPE A is not responsible for iNOS induction. The streptococcal supernatants alone did not induce significant nitrite production *in vitro*; neither did rmIFNγ alone. However, when the supernatants and rmIFNγ were added together, they acted synergistically to produce a highly significant increase in nitrite production. It is known from studies with normal mice that bacterial products often act in synergy with a variety of cytokines to cause the pathological effects seen, and that the individual components alone are relatively inactive [13]. IFNγ is a pivotal lymphokine regulating both non-specific inflammation and the specific immune response. It synergizes with many other cytokines, often exerting a disease promoting effect [14]. H22, a hamster monoclonal antibody to murine IFNγ has been shown to give significant protection against death in a model of *E. coli* sepsis in mice, and in addition, decreases the level of iNOS induction [15]. It is therefore not surprising that rmIFNγ is needed in order for iNOS to be induced.

With a monoclonal antibody to murine TNF-α, we observed that nitrite levels were reduced to those of medium alone in the presence of H250 culture supernatant and rmIFNγ. From clinical studies, we know that patients with both gram-positive and gram-negative infections have significantly raised circulating TNF-α levels [16, 17]. *In vitro*, sterile supernatants from gram-positive bacteria stimulate TNF-α production [12], and our results suggest that the stimulatory effect of the gram-positive supernatants on NO production is dependent, at least in part, on their ability to induce TNF-α.

SPE A is a member of a group of related superantigens, which although immunologically distinct, show significant sequence homology. Biochemically, these toxins are both heat resistant and of low molecular weight. SPE A is known to be a powerful mitogen and powerful inducer of cytokines. Our results show that *in vitro*, even in combination with IFNγ, SPE A only induces iNOS at concentrations >3.4 ng/mL. The anti-SPE A antibody did not affect the ability of H250 to induce nitrite *in vitro*, even in the presence of rmIFNγ. Finally, our results in the murine model failed to implicate SPE A in the pathogenesis of the condition. It is therefore unlikely that SPE A plays a role in the induction of iNOS in sepsis. The ability of other streptococcal toxins to induce iNOS needs to be investigated further.

Conclusions

What then is the role of NO in gram-positive sepsis? In gram-negative sepsis, induction of iNOS in vascular smooth muscle clearly plays a major role in the pathogenesis of circulatory collapse and shock. However, our results suggest that streptococcal supernatants do not induce iNOS in vascular smooth muscle (data not shown). In RAW cells (a macrophage cell line), there is induction of iNOS by streptococcal supernatants. NO is known to be involved in the non-specific cytotoxic activity of macrophages, and in recent studies in rodent models of gram-negative sepsis, we have shown increased levels of nitrosylated proteins (unpublished data), markers of cell damage by NO. It is therefore plausible that the increased production of NO in sepsis may be directly cytotoxic and in part responsible for the inflammatory damage and circulatory collapse seen.

Our results do not exclude the possibility that NO is implicated in the pathogenesis of gram-positive sepsis. For instance, other products of the gram-positive cell wall such as lipoteichoic acid are known to be capable of inducing cytokine production and upregulation of iNOS by macrophages [18, 19]. Furthermore, other gram-positive toxins, such as the TSST-1 produced by some strains of *Staphylococcus aureus*, are able to induce NO [20, 21]. Finally patients with gram-positive sepsis are known to have increased levels of TNF-α and plasma nitrate [9, 22], and it has been suggested that the use of non-specific inhibitors of NO activation such as methylene blue (one property of which is inhibition of soluble guanylyl cyclase), can reverse the loss of vascular responsiveness to norepinephrine seen in patients with gram-positive sepsis [23].

The pathophysiology of gram-positive shock is extremely complex. Further work is necessary to characterize the components of *S. pyogenes* that are capable of inducing iNOS, and the role of other cytokines needs to be investigated. It is clear that as is the case for gram-negative sepsis, acquiring a better understanding of the pathogenesis of gram-positive sepsis will allow the development of more effective treatment strategies.

References

1. The Working Group on Severe Streptococcal Infections (1993) Defining the group A streptococcal toxic shock syndrome. Rationale and consensus definition. JAMA 269: 390–391
2. Parrillo JE (1990) Septic shock in humans. Advances in the understanding of pathogenesis, cardiovascular dysfunction, and therapy. Ann Intern Med 113:227–242
3. Shaunak S, Wendon J, Monteil M, Gordon AM (1988) Septic scarlet fever due to *Streptococcus pyogenes* cellulitis. Q J Med 69:921–925
4. Cohen J, Donnelly JP, Worsley AM, Catovsky D, Goldman JM, Galton DAG (1983) Septicaemia caused by viridans streptococci in neutropenic patients with leukaemia. Lancet 2:1452–1454
5. Stevens DL, Tanner MH, Winship J, et al (1989) Severe group A streptococcal infections associated with a toxic shock-like syndrome and scarlet fever toxin A. N Engl J Med 321:1–7
6. Hoge CW, Schwartz B, Talkington DF, Breiman RF, MacNeill EM, Englender SJ (1993) The changing epidemiology of invasive group A streptococcal infections and the emergence of streptococcal toxic shock-like syndrome. A retrospective population-based study. JAMA 269:384–389
7. Moncada S, Palmer RM, Higgs EA (1991) Nitric oxide: Physiology, pathophysiology, and pharmacology. Pharmacol Rev 43:109–142
8. Nathan C (1992) Nitric oxide as a secretory product of mammalian cells. FASEB J 6: 3051–3064
9. Evans TJ, Carpenter A, Kinderman H, Cohen J (1993) Evidence of increased nitric oxide production in patients with the sepsis syndrome. Circ Shock 41:77–81
10. Finkel MS, Oddis CV, Jacob TD, Watkins SC, Hattler BG, Simmons RL (1992) Negative inotropic effects of cytokines on the heart mediated by nitric oxide. Science 257: 387–389
11. Stevens DL (1992) Invasive group A streptococcus infections. Clin Infect Dis 14:2–13
12. Bayston KF, Tomlinson M, Cohen J (1992) In-vitro stimulation of TNF-α from human whole blood by cell-free supernatants of gram-positive bacteria. Cytokine 4: 397–402
13. Rothstein JL, Schreiber H (1988) Synergy between tumor necrosis factor and bacterial products causes hemorrhagic necrosis and lethal shock in normal mice. Proc Natl Acad Sci USA 85:607–611

14. Heremans H, Van Damme J, Dillen C, Dijkmans R, Billiau A (1990) Interferon gamma, a mediator of lethal lipoloysaccharide-induced Shwartzman-like shock reactions in mice. J Exp Med 171:1853–1869
15. Evans TJ, Carpenter A, Silva AT, Cohen J (1992) Differential effects of monoclonal antibodies to tumor necrosis factor α and interferon gamma on induction of hepatic nitric oxide synthase in experimental gram-negative sepsis. Infect Immun 60: 4133–4139
16. Marks JD, Marks CB, Luce JM, et al (1990) Plasma tumor necrosis factor in patients with septic shock. Mortality rate, incidence of adult respiratory distress syndrome, and effects of methyl prednisolone administration. Am Rev Respir Dis 141:94–97
17. Calandra T, Baumgartner JD, Grau GE, et al (1990) Prognostic values of tumor necrosis factor/cachectin, interleukin-1, interferon-alpha, and interferon-gamma in the serum of patients with septic shock. J Infect Dis 161:982–987
18. Longchampt MO, Auguet M, Delaflotte S, Goulin-Schulz J, Chabrier PE, Braquet P (1992) Lipoteichoic acid: A new inducer of nitric oxide synthase. J Cardiovasc Pharmacol 20 (Suppl 12):S145–S147
19. Timmerman CP, Mattsson E, Martinez-Martinez L, et al (1993) Induction of release of tumor necrosis factor from human monocytes by staphylococci and staphylococcal peptidoglycans. Infect Immun 61: 4167–4172
20. Zembrowicz A, Vane JR (1992) Induction of nitric oxide synthase activity by toxic shock syndrome toxin 1 in a macrophage-monocyte cell line. Proc Natl Acad Sci USA 89:2051–2055
21. Cunha FQ, Moss DW, Leal LM, Moncada S, Liew FY (1993) Induction of macrophage parasiticidal activity by *Staphylococcus aureus* and exotoxins through the nitric oxide synthesis pathway. Immunology 78:563–567
22. Baud L, Cadranel J, Offenstadt G, Luquel L, Guidet B, Amstutz P (1990) Tumor necrosis factor and septic shock. Crit Care Med 18:349–350
23. Schneider F, Lutun P, Runge I, Couchot A, Tempe JD (1992) Effects of soluble guanylyl cyclase inhibition by methylene blue on the loss of vascular responsiveness to norepinephrine in human beings with either gram-negative or gram-positive septic shock. Clin Intensive Care 3 (Suppl):11 (Abst)

Effects of Nitric Oxide on various Cell Types and Organs

Nitric Oxide and the Heart in Sepsis

A. Kumar and J. E. Parrillo

Introduction

The physiologic and pathophysiologic roles of nitric oxide (NO), an endogenously produced molecule through which numerous local and systemic mediators produce their effects, is one of the most intensely studied fields of research in the health sciences today.

In the physiologic state, NO appears to be integral in normal neurotransmission, platelet-vessel wall interactions, vasomotor regulation, and immune cell function. In pathophysiologic conditions, NO can cause severe vasodilatation (in septic shock), contribute to inflammation caused by neutrophils and macrophages (in local infections), and may contribute to myocardial reperfusion injury (post-myocardial infarction). The role of NO in the heart is similarly diverse. Recent evidence suggests that NO may play a substantial part in physiologic regulation of cardiac contractility. In sepsis, however, physiologic control of NO generation in the heart appears to give way to pathophysiologic NO production. This chaper will review the nature of human myocardial dysfunction in sepsis and the recently defined role of NO and cyclic GMP (cGMP) in its development.

Myocardial Depression in Sepsis: Myocardial Hypoperfusion

The question of whether humans exhibit true myocardial depression (i.e. independent of loading conditions) in sepsis and septic shock had existed since the early studies of experimental endotoxic shock in animals almost 30 years ago [1, 2]. Utilizing a novel application of nuclear imaging technology, Parker and colleagues [3] were finally able, in the early 1980s, to definitively demonstrate the existence of human septic myocardial depression (decreased biventricular ejection fraction despite increased cardiac output). However, the etiology of such depression remained in doubt. Two theories regarding the nature of septic myocardial depression in humans and animals existed.

The first suggested that sepsis was associated with globally decreased myocardial perfusion resulting in ischemic injury to the septic heart. This position had been supported over the years by a number of investigators based largely on evidence of global myocardial hypoperfusion in canine models of endotoxic shock [4, 5]. However, as it became apparent that endotoxic shock is not an appropriate model through which to study septic myocardial dysfunction, another theory rose to prominence. The origins of the second major theory regarding the nature of myocardial depression in septic shock are found in Wigger's classic 1947 report [6] proposing the presence of a myocardial depressant factor in experimental hemorrhagic shock. By the 1960s, others had suggested that a similar circulating depressant substance was responsible for septic myocardial depression in humans with septic shock [7, 8].

Only recently have two studies definitively resolved the question [9, 10]. The first of these studies, performed by Cunnion and colleagues [9], involved placement of thermodilution coronary sinus catheters and serial measurements of coronary flow and metabolism in patients with acute septic shock. Coronary blood flow was measured and found to be either normal or increased (depending on the heart rate) compared to normal control subjects with comparable heart rates (Fig. 1). No patient with septic shock was found to have decreased myocardial blood flow. Both the patients who developed myocardial depression and those who did not had comparable myocardial flows. Further, no patient exhibited net myocardial lactate production.

Dhainaut and colleagues [10] subsequently confirmed the same phenomena of coronary blood flow and myocardial metabolism in patients with septic

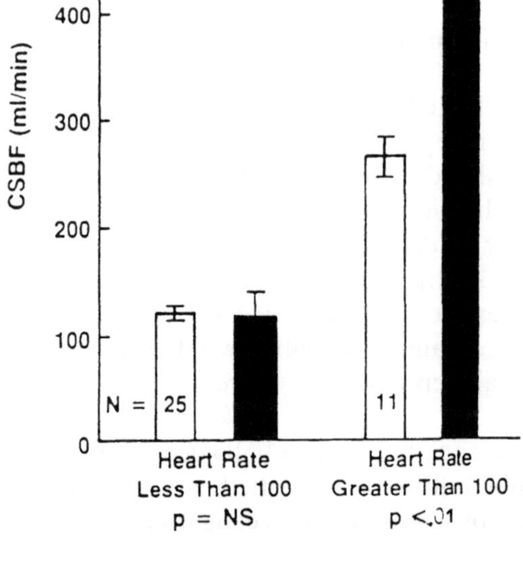

Fig. 1. Mean coronary sinus blood flow in 7 patients with septic shock compared with normal subjects. Flow measurements were stratified into heart rates above and below 100 beats/min. Coronary blood flow in septic shock patients equaled (heart rate <100/min) or exceeded (heart rate >100/min) coronary blood flow in control patients. CSBF = coronary sinus blood flow. (From [9] with permission)

shock using similar coronary sinus catheters. In addition, Goldfarb and colleagues [11, 12] supported the results in a porcine chronic endotoxemia model which closely mimicked human sepsis. Other recent studies [13, 14], some using nuclear magnetic resonance technology, have now shown that myocardial high energy phosphates are preserved in sepsis and septic shock suggesting that aberrant microvascular flow pattern do not result in focal myocardial ischemia. In sum, these studies argue strongly against either global or focal myocardial hypoperfusion as a cause of septic myocardial depression in humans.

Myocardial Depression in Sepsis: Circulating Depressant Substances

Clinical studies using central venous pressure catheters to measure preload in the early 1970s suggested that septic shock was characterized by a hypodynamic circulatory state with reduced cardiac output [15, 16]. These studies would later be determined to be faulty due to their dependence on central venous pressure for assessment of adequate preload, an approach now known to be unreliable in the critically ill. At the same time, however, Lefer and colleagues [8] were able to support these early clinical observations by documenting the existence of a myocardial depressant factor in the blood of dogs subjected to hypodynamic endotoxic shock. By providing a potential explanation for the results of these early (but somewhat misleading) clinical studies, Lefer and his colleagues helped to initiate the subsequent search for the nature of myocardial depressant substances in sepsis and septic shock.

Unfortunately, efforts to isolate the factor or factors (in both human and animal sepsis and septic shock models) have not been entirely successful (Table 1). Lefer and colleagues [8, 17], using cat papillary muscle to examine serum from experimental canine endotoxic shock and spontaneous human hemorrhagic and septic shock, suggested that myocardial depressant activity existed in a peptide of about 800 to 1000 daltons originating in the pancreas [8, 17]. The identity of this substance, however, has not been elucidated. McConn and coworkers [18] infused human septic plasma into canine coronary arteries *in vivo* to demonstrate the existence of a circulating myocardial depressant factor (Table 1). They suggested that depressant activity existed in at least two molecular weight fractions. The first fraction (under 1 kD) appeared to account for early/immediate depressant effects. A second fraction (between 1 and 10 kD) appeared to account for late/delayed depressant activity. However, there was no attempt to specifically isolate these substances.

Most recently, Mink and colleagues [19, 20] have attempted to isolate a factor or factors involved in canine endotoxic shock (live *E. coli* infusion). They have characterized but not identified two factors which are able to depress the contractility of canine trabecular preparations. The first factor may be analogous to the depressant substances described by Lefer [8, 17], since it appears to be under 1 kD in mass (personal communication) and is

Table 1. Myocardial depressant substances in endotoxic and septic shock: Physical and biochemical characterization

Authors	Ref.	Subject	Model	Assay	Control	*In vivo* correlation	Isolation
Wangensteen et al	[8]	canine	endotoxin bolus	cat papillary muscle	none	none	MW 800–1000 d ? peptide
Lovett et al	[17]	human	mixed septic and hemorrhagic shock	cat papillary muscle	serum from peri-operative patients	none	MW 800–1000 d ? peptide
McConn et al	[18]	human	spontaneous sepsis	*in vivo* canine cardiac performance during coronary perfusion with septic plasma	yes, but non-septic plasma caused depression	no formal comparison	multiple fractions MW <1000 d MW 1000–10000 d
Carli et al Benassayag et al	[22] [23]	human	spontaneous septic shock	rat cardiomyocytes	serum from healthy donors	none	MW <1000 d heat stable, lipid soluble ? estrogenic compound
Gomez et al Jha et al	[19] [20]	canine	live *E. coli* infusion	canine right ventricular trabeculae	plasma from non-endotoxic hypotensive dogs	change in *in vivo* pressure/volume curve correlated with *in vitro* trabecular depression	MW 10000–30000 d heat labile, acetone insoluble proteinase sensitive probable protein
Parrillo et al Reilly et al Kumar et al	[24] [25] [130]	human	spontaneous septic shock	rat cardiomyocytes	serum from multiple controls including critically ill non-septic patients	degree of myocyte depression correlated to LVEF	min 2 fractions MW 500–5000 d MW >10000 d heat labile, ethylacetate insoluble probable protein; activity eliminated by immunoabsorption of TNF and IL-1; NO and cGMP dependent

MW: Molecular weight, LVEF: left ventricular ejection fraction both heat stable and proteinase resistant. However, this factor was present in both normal and septic plasma. A second factor was found in only the 10–30 kD fraction of septic animals and found to be heat-labile, proteinase-sensitive and acetone-insoluble (consistent with a protein or peptide) (Table 1).

An alternative technical approach to the assessment of myocardial depressant activity originated with a 1960 report by Harary and Farley [21] who described the preparation of a cell culture of spontaneously beating cardiac myocytes derived from neonatal rats. Using this model, Carli and coworkers [22, 23] demonstrated that human serum collected during early septic shock depressed the contractility of myocytes in comparison to serum from healthy controls (Table 1). However, no *in vivo* correlation was offered nor were there other controls using serum from non-septic critically ill patients. The substance they described was heat stable, lipid soluble, had a molecular weight under 1 kD (similar to that described by Lefer [8, 17], and most recently by Mink et al [19, 20]. The authors suggested that it might represent an estrogenic compound.

A series of reports by Parrillo and colleagues in the mid to late 1980s were important in conclusively linking clinical myocardial depression during septic shock in humans to the effect of a myocardial depressant substance (MDS) in the patient's own serum. These reports also provided the best available data on the characteristics of the human form of MDS. These studies provide the strongest evidence that the myocardial depressant activity of serum or plasma from acutely septic patients and the ventricular myocardial depression (decreased ejection fraction) in spontaneous human sepsis are convincingly linked (Table 1) [24].

The investigators initially identified a subpopulation of septic shock patients with reversible depression of left ventricular ejection fraction (LVEF) as determined by radionuclide cineangiogram (RNCA) [3]. To determine the mechanism of this myocardial depression, an *in vitro* model employing a primary culture of spontaneously beating neonatal rat cardiac myocytes was established. Using phase contrast microscopy and an electronic video area detector, changes in the area of the field occupied by the rhythmically contracting cell (which represented the maximum extent of cell shortening) was determined (Fig. 2). The first derivative of maximum extent of shortening (i.e. peak velocity) was simultaneously electronically derived. Typical, reversible depression of extent and velocity of *in vitro* myocardial cell contraction on exposure to septic shock serum is shown in Fig. 3. The group of interest consisted of patients with acute septic shock and evidence of septic myocardial depression. The mean RNCA-determined LVEF for this group was 0.33. Serum obtained from these patients induced a highly significant 33% decrease in extent (Fig. 4) and 25% decrease in velocity of myocardial cell shortening. The decrease of the RNCA-determined LVEF in each of these patients in the acute phase of septic shock correlated significantly with the decrease in extent of myocardial cell shortening *in vitro* (Fig. 5). Left ventricular end-diastolic volume index, which was increased in patients with septic

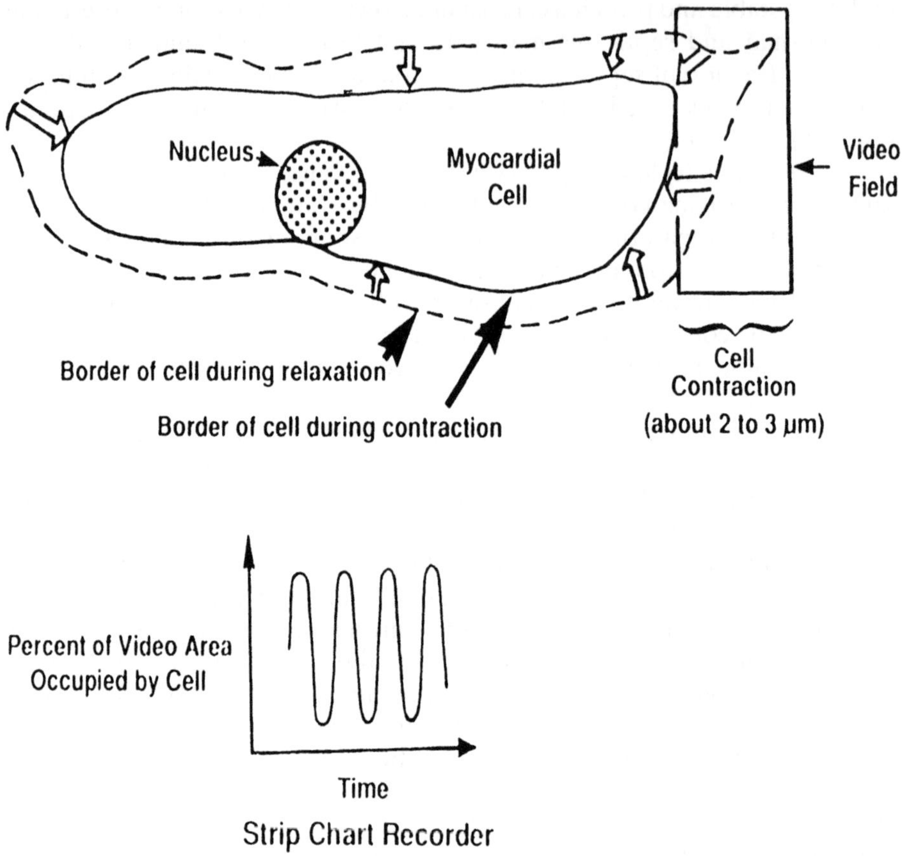

Fig. 2. A schematic diagram of the variable area method of detecting cell movement. The beating edge of a cell was placed in the video field, which measured the area within the field above a specific density threshold. The myocyte density was above this threshold and all the area surrounding the cell was below the threshold. The output was recorded on a strip chart recorder which displayed the percent of the video field which was occupied by the cell at any given moment. (From [24] with permission)

shock, was similarly correlated with the degree of depression in the *in vitro* assay. This established a strong link between *in vivo* cardiac function of patients with septic shock and the myocardial depressant activity of a substance circulating in their blood. Cardiac index (CI) and stroke volume index (SVI), however, did not demonstrate a similar correlation with *in vitro* serum depressant activity. As a control, serum from many of the same patients obtained before septic shock or after recovery induced no significant change in extent or velocity of myocardial cell shortening (Fig. 4). Serum from healthy laboratory personnel, patients with structural cardiovascular disease, and critically ill, non-septic ICU patients (who comprised 3 additional control groups) caused no significant change in maximum extent or velocity of cardiomyocyte shortening (Fig. 4).

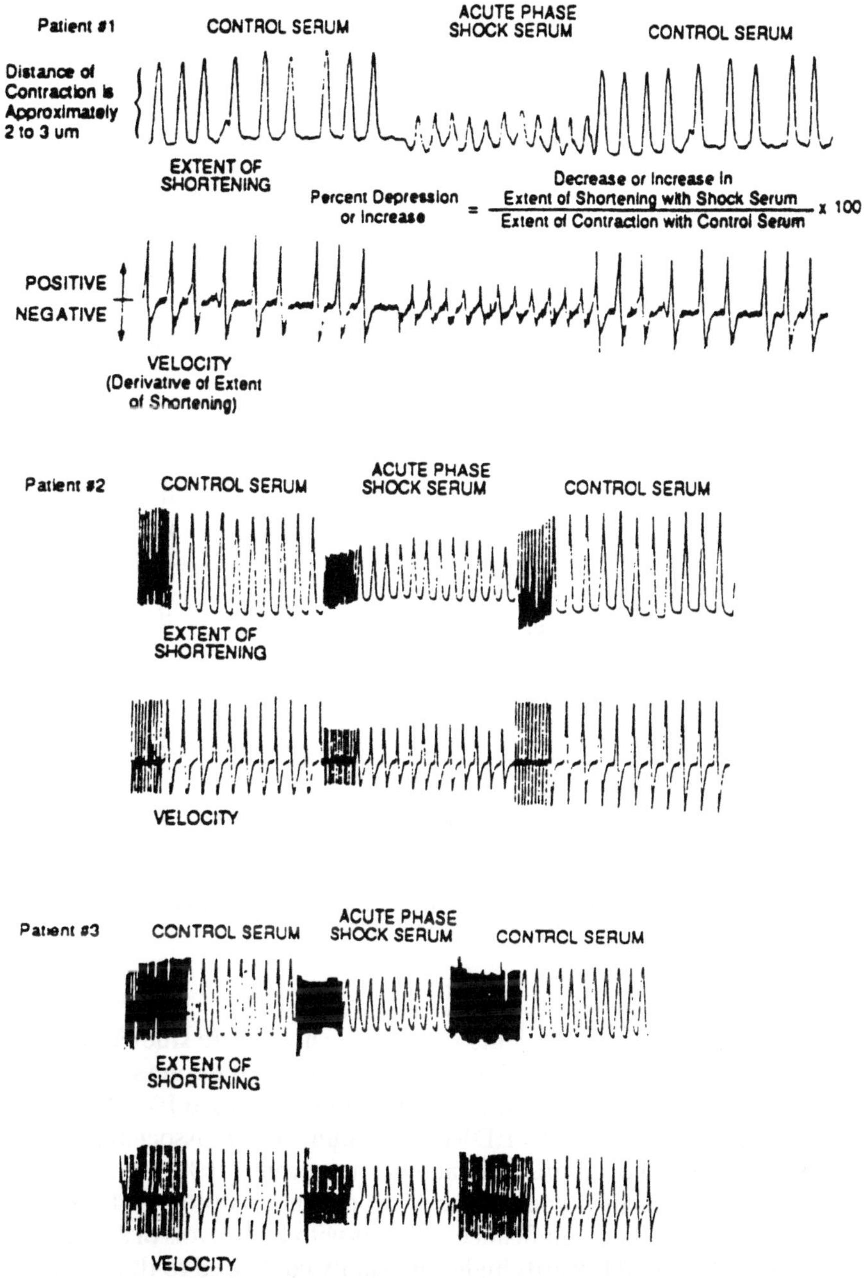

Fig. 3. Representative examples of the effect of serum from three different patients with septic shock. For each patient, the upper panel displays the extent of shortening and the lower panel displays the velocity of shortening. (From [24] with permission)

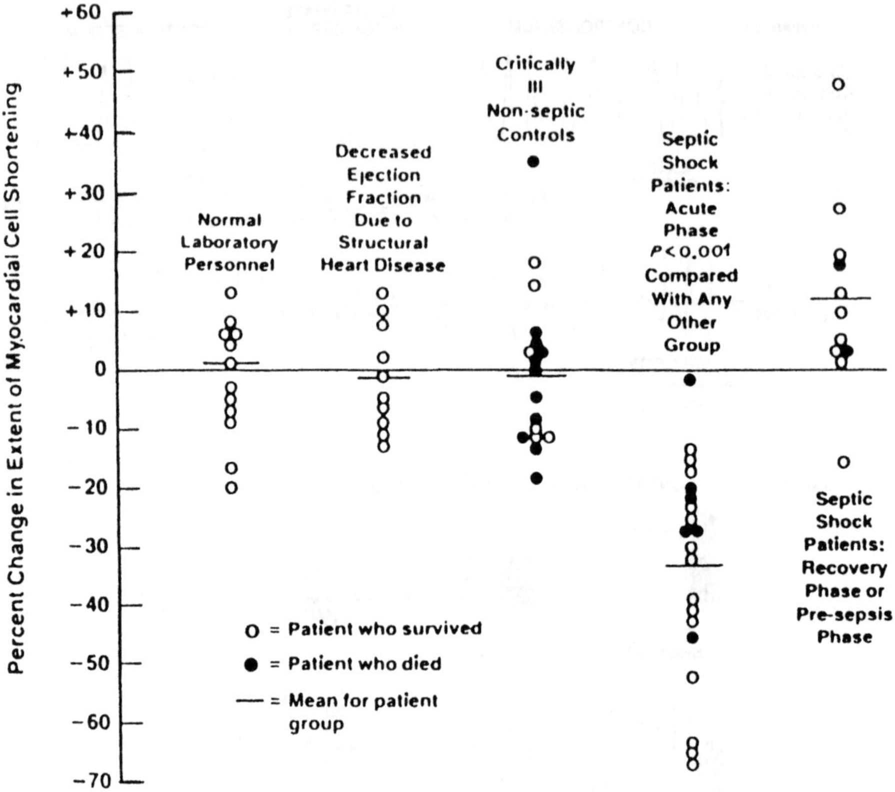

Fig. 4. The effect of serum from septic shock patients on the extent of myocardial cell shortening of spontaneously beating rat heart cells *in vitro*. Septic shock patients during the acute phase demonstrated a statistically significant lower extent of shortening ($p<0.001$) compared with any control group. Open circles = survivors, closed circles = non-survivors, horizontal line = mean for each group. (From [24] with permission)

In a subsequent study, patients with septic shock were studied for the presence of a circulating MDS [25]. Clinical myocardial depression as manifested by an elevated ventricular filling pressure (increased PWP) with ventricular dilation (increased LVEDVI) was found to be associated with high levels of myocardial depressant activity in serum (20% or greater mean decrease of extent of myocardial cell shortening). Although the study size was too small to reach significance, patients with high levels of serum depressant activity trended towards higher mortality compared to those with lower or absent activity. This may suggest a central pathophysiologic role for MDS in the cardiovascular dysfunction of septic shock.

The quantitative and temporal relationship between septic shock-associated reductions of *in vivo* ventricular function and septic shock serum-induced cardiac myocyte depression *in vitro* argues strongly in favor of an important pathophysiologic role for a circulating MDS during acute septic shock in

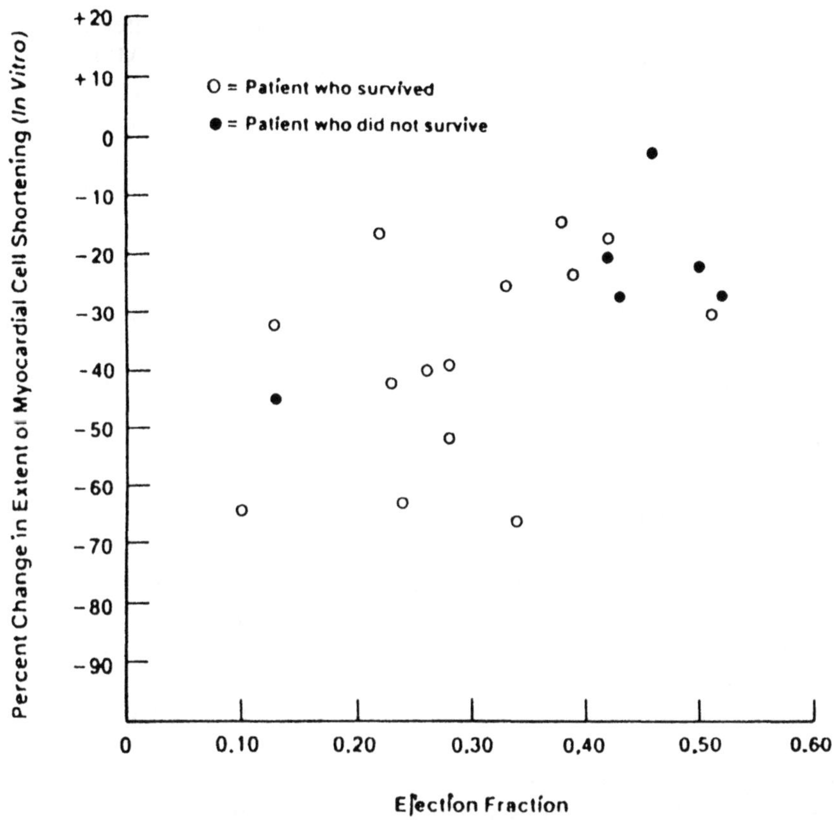

Fig. 5. Correlation of RNCA-determined LVEF and the extent of myocardial cell shortening *in vitro* in patients during the acute phase of septic shock. Spearman rank correlation revealed an $r = +0.52$ and $p < 0.01$. Open circles = survivors, closed circles = non-survivors. (From [24] with permission)

humans [24]. This possibility is further supported by trend towards an association between degree of septic shock serum-induced *in vitro* myocyte depression and mortality in the second study [25].

Filtration studies suggested that the MDS activity was present in at least two fractions of 0.5–5 kD and >10 kD. In addition, the substance was soluble in water but not ethylacetate and was heat labile [24, 25]. This combination of characteristics is most consistent with polypeptides or proteins.

Myocardial Depression in Sepsis: Potential Mediators

Although the previously mentioned studies were able to clarify some of the physical characteristics of MDS, it's molecular structure has never been fully characterized. An alternative approach to the identification of the myocardial depressant substance circulating during sepsis involves assessment of potential circulating factors known or suspected to be elevated during septic shock which match the known characteristics of human MDS (molecular weight >10 kD, heat labile, probably a polypeptide or protein).

The most obvious candidate is endotoxin (lipopolysaccharide, LPS), a component of the cell wall of gram-negative bacteria. Endotoxin is integrally involved in the development of septic shock associated with gram-negative infections. Suffredini and coworkers [26] have demonstrated that the hemodynamic effects of intravenous endotoxin administration qualitatively mimics those of spontaneous septic shock in human volunteers monitored over an 8 h period using pulmonary artery catheters and RNCA. Numerous animal studies have similarly demonstrated that administration of endotoxin (particularly chronic infusion) can mimic the cardiovascular manifestations of sepsis in humans [11, 12]. Danner and colleagues [27] measured serial blood levels of endotoxin in a large consecutive series of patients with septic shock. 43% of patients (including a number with negative blood cultures) had elevated levels of endotoxin on at least one determination. Patients with detectable endotoxemia, compared to patients without detectable endotoxemia, had a lower mean LVEF, lower SVR and higher plasma lactate level. In septic shock patients with positive blood cultures, endotoxemia was significantly associated with a decreased LVEF, increased vasopressor requirement, and higher mortality. Thus, evidence is compelling that endotoxin is an important mediator in the pathogenesis of gram-negative sepsis.

The evidence that endotoxin is a direct mediator of cardiovascular dysfunction in sepsis and septic shock is far less compelling. The occurrence of septic shock (with a hyperdynamic circulatory state and myocardial depression) can occur without gram-negative infection. Another mediator would have to be present in septic shock associated with gram-positive bacteria or fungi. Further, endotoxin does not fit the physical characteristics of MDS. LPS is a heat-stable glycolipid while available evidence suggests MDS is a heat-labile polypeptide. Finally, application of endotoxin by itself does not directly damage or otherwise affect myocardial tissues. Parker and Adams [28] demonstrated a lack of myocardial tissue contractility depression when 1 mg/mL of purified endotoxin was added directly to a guinea pig atrial muscle preparation while 100 fold lower concentrations of endotoxin administered to rats and pigs *in vivo* resulted in depressed isovolemic pressure-volume curves from the excised perfused hearts [29, 30]. We have shown that 1 mg/mL endotoxin does not directly depress cardiomyocyte contractility *in vitro* while far lower concentrations (0.1 ng/mL), when co-incubated with PMA-activated macrophages, result in media which contains marked depressant activity [31, 32].

That endotoxin generates significant hemodynamic depressant effects when infused *in vivo* into animal models while cardiovascular tissue exposed to endotoxin directly *in vitro* demonstrates marked resistance suggests that endogenously produced factors probably mediate *in vivo* responses to endotoxin. This proposal is supported by the demonstration that co-incubation of endotoxin with activated macrophages/monocytes produces supernatants with significant vascular and myocardial depressant activity [31, 33–37]. The list of potential cytokine inflammatory mediators which could contribute to cardiovascular depressant activity (including MDS) embraces, among others, IL-2, IL-4, IL-6, IL-8, IL-10, and interferon gamma (IFNγ) and, in particular, both tumor necrosis factor-α (TNF-α) and IL-1β.

IL-4, a highly pleiotropic molecule with both pro- and anti-inflammatory effects [38, 39], IL-8, documented to be elevated in human sepsis and known for its neutrophil chemotactic activity [40–42], and IL-10, a potent anti-inflammatory peptide which suppresses expression of proinflammatory peptides by macrophages [43, 44], are a varied group of cytokines each of which is thought to potentially have a role in septic shock. Although each shares characteristics with the larger of the myocardial depressant substances described by recent investigators (heat labile, molecular weight >10 kD), none causes significant hemodynamic compromise when infused into experimental animal models. Furthermore, none of these cytokines have been demonstrated to induce depression of myocardial (or vascular) tissue *in vitro*.

IL-6 is well known as the primary factor responsible for the induction of the hepatic acute phase protein response to systemic injury or inflammation [45, 46]. In that context, it appears to represent more a marker of inflammation and sepsis than its mediator or trigger. Although elevation of IL-6 in human septic shock is a strong predictor of outcome [47–49], it does not cause hemodynamic instability in animal models [50]. On the other hand, some investigators have suggested that IL-6 can cause direct *in vitro* depression of myocardial tissue in some models [51, 52].

Although possessing minimal cytotoxic activity alone, IFNγ, appears to synergize in *in vitro* cytotoxicity assays with endotoxin and other cytokines (including TNF and IL-1) thought to produce adverse effects in septic shock [40, 53]. Similarly, it exerts little cardiovascular effect by itself in *in vivo* animal models but does appear to augment the hemodynamic effects of endotoxin and TNF [54–56]. IFNγ, while inconsistently documented to be elevated in human sepsis, may still serve to amplify the adverse hemodynamic effects of TNF, IL-1 and various other proinflammatory factors involved in septic shock [40, 57]. No direct IFNγ-induced *in vitro* myocardial depressant effects have been demonstrated in the past.

The role of IL-2 in the hemodynamic and metabolic manifestations of sepsis and septic shock is somewhat unclear. IL-2 does qualitatively reproduce many of the hemodynamic manifestations of septic shock when administered to humans [58, 59]. However, given the fact that studies have generally failed to demonstrate a systemic elevation of IL-2 in humans with sepsis and septic shock [60, 61], this could potentially be explained by the fact that IL-2 is also known to mediate the release of other cytokines, particularly TNF and IL-1. *In vitro*, the effect of IL-2 on myocardial tissue is similarly confusing with some studies supporting *in vitro* myocardial depressant activity but others failing to do so [51, 62, 63].

As described earlier, TNF and IL-1 represent the two cytokines known to have the most significant effects on the cardiovascular system in both animals and humans. *In vivo*, TNF is produced during experimental animal endotoxic shock and septic shock [64, 65]. In addition, TNF is elevated in humans administered small quantities of endotoxin and in human septic shock [57, 66]. In animal models, administration of recombinant TNF causes fever, lactic acidosis, disseminated intravascular coagulation and death [65, 67]. The cardiovascular pattern of such animals is similar to that of spontaneous human septic shock and is characterized by hypotension, increased cardiac output, low systemic vascular resistance and decreased ejection fraction [68]. Evidence of myocardial depression can be demonstrated within 1 h of TNF infusion [69]. When given prior to challenge, antibodies to TNF protect mice [70] and rabbits [71] from endotoxic shock as well as protecting baboons from a lethal parenteral infusion of *E. coli* [72]. A recent report has suggested that application of anti-TNF monoclonal therapy to humans with septic shock improved left ventricular function even though it failed to improve mortality [73, 74]. As with TNF, experimental human and animal models of septic shock exhibit increased serum levels of IL-1 [75, 76]. Spontaneous human septic shock is also associated with elevations of measured circulating IL-1 [48, 57]. Infusion of IL-1 into animals qualitatively reproduces many of the hemodynamic manifestations of septic shock [77]. Although application of IL-1 receptor antagonist (an anti-IL-1 therapeutic strategy) has failed to improve overall mortality in human septic shock [78], amelioration of hemodynamic and metabolic manifestations of endotoxic shock in animals and spontaneous septic shock in humans [77, 79–81] has been demonstrated. Thus, TNF and IL-1 may both be integral to the development of septic shock and its cardiovascular manifestations in humans.

We have had the opportunity to examine a number of these factors in a cardiomyocyte assay system which employs a closed-loop video system to track the movement of latex microbeads attached to the myocyte cell membranes (Fig. 6) [25, 82]. We have been able to demonstrate that endotoxin, IL-2, IL-4, IL-6, IL-8, and IL-10 fail to induce a concentration-dependent depression of maximum extent or peak velocity of cardiomyocyte shortening within the 30 min of exposure up to concentrations as high as 500 to 1000 ng/mL (concentrations 100 × or more in excess of the highest serum levels documented during sepsis and septic shock) [32]. Similar results have been reported for IL-2 by some authors but others have documented an early depressant effect [51, 62, 63]. Similarly, other authors have suggested that IL-6 does cause early depression of guinea pig papillary muscle while we have been unable to demonstrate such an effect [51, 52]. This may relate to differences in models employed or may reflect a more basic difference between species in the precise actions of each cytokine.

In contrast to the other cytokines tested in our system, TNF and IL-1 both appear to cause significant concentration-dependent depression of cardiomyocyte contraction (both maximum extent and peak velocity of shortening) beginning within 10 min of exposure and increasing through to 30 min. TNF does so starting at concentrations at the upper extreme of those documented during human septic shock (0.8/mL) [83]. Cardiomyocyte

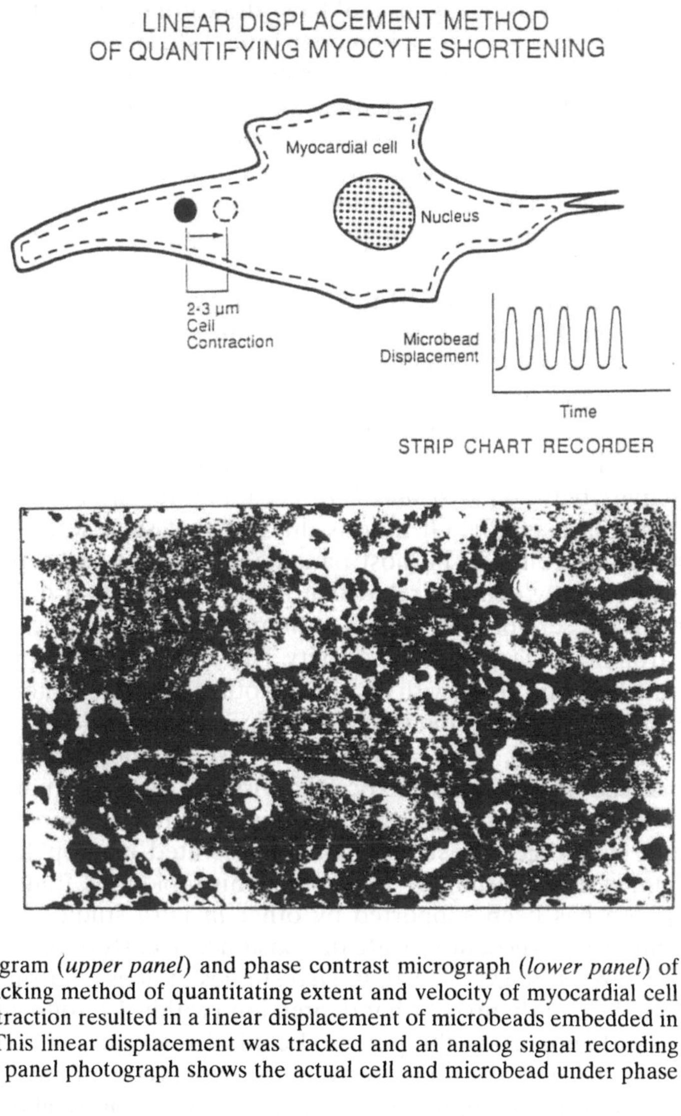

Fig. 6. Schematic diagram (*upper panel*) and phase contrast micrograph (*lower panel*) of closed loop video tracking method of quantitating extent and velocity of myocardial cell contraction. Cell contraction resulted in a linear displacement of microbeads embedded in the cell membrane. This linear displacement was tracked and an analog signal recording produced. The lower panel photograph shows the actual cell and microbead under phase contrast microscopy

depression induced by IL-1 begins at concentrations 4–10 × higher (between 8 and 32 ng/mL) than those typically seen during human septic shock [32]. Similar early depression of myocardial tissue contractility has been documented in other studies [51, 84].

TNF and IL-1 have been shown to exert synergistic hemodynamic effects in a variety of *in vivo* models of shock. Waage and colleagues [85] demonstrated that IL-1 synergistically potentiated the lethal effect of TNF administered to mice. Okusawa et al. [77] have shown that low doses of IL-1 in combination with TNF produced a shock-like state in rabbits. More recently,

Weinberg and colleagues [86] have shown that doses of endotoxin, TNF, and IL-1, too low to cause a hemodynamic disturbance individually, can act synergistically to produce hypotension in unanesthetized rabbits. Data also exists to show that supernatants of activated macrophages (containing cytokines such as TNF, IL-1, and IL-6) can produce depression of myocardial tissue at relatively low concentrations of the individual cytokines [31, 33, 34]. This suggests that combinations of cytokines may be able to generate *in vitro* (or *in vivo*) myocardial depression through additive or, possibly, synergistic effects.

We [87] have assessed the possibility of synergistic interactions between TNF and IL-1 in the generation of cardiomyocyte depression *in vitro*. Our data shows that in combination, TNF and IL-1 cause cardiomyocyte depression at concentrations which are inconsistent with simple additive effects. The generation of cardiomyocyte depression at these concentrations (50 to 100× lower than required individually) suggests synergism of depressant activity between IL-1 and TNF occurs in our model and may occur during human sepsis [87]. Of note is the fact that such serum concentrations are relatively common in most patients with sepsis and septic shock. We have also recently demonstrated that immunoabsorption of both TNF and IL-1 from the sera of 5 individual patients with acute septic shock eliminated its cardiomyocyte depressant activity. This data strongly supports the suggestion that TNF and IL-1 (with or without other cytokine factors) may cause septic myocardial depression via synergistic interactions [87].

Interestingly, there appear to be two distinct time frames of cytokine-induced depression of myocardial tissue. Our own studies have shown cardiomyocyte depression occurs within 10 min of exposure to TNF, IL-1, TNF and IL-1 together, supernatant of activated macrophages/monocytes, and human septic serum [31, 83, 87–89]. This rapid time frame for depressant response has been supported by other *in vitro* studies of myocardial tissue which similarly demonstrate the existence of early depression following cytokine exposure [51, 52, 84]. *In vivo* studies of dogs confirm the ability of TNF to induce early myocardial depression (within 1 h following initiation of infusion) supporting the potential clinical relevance of the *in vitro* phenomenon [68, 69]. This relatively early expression of myocardial depressant activity is too rapid to be explained by a mechanism which involves *de novo* protein synthesis. Parrillo and colleagues [24] have previously demonstrated that the degree of cardiomyocyte depression produced by human septic serum *in vitro* is proportional to the decrease of left ventricular ejection in the same humans during acute septic shock.

In contrast, other studies have suggested that TNF, IL-1 and supernatants of activated macrophages/monocytes may also induce a delayed and prolonged depressant effect on *in vitro* myocardial tissue beginning more than 1 h after exposure and potentially persisting for days [34, 90–92]. There is some evidence to suggest this delayed expression of depressant activity involves *de novo* protein synthesis and represents a somewhat different biochemical pathway than the early depression we have described.

The Physiologic Role of NO and cGMP in the Cardiovascular System

The role of NO (previously called endothelium-derived relaxing factor, EDRF) and cGMP is well established in the vasculature. In the normal vascular endothelium, a calcium/calmodulin-dependent constitutive NO synthetase (cNOS) generates picomolar quantities of NO and L-citrulline from L-arginine in response to stimulation by endothelial shear stress or in response to vasorelaxant mediators such as acetylcholine, bradykinin, or histamine [93, 94]. NO, a short-lived (half-life 6–10 sec) but highly diffusible molecule, enters the cytosol of the adjacent vascular smooth muscle myocytes where it activates soluble guanylate cyclase producing cGMP. cGMP mediates vascular smooth muscle relaxation by stimulating cGMP-dependent protein kinase which promotes resequestration of calcium into the sarcoplasmic reticulum by phospholamban resulting in decreased cytosolic calcium [94–96]. This mechanism appears to be a required element of normal vasomotor regulation. Decreased endothelial generation of NO has been associated with essential hypertension and atherogenesis.

Until recently, little data has been available regarding the potential role of NO and cGMP in the heart. In retrospect, the first suggestion that NO may play a normal physiologic role in the heart probably came from experiments in the 1960s which showed that acetylcholine, which is known to cause NO and cGMP-dependent vascular relaxation, depressed cardiac tissue *in vitro* through a cGMP-dependent pathway [97, 98]. Membrane permeable forms of cGMP directly depresses *in vitro* myocardial tissue contractility [99–101]. cGMP can also antagonize the inotropic properties of increased cAMP in myocardial tissue [100, 102, 103].

These effects appear to occur through at least three discrete mechanisms (Fig. 7). First, cGMP appears to reduce sarcolemmal Ca^{2+} influx through L-type channels via activation of a cGMP-dependent cAMP phosphodiesterase which reduces myocardial cAMP levels [102, 104, 105]. Reduced cAMP results in decreased cAMP-dependent phosphorylation of the L-type channel or an associated protein which would otherwise allow increased Ca^{2+} movement through the channel. Calcium influx into the cytosol through the L-type channel, one of the major determinants of degree of myocyte contraction, can also be reduced through a direct effect of a cGMP-dependent protein kinase [106]. Finally, another cGMP-dependent protein kinase may be responsible for reducing myocardial myofilament response to Ca^{2+} [101].

The role of NO is evident from other studies. Perfusion of isolated rat hearts with nitroprusside (a direct NO donor) and superfusion of *in vitro* cardiomyocytes with nitroprusside or NO in solution results in depression of contractility [100, 107–109]. This effect is associated with generation of cGMP and a reduction of L-type calcium current. Methylene blue, an inhibitor of guanylate cyclase, blocks both cGMP generation and L-type calcium channel reductions [100]. Recently, direct nitroprusside infusion into the

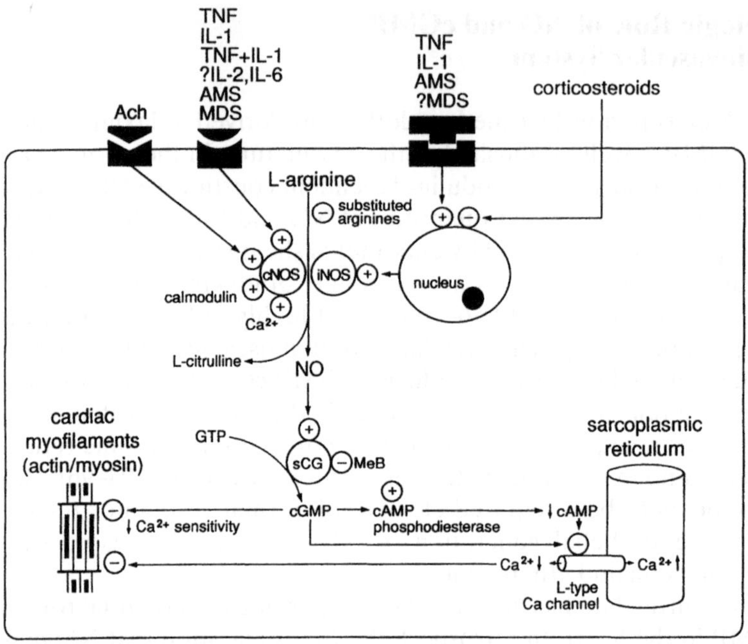

Fig. 7. Schematic diagram of myocardial tissue/cardiac myocyte. TNF = tumor necrosis factor-α, IL-1 = interleukin-1β, IL-2 = interleukin-2, IL-6 = interleukin-6, AMS = activated macrophage supernatants, MDS = myocardial depressant substance, Ach = acetylcholine, NO = nitric oxide, cNOS = constitutive NO synthase, iNOS = inducible NO synthase, sGC = soluble guanylate cyclase (see text for explanation)

coronary arteries of humans has been shown to depress intraventricular pressure development and improve diastolic relaxation and distensibility [110].

The existence of guanylate cyclase in the myocardium is generally acknowledged. More recently, a number of studies have elegantly demonstrated that an endogenous constitutive NO and cGMP signalling system must be involved in muscarinic cholinergic signal transduction in both rat and guinea pig cardiac myocytes [100, 111] (Fig. 7). A calcium/calmodulin-dependent cNOS has been detected in human and rat myocardium [112–114]. The cNOS of adult rat myocardium has been cloned and appears to bear substantial homology to the endothelial isoform [115]. The precise mechanism(s) by which cNOS are stimulated (apart from their dependence on Ca^{2+} and calmodulin) are not well defined. Although they are phosphorylated by a number of different protein kinases, phosphorylation has not yet been shown to regulate function of the enzyme [94]. This data suggests that NO and cGMP, in parallel to their function in the vasculature, play a central role in physiologic regulation of cardiac contractility.

The Role of NO and cGMP in Septic Cardiovascular Dysfunction

Available evidence supports a substantial role for aberrant generation of NO and cGMP in sepsis and septic shock in humans. NO is too short-lived to be measured directly in serum. However, nitrite/nitrate metabolites have been shown to be increased during human sepsis [116]. Similarly, serum cGMP has been shown to be elevated during endotoxic shock in animals [117]. However, the total quantity of cNOS in the human body is unlikely to be able to produce the quantities of NO which appear to be generated during septic shock.

In contrast to cNOS which are calcium/calmodulin-dependent and are limited to the generation of picomolar quantities of NO, inducible isoforms of NOS (iNOS), found in smooth muscle, macrophages, and, most recently, the heart, are calcium/calmodulin-independent and able to produce nanomolar quantities of NO. Generation of NO by these isoforms is primarily dependent on the quantity of induced enzyme. Because of their greater capacity to produce NO, once induction of the inducible enzyme is established, these isoforms may represent the dominant source of pathological NO production.

Although the activity of the endothelial cNOS has been shown to be rapidly augmented by endotoxin (and, potentially, other mediators) [118], a smooth muscle iNOS is clearly involved in septic vascular dysfunction. Endotoxin [35, 36], supernatant from activated macrophages [119], various cytokines (including TNF, IL-1, and IFNγ) [120–124], and sepsis [124–127] have been shown to contribute to septic vascular dilation through the generation of a vascular smooth muscle iNOS. The presence of an iNOS in affected vascular smooth muscle has been confirmed *in vitro* by direct identification of the enzyme or corresponding mRNA and by the demonstration that pathological vascular relaxation can be reversed by either substituted analogs of L-arginine (competitive NOS inhibitors) or by methylene blue (a guanylate cyclase inhibitor) [35, 93, 121, 123]. The potential clinical relevance of NO in sepsis was initially supported by the demonstration that the vascular effects of TNF *in vivo* can be substantially attenuated by treatment with NOS inhibitors [120–121]. More recent studies, further supporting a role for NO and cGMP in the pathophysiology of human septic shock, have shown that competitive NOS inhibitors (such as N^G-monomethyl-L-arginine, L-NMMA) and guanylate cyclase inhibitors (such as methylene blue) result in a prominent pressor response when administered to patients with septic shock [128, 129].

With respect to the role of NO in sepsis-associated myocardial dysfunction, we have shown that a number of cytokines, supernatants of activated macrophages, and serum from humans with septic shock can induce depression of *in vitro* cardiac tissue within 10 min of exposure [31, 83, 87, 88]. The mechanism involved in such depression appears to be sequential NO and cGMP generation [130–132], although the short time (10 min) to its manifestation suggests that stimulation of cNOS rather than *de novo* synthesis of iNOS must be responsible (Fig. 7). The precise mechanism by which cytokines, supernatants of activated macrophages, and septic serum stimulate the myocardial cNOS remains undefined at this time.

Our data demonstrate that cardiomyocyte depression induced by TNF-α, IL-1β, and human septic serum can be inhibited by NOS inhibitors such as L-NMA and re-established by L-arginine in excess [130–132]. Methylene blue, an inhibitor of guanylate cyclase also blocks depression [130–132]. In addition, we have shown that cardiomyocyte cGMP increases with exposure to increasing concentrations of TNF or IL-1 [131, 132]. Similarly, serum drawn from humans with septic shock induces an increase of intracellular cGMP concentration concomittant with decreases in cardiomyocyte contractility both of which can be blocked by the competitive NOS inhibitor, N-methylarginine, and re-established by L-arginine in excess [130]. Regarding potential interactions between cAMP and cGMP, TNF-induced attenuation of cAMP generation and contractile hyporesponsiveness to adrenergic stimulation was not reversed by NOS inhibition in this model suggesting that a second NO-independent source of early myocardial depression may be present in sepsis [133].

Finkel and colleagues have also shown that TNF-induced early depression of guinea pig papillary muscle was prevented by L-NMMA and re-established with an excess of L-arginine [51]. In addition, they showed that nitrites were significantly elevated in the course of depression. In contrast to our findings, their study suggested that IL-2 and IL-6 (at very high concentrations) caused similar depression in guinea pig papillary muscle (Fig. 7).

A number of investigators have shown that exposure of myocardial tissue *in vitro* to TNF, IL- 1, and supernatants from activated macrophages (which contain cytokines including TNF and IL-1) results in depression of myocardial tissue contractility via the production of iNOS, NO, and cGMP only after a minimum of several hours [33, 90, 112]. Others have shown that *in vivo* endotoxemia results in the production of iNOS and NO in the myocardium which is associated with a decrease in contractility of excised myocardial tissue [112, 134]. In these studies, production of iNOS is associated with a decrease in baseline and isoproterenol-stimulated contractility of myocardial preparations (also over a minimum of several hours) which can be prevented by dexamethasone and by NOS inhibitors (Fig. 7) [33, 90, 112].

A study by Kinugawa and colleagues [52] may help to reconcile these divergent findings of both early and late depressant activity. In their avian cardiac myocyte model, they were able to demonstrate that high concentrations of IL-6 induced both early (<30 min) and late (24 h) cardiac myocyte depression with an associated decrease in $[Ca^{2+}]_i$ (intracellular calcium transients). Early depression was associated with an increase of intracellular cGMP. Depression could be prevented by substituted arginine analogs or methylene blue and re-established by L-arginine. Depression was also blocked by pretreatment with EGTA, a calcium chelator suggesting that the myocardial calcium/calmodulin-dependent cNOS was involved. Late depression, on the other hand, was similarly associated with increased cGMP but could not be blocked by EGTA supporting a role for the calcium/calmodulin-independent iNOS. This study clearly suggests the possibility that cytokine (or sepsis) could stimulate the heart to sequentially produce NO via both cNOS and iNOS. Failure of some studies to show early depression could potentially be species or model specific.

The cellular origin of myocardial NOS (constitutive or inducible) has been the subject of controversy. Possible sources include myocardial vascular endothelium, cardiac myocytes, and non-myocytes (including fibroblasts). Although myocardial and intramyocardial vascular endothelium clearly have the constitutive ability to generate NO, whether they can produce sufficient quantities to account for septic myocardial dysfunction is questionable [108]. Recent studies suggest that endotoxin or IL-1 stimulation results in generation of iNOS within cardiac myocytes but not fibroblasts [135]. Another study suggested limited IL-1-stimulated iNOS activity in cardiac non-myocytes compared with myocytes [136].

Therapeutic Implications

Therapeutic approaches based on our knowledge of the role of NO and cGMP in sepsis and septic shock are in their infancy. While inhibition of NO or cGMP synthesis could theoretically correct both vascular and myocardial abnormalities in sepsis, there are also significant risks for adverse effects. NO and cGMP appear to be central to normal vasomotor control of appropriate distribution of perfusion. Inhibition of NO or cGMP generation could result in more inappropriate distribution of perfusion in sepsis. In addition, administration of substituted arginine inhibitors of NO synthesis have been shown to increase oxygen consumption (despite reducing total perfusion) in awake dogs, possibly through a direct effect on mitochondria [137].

In practice, both NOS and guanylate cyclase inhibitors have been shown to exhibit mixed hemodynamic effects in animal models of endotoxic or TNF-induced shock with increases of blood pressure [120, 125, 137]. However, limited, uncontrolled human studies to date suggest adverse effects may be dominant in clinical situations. Inhibition of NO synthesis using a substituted arginine analog in human septic shock resulted in increased blood pressure, increased vasopressor sensitivity, and increased vascular resistance but decreased cardiac output [129]. Administration of methylene blue (the guanylate cyclase inhibitor) to humans with septic shock similarly produced a vasopressor response with potentially adverse effects on cardiac output [128].

Given these data, it appears unlikely that widespread study of the clinical applications of NOS and guanylate cyclase inhibitors will occur until drugs of greater specificity and ease of titration become available.

Conclusion

Data accumulated over the last decade suggests that the human heart does exhibit myocardial depression during sepsis and septic shock. This depression is demonstrated by biventricular dilatation, decreased ejection fraction, and decreased response to fluid resuscitation. Evidence to date suggests that human septic myocardial depression is not mediated by myocardial hypoperfusion but rather by the effects of a circulating myocardial depressant substance or substances (MDS). This substance may represent low concentrations of TNF and IL-1 acting in synergy on the myocardium through NO.

The heart, like the vasculature, can be the target of NO generated by two distinct pathways. A constitutive NOS appears to be integral to physiologic parasympathetic regulation of cardiac contractility. This cNOS can also be stimulated by TNF, IL-1, possibly IL-2 and IL-6 (in some species), supernatants of activated macrophages, and MDS in septic serum to produce depression of cardiac myocyte contractility within minutes of exposure. This mechanism may play a clinical role in early human septic myocardial depression. When the inflammatory stimulus of sepsis is sustained, a higher capacity inducible NOS is generated within the myocardium. Clinically, this inducible NOS may be responsible for more prolonged septic myocardial depression. Both pathways act via generation of NO within the myocardial tissue. NO, whether generated by cytokine-stimulated cNOS or cytokine-induced iNOS, activates soluble guanylate cyclase in the myocardium producing cGMP. Decreased influx of calcium from the sarcoplasmic reticulum through the L-type calcium channel and decreased myofilament response to Ca^{2+} induced by cGMP result in loss of cardiac myocyte contractility and, clinically, septic myocardial depression.

Similar mechanisms may play a role in a variety of pathologic inflammatory cardiac conditions including myocarditis and post-infarction reperfusion injury. Serious clinical application of NOS and guanylate cyclase inhibitors will probably require the development of drugs with greater specificity.

References

1. Weil MH, MacLean LD, Visscher MD, Spink W (1956) Studies on the circulatory changes in the dog produced by endotoxin from gram-negative microorganisms. J Clin Invest 35:1191–1198
2. Solis RT, Downing SE (1966) Effects of E. coli endotoxemia on ventricular performance. Am J Physiol 211:307–313
3. Parker MM, Shelhamer JH, Bacharach SL, et al (1984) Profound but reversible myocardial depression in patients with septic shock. Ann Intern Med 100:483–490
4. Sibbald WJ, Paterson NAM, Holliday RL, Anderson RA, Lobb TR, Duff JH (1978) Pulmonary hypertension in sepsis: Measurement by the pulmonary artery diastolic-pulmonary wedge pressure gradient and the influence of passive and active factors. Chest 73:583–591
5. Peyton MD, Hinshaw LB, Greenfield LJ, Elkins RC (1976) The effects of coronary vasodilatation on cardiac performance during endotoxin shock. Surg Gynecol Obstet 143:533–538

6. Wiggers CJ (1947) Myocardial depression in shock. A survery of cardiodynamic studies. Am Heart J 33:633–650
7. Lefer AM, Martin J (1970) Origin of myocardial depressant factor in shock. Am J Physiol 218:1423–1427
8. Wangensteen SL, Geissenger WT, Lovett WL, Glenn TM, Lefer AM (1971) Relationship between splanchnic blood flow and a myocardial depressant factor in endotoxin shock. Surgery 69:410–418
9. Cunnion RE, Schaer GL, Parker MM, Natanson C, Parrillo JE (1986) The coronary circulation in human septic shock. Circulation 73:637–644
10. Dhainaut JF, Huyghebaert MF, Monsallier JF, et al (1987) Coronary hemodynamics and myocardial metabolism of lactate, free fatty acids, glucose, and ketones in patients with septic shock. Circulation 75:533–541
11. Goldfarb RD, Nightingale LM, Kish P, Weber PB, Loegering DJ (1986) Left ventricular function during lethal and sublethal endotoxemia in swine. Am J Physiol 251: H364–H373
12. Lee K, van der Zee H, Dziuban SW Jr, Goldfarb RD (1988) Left ventricular function during chronic endotoxemia in swine. Am J Physiol 254:H324–H330
13. Hotchkiss RS, Karl IE (1992) Reevaluation of the role of cellular hypoxia and bioenergetic failure in sepsis. JAMA 267:1503–1510
14. Solomon MA, Correa R, Alexander HR, et al (1994) Myocardial energy metabolism and morphology in a canine model of sepsis. Am J Physiol 266:H757–H768
15. Nishijima H, Weil MH, Shubin H, Cavanilles J (1973) Hemodynamic and metabolic studies on shock associated with gram-negative bacteremia. Medicine (Baltimore) 52:287–294
16. Shoemaker WC (1971) Cardiorespiratory patterns in complicated and uncomplicated septic shock. Ann Surg 174:119–125
17. Lovett WL, Wangensteen SL, Glenn TM, Lefer AM (1971) Presence of a myocardial depressant factor in patients with circulatory shock. Surgery 70:223–231
18. McConn R, Greineder JK, Wasserman F, Clowes GHA (1979) Is there a humoral factor that depresses ventricular function in sepsis? Circ Shock 1:9–22
19. Gomez A, Wang R, Unruh H, et al (1990) Hemofiltration reverses left ventricular dysfunction during sepsis in dogs. Anesthesiology 73:671–685
20. Jha P, Jacobs H, Bose D, et al (1993) Effects of E. coli sepsis and myocardial depressant factor on interval-force relations in dog ventricle. Am J Physiol 264: H1402–H1410
21. Harary I, Farley B (1960) In vitro studies of single isolated beating rat heart cells. Science 131:1674–1675
22. Carli A, Auclair MC, Vernimmen C, Jourdon P (1979) Reversal by calcium of rat heart cell dysfunction induced by human sera in septic shock. Circ Shock 6:147–157
23. Benassayag C, Christeff N, Autclair MC, et al (1984) Early released lipid-soluble cardiodepressant factor and elevated oestrogenic substances in human septic shock. Eur J Clin Invest 14:288–294
24. Parrillo JE, Burch C, Shelhamer JH, Parker MM, Natanson C, Schuette W (1985) A circulating myocardial depressant substance in humans with septic shock. Septic shock patients with a reduced ejection fraction have a circulating factor that depresses in vitro myocardial cell performance. J Clin Invest 76:1539–1553
25. Reilly JM, Cunnion RE, Burch-Whitman C, Parker MM, Shelhamer JH, Parrillo JE (1989) A circulating myocardial depressant substance is associated with cardiac dysfunction and peripheral hypoperfusion (lactic acidemia) in patients with septic shock. Chest 95:1072–1080
26. Suffredini AF, Fromm RE, Parker MM, et al (1989) The cardiovascular response of normal humans to the administration of endotoxin. N Engl J Med 321:280–287
27. Danner RL, Elin RJ, Hosseini JM, Wesley RA, Reilly JM, Parrillo JE (1991) Endotoxemia in human septic shock. Chest 99:169–175
28. Parker JL, Adams HR (1979) Myocardial effects of endotoxin shock: Characterization of an isolated heart muscle model. Adv Shock Res 2:163–175

29. Parker JL, Adams HR (1981) Contractile dysfunction of atrial myocardium from endotoxin-shocked pigs. Am J Physiol 240:H954–H962
30. Parker JL, Adams HR (1985) Development of myocardial dysfunction in endotoxin shock. Am J Physiol 248:H818–H826
31. Kumar A, Kosuri R, Ginsburg B, et al (1994) Myocardial cell contractility is depressed by supernatants of endotoxin stimulated THP-1 cells. Crit Care Med 22:A118 (Abst)
32. Kumar A, Kosuri R, Kandula P, et al (1994) Interleukin-1 beta but not endotoxin, interferon-gamma, or interleukin-6 depresses myocardial cell contractility in vitro. Clin Res 42:168A (Abst)
33. Balligand J, Ungureanu D, Kelly RA, et al (1993) Abnormal contractile function due to induction of nitric oxide synthesis in rat cardiac myocytes follows exposure to activated macrophage-conditioned medium. J Clin Invest 91:2314–2319
34. Gulick T, Chung MK, Pieper SJ, et al (1989) Interleukin 1 and tumor necrosis factor inhibit cardiac myocyte adrenergic responsiveness. Proc Natl Acad Sci 86:6753–6757
35. Rees DD, Cellek S, Palmer RMJ, Moncada S (1990) Dexamethasone prevents the induction by endotoxin of a nitric oxide synthase and the associated effects on vascular tone: An insight into endotoxic shock. Biochem Biophys Res Commun 173: 541–547
36. Julou-Schaeffer G, Gray GA, Fleming I, et al (1990) Loss of vascular responsiveness induced by endotoxin involves L-arginine pathway. Am J Physiol 259:H1038–H1043
37. McKenna TM, Reusch DW, Simpkins CO (1988) Macrophage conditioned medium and interleukin-1 suppress vascular contractility. Circ Shock 25:187–196
38. Powrie F, Coffman RL (1993) Inhibition of cell-mediated immunity by IL-4 and IL-10. Res Immunol 144:639–643
39. Butcher EC (1991) Leukocyte-endothelial cell recognition: Three (or more) steps to specificity and diversity. Cell 67:1033–1039
40. Billiau A, Vandekerckhove F (1991) Cytokines and their interactions with other inflammatory mediators in the pathogenesis of sepsis and septic shock. Eur J Clin Invest 21:559–573
41. Van Zee K, DeForge LE, Fischer E, et al (1991) IL-8 in septic shock, endotoxemia, and after IL-1 administration. J Immunol 146:3478–3482
42. Danner RL, Suffredini AF, Van Dervort AL, et al (1990) Detection of interleukin-6 (IL-6) and interleukin-8 (IL-8) during septic shock in humans. Clin Res 38:352A (Abst)
43. Moore KW, O'Garra A, de Waal Malefyr R, Vieira P, Mosmann TR (1993) Interleukin-10. Annu Rev Immunol 11:165–190
44. Albright JF, Oppenheim JJ (1992) Contemporary topics in immunology: Approaches to immunotherapy emerging from basic research. FASEB J 6:2890–2894
45. Le J, Vilcek J (1989) Interleukin-6: A multifunctional cytokine regulating immune reactions and the acute phase protein response. Laboratory Investigations 61: 588–602
46. Heinrich PC, Castell JV, Andus T (1990) Interleukin-6 and the acute phase response. Biochem J 265:621–636
47. Casey LC, Balk RA, Bone RC (1993) Plasma cytokine and endotoxin levels correlate with survival in patients with the sepsis syndrome. Ann Intern Med 119:771–778
48. Waage A, Brandtzaeg P, Halstensen A, Kierulf P, Espevik T (1989) The complex pattern of cytokines in serum from patients with meningococcal septic shock. J Exp Med 169:333–338
49. Hack CE, De Groot ER, Felt-Bersma RJ, et al (1989) Increased plasma levels of interleukin-6 in sepsis. Blood 74:1704–1710
50. Preiser JC, Schmartz D, Van der Linden P, et al (1991) Interleukin-6 administration has no acute hemodynamic or hematologic effect in the dog. Cytokine 3:1–4
51. Finkel MS, Oddis CV, Jacobs TD, Watkins SC, Hattler BG, Simmons RL (1992) Negative inotropic effects of cytokines on the heart mediated by nitric oxide. Science 257:387–389

52. Kinugawa K, Takahashi T, Kohmoto O, et al (1994) Nitric oxide-mediated effects of interleukin-6 on [Ca2+]i and cell contraction in cultured chick ventricular myocytes. Circ Res 75:285–295
53. Stolpen AH, Guinan EC, Fiers W, Pober JS (1986) Recombinant tumor necrosis factor and immune interferon act singly and in combination to re-organize human vascular endothelial cell monolayers. Am J Pathol 123:16–24
54. Heremans H, Van Damme J, Dillen C, Dijkmans R, Billiau A (1990) Interferon-gamma, a mediator of lethal lipopolysaccharide-induced Shwartzman-like reactions in mice. J Exp Med 171:1853–1869
55. Billiau A, Heremans H, Vanderkerckhove F, Dillen C (1987) Anti-interferon-gamma protects mice against the generalized Shwartzman reaction. Eur J Immunol 17:1851–1854
56. Silva AT, Cohen J (1992) Role of interferon-gamma in experimental gram-negative sepsis. J Infect Dis 166:331–335
57. Girardin E, Grau GE, Dayer JM, Roux-Lombard P, Lambert PH (1989) Plasma tumor necrosis factor and interleukin-1 in the serum of children with severe infectious purpura. N Engl J Med 319:397–400
58. Ognibene FP, Rosenberg SA, Lotze M, et al (1988) Interleukin-2 administration causes reversible hemodynamic changes and left ventricular dysfunction similar to those seen in septic shock. Chest 94:750–754
59. Gaynor ER, Vitek L, Sticklin L, et al (1988) The hemodynamic effects of treatment with interleukin-2 and lymphokine-activated killer cells. Ann Intern Med 109: 953–958
60. Pinsky MR, Vincent JL, Deviere J, Alegre M, Kahn RJ, Dupont E (1993) Serum cytokine levels in human septic shock: Relation to multiple system organ failure and mortality. Chest 103:565–575
61. Fong Y, Moldawer LL, Shires GT, Lowry SF (1994) The biologic characteristics of cytokines and their implication in surgical injury. Surg Gynecol Obstet 170:363–378
62. Sobotka PA, McMannis J, Fisher RI, Stein DG, Thomas JX (1990) Effects of interleukin-2 on cardiac function in the isolated rat heart. J Clin Invest 86:845–850
63. Weisensee D, Bereiter-Hahn J, Schoeppe W, Low-Friedrich I (1993) Effects of cytokines on the contractility of cultured cardiac myocytes. Int J Immunopharmacol 15: 581–587
64. Natanson C, Fink MP, Ballantyne HK, MacVittie TJ, Conklin JJ, Parrillo JE (1986) Gram-negative bacteremia produces both severe systolic and diastolic cardiac dysfunction in a canine model that simulates human septic shock. J Clin Invest 78: 259–270
65. Natanson C, Eichenholz PW, Danner RL, et al (1989) Endotoxin and tumor necrosis factor challenges in dogs simulate the cardiovascular profile of human septic shock. J Exp Med 169:823–832
66. Michie HR, Manogue KR, Spriggs DR, et al (1988) Detection of circulating tumor necrosis factor after endotoxin administration. N Engl J Med 318:1481–1486
67. Beutler B, Cerami A (1986) Cachectin and tumor necrosis factor as two sides of the same biological coin. Nature 320:584–586
68. Eichenholz PW, Eichacker PQ, Hoffman WD, et al (1992) Tumor necrosis factor challenges in canines: Patterns of cardiovascular dysfunction. Am J Physiol 263: H668–H675
69. Walley KR, Hebert PC, Wakai Y, Wilcox PG, Road JD, Cooper DJ (1994) Decrease in left ventricular contractility after tumor necrosis factor-alpha infusion in dogs. J Appl Physiol 76:1060–1067
70. Beutler B, Milsark IW, Cerami A (1985) Passive immunization against cachectin/tumor necrosis factor protects mice from the lethal effect of endotoxin. Science 229: 869–871
71. Mathison JC, Wolfson E, Ulevitch RJ (1988) Participation of tumor necrosis factor in the mediation of gram-negative bacterial lipopolysaccharide-induced injury in rabbits. J Clin Invest 81: 1925–1937

72. Tracey KJ, Fong Y, Hesse DG, et al (1987) Anti-cachectin/TNF monoclonal antibodies prevent septic shock during lethal bacteremia. Nature 330:662–664
73. Vincent JL, Bakker J, Marecaux G, Schandene L, Kahn RJ, Dupont E (1992) Administration of anti-TNF antibody improves left ventricular function in septic shock patients: Results of a pilot study. Chest 101:810–815
74. Wherry J, Wenzel R, Wunderink R, et al (1993) Monoclonal antibody to human tumor necrosis factor (TNF MAb): Multicenter efficacy and safety study in patients with the sepsis syndrome. Internat Conf Antimicrob Agents Chemother, Am Soc Microbiol 33:246 (Abst)
75. Wakabayashi G, Gelfand JA, Jung WK, Connolly RJ, Burke JF, Dinarello CA (1991) Staphylococcus epidermidis induces complement activation, tumor necrosis factor and interleukin-1, a shock-like state and tissue injury in rabbits without endotoxemia. Comparison to Escherichia coli. J Clin Invest 87:1925–1935
76. Hesse DG, Tracey KJ, Fong Y, et al (1988) Cytokine appearance in human endotoxemia and primate bacteremia. Surg Gynecol Obstet 166:147–153
77. Okusawa S, Gelfand JA, Ikejima T, Connolly RJ, Dinarello CA (1988) Interleukin-1 induces a shock-like state in rabbits. J Clin Invest 81:1162–1172
78. Fisher CJ Jr, Dhainaut JF, Opal SM, et al (1994) Recombinant human interleukin-1 receptor antagonist in the treatment of patients with the sepsis syndrome: Results from a randomized, double-blind, placebo-controlled trial. JAMA 271:1836–1843
79. Ohlsson K, Bjork P, Bergenfeldt M, Hageman R, Thompson RC (1990) Interleukin-1 receptor antagonist reduces mortality from endotoxin shock. Nature 348:550–552
80. Fischer E, Marano MA, Van Zee KJ, et al (1992) Interleukin-1 receptor blockade improves survival and hemodynamic performance in Escherichia coli septic shock but fails to alter host responses to sublethal endotoxemia. J Clin Invest 89:1551–1557
81. Fisher CJ Jr, Slotner GJ, Opal SM, et al (1994) Initial evaluation of human recombinant interleukin-1 receptor antagonist in the treatment of sepsis syndrome: A randomized, open-label, placebo-controlled multicenter trial. Crit Care Med 22:12–21
82. Schuette WH, Burch C, Roach PO, Parrillo JE (1987) Closed loop television tracking of beating heart cells in vitro. Cytometry 8:101–103
83. Kumar A, Dimou C, Snell RJ, et al (1992) Tumor necrosis factor produces depression of myocardial cell contraction in vitro. Crit Care Med 20:S52 (Abst)
84. Yokoyama T, Vaca L, Rossen RD, Durante W, Hazarika P, Mann DL (1993) Cellular basis for the negative inotropic effects of tumor necrosis factor-alpha in the adult mammalian heart. J Clin Invest 92:2303–2312
85. Waage A, Espevik T (1988) Interleukin-1 potentiates the lethal effect of tumor necrosis factor alpha/cachectin in mice. J Exp Med 167:1987–1992
86. Weinberg JR, Boyle P, Meager A, Guz A (1992) Lipopolysaccharide, tumor necrosis factor, and interleukin-1 interact to cause hypotension. J Lab Clin Med 120:205–211
87. Kumar A, Kosuri R, Kandula P, et al (1994) Human septic serum-induced in vitro cardiomyocyte depression is caused by the synergistic action of TNF-alpha and IL-1 beta. Clin Res 42:168A (Abst)
88. Kumar A, Brar R, Lakshminarayanan L, et al (1995) IL-1 beta but not endotoxin, IL-2, IL-4, IL-6, IL-8, IL-10, or interferon-gamma depresses myocardial cell contractility in vitro. Crit Care Med 23:A168 (Abst)
89. Kumar A, Kosuri R, Kandula P, et al (1995) Human septic serum induces depression of cardiac myocyte contractility through the synergistic actions of TNF-alpha and IL-1 beta. Crit Care Med 23:A262 (Abst)
90. Evans HG, Lewis MJ, Shah AM (1993) Interleukin-1 beta modulates myocardial contraction via dexamethasone sensitive production of nitric oxide. Cardiovasc Res 27:1486–1490
91. Hosenpud JD, Campbell SM, Mendelson DJ (1989) Interleukin-1-induced myocardial depression in an isolated beating heart preparation. J Heart Transplant 8:460–464

92. DeMeules JE, Pigula FA, Mueller M, Raymond SJ, Gamelli RL (1992) Tumor necrosis factor and cardiac function. J Trauma 32:686–692
93. Vane JR, Anggard EE, Botting RM (1990) Regulatory functions of the vascular endothelium. N Engl J Med 323:27–36
94. Nathan C (1992) Nitric oxide as a secretory product of mammalian cells. FASEB J 6: 3051–3064
95. Lincoln TM, Cornwell TL (1991) Towards an understanding of the mechanism of action of cyclic AMP and cyclic GMP in smooth muscle relaxation. Blood Vessels 28: 129-137
96. Moncada S, Palmer RMJ, Higgs EA (1991) Nitric oxide: Physiology, pathophysiology and pharmacology. Pharmacol Rev 43:109–142
97. Nawrath H (1977) Does cyclic GMP mediate the negative inotropic effect of acetylcholine in the heart? Nature 267:72–74
98. George WJ, Polson JB, O'Toole AG, Goldberg ND (1970) Elevation of guanosine 3,5-cyclic phosphate in rat heart after perfusion with acetylcholine. Proc Natl Acad Sci 66:398–403
99. Han X, Shimoni Y, Giles WR (1994) An obligatory role for nitric oxide in autonomic control of mammalian heart rate. J Physiol 476:309–314
100. Levi RC, Alloatti G, Penna C, Gallo MP (1994) Guanylate-cyclase-mediated inhibition of cardiac ICa by carbachol and sodium nitroprusside. Pflugers Arch 426: 419–426
101. Shah AM, Spurgeon HA, Sollott SJ, Talo A, Lakatta EG (1994) 8-bromo-cGMP reduces the myofilament response to Ca2+ in intact cardiac myocytes. Circ Res 74: 970–978
102. Levi RC, Alloatti G, Fischmeister R (1989) Cyclic GMP regulates Ca-channel current in guinea pig ventricular myocytes. Pflugers Arch 413:685–687
103. Tohse N, Sperelakis N (1991) cGMP inhibits the activity of single calcium channels in embryonic chick heart cells. Circ Res 69:352–361
104. Mery PF, Pavoine C, Belhassen L, Pecker F, Fischmeister R (1994) Nitric oxide regulates cardiac Ca^{2+} current. Involvement of cGMP-inhibited and cGMP-stimulated phosphodiesterases through guanylate cyclase activation. J Biological Chem 268: 26286–26295
105. Fischmeister R, Hartzell HC (1987) Cyclic guanosine 3′,5′-monophosphate regulates the calcium current in single cells from frog ventricle. J Physiol 387:453–472
106. Mery P, Lohmann SM, Walter U, Fischmeister R (1991) Ca2+ current is regulated by cyclic GMP-dependent protein kinase in mammalian cardiac myocytes. Proc Natl Acad Sci 88:1197–1201
107. Fort S, Lewis MJ (1991) Regulation of myocardial contractile performance by sodium nitroprusside in the isolated perfused heart of the ferret. Br J Pharmacol 102:351P (Abst)
108. Brady AJ, Warren JB, Poole-Wilson PA, Williams TJ, Harding SE (1993) Nitric oxide attenuates cardiac myocyte contraction. Am J Physiol 265:H176–H182
109. Grocott-Mason R, Fort S, Lewis MJ, Shah AM (1994) Myocardial relaxant effect of exogenous nitric oxide in isolated ejecting hearts. Am J Physiol 266:H1699–H1705
110. Paulus WJ, Vantrimpont PJ, Shah AM (1994) Acute effects of nitric oxide on left ventricular relaxation and diastolic distensability in humans. Assessment by bicoronary sodium nitroprusside infusion. Circulation 89:2070–2078
111. Balligand J, Kelly RA, Marsden PA, Smith TW, Michel T (1993) Control of cardiac muscle cell function by an endogenous nitric oxide signaling system. Proc Natl Acad Sci 90:347–351
112. Schulz R, Nava E, Moncada S (1992) Induction and potential biological relevance of a Ca (2+)-independent nitric oxide synthase in the myocardium. Br J Pharmacol 105:575–580
113. Smith JA, Radomski MW, Schulz R, Moncada S, Lewis MJ (1993) Porcine ventricular endocardial cells in culture express the inducible form of nitric oxide synthase. Br J Pharmacol 108:1107–1110

114. De Belder AJ, Radomski MW, Why HJF, et al (1993) Nitric oxide synthase activities in the human myocardium. Lancet 341:84–85
115. Balligand J, Kelly RA, Smith TW, Michel T (1994) Identification of a constitutive isoform of NO synthase in adult rat cardiac myocytes. Circulation 90:I193 (Abst)
116. Ochoa JB, Udekwu AO, Billiar TR, et al (1991) Nitrogen oxide levels in patients after trauma and during sepsis. Ann Surg 214:621–626
117. Rosenberg RB, Broner CW, O'Dorisio MS (1994) Modulation of cyclic guanosine monophosphate production during Escherichia coli septic shock. Biochem Med Metab Biol 51:149–155
118. Salvemini D, Korbut R, Anggard E, Vane JR (1990) Immediate release of a nitric oxide-like factor from bovine aortic endothelial cells by Escherichia coli lipopolysaccharide. Proc Natl Acad Sci 87:2593–2597
119. Bernard C, Szekely B, Philip I, Wollman E, Payen D, Tedgui A (1992) Activated macrophages depress the contractility of rabbit carotids via a L-arginine/nitric oxide-dependent effector mechanism. J Clin Invest 89:851–860
120. Kilbourn RG, Gross SS, Jubran A, et al (1990) N-methyl-L-arginine inhibits tumor necrosis factor-induced hypotension: Implications for the involvement of nitric oxide. Proc Natl Acad Sci USA 87:3629–3633
121. Busse R, Mulsch A (1990) Induction of nitric oxide synthase by cytokines in vascular smooth muscle cells. FEBS 275:87–90
122. Radomski MW, Palmer RMJ, Moncada S (1990) Glucocorticoids inhibit the expression of an inducible, but not the constitutive, nitric oxide synthase in vascular endothelial cells. Proc Natl Acad Sci 87:10043–10047
123. French JF, Lambert LE, Dage RC (1991) Nitric oxide synthase inhibitors inhibit interleukin-1 beta-induced depression of vascular smooth muscle. J Pharm Exp Therap 259:260–262
124. Kilbourn RG, Griffith OW (1992) Overproduction of nitric oxide in cytokine-mediated and septic shock. J Natl Cancer Inst 84:827–831
125. Lorente JA, Landin L, Renes E, et al (1993) Role of nitric oxide in the hemodynamic changes of sepsis. Crit Care Med 21:759–767
126. Petros A, Bennett D, Vallance P (1991) Effect of nitric oxide synthase inhibitors on hypotension in patients with septic shock. Lancet 383:1557–1558
127. Hollenberg SM, Cunnion RE, Zimmerberg J (1993) Nitric oxide synthase inhibition reverses arteriolar hyporesponsiveness to catecholamines in septic rats. Am J Physiol 33:H660–H663
128. Schneider F, Lutun P, Hasselmann M, Stoclet JC, Tempe JD (1992) Methylene blue increases systemic vascular resistance in human septic shock. Preliminary observations. Intensive Care Med 18:309–311
129. Petros A, Lamb G, Leone A, Moncada S, Bennett D, Vallance P (1994) Effects of a nitric oxide synthase inhibitor in humans with septic shock. Cardiovasc Res 28:34–39
130. Kumar A, Kosuri R, Thota V, et al (1993) Nitric oxide and cyclic GMP generation mediates human septic serum-induced in vitro cardiomyocyte depression. Chest 104: 12S (Abst)
131. Kumar A, Kosuri R, Thota V, et al (1993) Tumor necrosis factor-induced myocardial cell depression is mediated by nitric oxide and cyclic GMP generation. Circulation 883:I617 (Abst)
132. Kumar A, Kosuri R, Kandula P, et al (1995) Interleukin-1-beta-induced myocardial cell depression is mediated by nitric oxide and cyclic GMP generation. Crit Care Med 23:A149 (Abst)
133. Kumar A, Thota V, Kosuri R, et al (1995) Tumor necrosis factor impairs epinephrine-stimulated cardiomyocyte contractility and cyclic AMP response through a nitric oxide-independent mechanism. Crit Care Med 23:A148 (Abst)
134. Brady AJ, Poole-Wilson PA, Harding SE, Warren JB (1992) Nitric oxide production within cardiac myocytes reduces their contractility in endotoxemia. Am J Physiol 263:H1963–H1966

135. Shindo T, Ikeda U, Ohkawa F, et al (1994) Nitric oxide synthesis in rat cardiac myocytes and fibroblasts. Life Sci 55:1101–1108
136. Tsujino M, Hirata Y, Imai T, et al (1994) Induction of nitric oxide synthetase gene by interleukin-1 beta in cultured rat cardiocytes. Circulation 90:375–383
137. Shen W, Xu X, Ochoa M, Zhao G, Wolin MS, Hintze TH (1994) Role of nitric oxide in regulation of oxygen consumption in conscious dogs. Circ Res 75:1086–1095

Influence of Nitric Oxide on Cardiac Systolic and Diastolic Function

A. M. Shah

Introduction

The highly reactive gas, nitric oxide (NO), is now recognized to be a ubiquitous inter- and intra-cellular signaling molecule which has pivotal roles in many physiological processes, e.g. endothelial regulation of vascular smooth muscle tone and smooth muscle cell proliferation, inhibition of platelet adhesion and aggregation, peripheral and central neurotransmission, macrophage cytotoxicity, etc. (for reviews, see [1, 2]). In addition, an increasing body of data implicates abnormalities of NO pathway in several pathophysiological processes. The present chapter focuses on recent work which indicates that NO also has important physiological and pathophysiological influences on myocardial contractile function.

NO is generated from the amino acid L-arginine by a family of NO synthase (NOS) enzymes in a process that requires molecular oxygen and reduced nicotinamide adenine dinucleotide phosphate (NADPH) as co-substrates, and tetrahydrobiopterin as a co-factor. At least 3 isoforms of NOS, coded for by 3 separate genes, have been identified: constitutive Ca^{2+}-dependent endothelial and neuronal enzymes (eNOS and nNOS, respectively; or collectively cNOS), and a Ca^{2+}-independent inducible NOS (iNOS) which can be expressed in several different tissues, particularly following cytokine stimulation [2]. NO production by cNOS occurs in a rapid and tightly regulated fashion, with small puffs being released in response to specific mechanical and biochemical stimuli such as flow-induced shear stress (probably the most important physiological stimulus), moderate hypoxia, and a large number of agonists that act via specific membrane receptors, e.g. substance P, bradykinin, acetylcholine, excitatory amino acids, thrombin. By contrast, iNOS enzymes are generally activated by induction of transcription (e.g. by interferons, interleukins, lipopolysaccharide) and subsequently produce large amounts of NO in an unregulated manner. Inhibition of NO production by cNOS or iNOS may be achieved pharmacologically by the use of L-arginine analogs, while NO activity can be inhibited by hemoglobin, which binds to and inactivates it. Induction of iNOS isoforms can be prevented by corticosteroids and certain cytokines.

NO has a very brief half-life in biological systems so that its activity is generally restricted to sites close to the source of release. The main subcellular actions of NO in many tissues are mediated through stimulation of soluble guanylyl cyclases and consequent elevation of intracellular cyclic guanosine 3′,5′ monophosphate (cGMP), in a process that involves the binding of NO to the heme moiety of the enzyme. However, there is increasing evidence that NO has several additional molecular targets, and can therefore exert a variety of cGMP-independent actions [3].

NO and Cardiac Function

NO-mediated changes in peripheral and coronary vascular tone have the potential to indirectly influence cardiac function. For example, alterations in peripheral arterial tone will result in altered cardiac loading. Changes in venous tone may alter capacitance and venous return and may thus modify ventricular contractile function through the Frank-Starling mechanism (i.e. secondary to changes in ventricular filling volumes). Modulation of coronary vascular tone will influence coronary perfusion, and thereby influence cardiac function particularly under pathophysiological situations. In addition to these *indirect* effects on the heart it is now apparent that NO also exerts *direct* effects on myocardium which are both physiologically and pathophysiologically relevant. At least two cellular sources of NO are important in this respect, cardiac endothelial cells and the cardiac myocytes themselves. It is possible that NO derived from other cell types, e.g. nerve tissue, may also exert significant indirect and/or direct effects on cardiac function, but this aspect is relatively unexplored as yet. The discussion below is restricted to the effects of NO on myocardial *contractile* function, although it should be noted that evidence is emerging to suggest modulation of other aspects of cardiac function.

Basal Systolic and Diastolic Function

Isolated Papillary Muscle Studies

The first studies to investigate the effects of endothelium-derived NO on myocardial contraction were performed in isolated mammalian cardiac papillary muscle preparations. Stimulation of NO release from the endocardial endothelium in ferret papillary muscles, using the agonist substance P, resulted in enhanced myocardial relaxation and a slight reduction in peak isometric force, but no change in the early phase of contraction, e.g. in the maximal rate of force development (Fig. 1) [4]. These effects were accompanied by an elevation of myocardial cGMP, and could be abolished either by denudation of the endocardial endothelium, by inhibitors of NO activity (hemoglobin), or by L-arginine analogs, confirming that they were the result of NO release from endothelial cells. Similar effects were also observed with the exogenous NO donor, sodium nitroprusside, or with a lipid-soluble

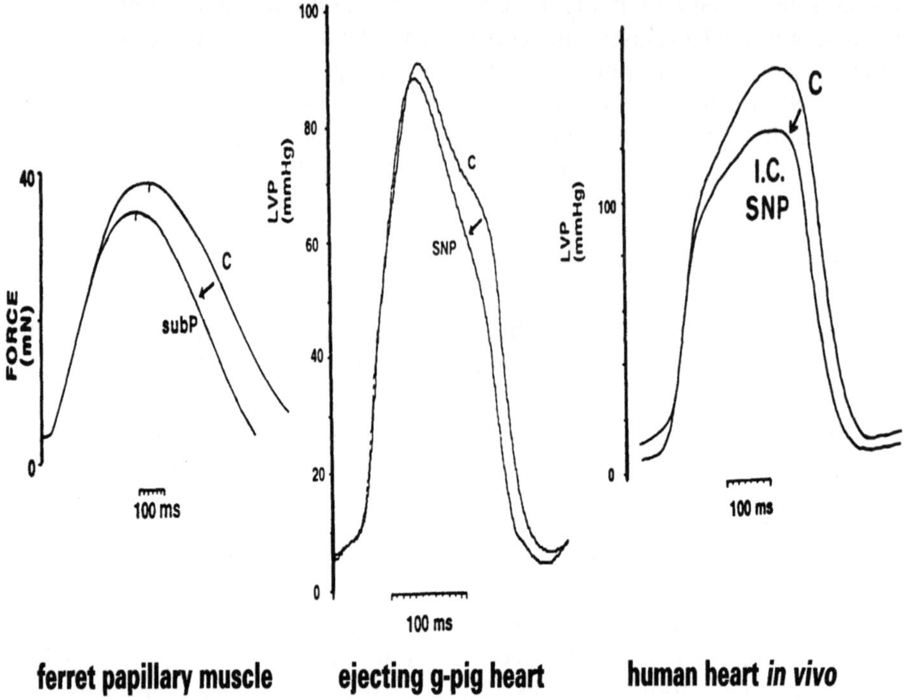

Fig. 1. Representative examples of the basal effect of NO in different cardiac preparations and species. Left panel: effect of substance P (subP) on isometric twitch contraction of an isolated ferret papillary muscle preparation. Middle panel: effect of sodium nitroprusside (SNP) on left ventricular pressure in the isolated ejecting guinea-pig heart. Right panel: effect of low-dose bicoronary SNP infusion on left ventricular pressure in a normal human subject undergoing diagnostic cardiac catheterization. C = control

cGMP analog, 8-bromo-cGMP [4, 5]. This pattern of myocardial activity, affecting mainly late systole and relaxation, is rather unusual and differs from most conventional therapeutic agents, e.g. Ca^{2+}-antagonists, β-adrenergic blockers, phosphodiesterase inhibitors, which generally alter both early systolic function and relaxation.

Isolated Heart Studies

In the isolated papillary muscle preparation, most of the myocardium is readily accessible to factors released by the endocardial endothelium. In the intact heart, however, it seems unlikely that NO released at the endocardial (cavitary) surface could significantly influence contraction of the total myocardial mass, in view of its brief half-life. On the other hand, no cardiac myocyte is more than approximately 5 μ from a coronary vascular endothelial cell. It was therefore an important question whether NO derived from coronary vascular endothelial cells could also modulate myocardial contraction in a similar manner to that observed with endocardial NO in papillary muscles. We addressed this issue in studies in the isolated ejecting (working) guinea-pig heart preparation, where filling pressure (preload), afterload and heart rate can be maintained constant. The effects of coronary endothelium-derived factors or of other interventions on contractile function can therefore be assessed independently of changes in loading and heart rate. High-fidelity left ventricular (LV) pressure data obtained with a 2F tip-micromanometer inserted directly into the ventricular cavity allow accurate quantification of changes in LV pressure and in LV relaxation [6].

Both the NO donor, sodium nitroprusside, and agents that release NO from coronary vascular endothelial cells (e.g. substance P or bradykinin) induced a dose-dependent earlier onset and acceleration of LV relaxation, without significant effect on peak LV dP/dt, similar to the relaxant effects observed in isolated papillary muscles (Fig. 1) [6, 7]. These relaxant effects were independent of increases in coronary flow and were not reproduced by a cGMP-independent vasodilator, the Ca^{2+}-antagonist nicardipine, which increased coronary flow to a similar extent [6]. The actions of sodium nitroprusside and of substance P or bradykinin were inhibited in the presence of hemoglobin, supporting the hypothesis that they were mediated by NO. In addition, neither substance P nor bradykinin had direct myocardial effects in endothelium-denuded guinea-pig papillary muscles [7]. Thus, endothelium-derived NO selectively modulates LV relaxation in the whole heart without major effect on early systolic pressure development.

Isolated Myocyte Studies

The effects of exogenous NO or of cGMP analogs have also been studied in isolated myocyte preparations. In isolated single rat cardiac myocytes, the cGMP analog, 8-bromo-cGMP, induced similar effects to those described above, i.e. a reduction in amplitude of isotonic twitch shortening, an earlier onset of twitch relaxation, and a significant increase in myocyte diastolic length [8]. These effects were not accompanied by changes in the cytosolic Ca^{2+} transient, monitored using the fluorescent probe indo-1, or in the action potential, suggesting that they were mediated by a cGMP-induced reduction in the myofilament response to Ca^{2+} (Fig. 2). This mechanism of action was also supported by studies of the steady-state relationship between cell shortening and cytosolic Ca^{2+} in intact myocytes tetanized by high frequency stimulation following inhibition of the sarcoplasmic reticulum adenosine triphosphatase (ATPase); 8-bromo-cGMP decreased steady-state tetanic shortening without reducing steady-state tetanic elevation of Ca^{2+} [8]. The reduction in myofilament response to Ca^{2+} induced by 8-bromo-cGMP appears to be mediated via stimulation of cGMP-dependent protein kinase since it was inhibited by KT5823, a specific inhibitor of the kinase. The acute increase in myocyte diastolic length induced by 8-bromo-cGMP, which was also observed in quiescent (i.e. electrically unstimulated) myocytes, may

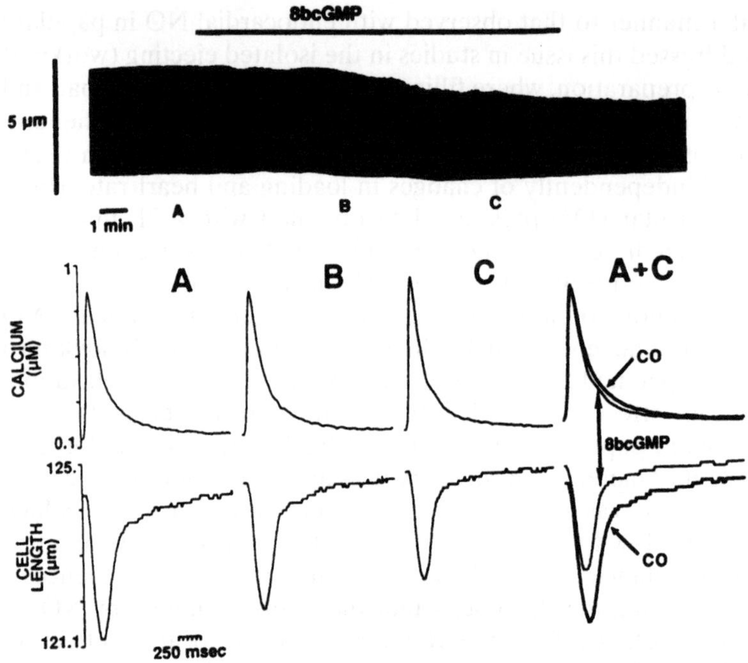

Fig. 2. Effect of 8-bromo-cGMP (8bcGMP, 50 µM) on an isolated rat cardiac myocyte loaded with the fluorescent probe indo-1. Above: Chart recording of myocyte shortening. Below: Expanded scale cytosolic Ca^{2+} transients and twitch contractions at time-points indicated (**A** baseline; **B** 4 min after, and **C** 11 min after 8bcGMP). The contractile changes produced by 8bcGMP were not accompanied by significant changes in the Ca^{2+} transient. (From [8] with permission)

reflect a reduction in intrinsic diastolic tone in mechanically (externally) unloaded myocytes, and is consistent with the increase in diastolic distensibility observed in clinical studies (see below) [9].

About 50% of the rat myocytes exposed to 8-bromo-cGMP in the above study demonstrated a transient positive inotropic response which was associated with an increase in amplitude of the cytosolic Ca^{2+} transient [8]. A recent preliminary study in isolated rat cardiac myocytes has also reported biphasic inotropic effects of exogenous NO-donor drugs, with an increase in twitch amplitude at low concentrations of the drugs and a decrease in twitch amplitude at higher concentrations [10]. By contrast, a study of the effects of NO-donor drugs or of authentic NO in solution in isolated guinea-pig cardiac myocytes reported only modest reduction (approximately 20%) in the amplitude of isotonic twitch shortening, but no positive inotropic effect [11]. The physiological relevance of possible biphasic effects of NO/cGMP remains to be established.

Clinical Studies

Recently, we have extended our studies on the cardiac effects of NO to human subjects [9,12]. The effects of low-dose (≤ 4 μg/min) bicoronary infusions of the exogenous NO donor, sodium nitroprusside, were studied in normal subjects undergoing cardiac catheterization for investigation of atypical chest pain [9]. None of the subjects had any evidence of cardiac disease. Measurements were made of high-fidelity tip-micromanometer LV pressure, dP/dt right atrial pressure, and LV volume (by sequential angiography). A 5 min bicoronary infusion of sodium nitroprusside resulted in a significant reduction in peak LV systolic pressure and an earlier onset of LV relaxation (i.e. a reduced time to dP/dt_{min}), but caused no change in LV dP/dt_{max} (Fig. 1). LV end-diastolic pressure was significantly reduced while LV end-diastolic volume increased. The latter, together with a downward shift in end-diastolic LV pressure-volume relations, indicated an increase in LV end-diastolic distensibility. These changes were not accompanied by significant reduction in ejection fraction or in stroke volume. Infusion of an identical dose of sodium nitroprusside into the right atrium failed to reproduce these effects, indicating that they could not be attributed to systemic vasodilatation. Thus, exogenous NO exerted potentially beneficial effects on LV relaxation and LV diastolic distensibility without reducing stroke volume (despite reduction in peak LV pressure). In further studies using a similar investigation protocol, we have examined the effects of bicoronary infusion of substance P (20 pg/min) in normal subjects and in transplant recipients free of rejection or graft vasculopathy being catheterized at annual follow-up [12]. We observed a similar effect of substance P on LV relaxation, peak LV systolic pressure and LV diastolic distensibility to that observed with sodium nitroprusside, again without reduction in ejection fraction or stroke volume. These studies provide the first clinical evidence for a paracrine effect of NO released from coronary vascular endothelium in modulating LV contractile function.

Physiological Significance

The general pattern of the effects of NO described above is a relatively selective action on LV relaxation and diastolic distensibility, without major depression of systolic function. *In vivo*, enhanced LV relaxation and increased end-diastolic distensibility secondary to NO release from coronary vascular endothelial cells may benefit 1) LV diastolic filling, both by reducing filling pressure and by increasing the diastolic interval; and 2) coronary perfusion, also by increasing the diastolic interval and by reducing extravascular compressive forces. The most relevant physiological stimuli for NO release from coronary vascular endothelium may be flow-induced shear stress and the cyclical mechanical deformation of the endothelium that occurs during the cardiac cycle [13,14]. A NO-dependent pathway may be envisaged which enables the coupling of the *pattern* of ventricular contractile function to changes in coronary flow and/or heart rate. Consider, for example, what might happen during an increase in heart rate. The increase in frequency and velocity of pulsatile coronary flow would increase NO release [13, 14] which would then enhance LV relaxation and improve diastolic distensibility [9, 12]. The subsequent improvements in cardiac filling and subendocardial perfusion, without compromise to stroke output, would be beneficial in the context of a reduced diastolic interval and a ventricle performing more work.

Modulation of β-adrenergic Inotropic Responses

As mentioned previously, constitutive production of NO by cardiac myocytes has been reported. Recent studies by Balligand et al. [15] suggest that this intra-myocyte signaling pathway may have a role in modulating the positive inotropic response to β-adrenergic stimulation, at least in rat heart. In adult rat ventricular myocytes, the inotropic response to isoproterenol was augmented by the NOS inhibitor N^{ω}-nitro-L-arginine although the inhibitor had no effect on basal contraction. These effects were associated with an increase in NO production detected by a reporter cell assay. It was thus suggested that basal NO production reduces the inotropic response to isoproterenol. The subcellular mechanism of these effects was not determined, but is likely to be different from the cGMP-mediated reduction in myofilament response to Ca^{2+} discussed above. Some insights are provided by a study in frog myocytes [16] in which cyclic adenosine monophosphate (cAMP)-stimulated sarcolemmal L-type Ca^{2+} current was inhibited by the NO exogenous NO donor, 3-morpholinosydnonimine (SIN-1), reportedly via activation of a cAMP-phosphodiesterase (i.e. a reduction in cAMP levels). It was also observed in this study that low (nanomolar) concentrations of SIN-1 increased cAMP-stimulated Ca^{2+} current, suggesting inhibition of a different cAMP-phosphodiesterase (i.e. an increase in cAMP levels). The latter observation is obviously similar to the reports of biphasic inotropic effects of NO donor drugs in rat myocytes discussed above.

Regulation of Heart Rate

An endogenous (i.e. intra-myocyte) NO pathway has been implicated in the regulation of heart rate. In spontaneously beating cultured rat neonatal cardiac myocytes, the negative chronotropic action of the muscarinic agonist carbachol could be blocked by NOS inhibitors and mimicked by cGMP analogs [15]. These actions were attributed to intracellular generation of NO by cNOS and a consequent elevation of cGMP. In another study in rat neonatal cardiac myocytes, induction of iNOS activity by IL-1β (which occurred after about 15 h) was accompanied by a reduction in the frequency of spontaneous beating; these effects could also be inhibited by NOS inhibitors [17]. Both the induction of iNOS activity and the effect on beating rate were prevented by transforming growth factor β (TGF-β). Whether NO exerts similar effects on beating rate in spontaneously beating *adult* pacemaker tissue or whether endothelium-derived NO exerts chronotropic activity in the intact heart requires further investigation. A recent paper reports that cholinergic inhibition of isoproterenol-stimulated Ca^{2+} current in spontaneously beating adult rabbit sino-atrial cells is mediated by NO, and concludes that NO may reduce heart rate by this mechanism [18]. However, no measurements of beating rate or of the effect of NO on basal (unstimulated) rate were reported.

Subcellular Mechanisms

Probably all the effects of cNOS referred to above involve the stimulation of guanylyl cyclase and subsequent production of cGMP, although the possibility of cGMP-independent actions should not be excluded. A detailed discussion of the subcellular mechanisms of action of cGMP is outside the scope of this review and has been covered elsewhere [19]. The main target proteins for cGMP in myocardium are thought to be cGMP-dependent protein kinases (PKG) and cGMP-regulated cyclic nucleotide phosphodiesterases. The former pathway has the potential to alter function via phosphorylation, whilst the latter modulates intracellular levels of cAMP. A cGMP-stimulated phosphodiesterase is postulated to decrease intracellular cAMP whereas a cGMP-inhibited phosphodiesterase will have the opposite effect. The relative importance of these different pathways in different parts of the heart (e.g. atrium or ventricle), in different species, and under different conditions of myocardial activation has not been worked out in detail. Further adding to the complexity of possible responses and mechanisms are 1) the multiple isoforms of cGMP target proteins, 2) cross-reactions between cAMP- and cGMP-activated kinases, 3) possible subcellular compartmentation of cGMP and/or its target proteins, and 4) the possible role of cGMP-activated phosphatases.

Actions attributed to cGMP in heart include 1) modulation of sarcolemmal Ca^{2+} influx (either an increase or a reduction), generally only in the presence of elevated cAMP, 2) changes in action potential characteristics, 3) reduction in Ca^{2+}-sensitivity of the myofilaments, 4) phosphorylation of sarcoplasmic reticular proteins, and 5) modulation of cell volume (reviewed in [8]). The effects of NO/cGMP on myocardial relaxation and diastolic tone discussed above appear to be the result of PKG-mediated reduction in myofilament response to Ca^{2+} [8, 20]. On the other hand, the modulation of β-adrenergic responses is likely to involve changes in cAMP levels mediated via cGMP-regulated phosphodiesterases [16, 18].

In contrast to NO derived from cNOS, production of large amounts of this reactive molecule by iNOS (see below) is likely to exert significant cGMP-independent as well as cGMP-dependent effects [2, 3]. It is recognized that

NO can react directly with mitochondrial respiratory enzymes and with a variety of nuclear proteins, with the potential to cause cellular damage by these mechanisms. Indeed, this appears to be the major mode of action of the cytotoxic effect of NO on parasites and tumor cells.

Pathophysiology

The possible relevance of these NO-mediated effects on myocardial function in disease states is an area that has received little attention so far. Most of the work to date has focused on abnormal NO production by iNOS, whereas the role of the cNOS pathways in myocardial disease states remains largely speculative.

NO Production by iNOS

Following the recognition that activation of iNOS contributes to the pathophysiology of immune-mediated conditions (e.g. endotoxic shock, allograft rejection, etc.), and that iNOS can be expressed in multiple tissues, attention has turned to the role of iNOS in heart. Biochemical evidence for Ca^{2+}-independent cardiac production of NO was first reported in isolated rat cardiac myocytes following treatment of the animals with endotoxin [21]. Subsequent studies have shown that a variety of cytokines as well as medium conditioned by activated macrophages can also cause expression of iNOS activity in myocardial tissue in several different species [22–25]. NO activity has been demonstrated by electron paramagnetic resonance studies in rejecting cardiac grafts in an experimental rat model of transplantation [26]. A subsequent study has confirmed the presence of iNOS mRNA and iNOS protein in rejecting rat cardiac allografts [27], and iNOS message and protein have been detected in cultured rat myocytes treated with IL-1β [28]. Interestingly, iNOS activity has also been reported in experimental myocardial infarction [29]. It should be noted that in addition to cardiac myocytes, iNOS may also be expressed in other cell types within the heart (e.g. endothelial cells, endocardial cells, vascular smooth muscle, and inflammatory cells) [1]. A recent study has now confirmed the potential for Ca^{2+}-independent NOS activity in endomyocardial biopsies obtained from patients with dilated cardiomyopathy [30].

The functional consequence of activation of iNOS in cardiac tissue appears to be myocardial depression. In isolated papillary muscles and in isolated cardiac myocytes, activation of iNOS was found to depress basal and/or β-adrenergic stimulated myocardial inotropic responses [22–25]. These abnormalities could be reversed by inhibitors of NOS (L-arginine analogs) and in some cases could be prevented by pretreatment with dexamethasone. These data raise the possibility that expression of cardiac iNOS may contribute to cardiac pathophysiology and dysfunction in disease states where an involvement of cytokines is implicated (e.g. endotoxic shock, cardiomyopathy, allograft rejection (? myocardial infarction)).

A rapid (within minutes) NO-mediated depression of myocardial contraction following administration of some cytokines has also been reported [25, 31]. This cannot be attributed to expression of iNOS but most likely results from cytokine-induced activation of cNOS. In cultured avian cardiac myocytes, such depression of contraction, mediated by IL-6, was associated with a depression in intracellular Ca^{2+} transients [25].

Role of cNOS-derived NO in Cardiac Disease States?

Endothelial dysfunction is well recognized in relation to the regulation of vascular tone (e.g. in hypertension, left ventricular hypertrophy, heart failure, hypercholesterolemia, ischemia-reperfusion) [1, 2]. Whether abnormalities of the NO pathway play a role in the pathophysiology of myocardial disease states is an important question. In view of the influences of endothelium-derived NO on myocardial relaxation and diastolic tone discussed above, a possible role of this pathway in the pathogenesis of diastolic dysfunction is of interest. Reduced NO activity might be postulated to result in delayed myocardial relaxation and decreased end-diastolic distensibility, which would compromise both cardiac filling and subendocardial coronary perfusion, and contribute to elevated filling pressures and pulmonary congestion - often an important feature in cardiac disease states [32]. On the other hand, agents that exert selective effects on myocardial relaxation and diastole (via NO/cGMP-dependent mechanisms) might be useful as part of the therapeutic approach to these conditions. In a recent study in an experimental rat cardiac myocyte model of acute "diastolic dysfunction" produced by brief hypoxia-reoxygenation, pretreatment with the cGMP analog 8-bromo-cGMP completely inhibited the abnormalities of relaxation and diastolic function; the mechanism appeared to involve an alteration of the myofilament response to Ca^{2+} [33]. Vatner and colleagues [34] have studied the effects of inhibition of NOS on myocardial stunning following brief (10 min) ischemia in conscious dogs. Inhibition of NOS enhanced the degree of myocardial depression in this model, implying that NO had a protective effect against myocardial stunning. NO also appears to protect isolated rat cardiac myocytes against the injury observed at reoxygenation following prolonged anoxia [35].

Conclusions

There is now convincing evidence that NO pathways exert physiologically important direct as well as indirect influences on myocardial contractile functions. At least two pathways appear to be important, a paracrine pathway involving endothelium-derived NO and an intracrine (or possibly autocrine) pathway involving NO production within cardiac myocytes. NO influences both basal myocardial function, particularly relaxation and diastolic tone, and β-adrenergic inotropic and chronotropic responses. The paracrine pathway may have a particular role in allowing crosstalk between the coronary circulation and the myocardium, thus facilitating adaptive matching of flow and function. The latter should be considered in the context of a broader endothelial paracrine pathway that involves several other cardioactive factors [32, 36–39] including endothelin [40–42], two unidentified but biologically characterized substances termed "myofilament desensitizing factor" [43] and "hypoxic factor" [44], and possibly other agents (adenine nucleotides, angiotensin, prostaglandins) (Fig. 3). Accumulating evidence also supports a role for intracardiac iNOS pathways in pathophysiological states where immune mechanisms or cytokines are implicated. It seems likely that abnormalities of cardiac cNOS pathways too will be relevant in disease states. The scene is thus set for an explosion of work in this field, with the potential for therapeutic advances.

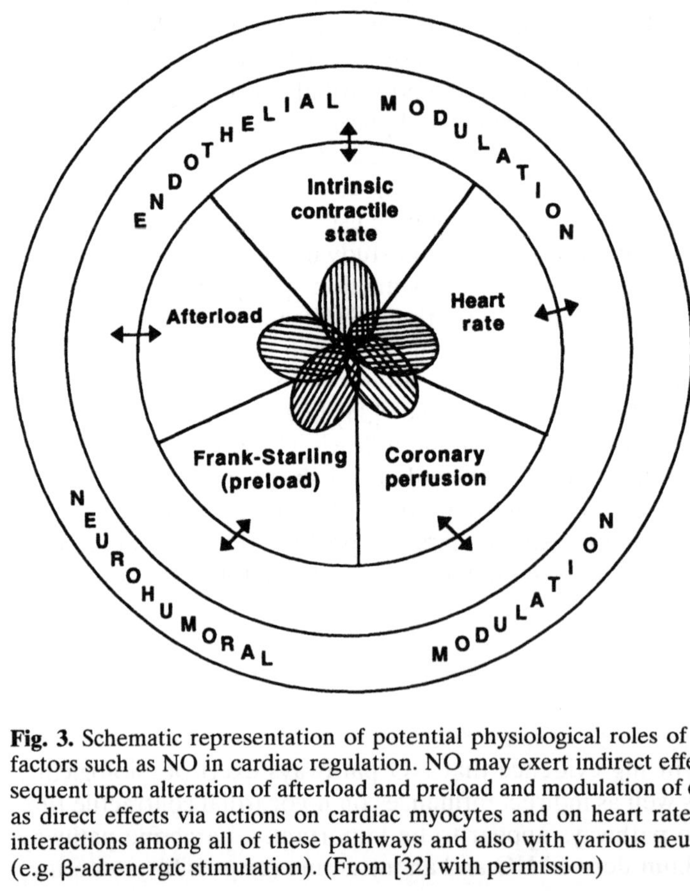

Fig. 3. Schematic representation of potential physiological roles of endothelium-derived factors such as NO in cardiac regulation. NO may exert indirect effects on the heart consequent upon alteration of afterload and preload and modulation of coronary flow, as well as direct effects via actions on cardiac myocytes and on heart rate. There are potential interactions among all of these pathways and also with various neurohumoral influences (e.g. β-adrenergic stimulation). (From [32] with permission)

Acknowledgements: This work was supported by the British Heart Foundation and the Welsh Scheme for the Development of Health and Social Sciences Research.

References

1. Moncada S, Palmer RMJ, Higgs HA (1991) Nitric oxide. Physiology, pathophysiology and pharmacology. Pharmacol Rev 43:109–142
2. Nathan C (1992) Nitric oxide as a secretory product of mammalian cells. FASEB J 6: 3051–3064
3. Stamler JS, Singel DJ, Loscalzo J (1993) Biochemistry of nitric oxide and its redox-activated forms. Science 258:1898–1902
4. Smith JA, Shah AM, Lewis MJ (1991) Factors released from the endocardium of the ferret and pig modulate myocardial contraction. J Physiol 439:1–14
5. Shah AM, Lewis MJ, Henderson AH (1990) Effects of 8-bromo-cyclic GMP on contraction and on inotropic response of ferret cardiac muscle. J Mol Cell Cardiol 23: 55–64
6. Grocott-Mason RM, Fort S, Lewis MJ, Shah AM (1994) Myocardial relaxant effect of exogenous nitric oxide in the isolated ejecting heart. Am J Physiol 266:HI699–H1705
7. Grocott-Mason RM, Anning PB, Evans H, Lewis MJ, Shah AM (1994) Modulation of left ventricular relaxation in isolated ejecting heart by endogenous nitric oxide. Am J Physiol 267:H1804–H1813
8. Shah AM, Spurgeon H, Sollott SJ, Talo A, Lakatta EG (1994) 8-bromo cyclic GMP reduces the myofilament response to calcium in intact cardiac myocytes. Circ Res 74: 970–978
9. Paulus WJ, Vantrimpont PJ, Shah AM (1994) Acute effects of nitric oxide on left ventricular relaxation and diastolic distensibility in man. Circulation 89:2070–2078
10. Kojda G, Kottenberg K, Nix P, Schluter KD, Piper HM (1994) Opposite responses of rat cardiac myocytes to different nitrovasodilators. Circulation 90:1–592 (Abst)
11. Brady AJB, Warren JB, Poole-Wilson PA, Williams TJ, Harding SE (1993) Nitric oxide attenuates cardiac myocyte contraction. Am J Physiol 265:H176–H182
12. Paulus WJ, Vantrimpont PJ, Shah AM (1994) Paracrine coronary endothelial control of left ventricular function in humans. Circulation 90:I–17 (Abst)
13. Hutcheson IR, Griffith TM (1991) Release of endothelium-derived relaxing factor is modulated both by frequency and amplitude of pulsatile flow. Am J Physiol 261: H257–H262
14. Lamontagne D, Pohl U, Busse R (1992) Mechanical deformation of vessel wall and shear stress determine the basal release of endothelium-derived relaxing factor in the intact rabbit coronary vascular bed. Circ Res 70:123–130
15. Balligand JL, Kelly RA, Marsden PA, Smith TW, Michel T (1993) Control of cardiac muscle function by an endogenous nitric oxide signaling system. Proc Natl Acad Sci USA 90:347–351
16. Mery PF, Pavoine C, Belhassen L, Pecker F, Fischmeister R (1994) Nitric oxide regulates cardiac Ca^{2+} current. J Biol Chem 268:26286–26295
17. Roberts AB, Vodovotz Y, Roche NS, Sporn MB, Nathan CF (1992) Role of nitric oxide in antagonistic effects of transforming growth factor-β and interleukin-1β on the beating rate of cultured cardiac myocytes. Mol Endocrinol 6:1921–1930
18. Han X, Shimoni Y, Giles WR (1994) An obligatory role for nitric oxide in autonomic control of mammalian heart rate. J Physiol 476:309–314
19. Lincoln TM, Cornwell TL (1993) Intracellular cyclic GMP receptor proteins. FASEB J 7:328–338
20. Pfitzer G, Rüegg JC, Flockerzi V, Hofmann F (1982) cGMP-dependent protein kinase decreases calcium sensitivity of skinned cardiac fibers. FEBS Lett 149:171–175

21. Schulz R, Nava E, Moncada S (1992) Induction and biological relevance of a Ca^{++}-independent nitric oxide synthase in the myocardium. Br J Pharmacol 105:575–580
22. Brady AJB, Poole-Wilson PA, Harding SE, Warren JB (1992) Nitric oxide within cardiac myocytes reduces their contractility in endotoxemia. Am J Physiol 263: H1963–H1966
23. Evans HG, Lewis MJ, Shah AM (1993) Interleukin-1β modulates myocardial contraction via dexamethasone-sensitive production of nitric oxide. Cardiovasc Res 27: 1486–1490
24. Balligand JL, Ungureanu D, Kelly RA, et al (1993) Abnormal contractile function due to induction of nitric oxide synthesis in rat cardiac myocytes follows exposure to activated macrophage-conditioned medium. J Clin Invest 91:2314–2319
25. Kinugawa K, Takahashi T, Kohmoto O, et al. (1994) Nitric oxide-mediated effects of interleukin-6 on $[Ca^{2+}]_i$ and cell contraction in cultured chick ventricular myocytes. Circ Res 75:285–295
26. Lancaster JR, Langrehr JM, Bergonia HA, Murase N, Simmons RL, Hoffman RA (1992) EPR detection of heme and non-heme iron-containing protein nitrosylation by nitric oxide during rejection of rat heart allograft. J Biol Chem 267:10994–10998
27. Yang X, Chowdhury N, Cai B, et al (1994) Induction of myocardial nitric oxide synthase by cardiac allograft rejection. J Clin Invest 94:714– 721
28. Tsujino M, Hirata Y, Imai T, et al (1994) Induction of nitric oxide synthase gene by interleukin-1β in cultured rat cardiocytes. Circulation 90:375–383
29. Dudek RR, Wildhirt S, Pinto V, Giesler G, Bing RJ (1994) Dexamethasone inhibits the expression of an inducible nitric oxide synthase in infarcted rabbit myocardium. Biochem Biophys Res Commun 202:1120–1126
30. De Belder AJ, Radomski MW, Why HJF, et al (1993) Nitric oxide synthase in human myocardium. Lancet 341:84–85
31. Finkel MS, Oddis CV, Jacob TD, Watkins SC, Hattler BG, Simmons RL (1992) Negative inotropic effects of cytokines on the heart mediated by nitric oxide. Science 257: 387–389
32. Shah AM (1994) Modulation of myocardial contractile function by endothelium. In: Haunso S, Kjeldsen K (eds) Proceedings of the International Society for Heart Research European Section Meeting. Monduzzi Editore, Bologna, pp 53–60
33. Shah AM, Silverman HS, Lakatta EG (1993) Cyclic GMP abolishes relaxation abnormalities at reoxygenation following brief anoxia in single cardiac myocytes. Circulation 88:I-3 (Abst)
34. Hasebe N, Shen YT, Vatner SF (1993) Inhibition of endothelium- derived relaxing factor enhances myocardial stunning in conscious dogs. Circulation 88:2862–2871
35. Schluter KD, Weber M, Schraven E, Piper HM (1994) NO donor SIN-1 protects against reoxygenation-induced cardiomyocyte injury by a dual action. Am J Physiol 267:H1461–H1466
36. Brutsaert DL, Andries LJ (1992) The endocardial endothelium. Am J Physiol 263: H985–H1002
37. Ramaciotti C, Sharkey A, McClellan G, Winegrad S (1992) Endothelial cells regulate cardiac contractility. Proc Natl Acad Sci 89:4033–4036
38. Shah AM, Lewis MJ (1993) Modulation of myocardial contraction by endocardial and coronary vascular endothelium. Trends Cardiovasc Med 3:98–103
39. Ramaciotti C, McClellan G, Sharkey A, Rose D, Wiseberg A, Winegrad S (1993) Cardiac endothelial cells modulate contractility of rat heart in response to oxygen tension and coronary flow. Circ Res 72:1044–1064
40. Mebazaa A, Mayoux E, Maeda K, Lakatta EG, Robotham JL, Shah AM (1993) Paracrine effects of endocardial endothelial cells on myocyte contraction mediated via endothelin. Am J Physiol 265:H1841–H1846
41. McClellan G, Weisberg A, Rose D, Winegrad S (1994) Endothelial cell storage and release of endothelin as a cardioregulatory mechanism. Circ Res 70:787–803
42. Evans HG, Lewis MJ, Shah AM (1994) Modulation of myocardial relaxation by basal release of endothelin from endocardial endothelium. Cardiovasc Res 28:1694–1699

43. Shah AM, Mebazaa A, Wetzel RC, Lakatta EG (1994) Novel cardiac myofilament desensitizing factor released by endocardial and vascular endothelial cells. Circulation 89:2492–2497
44. Shah AM, Mebazaa A, Cuda G, et al. (1994) Ca^{2+}-independent inhibition of actomyosin crossbridge cycling in cardiac myocytes by endothelial cell factors in response to hypoxia. J Physiol 475.P:80P (Abst)

Cellular Actions of Nitric Oxide on the Circulatory System

A. M. Lefer

Introduction

In recent years, nitric oxide (NO) has been widely studied by a variety of biomedical scientists including pharmacologists, physiologists, microbiologists, immunologists as well as those interested in circulatory shock states and other cardiovascular disorders including ischemia-reperfusion [1,2]. In these disorders, significant interest has focused on the potential toxic effects of NO, since NO is thought to kill bacteria and has been postulated to be the active principle responsible for the hypotension and impaired vascular responsiveness to vasoconstrictors in septic shock [3–5]. On the other hand, several others have obtained strong evidence that NO functions as an intercellular messenger and is essential in the normal physiology of the circulatory system [6–8]. Moreover, a large body of evidence has been obtained indicating that in endotoxin shock, NO concentrations, or the effectiveness of NO actually declines [9–11]. In this connection, infusion of low doses of NO or organic nitrates which release NO, actually improves circulatory shock states including endotoxic shock [12–15]. Indeed, this controversy has undergone spirited debate and the final decision has not been reached. In the past few years however, a significant shift in our thinking has occurred away from viewing NO as being the primary mediator of endotoxin or septic shock toward appreciating its role as a physiological regulator of circulatory homeostasis. The major purposes of this chapter are

1) to review the role of NO as a modulator of cell function in the major cell types which comprise the circulatory system, and
2) to relate these cellular actions to potential effects of NO in the setting of endotoxic or septic shock.

Major Effects of NO on Cells Within the Circulatory System

In order to evaluate the role of NO in the overall circulatory system, we first have to assess its direct effects on blood cells, on vascular cells, and on cardiac cells (Table 1). At low to moderate concentrations, most but not all of the effects of NO can be seen to be homeostatic, that is beneficial and physiological. Only the inhibitory effect of high concentrations of NO on the oxy-

Table 1. Effect of NO on blood cells and cells of the cardiovascular system

Cell Type	Acute effect (min)	Chronic effect (days)
Blood Cells		
– Erythrocytes	Impairs O_2 carrying capacity	?
– Leukocytes	Attenuates adherence to endothelium and quenches free radicals	?
– Platelets	Inhibits aggregation	?
Vascular Cells		
– Vascular smooth muscle cells	Relaxes	Inhibits proliferation
– Endothelial cells	Downregulates adhesion molecule Inhibits permeability leakiness Overrides superoxide modules	Stimulates proliferation
Cardiac Cells		
– Cardiac myocytes	Very small or no decrease in contractility	?
– Connective tissue cells	Not known	Not known

gen carrying capacity of erythrocytes is clearly deleterious. At even lower concentrations, NO exerts two physiologically important effects on the other formed elements of the blood namely it inhibits the adherence of leukocytes, particularly polymorphonuclear (PMN) leukocytes (i.e. neutrophils) to the vascular endothelium [16, 17] and it inhibits platelet aggregation.

Effects of NO on Leukocytes

Figure 1 illustrates the anti-adhesive effect of NO on neutrophil-endothelial interaction. In this experiment, cat neutrophils were stimulated with 100 nM leukotriene B_4 (LTB_4). This chemotactic agent markedly increased the adhesion of these activated PMNs for the endothelium of autologous cat coronary artery segments opened up to expose the endothelial surface. In the presence of 500 nm SPM-5185, a cysteine containing NO donor, the adherence of PMNs was markedly attenuated, whereas 500 nM SPM-5267, the same compound minus the NO group, was ineffective in this regard. This anti-neutrophil effect is probably related to inhibition of CD11/CD18 upregulation on the surface of the PMN. Inhibition of adhesion molecules appears to be important in preventing reperfusion injury [18, 19]. Interestingly, human PMNs and PMNs of other mammalian species lack NO synthase activity [20]. NO also inhibits the release of free radicals by activated PMNs, particularly the superoxide radical [9, 10]. These two effects constitute an important anti-inflammatory action of NO which is important in supporting the natural defense mechanisms of the body in ischemia-reperfusion and shock states [1, 20]. With regard to platelets, NO has been known for some time to inhibit platelet aggregation [21, 22]. More recently, NO has been shown to down-

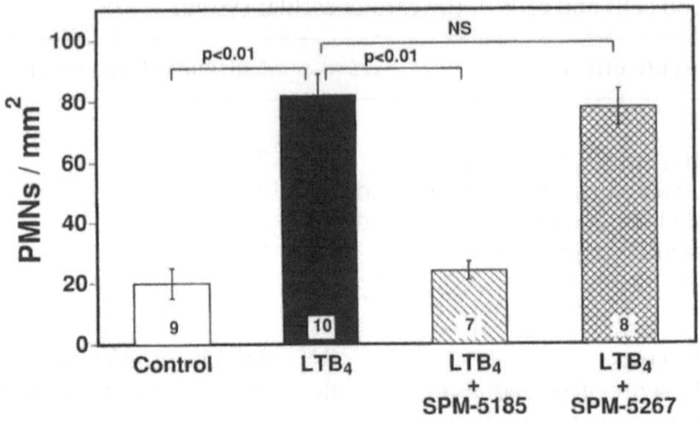

Fig. 1. Adherence of autologous cat neutrophils (PMNs) to sheets of cat coronary endothelium. Bar heights are numbers of PMNs adhering to the endothelium 20 min following stimulation with 100 nM leukotriene B_4 (LTB_4) or no stimulation (control). Brackets indicate ± SEM numbers in the bars are numbers of experiments. SPM-5185, a cysteine class of NO donor or its analog minus the NO, SPM-5267 were added to yield a final concentration of 500 nM. Only SPM-5185 significantly attenuated PMN adherence, providing evidence that NO exerts anti-neutrophil effects

regulate the adhesion glycoprotein P-selectin [23]. Since, P-selectin is found in the α-granules of platelets as well as the Weibel-Palade bodies of endothelial cells, this regulatory effect may inhibit potential interaction between platelets and neutrophils and neutrophil-endothelial interaction in ischemic and shock states [24].

Effects of NO on Smooth Muscle Cells

In the blood vessel wall, the two major cell types of greatest physiological importance are endothelial cells and vascular smooth muscle cells. One of the earliest and best characterized effects of NO relates to its vasodilator or vasorelaxant action. In fact, endothelium-derived relaxing factor (EDRF) was discovered precisely because of its vasorelaxing effect on blood vessels [25]. Later, when EDRF was identified as NO [6, 26], the vasorelaxing action of NO became widely appreciated. However, in addition to this rapidly activated vasorelaxing effect, which occurs in a few seconds and is mediated by the activation of cyclic GMP [27], there are other effects of NO on vascular smooth muscle cells which occur over periods of days. In this connection, NO has been shown to inhibit the proliferation of cultured vascular smooth muscle cells over 3 to 7 days [28, 29]. This anti-proliferative effect of NO on vascular smooth muscle cells appears to be a very important mechanism inhibiting neointimal proliferation (i.e. restenosis of an injured blood vessel) [29].

Effects of NO on Endothelial Cells

NO also exerts a variety of important physiologic effects in endothelial cells. Acutely, NO has been shown to reduce the adhesion of leukocytes to the endothelium [30, 31]. This occurs both *in vivo* and *in vitro*, and is at least partly related to the ability of NO to downregulate the expression of P-selectin and possibly intercellular adhesion molecule (ICAM-1) on the surface of endothelial cells. This has important beneficial actions in ischemia-reperfusion of the mesenteric circulation [23, 31], and of the heart [19, 32], as well as in the early stages of hypercholesterolemia and atherogenesis [33]. In other acute actions on endothelial cells, NO has been shown to inhibit the increased endothelial permeability frequently occurring in inflammatory states [34, 35]. This may be of benefit in shock states where hypovolemia is known to occur (e.g. hemorrhagic or endotoxemic shock). Additionally, NO is known to quench superoxide radicals, and thus decrease the oxidative stress on endothelial cells, particularly since endothelial cells are the probable initial source of much of the superoxide radical generation in the early stages of reperfusion injury [36, 37]. This may have particular importance in myocardial and mesenteric reperfusion injury [38, 39]. The probable mechanism of the generation of the superoxide radicals is by activation of xanthine oxidase from its xanthine dehydrogenase form [40].

Chronically, NO also exerts effects on endothelial cells. Thus, Granger et al. [41] first showed that NO stimulates the proliferation of cultured endothelial cells. This has been confirmed under more precise conditions by Guo et al [42]. Table 2 illustrates an example of the endothelial stimulatory effect of NO. Cultured rat aortic endothelial cells were brought to subconfluent cell monolayers from the fourth passage, and then made quiescent for 2 days with 0.1% fetal bovine serum (FBS). BrdU (5-bromo-2′-deoxyurudine) binding was determined microscopically by measuring deposition of anti-BrdU-antibody complexes. The furazan NO donor CAS-1009 concentration dependently stimulated BrdU incorporation (i.e. endothelial cell proliferation), whereas a control non-NO-donating compound (C-3934) was inactive in this regard. Counting of cell numbers agreed very closely with the BrdU values, thus confirming the NO stimulation of endothelial cell proliferation. Thus, NO is capable of stimulating a significant degree of endothelial cell prolife-

Table 2. BrdU incorporation in cultured rat aortic endothelial cells

Concentration (μg/mL)	CAS-1609 (active NO donor)	C-3934 (Inactive control compound)
0	42 ± 2	
2	52 ± 3*	42 ± 5
10	63 ± 4**	44 ± 3

All values are means ± SEM of three observations performed in triplicate; * $p < 0.05$ from control; ** $p < 0.02$ from control ration at concentrations of 2 to 10 μg/mL. These concentrations are below those which produce a vasodilator effect in isolated blood vessels or in the intact animal [42]. These cell proliferative effects may have great significance in regeneration of the endothelium following vascular injury (e.g. angioplasty, trauma, etc.) and may also contribute to the inhibition of restenosis brought about by NO donors [42]. In summary, all of the known effects of NO on vascular smooth muscle and endothelial cells are pro-homeostatic, and would be beneficial in a variety of circulatory disease states including circulatory shock or vascular injury [43].

Effects of NO on Cardiac Muscle Cells

The third general group of cardiovascular cell types are those cell types found in the heart. These are largely cardiac myocytes and connective tissue cells. Very little is known about the effects of NO on cardiac connective tissue cells (e.g. fibroblasts) since their responses to NO have not been actively studied. In contrast, there is a great deal of interest in the effects of NO on cardiac muscle. Unfortunately, this topic is replete with conflicting reports and controversies over the actual effects of NO on cardiac muscle contractility. This has significant implications for understanding the role of NO in circulatory shock states including septic shock. In hemorrhagic, septic, bowel ischemic, and cardiogenic shock, myocardial depression is a well recognized feature of the pathophysiology [44]. However, many factors are thought to

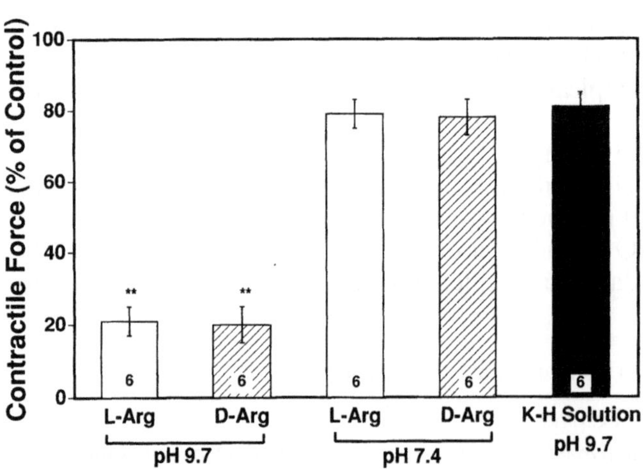

Fig. 2. Developed force (contractile force) responses expressed as % of initial control for 6 isolated electrically stimulated left ventricular rat papillary muscles. Means ± SEM. L-arginine and D-arginine at pH 9.7 significantly decreased contractile force from control values ($p<0.01$). However at pH 7.4, the effects were only minimal compared with a time matched control of 85%. These data do not support the role of NO as a direct negative inotropic agent

play a role in this impaired cardiac contractility including alterations in coronary blood flow and circulating cardiodepressant factors including myocardial depressant factor (MDF) [45].

Recently, NO has been suggested as an important negative inotropic agent in septic shock [46]. This was based on the finding that L-arginine exerted a large decrease in the contractility of isolated guinea pig papillary muscles. Unfortunately, D-arginine was not studied, in these experiments, and the pH of the arginine solution was not measured. More recently, others have reported a much more modest negative inotropic effect of NO which occurs primarily in the presence of adrenergic stimulation [47, 48]. This negative inotropic effect is on the order of a 7% decrease in cardiac contractility, a value comparable to the 5–6% decrease in contractility observed in response to sodium nitroprusside in isolated ferret hearts [49], and certainly too small to account for the cardiac depression observed in shock states.

Our laboratory recently addressed the question of the role of NO in cardiac muscle isolated from rat and cat hearts [50]. We wanted to: 1) test more than one mammalian species in order to eliminate idiosyncratic responses of a single species, 2) rigorously determine the effects of the amino acid arginine on cardiac muscle contractility, and 3) test the effects of authentic NO gas on cardiac muscle contractility, under conditions allowing us to measure the NO concentration of the medium.

Figure 2 summarizes the effects of arginine on isolated electrically stimulated left ventricular rat papillary muscles. Arginine, at concentrations of 1 to 25 mM exerted no detectable effect on the contractility of papillary muscles of rats or cats (data not shown). However, 50 mM L-arginine, the concentration employed found by Finkel et al [46], exerted a marked negative inotropic effect decreasing the contractility of the rat papillary muscles by 78% (Fig. 2). Although this seems to be straight forward, there are several problems with this finding. Firstly, D-arginine, at 50 mM, produced the same degree of cardiac depression as did L arginine (Fig. 2) [50], so that the cardiac depression is clearly unrelated to generation of NO. Secondly, the pH of the L-arginine solution was measured and found to be 9.7. When the arginine solutions were titrated to pH 7.4, the negative inotropic effect declined to only 18%. Moreover, D-arginine alone at pH 7.4 or Krebs-Henseleit buffer at pH 9.7 also exerted a negative inotropic effect of about 18% (Fig. 2). Thus, the alleged negative inotropic effect of extremely high concentrations of arginine is not due to generation of NO. Supporting this contention are additional findings that 50 mM L-aspartic also exerts marked decreases in contractility of rat papillary muscles [50].

Additional experiments were conducted in isolated cat papillary muscles employing NO gas in oxygen purged Krebs-Henseleit solution. Concentrations of NO up to 500 nM exerted no detectable negative inotropic effect (Fig. 3). To put this in perspective, complete vasodilation of the entire arterial system would occur at 100 to 200 nM NO, resulting in a mean arterial blood pressure equal to the mean circulatory pressure (i.e. 7 mm Hg). Thus, 500 nM is a very high concentration of NO and yet was without significant

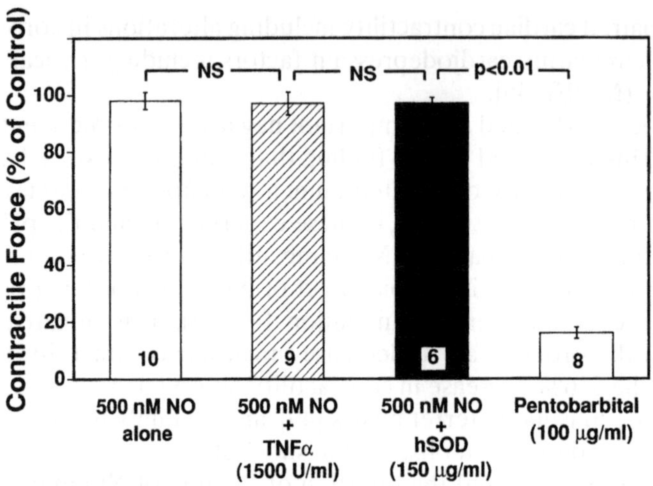

Fig. 3. Developed force (contractile force) of isolated electrically stimulated right ventricular cat papillary muscles. Bar heights are means, brackets are ± SEM, and numbers in the bars are numbers of papillary muscles studied. NO gas at 500 nM, measured with a NO electrode, did not significantly alter control contractile force. Neither did NO in the presence of TNF-α (1500 U/mL) or human recombinant superoxide dismutase (150 μg/mL), the superoxide scavenger. Only sodium pentobarbital (100 μg/mL) significantly depressed the contractility of the cat papillary muscle providing evidence that the papillary muscles are responsive to inotropic agents

inotropic activity. Fig. 3 also shows that 500 nM NO in the presence of 1500 U/mL tumor necrosis factor-α (TNFα), a cytokine which is produced in septic shock, still does not produce a significant cardiodepressant effect. In further efforts to unmask any potential inotropic effects of NO, we attempted to reduce the degradation of NO by superoxide radicals, the major inactivator of NO. We tested 500 nM NO in the presence of 150 μg/mL of recombinant human superoxide dismutase (hSOD). Although this is a high concentration of hSOD, it still failed to reveal a significant negative inotropic effect of NO (Fig. 3). To ensure that the papillary muscles were indeed responsive to negative inotropic agents, we added sodium pentobarbital at 100 μg/mL. This decreased contractility by 86% (Fig. 3), and verified that these papillary muscles were responsive to negative inotropic agents. Thus, we are forced to conclude that NO does not exert any physiologically important negative inotropic effect in these unstimulated preparations of cardiac muscle. We then tested the effects of NO in stimulated cardiac muscle. Weyrich et al. [50] found that 500 nM NO, in the presence of 5 μM norepinephrine, decreased contractile force by only 6%, a finding similar to that of others [48]. Although this negative inotropic effect was statistically significant, it clearly is a physiological minor event which could not account for the cardiac depression observed in endotoxic or septic shock.

Regarding cardiac fibroblasts or other connective tissue cells, there are no known effects of NO either of a stimulatory or an inhibitory nature. NO also acts as a neurotransmitter in the central nervous system, however there are no reported new effects of NO that would appear to impact on either the vasculature or the heart.

Conclusion

NO exerts a wide variety of effects on the circulatory system. The vast majority of these effects are physiologic, and promote cardiovascular homeostasis. The only important anti-homeostatic effect of NO is its well known action binding to the heme of hemoglobin to restrict the oxygen carrying capacity of erythrocytes. It is possible that extremely high concentrations of NO could play a role in the pathophysiology of septic shock, but no compelling evidence for this currently exists. Initially, NO was postulated as a major mediator of the hypotension of septic shock [4, 54]. In this regard, NO synthase inhibitors severely constrict the arterial system resulting in reduced blood flow to vital organs (i.e. gastrointestinal tract, kidneys), increasing total peripheral resistance, and effecting a reduction in cardiac output. These consequences of reduced NO merely aggravate the hemodynamic state of septic shock and further compromise the ability of the organism to respond to the shock insult [55]. This situation reminds one of the controversy surrounding the use of norepinephrine to treat shock 30 to 40 years ago. At that time, it was argued that norepinephrine should be given to increase arterial blood pressure and thus alleviate the hypotension of shock states. However, it became clear that norepinephrine reduced blood flow to vital organs even further than that occurring in shock, leading to aggravation of the shock condition. Eventually, α-adrenergic blockers along with fluids became the recognized approach to treat circulatory shock. We should not forget this illuminating example since the current controversy over NO appears to be a very similar situation.

Acknowledgements: Supported in part by Research Grant # GM-45 434 from the National Institutes of Health, Bethesda, MD.

References

1. Lefer AM, Lefer DJ (1993) Pharmacology of the endothelium in ischemia-reperfusion and circulatory shock. Annu Rev Pharmacol Toxicol 33:71–90
2. Kovach AGB, Lefer AM (1993) Endothelial dysfunction in shock states. News Physiol Sci 8:145–148
3. Petros A, Bennett D, Vallance P (1991) Effect of nitric oxide synthase inhibitors on hypotension in patients with septic shock. Lancet 338:1557–1558
4. Kilbourn RG, Jubran A, Gross SS, et al (1990) Reversal of endotoxin-mediated shock by N^G-methyl-L-arginine, an inhibitor of nitric oxide synthesis. Biochem Biophys Res Commun 172:1132–1138
5. Marletta MA (1989) Nitric oxide: Biosynthesis and biological significance. Trends Biochem Sci 14:488–492

6. Palmer RMJ, Ferrige AG, Moncada S (1987) Nitric oxide release accounts for the biological activity of endothelium-derived relaxing factor. Nature (Lond) 327:524–526
7. Johnson G III, Tsao PS, Lefer AM (1991) Cardioprotective effects of authentic nitric oxide in myocardial ischemia and reperfusion. Crit Care Med 19:244–252
8. Kelm M, Schrader J (1990) Control of coronary vascular tone by nitric oxide. Circ Res 66:1561–1575
9. Graier WF, Myers PR, Rubin LJ, Adams HR, Parker JL (1994) Escherichia coli endotoxin inhibits agonist-mediated cytosolic Ca^{2+} mobilization and nitric oxide biosynthesis in cultured endothelial cells. Circ Res 75:659–668
10. Wylam M, Samsel R, Umans J, Mitchell R, Leff A, Schumacker P (1990) Endotoxin in vivo impairs endothelium-dependent relaxation of canine arteries in vitro. Am Rev Respir Dis 142:1263–1267
11. Siegfried MR, Ma XL, Lefer AM (1992) Splanchnic vascular endothelial dysfunction in rat endotoxemia: Role of superoxide radicals. Eur J Pharmacol 212:171–176
12. Aoki N, Johnson G III, Lefer AM (1990) Beneficial effects of two forms of NO administration in feline splanchnic artery occlusion shock. Am J Physiol 258:G275–G281
13. Carey C, Siegfried M, Ma XL, Weyrich AS, Lefer AM (1992) Antishock and endothelial actions of a NO donor in mesenteric ischemia and reperfusion. Circ Shock 38:209–216
14. Hutcheson IR, Whittle BJR, Boughton-Smith NK (1990) Role of nitric oxide in maintaining vascular integrity in endotoxin-induced acute intestinal damage in the rat. Br J Pharmacol 101:815–820
15. Wright EE, Rees DD, Moncada S (1992) Protective and pathological roles of nitric oxide in endotoxin shock. Cardiovasc Res 26:48–57
16. Gaboury J, Woodman RC, Granger DN, Reinhardt P, Kubes P (1993) Nitric oxide prevents leukocyte adherence: Role of superoxide. Am J Physiol 265:H862–H867
17. Clancy RM, Leszcynska-Piziak J, Abramson SB (1992) Nitric oxide, an endothelial cell relaxation factor, inhibits neutrophil superoxide anion production via a direct action on the NADPH oxidase. J Clin Invest 90:1116–1121
18. Ma XL, Weyrich AS, Lefer DJ, Lefer AM (1993) Diminished basal nitric oxide release after myocardial ischemia and reperfusion promotes neutrophil adherence to coronary endothelium. Circ Res 72:403–412
19. Siegfried MR, Carey C, Ma XL, Lefer AM (1992) Beneficial effects of SPM-5185, a cysteine-containing NO donor in myocardial ischemia-reperfusion. Am J Physiol 263: H771–H777
20. Yan L, Vandivier RW, Suffredini AF, Danner RL (1994) Human poly-morphonuclear leukocytes lack detectable nitric oxide synthase activity. J Immunol 153:1825–1834
21. Radomski MW, Palmer RMJ, Moncada S (1987) The anti-aggregating properties of vascular endothelium: Interactions between prostacyclin and nitric oxide. Br J Pharmacol 92:639–644
22. Stamler JS, Loscalzo J (1991) The antiplatelet effects of organic nitrates and related nitroso compounds in vitro and in vivo and their relevance to cardiovascular disorders. J Am Coll Cardiol 18:1529–1536
23. Gauthier TW, Davenpeck KL, Lefer AM (1994) Nitric oxide attenuates leukocyte-endothelial interaction via P-selectin in splanchnic ischemia-reperfusion. Am J Physiol 267:G562–G568
24. Weyrich AS, Ma XL, Lefer DJ, Albertine KH, Lefer AM (1993) In vivo neutralization of P-selectin protects feline heart and endothelium in myocardial ischemia and reperfusion injury. J Clin Invest 91:2620–2629
25. Furchott RF, Zawadzki JV (1980) The obligatory role of endothelial cells in the relaxation of arterial smooth muscle by acetylcholine. Nature 288:373–376
26. Ignarro LJ, Buga GM, Wood KS, Byrns RE, Chaudhuri G (1987) Endothelium-derived relaxing factor produced and released from artery and vein is nitric oxide. Proc Natl Acad Sci USA 84:9265–9269
27. Murad F, Mittal CK, Arnold WP, Katsuki S, Kimura H (1978) Guanylate cyclase: Activation by azide, nitro compounds, nitric oxide and hydroxyl radical and inhibition by hemoglobin and myoglobin. Adv Cyclic Nucleotide Res 9:145–158

28. Garg UC, Hassid A (1989) Nitric oxide-generating vasodilators and 8-bromo-cyclic guanosine monophosphate inhibit mitogenesis and proliferation of cultured rat vascular smooth muscle cells. J Clin Invest 83:1774–1777
29. Guo JP, Milhoan KA, Tuan RS, Lefer AM (1994) Beneficial effect of SPM-5185, a cysteine-containing nitric oxide donor, in rat carotid artery intimal injury. Circ Res 75: 77–84
30. Kubes P, Suzuki M, Granger DN (1991) Nitric oxide: An endogenous modulator of leukocyte adhesion. Proc Natl Acad Sci USA 88:4651–4655
31. Davenpeck KL, Gauthier TW, Lefer AM (1994) Inhibition of endothelial-derived nitric oxide promotes P-selectin expression and actions in the rat microcirculation. Gastroenterology 107:1050–1058
32. Siegfried MR, Erhardt J, Rider T, Ma XL, Lefer AM (1992) Cardioprotection and attenuation of endothelial dysfunction by organic nitric oxide donors in myocardial ischemia-reperfusion. J Pharmacol Exp Ther 260:668–675
33. Lefer AM, Ma XL (1993) Decreased basal nitric oxide release in hypercholesterolemia increases neutrophil adherence to rabbit coronary artery endothelium. Arterioscl Thromb 13:771–776
34. Kurose I, Kubes P, Wolf R, et al (1993). Inhibition of nitric oxide production: Mechanisms of vascular albumin leakage. Circ Res 73:164–171
35. Kubes P, Grisham MB, Barrowman JA, Gaginella T, Granger DN (1991) Leukocyte-induced vascular protein leakage in cat mesentery. Am J Physiol 261:H1872–H1879
36. Zweier J, Kuppusamy P, Lutty GA (1988) Measurement of endothelial cell free radical generation: Evidence for a central mechanism of free radical injury in post-ischemic tissues. Proc Natl Acad Sci USA 85:4046–4050
37. Tsao PS, Lefer AM (1990) Time course and mechanism of endothelial dysfunction in isolated ischemic and hypoxic perfused rat hearts. Am J Physiol 259:H1660– H1666
38. Lefer AM, Tsao PS, Lefer DJ, Ma XL (1991) Role of endothelial dysfunction in the pathogenesis of reperfusion injury following myocardial ischemia. FASEB J 5: 2029–2034
39. Tsao PS, Aoki N, Lefer DJ, Johnson G III, Lefer AM (1990) Time course of endothelial dysfunction and myocardial injury during myocardial ischemia and reperfusion in the cat. Circulation 82:1402–1412
40. Bulkley GB (1989) Mediators of splanchnic organ injury: Overview and perspective. In: Marston A, Bulkley GB, Fiddian-Green RG, Haglund UH (eds) Splanchnic ischemia and multiple organ failure. Edward Arnold, a division of Hodder and Stoughton, pp 183–191
41. Ziche M, Morbidelli L, Masini E, et al (1994) Nitric oxide mediates angiogenesis in vivo and endothelial cell growth and migration in vitro promoted by substance P. J Clin Invest 94:2036–2044
42. Guo JP, Panday MM, Consigny PM, Lefer AM (1995) CAS 1609, a novel NO donor inhibits neointimal formation and accelerates endothelial regeneration following rat carotid artery injury. Am J Physiol (In Press)
43. Lefer AM, Lefer DJ (1994) Therapeutic role of nitric oxide donors in the treatment of cardiovascular disease. Drugs of the Future 19:665–672
44. Lefer AM (1974) Myocardial depressant factor and circulatory shock. Klin Wochenschr 52:358–370
45. Lefer AM (1987) Interaction between myocardial depressant factor and vasoactive mediators with ischemia and shock. Am J Physiol 252:R193–R205
46. Finkel MS, Oddis CV, Jacob TD, Hattler BG, Simmons RL (1992) Negative inotropic effects of cytokines on the heart mediated by nitric oxide. Science 257:387–389
47. Brady AJB, Warren JB, Poole-Wilson PA, Williams TJ, Harding SE (1993) Nitric oxide attenuates cardiac myocyte contraction. Am J Physiol 265:H176–H182
48. Balligand JL, Kelly RA, Marsden PA, Smith TW, Michel T (1993) Control of cardiac muscle cell function by an endogenous nitric oxide signaling system. Proc Natl Acad Sci USA 90:347–351

49. Fort S, Lewis ML (1991) Regulation of myocardial contractile performance by sodium nitroprusside in the isolated perfused heart of the ferret. Br J Pharmacol 102:351P (Abst)
50. Weyrich AS, Ma XL, Buerke M (1994) Physiological concentrations of nitric oxide do not elicit an acute negative inotropic effect in unstimulated cardiac muscle. Circ Res 75:692–700
51. Rubanyi GM, Vanhoutte PM (1986) Oxygen-derived free radicals, endothelium, and responses of vascular smooth muscle. Am J Physiol 250:H815–H824
52. Gryglewski RJ, Palmer RMJ, Moncada S (1986) Superoxide anion is involved in the breakdown of endothelium-derived relaxing factor. Nature (Lond) 320:454–456
53. Stewart DJ, Pohl U, Bassenge E (1988) Free radicals inhibit endothelium-dependent dilation in the coronary resistance bed. Am J Physiol 255:H765–H769
54. Julou-Schaeffer G, Gray GA, Fleming I, Schott C, Parratt JR, Stoclet JC (1990) Loss of vascular responsiveness induced by endotoxin involves the L-arginine pathway. Am J Physiol 259:H1038–H1043
55. Lefer AM (1994) Endotoxin, cytokines, and nitric oxide in shock: An editorial comment. Shock 1:79–80

Role of Nitric Oxide in Cerebral Ischemia

J. R. Kirsch and R. J. Traystman

Introduction

In this chapter, we will discuss the role of nitric oxide (NO) as an important mediator of cerebral vascular tone and in the mechanism of cerebral blood flow (CBF) changes during cerebral ischemia and reperfusion and during reperfusion. We also address the possibility that NO may be an important mediator of brain injury in the setting of focal ischemia.

NO and Control of CBF

Endothelium-derived relaxation and constriction factors have been implicated as important mediators of vascular tone [1]. In brain, removal of or damage to endothelium results in the loss of cerebral vasodilation to acetylcholine [2] but dilation remains intact to administration of nitro-vasodilators [3]. NO and/or a NO-containing substance (e.g. nitrosothiol) [4] have been demonstrated to be important mediators of endothelium-derived relaxation [5]. NO synthase (NOS) produces NO in vascular endothelium during conversion of L-arginine to citrulline [6]. NOS also has been found in perivascular neurons [7], and astrocytes [8].

Rosenblum et al. [9] have demonstrated that mouse pial arterioles *in vivo* dilate in response to L-arginine (which presumably increases the amount of NO) and constrict in response to NOS inhibition with N^G-monomethyl-L-arginine (L-NMMA). Likewise, in piglets, local application of mono-L-arginine containing compounds produces dilation of pial vessels which can be blocked by NOS inhibition [10]. Tone of large cerebral vessels also appears to be altered by endothelium-related processes *in vivo* [2].

Several investigators [11] have reported a reduction in local CBF in adult animals after inhibition of NOS. In immature piglets, NOS inhibition causes prompt reduction in CBF with a concomitant increase in cerebrovascular resistance [12]. Importantly, this effect on cerebral vascular tone is reversible by administration of L-arginine hydrochloride. Regions with poor blood brain barrier function and regions with high NOS activity [13] were most affected by L-NMMA administration. The decrease in CBF during NOS inhibition was not associated with a change in cerebral oxygen consumption

($CMRO_2$) and, therefore, the mechanism for the observed reduction in CBF is not related to a decrease in $CMRO_2$.

In the presence of a less intact (from injury) or less mature blood brain barrier, the cerebral parenchyma may be more affected by NOS inhibition with a resultant greater fall in $CMRO_2$ than was observed in our study [12]. The greater fall in CBF with NOS inhibition in areas of less developed blood brain barrier has been reported. For example, in thiopental anesthetized rabbits, choroid plexus blood flow falls to a greater extent compared with forebrain flow during NOS inhibition [14]. Likewise, NOS inhibition causes a greater reduction in flow to pituitary compared to other regions [15].

When comparing studies from different laboratories, it is important to have an accurate assessment of degree of NOS inhibition in each study. NOS activity can be assessed by either measuring enzyme activity in brain biopsy samples or by measuring the physiologic response to an intervention which is known to work by a NO-mediated mechanism. Although measuring NOS activity in whole brain biopsy samples is now a standard technique to assess efficacy of enzyme inhibition, these samples represent mainly brain parenchyma and, as a result enzyme activity in these samples may not accurately assess the degree of NOS inhibition in endothelium. The ability to differentiate the degree of NOS inhibition in parenchyma and endothelium will become increasingly important as more specific inhibitors of brain and endothelial NOS are developed.

NO and Control of CBF during Reperfusion

Following a period of ischemia, there is an initial period of cerebral hyperemia which is quickly followed by a period of delayed hypoperfusion [16]. It has been hypothesized that these significant changes in CBF which occur during reperfusion may be important in the mechanism of post-ischemic brain injury. The mechanism(s) responsible for this hyperemia is the subject of considerable controversy. Metabolites which accumulate in ischemic brain may cause vasodilation. Such metabolites include lactic acid, adenosine, endogenous opioids, neurotransmitters, and free fatty acids. We hypothesized that NO was an important mediator of post-ischemic hyperperfusion [17]. We found that NOS inhibition with intravenous N^{ω}-nitro-L-arginine methyl ester (L-NAME) attenuates (but did not ablate) post-ischemic hyperemia in a dose-related fashion. These findings support the hypothesis that release of NO, and/or a subsequently produced NO-containing substance, participates in the mechanism for post-ischemic hyperemia in pentobarbital anesthetized piglets early in reperfusion. We also found regionally dependent amelioration of hyperemia which presumably paralleled the localization of NOS activity in pig brain.

Our data suggest that there may be an increased NO production associated with global cerebral ischemia/reperfusion. The mechanism for NO release and action was not delineated in this study. NO may originate from neurons [18], endothelium [6], astrocytes [8] or leukocytes [19]. It has been suggested that during ischemia there is a diminished amount of substrate (i.e. arginine) for NOS [20]. Under these conditions, NOS may lead to the production of superoxide anion [21]. Oxygen radicals have been shown to produce vasoparalysis and could account for the vasodilation and hyperemia following ischemia. However, we have been unable to affect post-ischemic hyperemia in a similar model of ischemia using polyethylene glycol conjugated superoxide dismutase (PEG-SOD) [22]. It is unclear if the PEG-SOD could have reached the site of superoxide anion production during the time course of the experiment [23].

Moskowitz et al. [24] have demonstrated that post-ischemic hyperemia can be attenuated with chronic trigeminal ganglionectomy. These authors have postulated that the reduced hyperemia is due to decreased release of mediators which, under normal conditions, can cause stimulation of NOS and increased release of NO. Therefore, under conditions of trigeminal ganglionectomy, decreased mediator release may result in decreased stimulation of NOS and decreased production of NO. This would then attenuate post-ischemic hyperemia.

Nitric Oxide and Delayed Hypoperfusion

Reperfusion following transient global cerebral ischemia is commonly characterized by a period of delayed hypoperfusion, when $CMRO_2$ and CBF fall to below pre-ischemic values [25]. The significance and mechanism for this alteration in blood flow remain unclear, but it is likely that damage to vascular endothelium and consequent reduced vasodilatory capacity are important factors. In this regard, our laboratory has addressed the possibility that delayed hypoperfusion is due to cessation of tonically produced NO or to an abnormality of stimulated production of NO. Following cerebral ischemia, oxygen radicals produced during reperfusion [26] may result in brain injury and impaired NO synthesis [27]. In addition, superoxide anion avidly reacts with NO resulting in its deactivation. Thus, increased production of superoxide anion during reperfusion may prevent the vasodilator effect of NO. Therefore, ischemia-induced brain damage may lead to a decreased tonic production of NO or an alteration in the NO signal transduction pathway, both of which would result in impaired tonic NO-mediated cerebral vasodilation. We hypothesized that tonic production of NO is impaired during delayed hypoperfusion. We tested this hypothesis by determining if the decrease in CBF which occurs after L-NAME administration is attenuated during delayed hypoperfusion after transient global ischemia [28].

Our study demonstrated that during post-ischemic delayed hypoperfusion, L-NAME administration induced a marked CBF reduction with increased cerebral vascular resistance. These effects were probably due to NOS activity inhibition since they were reversed by the administration of L-arginine [28]. Thus, while CBF was profoundly reduced during delayed hypoperfusion, NOS inhibition resulted in a further decrease of flow. These results suggest that tonic NO vasodilation persists in cats despite post-ischemic delayed hypoperfusion. Consistent with our study, Prado et al. [29] have found that inhibition of NOS in rats prior to induction of transient forebrain ischemia resulted in more severe delayed cortical hypoperfusion as compared to rats treated with saline.

Reduction of flow following L-NAME may also be due to microvascular stasis via cell aggregation as well as increased vasoconstriction. L-NAME does not inhibit neutrophil NOS or produce leukocyte aggregation [30]. However, intravascular stasis is observed following intra-arterial administration of L-NAME [31], and topical NOS inhibition enhances platelet aggregation in pial vessels at the site of endothelial damage [32]. Therefore, some proportion of the CBF reduction by L-NAME can be attributed to vasoconstriction and another proportion to microcirculatory stasis from platelet aggregation.

With regard to stimulated release of NO during delayed hypoperfusion, several reports indicate that pial vasodilatory responses to topical agents such as acetylcholine are abnormal *in vivo* for varying intervals after the ischemic insult [33]. However, others [34] have demonstrated normal endothelium-dependent and independent relaxation of basilar artery after ischemia/reperfusion, suggesting a size-dependence or vessel specificity in vasodilatory abnormalities. Further, delayed hypoperfusion may be the CBF consequence of a decrease in endogenous vasodilator production, whether under tonic conditions or in response to agonist stimulation.

Using a series of agonists applied to cat pial microvessels *in situ*, we sought to localize the defect in microcirculatory signal-response to acetylcholine in hopes of achieving a better understanding of the mechanisms involved in delayed hypoperfusion [35]. In addition, intravenous oxotremorine (a muscarinic receptor agonist which results in endothelial-dependent NO release [36]) was employed in order to determine the effect of muscarinic receptor stimulation on blood flow within deeper brain regions.

Our study [35] demonstrated that pial vasodilation to both the muscarinic receptor agonist, acetylcholine, and the receptor-independent, calcium ionophore, A23187, was impaired during post-ischemic hypoperfusion. This finding suggests that membrane bound receptor dysfunction does not solely explain altered cerebrovascular responses to muscarinic stimulation. We also found that pial vasodilation in response to each agonist was lost or attenuated in animals that did not demonstrate post-ischemic hypoperfusion. This finding suggests that the presence of persistent vasodilation during reperfusion, without CBF reduction, was associated with impaired vascular smooth muscle relaxation to NO, as well as endothelial dysfunction. In addition, defects in agonist-induced vasodilatory responses in the presence of hypoperfusion were less pronounced in large pial arterioles (201 ± 11 μm) than in medium (66 ± 2 μm) or small (35 ± 2 μm) arterioles, emphasizing that small vessel endothelial pathology is important in post-ischemic hypoperfusion. Lastly, increases in CBF in response to systemic muscarinic stimulation via intravenous oxotremorine were largely attenuated after ischemia. This finding indicates that altered cerebrovascular responses are unlikely to be limited to the pial circulation and occur throughout forebrain.

It may be that post-ischemic hypoperfusion is associated with a global disruption of endothelium-mediated vasoresponsivity and is not specific to the L-arginine-NO synthetic pathway. If A23187 produces dilation by releasing an endothelium derived relaxant factor (EDRF) other than NO, then additional dilator signals may be lost. If, as previously suggested, A23187-induced vasodilation is accomplished via NO, then the present data indicate that NO synthesis or availability is aberrant. NO protection from deactivation by superoxide anion may be an important mechanism in the efficacy of antioxidant therapies which restore vasodilator responses [37]. Nelson et al. [37] demonstrated that application of superoxide dismutase and catalase or deferoxamine topically within a cranial window restored pial arteriolar reactivity to acetylcholine [37].

In cat, nitroprusside causes dilation by a mechanism which involves release of calcitonin gene-related peptide from perivascular neurons which in turn activates guanylate cyclase in vascular smooth muscle [38]. Arteriolar dilation to nitroprusside was not impaired during delayed hypoperfusion, suggesting normal guanylate cyclase activity within vascular smooth muscle.

NO and Mechanism of Brain Injury during Ischemia

Although in normal brain NO seems to be a non-toxic mediator of cerebral vasodilation [39], recent data suggests that NO may have neurotoxic effects if present in abnormally high concentrations [40]. In endothelium and brain, ionized calcium is the intracellular messenger initiating the NADPH-dependent oxidation of arginine to produce NO [41]. It is well known that cerebral ischemia causes an increase in extracellular concentration of glutamate and aspartate, and these excitatory amino acids bind to and stimulate N-methyl-D-aspartate (NMDA) receptors in brain [42]. Excessive activation of the NMDA receptors allows influx of ionized calcium into neurons, which may stimulate production of superoxide anion via a prostaglandin pathway, and NO via stimulation of NO synthase. Stimulation of non-NMDA glutamate receptors is also thought to generate NO. Moreover, inhibition of NOS in neuronal culture ameliorates glutamate toxicity [40]. One postulated mechanism of NO toxicity is that at high concentrations, superoxide anion and NO may react to form the peroxynitrite anion, which decomposes at acidic pH into strong oxidants [43]. Thus, excessive NO generation might act as an agent of cell death in stroke by reacting with superoxide.

We tested the hypothesis that inhibiting NOS with L-NAME could alleviate ischemic brain injury following 4 h focal cerebral ischemia in halothane anesthetized cats [20]. Our study demonstrated that intravenous administration of L-NAME prior to prolonged middle cerebral artery (MCA) occlusion, in cats, results in amelioration of injury. Protection from injury did not appear to be due to a favorable redistribution of CBF during MCA occlusion in spite of increased arterial pressure.

In rat, others [44] have demonstrated that L-NAME (3 mg/kg but not 1 mg/kg) reduced the amount of brain injury resulting from permanent MCA occlusion. Likewise, in mouse, systemic administration of N^{ω}-nitro-L-arginine (L-NA) resulted in reduction in cerebral infarction from permanent MCA occlusion [45]. However, in both studies, efficacy of NOS inhibition was not determined nor was there any attempt to control arterial blood gases which affect CBF.

L-NAME has recently been found to be a muscarinic receptor antagonist. However, this effect is not likely the cause of its protection. Rather this effect may actually be detrimental to neurologic recovery since others have found that parasympathetic denervation increases infarction volume following MCA occlusion [46] and electrical stimulation of the cerebellar fastigial nucleus, which increases CBF by a muscarinic mechanism reduces ischemic infarction elicited by MCA occlusion in rat [47]. We speculate that the mechanism for protection by NO-synthase inhibitors results from inhibition of NO synthesis but without a beneficial effect on blood flow. Possible sites of NO production would include vascular endothelium [6], perivascular neurons [13], or astrocytes [8]. The site of protection for NOS inhibition is most likely in neurons since mice deficient in neuronal NOS are resistant to injury by ischemia [48]. However, when these mice are administered a NOS inhibitor to inhibit endothelial NOS, they experience more brain damage during ischemia. These data underscore the need to develop more specific pharmacologic inhibitors of neuronal NOS prior to considering clinical trials of agents which are non-specific NOS inhibitors.

Although it is possible that NO may be directly toxic to cells by reacting with non-heme iron-sulfur complex [49], this seems unlikely because NO is also a natural mediator of vasodilation for agents which act via an endothelial-dependent process. Nonetheless, both NO and excitatory amino acids appear to be linked to the mechanism of injury following focal cerebral ischemia [50].

An alternative explanation for protection by L-NAME may be via an effect on cerebral metabolism. For example, other drugs (e.g. barbiturates) which decrease cerebral metabolism have been demonstrated to protect brain during focal ischemia [51]. We believe this is an unlikely mechanism for the beneficial effects of L-NAME because we have previously demonstrated reduced CBF during inhibition of NOS activity without a decrease in $CMRO_2$ [52].

Inhibition of NOS activity may not be protective in all types of ischemia. Indeed, in gerbil, systemic administration of L-NAME resulted in delayed electrocortical recovery following transient forebrain ischemia [53]. However, efficacy of NOS inhibition was not determined and physiologic variables were not recorded. These authors also did not determine the effect of L-NAME on electrocortical activity in control animals to determine baseline effects.

More recently, others [54] have demonstrated that administration of L-NA to spontaneously hypertensive rats was associated with a worsening of cortical injury volume following MCA occlusion. Furthermore, in this study [54], the detrimental effect of NOS inhibition could be prevented with L-arginine but not with D-arginine. Unlike normotensive rats in other studies, inhibition of NOS in hypertensive rats [54] did not appear to reduce CBF. This observation suggests a different effect of NOS inhibition on brain in spontaneously hypertensive rats.

NO and Transient Focal Ischemia

If inhibitors of NOS are to be effective from a clinical perspective, they must show clinical efficacy in a paradigm of delayed administration. Therefore, we tested the hypothesis that administration of L-NAME before or at the end of transient focal ischemia alleviates early brain injury. In order to assess the question of specificity of the L-NAME response for a NO-mediated mechanism, we also examined whether administration of L-arginine, the substrate for NOS, reverses any beneficial effect of L-NAME on brain injury [55]. Regional CBF was measured during cerebral ischemia and reperfusion to determine if these changes could account for any beneficial effect detected with drug treatment.

L-NAME administration, either before or at the end of transient MCA occlusion, resulted in partial protection from injury [55]. Protection by L-NAME administered at the end of ischemia was prevented by co-administration of L-arginine. Protection from injury was not due to a favorable redistribution of blood flow during MCA occlusion despite increased arterial pressure.

Malinski et al [56] have demonstrated increased accumulation of NO during the initial period of transient focal ischemia and the initial period of reperfusion. The mechanism for increased accumulation of NO was presumably the result of increased NOS activity rather than decreased NO clearance. We found that treatment with L-NAME at the end of ischemia resulted in a similar degree of protection in caudate as compared to pretreatment. This suggests that NO produced during reperfusion is important in the pathophysiology of brain injury produced following transient focal ischemia. Enhanced NO production during reperfusion may be attributable to persistent elevation of excitatory amino acids during reperfusion [57] at a time when oxygen, which is necessary for production of NO by NOS would be available. Evidence that excitatory amino acids play a role in post-ischemic brain injury comes from the finding that post-ischemic administration of NMDA glutamate receptor antagonists [58] prevented brain injury following transient focal ischemia. In addition, activity of NOS is primarily regulated by intracellular calcium, which presumably remains elevated during early reperfusion.

In order to determine the specificity of L-NAME in our study [55], we administered a large dose of the L-arginine substrate for the enzymatic reaction to competitively reverse the effects of L-NAME. Because the beneficial effect of L-NAME in our study was prevented by co-administration of L-arginine, our data indicate a specific role for NO in the mechanism of early brain injury following transient focal ischemia.

NO as a Mechanism of Injury in Stroke-Prone Rats

Endothelium-dependent NO-mediated cerebral vasodilation is impaired in spontaneously hypertensive stroke-prone rats (SHR-SP) [59]. In addition, NO inhibits platelet adhesion and aggregation, and platelet and monocyte activation appears closely related to the occurrence of cerebrovascular occlusions in SHR-SP. Because of the high activity of NOS found within brain [13], we hypothesized that there may be a genetic defect or altered enzyme expression that would lead to altered NOS activity in brain which could lead to predisposition for stroke in SHR-SP. We measured brain NOS activity in both SHR and SHR-SP strains, as well as in the normotensive genetically-related Wistar-Kyoto (WKY) strain [60]. Further, because these isoforms of NOS are Ca^{2+}-dependent and dietary Ca^{2+} supplementation offers partial protection from salt-induced stroke in SHR-SP [61], we determined whether $[Ca^{2+}]$-dependence of the enzyme differed between the 3 rat strains. In all rats tested, high levels of brain NOS activity was demonstrated, the highest levels occurring in the cerebellum, in agreement with previous anatomic observations [13]. However, SHR-SP as well as SHR rats showed no difference in regional brain NOS activity as compared to their normotensive controls, WKY. Therefore, these results do not support the hypothesis that the genetic predisposition to stroke in SHR-SP rats is due to an altered ability of brain NOS to produce NO. These findings are consistent with the conclusion of Mayhan [62] that the mechanism for altered endothelium-dependent vasodilation during chronic hypertension is due to activation of the prostaglandin H_2-thromboxane A_2 receptor rather than decreased release of an endothelium-dependent relaxation factor.

Our study could not rule out a specific alteration in endothelial NO production with normal neuronal NOS activity because the major fraction of NOS measured in our samples comes from neurons and glia rather than endothelium [13]. In favor of the hypothesis of a specific alteration in endothelium NOS is the finding of morphological alterations of cerebral vessel wall, including endothelium, in SHR-SP [63].

Alternatively, *in vivo* brain NOS activity may be impaired despite normal *in vitro* findings if the availability of any of the cofactors necessary for brain NOS activity is limited in the *in vivo* environment. For example, brain NOS activity is calcium-dependent. However, our data do not suggest a different calcium requirement for brain NOS activity between the 3 rat strains. However, alterations in intracellular calcium concentration and activation of calcium channel currents have been described in SHR vascular smooth muscle, blood cells [64], and brain synaptosomes [65]. An abnormal movement of calcium may thus occur in SHR-SP brain which could be responsible for an altered *in vivo* NOS activity. Similarly, a lack of endogenous L-arginine, the natural substrate for NOS , has been described in some experimental models of hypertension [66]. Yet dietary L-arginine supplementation does not reduce the stroke rate in SHR-SP [67] nor the hypertension in SHR [66]. Finally, endogenous inhibitors may be present *in vivo*, counteracting the effect of an otherwise normally functioning NOS. Consistent with this is the observation of the accumulation of an endogenous NOS inhibitor in chronic renal failure [68], and that renal function impairment occurs in parallel with stroke in SHR-SP rats [69].

It is also possible that decreased NO production or increased NO consumption was the cause of increased incidence of stroke in SHR-SP. If this were the case, we would expect less reduction in CBF after administration of a pharmacologic NOS inhibitor. To test this hypothesis, we measured sensitivity of brain NOS to inhibition by measuring the effect of intravenous L-NMMA and L-NA on CBF and maximal NOS activity measured in brain biopsy samples under optimal conditions *in vitro* [70]. We found that although salt loading and feeding SHR-SP a Japanese style diet predisposes these animals to stroke [71], it does not change cortical NOS activity. In addition, there was no difference between strains in sensitivity of NOS for inhibition by L-NMMA or L-NA. Our finding of similar NOS inhibition and similar alteration of CBF and cerebral vascular resistance between strains with L-NMMA and L-NA suggests that it is unlikely that an alteration in sensitivity of whole brain NOS to pharmacologic inhibition accounts for predisposition to stroke in SHR-SP

Conclusions

Existing data strongly support a role for NO in the maintenance of normal cerebral vascular tone. In addition, NO produced by neurons during ischemia appears to be an important mediator of brain injury. However, NO produced by endothelium may serve a protective role because its inhibition results in greater brain injury during focal ischemia. Post-ischemic hyperemia is at least partially due to an excess availability of NO during the initial period of reperfusion. Delayed hypoperfusion is associated with reduced agonist-induced cerebral vasodilation while vasodilation due to direct application of NO remains robust.

Acknowledgements: This research was supported by USPHS NIH grant #NS20020.

References

1. Furchgott RF (1983) Role of endothelium in responses of vascular smooth muscle. Circ Res 53:557–573
2. Faraci FM (1990) Role of nitric oxide in regulation of basilar artery tone in vivo. Am J Physiol 259:H1216–H1221
3. Rosenblum WI, Nelson GH, Povlishock JT (1987) Laser-induced endothelial damage inhibits endothelium-dependent relaxation in the cerebral microcirculation of the mouse. Circ Res 60:169–176
4. Myers PR, Minor RL Jr, Guerra R Jr, Bates JN, Harrison DG (1990) Vasorelaxant properties of the endothelium-derived relaxing factor more closely resemble S-nitrosocysteine than nitric oxide. Nature 345:161–163
5. Moncada S, Radomski MW, Palmer RM (1988) Endothelium-derived relaxing factor. Identification as nitric oxide and role in the control of vascular tone and platelet function. Biochem Pharmacol 37:2495–2501
6. Palmer RM, Ashton DS, Moncada S (1988) Vascular endothelial cells synthesize nitric oxide from L-arginine. Nature 333:664–666
7. Toda N, Okamura T (1990) Mechanism underlying the response to vasodilator nerve stimulation in isolated dog and monkey cerebral arteries. Am J Physiol 259: H1511–H1517
8. Murphy S, Minor RL Jr, Welk G, Harrison DG (1990) evidence for an astrocyte-derived vasorelaxing factor with properties similar to nitric oxide. J Neurochem 55: 349–351
9. Rosenblum WI, Nishimura H, Nelson GH (1990) Endothelium-dependent L-Arg- and L-NMMA-sensitive mechanisms regulate tone of brain microvessels. Am J Physiol 259:H1396–H1401
10. Busija DW, Leffler CW, Wagerle LC (1990) Mono-L-arginine-containing compounds dilate piglet pial arterioles via an endothelium-derived relaxing factor-like substance. Circ Res 67:1374–1380
11. Tanaka K, Gotoh F, Gomi S, et al (1991) Inhibition of nitric oxide synthesis induces a significant reduction in local cerebral blood flow in the rat. Neurosci Lett 127:129–132
12. Greenberg RS, Helfaer MA, Kirsch JR, Moore LE, Traystman RJ (1994) Nitric oxide synthase inhibition with N^G-mono-methyl-L-arginine reversibly decreases cerebral blood flow in piglets. Critical Care Med 22:384–392
13. Bredt DS, Hwang PM, Snyder SH (1990) Localization of nitric oxide synthase indicating a neural role for nitric oxide. Nature 347:768–770
14. Faraci FM, Heistad DD (1992) Does basal production of nitric oxide contribute to regulation of brain-fluid balance? Am J Physiol 262:H340–H344
15. Kovach AGB, Szabo C, Benyo Z, Csaki C, Greenberg JH, Reivich M (1992) Effects of N^G-nitro-L-arginine and L-arginine on regional cerebral blood flow in the cat. J Physiol (Lond) 449:183–196
16. Kirsch JR, Helfaer MA, Blizzard K, Toung TJ, Traystman RJ (1990) Age-related cerebrovascular response to global ischemia in pigs. Am J Physiol 259:H1551–H1558
17. Greenberg RS, Helfaer MA, Kirsch JR, Traystman RJ (1995) The effect of nitric oxide synthase inhibition upon post-ischemic cerebral hyperemia. Am J Physiol (In Press)
18. Nozaki K, Moskowitz MA, Maynard KI, et al (1993) Possible origins and distribution of immunoreactive nitric oxide synthase-containing nerve fibers in cerebral arteries. J Cereb Blood Flow Metab 13:70–79
19. McCall TB, Boughton Smith NK, Palmer RM, Whittle BJ, Moncada S (1989) Synthesis of nitric oxide from L-arginine by neutrophils. Release and interaction with superoxide anion. Biochem J 261:293–296
20. Nishikawa T, Kirsch JR, Koehler RC, Bredt DS, Snyder SH, Traystman RJ (1993) Effect of nitric oxide synthase inhibition on cerebral blood flow and injury volume during focal ischemia in cats. Stroke 24:1717–1724
21. Faraci FM, Brian JEJ (1994) Nitric oxide and the cerebral circulation. Stroke 25: 692–703

22. Helfaer MA, Kirsch JR, Haun SE, Moore LE, Traystman RJ (1991) Polyethylene glycol conjugated superoxide dismutase fails to blunt post-ischemic reactive hyperemia. Am J Physiol 261:H548–H553
23. Haun SE, Kirsch JR, Helfaer MA, Kubos KL, Traystman RJ (1991) Polyethylene glycol-conjugated superoxide dismutase fails to augment brain superoxide dismutase activity in piglets. Stroke 22:655–659
24. Moskowitz MA, Sakas DE, Wei EP, et al (1989) Postocclusive cerebral hyperemia is markedly attenuated by chronic trigeminal ganglionectomy. Am J Physiol 257: H1736–H1739
25. Snyder JV, Nemoto EM, Carroll RG, Safar P (1975) Global ischemia in dogs: Intracranial pressures, brain blood flow and metabolism. Stroke 6:21–27
26. Kontos CD, Wei EP, Williams JI, Kontos HA, Povlishock JT (1992) Cytochemical detection of superoxide in cerebral inflammation and ischemia. Am J Physiol 263: H1234–H1242
27. Wei EP, Christman CW, Kontos HA, Povlishock JT (1985) Effects of oxygen radicals on cerebral arterioles. Am J Physiol 248:H157–H162
28. Clavier N, Kirsch JR, Hurn PD, Traystman RJ (1994) Cerebral blood flow is reduced by N^{ω}-nitro-L-arginine methyl ester during delayed hypoperfusion in cats. Am J Physiol 267:H174–H181
29. Prado R, Watson BD, Wester P (1993) Effects of nitric oxide synthase inhibition on cerebral blood flow following bilateral carotid artery occlusion and recirculation in the rat. J Cereb Blood Flow Metab 13:720–723
30. McCall TB, Feelisch M, Palmer RM, Moncada S (1991) Identification of N-iminoethyl-L-ornithine as an irreversible inhibitor of nitric oxide synthase in phagocytic cells. Br J Pharmacol 102:234–238
31. Prado R, Watson BD, Kuluz J, Dietrich WD (1992) Endothelium-derived nitric oxide synthase inhibition effects on cerebral blood flow, pial artery diameter, and vascular morphology in rats. Stroke 23:1118–1124
32. Nishimura H, Rosenblum WI, Nelson GH, Boynton S (1991) Agents that modify EDRF formation alter antiplatelet properties of brain arteriolar endothelium in vivo. Am J Physiol 261:H15–H21
33. Macfarlane R, Moskowitz MA, Tasdemiroglu E, Wei EP, Kontos HA (1991) Postischemic cerebral blood flow and neuroeffector mechanisms. Blood Vessels 28:46–51
34. Katusic ZS, Michenfelder JD, Milde JH (1991) Ischemia-reperfusion does not affect reactivity of isolated canine basilar artery. J Cereb Blood Flow Metab 11:824–828
35. Clavier N, Kirsch JR, Hurn PD, Traystman RJ (1994) Effect of postischemic hypoperfusion on vasodilatory mechanisms in cat. Am J Physiol 267:H2012–H2018
36. Wang Q, Pelligrino DA, Koenig HM, Albrecht RF (1994) The role of endothelium and nitric oxide in rat pial arteriolar dilatory responses to CO_2 in vivo. J Cereb Blood Flow Metab 14:944–951
37. Nelson CW, Wei EP, Povlishock JT, Kontos HA, Moskowitz MA (1992) Oxygen radicals in cerebral ischemia. Am J Physiol 263:H1356–H1362
38. Wei EP, Moskowitz MA, Boccalini P, Kontos HA (1992) Calcitonin gene-related peptide mediates nitroglycerin and sodium nitroprusside-induced vasodilation in feline cerebral arterioles. Circ Res 70:1313–1319
39. Faraci FM (1991) Role of endothelium-derived relaxing factor in cerebral circulation: Large arteries vs. microcirculation. Am J Physiol 261:H1038–H1042
40. Dawson VL, Dawson TM, London ED, Bredt DS, Snyder SH (1991) Nitric oxide mediates glutamate neurotoxicity in primary cortical culture. Proc Natl Acad Sci USA 88:6368–6371
41. Moncada S, Palmer RM, Higgs EA (1989) Biosynthesis of nitric oxide from L-arginine. A pathway for the regulation of cell function and communication. Biochem Pharmacol 38:1709–1715
42. Benveniste H, Drejer J, Schousboe A, Diemer NH (1984) Elevation of the extracellular concentrations of glutamate and aspartate in rat hippocampus during transient cerebral ischemia monitored by intracerebral microdialysis. J Neurochem 43:1369–1374

43. Beckman JS, Beckman TW, Chen J, Marshall PA, Freeman BA (1990) Apparent hydroxyl radical production by peroxynitrite: Implications for endothelial injury from nitric oxide and superoxide. Proc Natl Acad Sci USA 87:1620–1624
44. Buisson A, Plotkine M, Boulu RG (1992) The neuroprotective effect of a nitric oxide inhibitor in a rat model of focal cerebral ischemia. Br J Pharmacol 106:766–767
45. Nowicki JP, Duval D, Poignet H, Scatton B (1991) Nitric oxide mediates neuronal death after focal cerebral ischemia in the mouse. Eur J Pharmacol 204:339–340
46. Kano M, Moskowitz MA, Yokota M (1991) Parasympathetic denervation of rat pial vessels significantly increases infarction volume following middle cerebral artery occlusion. J Cereb Blood Flow Metab 11:628–637
47. Reis DJ, Berger SB, Underwood MD, Khayata M (1991) Electrical stimulation of cerebellar fastigial nucleus reduces ischemic infarction elicited by middle cerebral artery occlusion in rat. J Cereb Blood Flow Metab 11:810–818
48. Huang Z, Huang PL, Panahian N, Dalkara T, Fishman MC, Moskowitz MA (1995) Effects of focal ischemia in mice deficient in neuronal nitric oxide synthase. Stroke 26: 13
49. Dawson TM, Dawson VL, Snyder SH (1992) A novel neuronal messenger molecule in brain: The free radical, nitric oxide. Ann Neurol 32:297–311
50. Trifiletti RR (1992) Neuroprotective effects of N^G-nitro-L-arginine in focal stroke in the 7-day old rat. Eur J Pharmacol 218:197–198
51. Nehls DG, Todd MM, Spetzler RF, Drummond JC, Thompson RA, Johnson PC (1987) A comparison of the cerebral protective effects of isoflurane and barbiturates during temporary focal ischemia in primates. Anesthesiology 66:453–464
52. Greenberg RS, Kirsch JR, Helfaer MA, Moore LE, Traystman RJ (1991) N-monomethyl-L-arginine decreases cerebral blood flow in piglets. FASEB J 5:A375 (Abst)
53. Sancesario G, Iannone M, D'Angelo V, Nistico G, Bemardi G (1992) N^ω-nitro-L-arginine-methyl ester inhibits electrocortical recovery subsequent to transient global brain ischemia in mongolian gerbil. Funct Neurol 7:123–127
54. Yamamoto S, Golanov EV, Berger SB, Reis DJ (1992) Inhibition of nitric oxide synthesis increases focal ischemic infarction in rat. J Cereb Blood Flow Metab 12:717–726
55. Nishikawa T, Kirsch JR, Koehler RC, Miyabe M, Traystman RJ (1994) Nitric oxide synthase inhibition reduces caudate injury following transient focal ischemia in cats. Stroke 25:877–885
56. Malinski T, Bailey F, Zhang ZG, Chopp M (1993) Nitric oxide measured by a porphyrinic microsensor in rat brain after transient middle cerebral artery occlusion. J Cereb Blood Flow Metab 13:355–358
57. Baker AJ, Zornow MH, Scheller MS, et al (1991) Changes in extracellular concentration of glutamate, aspartate, glycine, dopamine, serotonin, and dopamine metabolites after transient global ischemia in the rabbit brain. J Neurochem 57:1370–1379
58. Nishikawa T, Kirsch JR, Koehler RC, Miyabe M, Traystman RJ (1994) Competitive N-methyl-D-aspartate receptor blockade reduces brain injury following transient focal ischemia in cats. Stroke 25:2258–2264
59. Yang ST, Mayhan WG, Faraci FM, Heistad DD (1991) Endothelium-dependent responses of cerebral blood vessels during chronic hypertension. Hypertension 17:612–618
60. Clavier N, Tobin J, Kirsch JR, Izuta M, Traystman RJ (1994) Brain nitric oxide synthase activity in normal, hypertensive, and stroke prone rats. Stroke 25:1674–1678
61. Peuler JD, Schelper RL (1992) Partial protection from salt-induced stroke and mortality by high oral calcium in hypertensive rats. Stroke 23:532–538
62. Mayhan WG (1992) Role of prostaglandin H_2-thromboxane A2 in responses of cerebral arterioles during chronic hypertension. Am J Physiol 262:H539–H543
63. Tagami M, Yamori Y (1988) Morphological analysis of the pathogenesis of hypertensive cerebrovascular lesions: Role of monocytes and platelets in intracerebral vessel occlusions. Jpn Circ J 52:1351–1356
64. Wuorela H, Porsti I, Arvola P, Makynen H, Vapaatalo H (1992) Three levels of dietary calcium-effects on blood pressure and electrolyte balance in spontaneously hypertensive rats. Naunyn Schmiedebergs Arch Pharmacol 346:542–549

65. Honda H, Shibuya T, Salafsky B (1990) Brain synaptosomal Ca^{2+} uptake: Comparison of Sprague-Dawley, Wistar-Kyoto and spontaneously hypertensive rats. Comp Biochem Physiol [B] 95:555–558
66. Chen PY, Sanders PW (1991) L-arginine abrogates salt-sensitive hypertension in Dahl/Rapp rats. J Clin Invest 88:1559–1567
67. Stier CT Jr, Sim GJ, Levine S (1991) Dietary arginine falls to protect against cerebrovascular damage in stroke-prone hypertensive rats. Brain Res 549:354–356
68. Vallance P, Leone A, Calver A, Collier J, Moncada S (1992) Accumulation of an endogenous inhibitor of nitric oxide synthesis in chronic renal failure. Lancet 339:572–575
69. Ogiku N, Sumikawa H, Hashimoto Y, Ishida R (1993) Prophylactic effect of imidapril on stroke in stroke-prone spontaneously hypertensive rats. Stroke 24:245–252
70. Izuta M, Clavier N, Kirsch JR, Traystman RJ (1995) Cerebral blood flow during inhibition of brain nitric oxide synthase activity in normal, hypertensive, and stroke prone rats. Stroke (In press)
71. Yamori Y, Horie R, Tanase H, Fujiwara K, Nara Y, Lovenberg W (1984) Possible role of nutritional factors in the incidence of cerebral lesions in stroke-prone spontaneously hypertensive rats. Hypertension 6:49–53

Platelet Regulation and Damage in Vascular Thrombotic and Septic Disorders

M. W. Radomski and E. Salas

Introduction

The discovery of the endogenous pathway for biosynthesis of nitric oxide (NO) from L-arginine and the appreciation that this mediator regulates not only vascular tone but also platelet function have resulted in revisiting the physiological regulation of vascular hemostasis [1]. In this chapter, we will describe the basis for this regulation in health and review how the changes in the release, metabolism and action of NO may be involved in the pathogenesis of vascular pathologies including septicemia, thrombosis and atherosclerosis.

Platelet Activation Response

Vascular homeostasis modulates both the vessel wall contractility and hemostasis. *Hemostasis* is often defined as complex regulatory mechanisms resulting from accidental injury of the vessel wall which lead to the restoration of vascular integrity. *Thrombosis* is believed to form a pathological extension of hemostasis and occurs when the hemostatic repair is exaggerated or ill-controlled. Thrombosis often compromises the tissue perfusion and precipitates ischemia and cellular damage. It is difficult to draw a clear-cut line separating thrombosis from hemostasis. For this reason, the term of *hemostatic-thrombotic balance* has been coined to denote these sophisticated, regulating mechanisms [2].

The platelets play a pivotal role in maintaining hemostatic-thrombotic balance. They are anucleate, small blood elements that are formed by fragmentation of large mother cells, megakaryocytes. The presence of plasma granules is the most important feature of platelet ultrastructure. The granules contain potent activator and proliferating agents whose release is vital for the platelet activation response. Under physiological conditions, platelets remain in a close contact with the endothelium, yet these interactions do not cause activation. However, when a blood vessel is damaged and the endothelium is disrupted, platelets are quickly recruited from plasma to the place of injury where they form an occlusive plug. These reactions are initiated by platelet contact and spreading on the subendothelial matrix (*platelet ad-*

hesion), and lead to the formation of fibrinogen-derived bridges between platelet themselves (*platelet aggregation*), reinforcement of the primary plug with fibrin and finally recanalization of the affected vessel due to activation of the fibrinolytic and tissue-repairing processes (Fig. 1).

Like all cells platelets possess the transduction system to recognize, analyze and respond to the changes in their microenvironment. During adhesion platelets extend surface membrane receptors that are capable of anchoring them to specific matrix molecules such as von Willebrand factor, thrombospondin, fibronectin and collagen. These are glycoprotein receptors mostly belonging to the integrin superfamily of cell adhesion receptors. Platelets also possess a number of receptors that are recognized by soluble factors which circulate in plasma including epinephrine, adenosine diphosphate

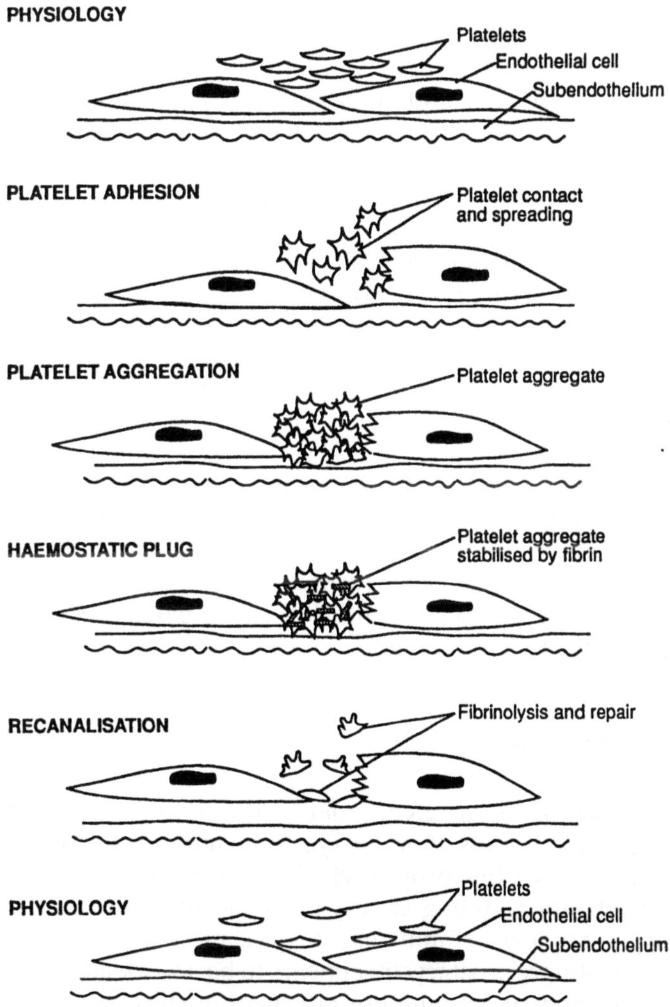

Fig. 1. Platelet hemostasis (From [2] with permission)

(ADP), vasopressin, serotonin and thrombin. These factors act in concert to amplify the signal initiated by adhesion to the subendothelial matrix. The amplification of the platelet membrane signaling system involves activation of lipid-metabolizing enzymes such as phospholipases A_2 and C. Phospholipase A_2 liberates arachidonic acid from cellular lipids which is further converted by cyclooxygenase to cyclic endoperoxides and by thromboxane synthase to thromboxane A_2 (TXA_2) which potentiate platelet activation. Phospholipase C converts membrane phosphatidylinositol diphosphate into diacyglicerol and inositol triphosphate that activate protein kinase C and the release of Ca^{2+} from the dense granules, respectively. Protein kinase C phosphorylates substrate proteins while Ca^{2+} triggers off a number of platelet reactions. These biochemical changes are accompanied by a dramatic reorganization of platelet cytoskeleton proteins leading to the change of platelet shape from disk to spheric structure protruding numerous pseudopodia. At the same time, platelet granules are grouped and secreted outside by the force generated by platelet contractile proteins (internal contraction). The secretion of Ca^{2+} and other granule constituents (the platelet release reaction) is clearly linked to platelet aggregation. The platelet releasate contains a number of agents that support aggregation and act via amplification of the release reaction (ADP and TXA_2) or by stabilizing loose bonds between adjacent platelets (fibrinogen and von Willebrand factor). The stabilization of platelet-platelet bridges is achieved by the conversion of platelet glycoprotein integrin receptor IIb/IIIa to the active conformation that binds tightly to fibrinogen. Finally, platelet secretion may facilitate also the formation of platelet-leukocyte aggregates by allowing alpha-granule protein P-selectin to be translocated to the platelet surface membrane and bind to the leukocyte ligand [2].

Physiological Regulation of Platelet Activation by Labile Mediators

The process of platelet activation is under tight regulatory control. The labile mediators (prostacyclin and NO) play a paramount role in regulation of the platelet activation cascade.

Prostacyclin

Prostacyclin is an eicosanoid synthesized from arachidonic acid by the sequential actions of cyclooxygenase and prostacyclin synthase enzymes [3]. Albeit, very labile prostacyclin is a potent vasodilator and inhibitor of platelet activation. Acting at very low concentrations (nM), prostacyclin inhibits aggregation and stimulates disaggregation of preformed platelet aggregates. Interestingly, prostacyclin appears to be a weak inhibitor of platelet adhesion [4]. These platelet-inhibitor effects of prostacyclin are mediated by stimulation of specific receptors linked to the adenylate cyclase system leading to an increase in cyclic AMP levels. cAMP-mediated responses result in the phosphorylation of specific platelet proteins followed by downregulation of several stages of the platelet transduction mechanism including inhibition of intraplatelet Ca^{2+} release and activation of platelet receptors [5].

Nitric Oxide

The hemostatic-thrombotic balance is maintained by the interactions between cellular and non-cellular components of the blood-vessel system. The platelets, endothelial cells, leukocytes and erythrocytes are the main cell types involved in this regulation. Interestingly, except erythrocytes (which remove NO by binding to hemoglobin), the remaining blood elements generate NO that exerts hemostatic influence in the system.

Nitric oxide synthase enzymes in hemostatic cells. In 1990, found that platelets generate NO using a constitutive, L-arginine-, Ca^{2+}-dependent NO synthase (NOS) mostly associated with the soluble fraction of these blood elements [6, 7]. Subsequently, a number of groups confirmed our findings [8–12]. Recently, a constitutive isoform of NOS (iNOS) has been purified to homogeneity from human platelet cytosolic fractions [13]. The native enzyme appeared to be dimeric, Ca^{2+}-, calmodulin- and biopterin-dependent and its molecular weight was approximately 150 kD.

The endothelial cells possess also a constitutive isoform of NOS (eNOS). The eNOS from human umbilical vein and bovine aorta endothelial cells have been cloned [14, 15]. This is a Ca^{2+}-, NADPH-, flavin- and biopterin-dependent enzyme which utilizes the guanido nitrogen atom of L-arginine and incorporate molecular oxygen to generate NO and L-citrulline [16]. Neutrophils synthesize NO through the activity of a constitutive NOS [17].

Regulation of Platelet Hemostasis by NO

In the resting platelets, the synthesis of NO is not detectable. However, the platelet NOS becomes activated during platelet aggregation induced by collagen, ADP and arachidonic acid [7]. Since this enzyme is strictly Ca^{2+}-dependent and platelet aggregation is associated with an increase in intraplatelet Ca^{2+}, it is possible that this cation controls the activation of the platelet NOS *in vivo*. However, our recent studies using direct electrochemical measurement of NO released from human platelets have shown that NO is released during collagen- but not thrombin-induced aggregation [18]. The reasons for the differential action of collagen and thrombin on NO release are not clear since both aggregating agents are known to mobilise Ca^{2+} from intraplatelet compartments. Since differential Ca^{2+} compartments have been described in human platelets [19], the activation of the platelet NOS may be linked to a collagen-sensitive but thrombin-insensitive pool. Alternatively, NO generated during platelet aggregation with thrombin is rapidly converted to a metabolite that is not detected by NO-selective electrochemical system.

The amounts of NO available for regulation of platelet function are supplemented by its production by the vascular endothelium. Studies with inhibitors of NOS suggest that eNOS generates NO constantly to provide the vasodilator tone [20]. The physiological stimuli for generation of NO by the endothelium are not yet fully understood, but flow and shear stress seem to stimulate the synthesis of NO via activation of the potassium K_{Ca} channel [21].

In 1987, it was also shown that cultured and fresh endothelial cells, when stimulated with bradykinin, release NO in quantities sufficient to inhibit platelet adhesion [4, 22, 23]. Moreover, the coronary and pulmonary vasculature release NO to inhibit platelet adhesion under constant flow conditions [24, 25].

Platelet aggregation *in vitro* induced by a variety of agonists is inhibited by NO released from fresh or cultured endothelial cells [26–33]. This NO also causes disaggregation of preformed platelet aggregates [27]. Moreover, basal [9, 34–37] or stimulated by cholinergic stimuli and substance P [38–40] release of NO results in inhibition of platelet aggregation induced by some aggregating agents or endothelial injury *in vivo*. Finally, the inhibition of NOS in healthy volunteers leads to increased platelet aggregation and granule release [41]. Thus, a concerted action of cellular NOS regulates platelet activation, causing inhibition of adhesion and aggregation, and induction of disaggregation.

The contribution of NO released from neutrophils [17] to the regulation of platelet function *in vivo* remains to be established. However, it is of interest that NO-cGMP system also inhibits the adhesion and chemotaxis of stimulated neutrophils [42, 43].

In vivo, the synthesis and release of a single inhibitor is unlikely to account for regulation of platelet aggregation. We have shown that NO and prostacyclin synergize with each other as inhibitors of platelet aggregation and inducers of disaggregation [27]. In addition, synergistic induction of platelet disaggregation has recently been demonstrated by combining glyceryl trinitrate (an NO donor), prostaglandin E_1 and tissue plasminogen activator, which act via cGMP, cAMP and plasmin-dependent mechanisms, respectively [44]. Furthermore, inhibition of NO generation *in vivo* abolishes antithrombotic activity of aspirin and ketanserin [35]. Thus, it is likely that platelet aggregation *in vivo* is regulated by the synergistic action of several inhibitors of platelet function.

Molecular Basis for the Regulatory Effects of NO on Hemostasis

The Soluble Guanylyl Cyclase

The stimulation by NO of the soluble guanylyl cyclase (GC-S) accounts for major regulatory actions of this mediator on cell-cell interactions in hemostasis. This activation leads to the conversion of magnesium guanosine 5′-triphosphate to guanosine 3′,5′-monophosphate (cGMP) which stimulates cGMP-dependent protein kinase. The biochemical effects of cGMP may result in inhibition of phosphorylation of myosin light chains and of protein kinase C, stimulation of phosphorylation of the β-subunit of glycoprotein I, and modulation of phospholipase A_2- and C-mediated responses [45]. cGMP-regulated responses also decrease intracellular Ca^{2+} [46]. All these effects may be important in regulation of platelet adhesins. Indeed, cGMP is likely to regulate the function of the fibrinogen receptor both during adhesion and aggregation [47, 48] and decrease the expression of P-selectin on platelet surface during platelet-platelet and platelet-leukocyte interactions [48, 49]. Not all platelet receptors appear to be influenced by cGMP and the fibronectin receptor may be an example of those adhesins that are not regulated by this cyclic nucleotide [47]. Some actions of NO on platelets may stem from a "cross-talk" between GC-S and the adenylate cyclase systems [50].

cGMP-Independent Regulation of Platelet Function by NO

In addition to GC-S, NO may also interact with other affinity targets including thiols, transition metal-containing clusters and secondary amines [51]. However, the biological significance of these interactions for the regulation of platelet hemostasis remains to be elucidated.

Role of NO in Vascular Pathology

Endotoxemia

The exposure of humans to bacterial endotoxin (lipopolysaccharide, LPS) initiates fever, cardiovascular shock and may precipitate severe disturbances in hemostatic-thrombotic balance including disseminated intravascular coagulation (DIC). Indeed, many of the deleterious effects of LPS result from the activation of plasma factors, blood cells and the endothelium. LPS causes the release of a wide range of cell-derived substances including cytokines [52] and procoagulants such as tissue factor [53]. In addition, cytokines increase the thrombogenicity of endothelial cells through induction of the expression of endothelial adhesins [52]. The changes in hemostatic-thrombotic balance underline the pathogenesis of LPS-induced DIC which has been originally characterized as consumption coagulopathy with thrombocytopenia secondary to the activation of the coagulation cascade [54]. A primary effect of LPS on platelets has been also studied. Interestingly, an important species difference in the reactivity of platelets to LPS has been found. Studies on isolated platelets have reported a proaggregatory effect of LPS on rabbit [55, 56] but not on human [57, 58] platelets. More recently, Hawiger et al. [59] demonstrated membrane changes and release of serotonin from platelets induced by LPS, while Saba and colleagues [60] reported that a variety of endotoxins, when added to human platelets, inhibited platelet aggregation initiated by several aggregating agents. Thus, the issue whether LPS stimulates or inhibits platelet function remains controversial.

Over the past few years, evidence has accumulated for the release of NO in mediating biological effects of LPS in the vascular system. It is now well recognized that the exposure of target cells to LPS and cytokines leads to the expression of the inducible NOS (iNOS). The iNOS, purified to homogeneity and cloned from murine macrophages [61], has been now identified in different hemostatic cells including the endothelium [62] and smooth muscle vascular cells [63]. The expression of iNOS requires protein synthesis and as such is inhibited by known inhibitors of this process including cycloheximide and actinomycin D. In addition, glucocorticoids also downregulate the expression of iNOS [62, 63]. The platelets, although anucleate, do synthesize proteins.

Recently, it has been shown that platelets contain small amounts of mRNA for iNOS and express iNOS activity when induced by cytokines [64]. It is, therefore, likely that the platelet iNOS may originate both from platelets and megakaryocytes since the latter cells (from which platelets derive) express iNOS upon cytokine stimulation and may pass it on to platelets [65]. LPS- and cytokine-mediated expression of iNOS is likely to have complex repercussions for vascular hemostasis. On one hand, inducible NO acts to promote hemostasis and inhibit thrombosis since the inhibition of its generation by NOS inhibitors greatly potentiated cytokine-stimulated platelet adhesion to cultured human endothelial cells [66] and precipitated renal glomerular thrombosis [67]. On the other hand, the exposure of endothelial cells to cytokines may result in cell toxicity and destruction [68]. There is growing evidence that the cellular toxicity associated with the expression of inducible NO is indirect and mediated via formation of peroxynitrite anion ($ONOO^-$) from NO and superoxide anion (O_2^-). Peroxynitrite is a potent and highly reactive oxidant which, when formed, can propagate the cascade of oxidation and cell damage [69]. Indeed, it has been suggested that $ONOO^-$ may play an important role in oxidant-induced endothelial injury [70]. Recently, we have found that $ONOO^-$ stimulated aggregation of human platelets and reversed inhibition of aggregation by NO and prostacyclin [71]. In addition, the exposure of the isolated perfused coronary vasculature to this oxidant led to a decrase in vascular reactivity to the vasodilators [72]. Interestingly, the effects of $ONOO^-$ on platelets were attenuated in the presence of thiols due to its conversion to S-nitrosothiols which inhibited platelet aggregation. In addition, $ONOO^-$-induced impairment in vascular reactivity was prevented by exogenous NO or prostacyclin. Thus, the net effect (cytoprotective or cytodestructive) of inducible NO is likely to be dependent on the microenvironment where this molecule is generated, reacts and acts.

Vascular Ischemic Disorders

Although it is well established that both the vasodilator and platelet-regulatory functions of endothelium are impaired during the course of atherosclerosis, essential hypertension, diabetes and coronary artery disease [73], the pathomechanism of these changes remains unclear. Since oxidative modification of low-density lipoproteins (LDL) has been suggested to play a key role in atherogenesis, a number of studies have examined the effects of native and oxidized LDL on NO-mediated vascular functions. In most of these studies lipoproteins have been found to decrease the bioactivity of NO [74]. Several hypotheses have been proposed to explain these effects of LDL including inhibition of NOS activity, direct inactivation of NO or changes in NO metabolism [75]. In addition, LDL may inhibit arginine uptake to platelets, and through this mechanism decrease NOS activity and promote thrombosis [76]. An impaired NO generation or action may also underline the pathomechanism of vasospastic and thrombotic changes of essential hypertension [10, 77], diabetes [78] and the coronary artery disease. Indeed, the endogenous NO has been shown to inhibit microthrombo-embolism in the ischemic heart and protect against intracoronary thrombosis [79, 80].

Interestingly, some researchers have reported increased generation of nitrogen oxides during atherogenesis that was associated with decreased bioactivity of these compounds [81]. It is, therefore, possible that conversion of

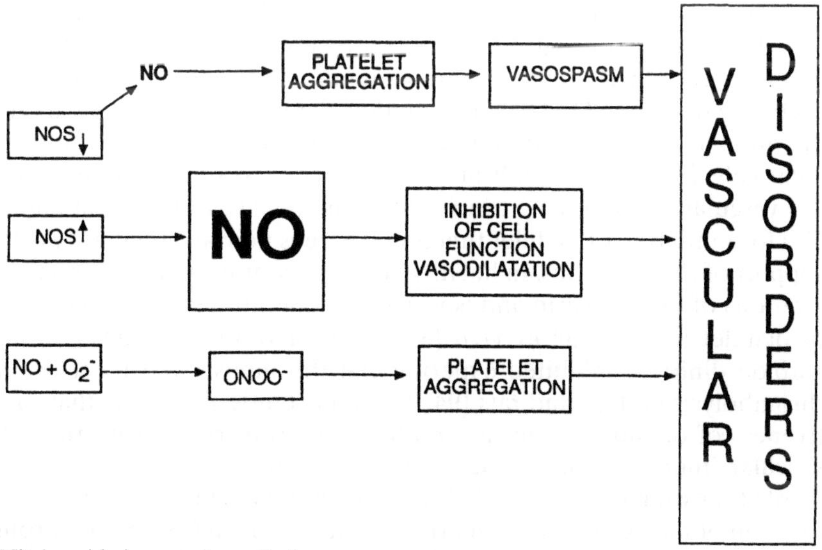

Fig. 2. Nitric oxide in vascular pathology

NO to the species such as $ONOO^-$ may play a role in the pathogenesis of atherogenesis and other vascular pathologies (Fig. 2) [71, 72].

Pharmacological Regulation of Platelet Function by NO Gas and NO Donor Drugs

NO gas is a potent inhibitor of platelet activation *in vitro*, however its anti-platelet activity is limited by short chemical (sec) and biological (1–4 min) half-lives [26]. Interestingly, inhalation of NO gas by healthy volunteers and experimental animals resulted in a longer-lasting (appr. 30 min) inhibition of platelet function as detected by prolongation of bleeding time [82]. This phenomenon may be explained either by accumulation and subsequent slow release of NO from lipid part of cell membrane [83] or formation of endogenous NO donors.

The mechanism of pharmacological activity of NO donors is related to their ability to release NO [84]. The main pharmacological advantage of these compounds is their longer than NO gas half-life. The NO donors can be classified according to their ability to release NO *in vitro*. The first group is comprised of the drugs that require metabolic transformation to release NO. Organic nitrates are the most prominent members of this class. *In vitro*, these compounds are poor spontaneous releasers of NO and require the presence of a thiol co-factor (e.g. N-acetylcysteine) for acceleration of this liberation [84]. *In vivo* in addition to thiols, the release of NO from organic nitrates is greatly enhanced by enzyme(s) which still remain to be identified. Vascular tissues show a remarkable capacity for enzymatic conversion of organic nitrates to NO [85], however this ability does not appear to be shared by platelets. Indeed, organic nitrates are weak inhibitors of aggregation of isolated platelets *in vitro* [86]. However, following stimulation of platelet aggregation by ADP *in vitro*, nitroglycerin has been shown to exert a potent disaggregating effect which may be due to a temporary elevation in platelet guanylyl cyclase activity to NO [87]. Nitrate-induced inhibition of platelet aggregation *in vitro* can be greatly potentiated in the presence of thiols or cultured vascular cells [85, 88]. This indicates that the conversion of organic nitrates by the vascular tissue *in vivo* can result in the release of sufficient amounts of NO for inhibition of platelet function. There is now an abundant evidence in support of these notions. In healthy volunteers, oral and intravenous administrations of nitroglycerin and isosorbide mononitrates resulted in inhibition of platelet aggregation *ex vivo* [89–93]. In addition, nitroglycerin and isosorbide dinitrate inhibited experimental thrombosis and reocclusion after thrombolysis in dogs and rats [94, 95]. Importantly, it appears that the effectiveness of organic nitrates as antithrombotics increases with the extent of vascular injury. In normal pigs, the deposition of platelets on arterial segments following injury using balloon angioplasty is inhibited by intravenous infusion of nitroglycerin when arterial injury is deep (extending through the internal elastic lamina) rather than mild (deendothelialization only) [96].

Furthermore, short- and long-lasting administration of nitroglycerin and isosorbide dinitrate to patients suffering from coronary artery disease and acute myocardial infarction resulted in a significant inhibition of platelet adhesion and aggregation [92, 97–99]. Of particular interest are interactions of organic nitrates with other inhibitors of platelet function. Indeed, organic nitrates synergize with prostacyclin or aspirin as inhibitors of platelet aggregation *in vitro* and *in vivo* [93, 100]. Also, isosorbide dinitrate has been shown to potentiate the antiplatelet activity of prostaglandin E_1 in patients with peripheral vascular disease [101]. Thus, all these observations clearly indicate that organic nitrates inhibit platelet function, and their antiplatelet properties are relevant to the mechanism of clinical effectiveness of these compounds in vascular disorders.

The second group of NO donors is composed of compounds which release NO without the need of metabolic activation. Sodium nitroprusside, molsidomine and SIN-1 are clinically used members of this group.

Sodium Nitroprusside, Molsidomine and SIN-1

Because of its powerful vasodilator action, sodium nitroprusside is often used to combat vascular emergencies associated with hypertensive crisis. Since this compound shows some antiplatelet activity both *in vitro* and *in vivo* [102, 103], its acute clinical effects may also be mediated, in part, through inhibition of platelet function.

Molsidomine and its active metabolite SIN-1 have been shown to inhibit experimental thrombosis and platelet aggregation in healthy volunteers and in patients suffering from acute myocardial infarction [104]. Interestingly, SIN-1 in addition to NO has been also found to generate O_2^- and $ONOO^-$ [105]. Since $ONOO^-$ has been shown to cause platelet aggregation and counteract the platelet inhibitory activity of NO [71], the formation of this radical may counteract the antiplatelet activity of NO released from SIN-1.

NO Donors as Inhibitors of Platelet Function

Despite almost 140 years of use, organic nitrates remain one of the safest and most widely applied cardiovascular drugs. There is, however, plenty of scope for improvement and the current pharmacological literature is overwhelmed by communications describing other families of NO donors. Since organic nitrates under some conditions are known to induce tachyphylaxis, there have been many attempts to synthesize tachyphylaxis-free NO donors. One of more promising groups of drugs has been recently synthesized by Schwarz Pharma. They are cysteine-containing nitrates. Interestingly, the incorporation of a cellular thiol cysteine to the structure of organic nitrate resulted in a high effectiveness of these compounds as inhibitors of platelet and leukocyte functions both *in vitro* and *in vivo* [106].

Another challenging problem is tissue selectivity. The platelet-inhibitory actions of organic nitrates cannot be separated from effects on vascular wall. The concept of platelet-selective NO donors has arisen from our experiments with S-nitrosoglutathione (GSNO). GSNO is a tripeptide S-nitrosothiol which is formed by S-nitrosylation of glutathione, the most abundant intracellular thiol. We have found that the intravenous administration of GSNO into conscious rats inhibits platelet aggregation at doses that have only small effect on the blood pressure [107]. Moreover, similar platelet/vascular differentiation is detected following intraarterial administration of GSNO into the circulation of human forearm [108]. Finally, we have infused GSNO into patients undergoing balloon angioplasty and found that this NO donor effectively protected platelets from activation at the site of angioplastic injury without altering blood pressure [109]. This differential effect of GSNO is likely to be due to the presence in platelets of enzyme(s) that metabolize GSNO and release NO [107]. Thus, these observations show that it is possible to design pharmacologically tissue-selective NO donors, and activated platelets and leukocytes are likely therapeutic targets.

Position of Organic Nitrates among other Inhibitors of Platelet Activation

The antiplatelet activities of organic nitrates add yet another dimension to the pharmacological profile of these compounds. How does the platelet-regulatory profile of organic nitrates compare to those of "classical" inhibitors of platelet function? Actelylsalicylic acid (aspirin) is by far the most widely used antiplatelet drug in clinical practice, and its benefits in terms of decreasing mortality due to reinfarction have been unequivocally demonstrated [110–113] while those of organic nitrates have not yet been established. However, it is important to realize that aspirin, a cyclooxygenase inhibitor, blocks only thromboxane-mediated platelet aggregation [112] leaving the remaining pathways of adhesion and aggregation unopposed. In contrast, NO inhibits the activation cascade of mediators generated by all known pathways of platelet aggregation [114] and some pathways of platelet adhesion to subendothelium [47]. Thus, future clinical studies will determine the place and the effectiveness of traditional organic nitrates and new NO donors in maintainig hemostatic-thrombotic balance and preventing the ischemic complications of thrombotic vascular disorders.

Conclusion

NO generated by endothelial cells, platelets and leukocytes has proved to be an important physiological and regulatory mediator ensuring vessel wall hemostasis and preventing thrombosis. The changes in its generation or metabolism are likely to play roles in the pathomechanism of vasospastic and thrombotic disorders. Understanding the regulatory and damaging properties of NO and its interaction with other hemostatic factors may help in improving current antithrombotic therapy.

Acknowledgements: This work was supported by establishment award from the Alberta Heritage Foundation for Medical Research.

References

1. Radomski MW, Moncada S (1993) Regulation of vascular homeostasis by nitric oxide. Thromb Haemost 70:36–41
2. Radomski MW (1994) Platelet regulation: Another dimension for NO and nitrates. Schwarz Pharma Scientific Forum, Schwarz Pharma Publisher, vol 7
3 Moncada S, Gryglewski RJ, Bunting S, Vane JR (1976) An enzyme isolated from arteries transforms prostaglandin endoperoxides to an unstable substance that inhibits platelet aggregation. Nature 263:663–665
4. Radomski MW, Palmer RMJ, Moncada S (1987) Endogenous nitric oxide inhibits human platelet adhesion to vascular endothelium. Lancet 2:1057–1058
5. Moncada S (1982) Biological importance of prostacyclin. Br J Pharmacol 76:3–31
6. Radomski MW, Palmer RMJ, Moncada S (1990) An L-arginine/nitric oxide pathway present in human platelets regulates aggregation. Proc Natl Acad Sci USA 87: 5193–5197
7. Radomski MW, Palmer RMJ, Moncada S (1990) Characterization of the L-arginine/nitric oxide pathway in human platelets. Br J Pharmacol 101:325–328
8. Pronai L, Ichimori K, Nozaki H, et al (1991) Investigation of the existence and biological role of L-arginine/nitric oxide pathway in human platelets by spin-trapping/EPR studies. Eur J Biochem 202:923–930
9. Yao SK, Ober JC, Krishnaswami A, et al (1992) Endogenous nitric oxide protects against platelet aggregation and cyclic flow variations in stenosed and endothelium-injured arteries. Circulation 86:1302–1309
10. Cadwgan TM, Benjamin N (1993) Evidence for altered platelet nitric oxide synthesis in essential hypertension. J Hypertension 11:417–420
11. Noris M, Benigni A, Boccardo P, et al (1993) Enhanced nitric oxide synthesis in uremia: Implications for platelet dysfunction and dialysis hypotension. Kidney Int 44: 445–450
12. Berkels R, Klaus W, Boller M, Rosen R (1994) The calcium modulator nifedipine exerts its antiaggregatory property via a nitric oxide mediated process. Thromb Haemost 72:309–312
13. Muruganandam A, Mutus B (1994) Isolation of nitric oxide synthase from human platelets. Biochem Biophys Acta 1200:1–6
14. Marsden PA, Shappert KT, Chen HS, et al (1992) Molecular cloning and characterisation of human endothelial nitric oxide synthase. FEBS Lett 307:287–293
15. Sessa WV, Harrison JK, Barber CM, et al (1992) Molecular cloning and expression of cDNA encoding endothelial cell nitric oxide synthase. J Biol Chem 267:15274–15276
16. Moncada S (1992) The 1991 Ulf von Euler Lecture. The L-Arginine: Nitric oxide pathway. Acta Physiol Scand 145:201–227
17. McCall TB, Boughton-Smith NK, Palmer RMJ, Whittle BJR, Moncada S (1989) Synthesis of nitric oxide from L-arginine by neutrophils. Biochem J 261:293–296
18. Malinski T, Radomski MW, Taha Z, Moncada S (1993) Direct electrochemical measurement of nitric oxide released from human platelets. Biochem Biophys Res Commun 194:960–965
19. Brune B, Volker U (1991) Different calcium pools in human platelets and their role in thromboxane A_2 formation. J Biol Chem 266:19232–19237

20. Rees DD, Palmer RMJ, Moncada S (1989) Role of endothelium-derived nitric oxide in the regulation of blood pressure. Proc Natl Acad Sci USA 86:3375–3378
21. Cooke JP, Rossitch E Jr, Andon NA, Loscalzo J, Dzau VJ (1991) Flow activates an endothelial potassium channel to release an endogenous nitrovasodilator. J Clin Invest 88:1663–1671
22. Radomski MW, Palmer RMJ, Moncada S (1987) The role of nitric oxide and cGMP in platelet adhesion to vascular endothelium. Biochem Biophys Res Commun 148: 1482–1489
23. Sneddon JM, Vane JR (1988) Endothelium-derived relaxing factor reduces platelet adhesion to bovine endothelial cells. Proc Natl Acad Sci USA 85:2800–2804
24. Venturini CM, Del Vecchio PJ, Kaplan JE (1989) Thrombin-induced platelet adhesion to endothelium is modified by endothelial derived relaxing factor (EDRF). Biochem Biophys Res Commun 159:349–354
25. Pohl U, Busse R (1989) EDRF increases cyclic GMP in platelets during passage through the coronary vascular bed. Circ Res 65:1798–1803
26. Radomski MW, Palmer RMJ, Moncada S (1987) Comparative pharmacology of endothelium-derived relaxing factor, nitric oxide and prostacyclin in platelets. Br J Pharmacol 92:181–187
27. Radomski MW, Palmer RMJ, Moncada S (1987) The anti-aggregating properties of vascular endothelium: Interactions between prostacyclin and nitric oxide. Br J Pharmacol 92:639–646
28. Furlong B, Henderson AH, Lewis MJ, Smith JA (1987) Endothelium-derived relaxing factor inhibits in vitro platelet aggregation. Br J Pharmacol 90:687–692
29. Busse R, Luckhoff A, Bassenge E (1987) Endothelium-derived relaxing factor inhibits platelet activation. Naunyn Schmiedeberg's Arch Pharmacol 336:566–571
30. Macdonald PS, Read MA, Dusting GJ (1988) Synergistic inhibition of platelet aggregation by endothelium-derived relaxing factor and prostacyclin. Thromb Res 49:437–449
31. Alheid U, Reichwehr I, Forstermann U (1989) Human endothelial cells inhibit platelet aggregation by separately stimulating platelet cyclic AMP and cyclic GMP. Eur J Pharmacol 164:103–110
32. Houston DS, Robinson P, Gerrard JM (1990) Inhibition of intravascular platelet aggregation by endothelium-derived relaxing factor: Reversal by red blood cells. Blood 76:953–958
33. Broekman MJ, Eiroa AM, Marcus AJ (1991) Inhibition of human platelet reactivity by endothelium-derived relaxing factor from human umbilical vein endothelial cells in suspension: Blockade of aggregation and secretion by an aspirin-insensitive mechanism. Blood 78:1033–1040
34. Rosenblum WI, Nelson GH, Povlishock JT (1987) Laser-induced endothelial damage inhibits endothelium-dependent relaxation in the cerebral microcirculation of the mouse. Circ Res 60:169–176
35. Golino P, Capelli-Bigazzi M, Ambrosio G, et al (1992) Endothelium-derived relaxing factor modulates platelet aggregation in an in vivo model of recurrent platelet activation. Circ Res 71:1447–1456
36. May GR, Crook P, Moore PK, Page CP (1991) The role of nitric oxide as an endogenous regulator of platelet and neutrophil activation within the pulmonary circulation. Br J Pharmacol 102:759–763
37. Herbaczynska-Cedro K, Lembowicz K, Pytel B (1991) N^G-monomethyl-L-arginine increases platelet deposition on damaged endothelium in vivo. A scanning electron microscopy study. Thromb Res 64:1–9
38. Bhardwaj R, Page CP, May GR, Moore PK (1988) Endothelium-derived relaxing factor inhibits platelet aggregation in human whole blood in vitro and in the rat in vivo. Eur J Pharmacol 157:83–91
39. Hogan JC, Lewis MJ, Henderson AH (1988) In vivo EDRF activity influences platelet function. Br J Pharmacol 94:1020–1022

40. Humphries RG, Tomlinson W, O'Connor SE, Leff P (1990) Inhibition of collagen- and ADP-induced platelet aggregation by substance P in vivo: Involvement of endothelium-derived relaxing factor. J Cardiovasc Pharmacol 16:292–297
41. Bodzenta-Lukaszyk A, Gabryelewicz A, Lukaszyk A, et al (1994) Nitric oxide synthase inhibition and platelet function. Thromb Res 75:667–672
42. Kubes P, Suzuki M, Granger DN (1991) Nitric oxide: An endogenous modulator of leukocyte adhesion. Proc Natl Acad Sci USA 88:4651–4655
43. Moilanen E, Vuorinen P, Metsa-Ketela T, Vapaatalo H (1993) Inhibition by nitric oxide donors of human polymorphonuclear leucocyte functions. Br J Pharmacol 109: 852–858
44. Stamler JS, Vaughan DE, Loscalzo J (1989) Synergistic disaggregation of platelets by tissue-type plasminogen activator, prostaglandin E_1 and glyceryl trinitrate. Circ Res 65:796–804
45. Walter U (1989) Physiological role of cGMP and cGMP-dependent protein kinase in the cardiovascular system. Rev Physiol Biochem Pharmacol 113.41–88
46. Nakashima S, Tohmatsu T, Hattori H, Okano Y, Nozawa Y (1986) Inhibitory action of cyclic GMP on secretion, phosphoinositide hydrolysis and calcium mobilization in thrombin-stimulated human platelets. Biochem Biophys Res Commun 135:1099–1104
47. Shahbazi T, Jones N, Radomski MW, Moro MA, Gingell D (1994) Nitric oxide donors inhibit platelet spreading on surfaces coated with fibrinogen but not fibronectin. Thromb Res 75:631–642
48. Salas E, Moro MA, Askew S, et al (1994) Comparative pharmacology of analogues of S-nitroso-N-acetyl-DL-penicillamine in platelets. Br J Pharmacol 112:1071–1076
49. Kurose I, Kubes P, Wolf R, Anderson DC, et al (1993) Inhibition of nitric oxide production. Mechanisms of vascular albumin leakage.Circ Res 73:164–171
50. Maurice DH, Haslam RJ (1990) Molecular basis of the synergistic inhibition of platelet function by nitrovasodilators and activators of adenylate cyclase: Inhibition of cyclic AMP breakdown by cyclic GMP. Mol Pharmacol 37:671–681
51. Hibbs JB Jr, Taintor RR, Vavrin Z, et al (1990) Synthesis of nitric oxide from a terminal guanidino nitrogen atom of L-arginine: A molecular mechanism regulating cellular proliferation that targets intracellular iron. In: Moncada S, Higgs EA (eds) Nitric Oxide from L-Arginine: A Bioregulatory System. Elsevier, Amsterdam, pp 189–223
52. Mantovani A, Dejana E (1989) Cytokines as communication signals between leukocytes and endothelial cells. Immunol Today 10:370–375
53. Rickles FR, Levin J, Hardin JA, Conrad ME Jr (1977) Tissue factor generation by human mononuclear cells: Effects of endotoxin and dissociation of tissue factor generation from mitogenic response. J Lab Clin Med 89:792–803
54. Corrigan JJ Jr, Ray W, May N (1968) Changes in the blood coagulation system associated with septicemia. N Engl J Med 279:851–856
55. Des Pres RM, Horowitz HI, Hook EW (1967) Effects of bacterial endotoxin on rabbit platelets. I. Platelet aggregation and release of platelet factors in vitro. J Exp Med 114:857–873
56. Semeraro N, Colucci M, Fumarola D, Vermylen J (1978) Platelets and endotoxins: Complement-dependent and complement-independent interactions. In: di Gaetano G, Garatini S (eds) Platelets: A multidisciplinary approach. Raven Press, New York, pp 292–301
57. Corn M (1968) Effect of thrombin, ADP, connective tissue and endotoxin on platelet glycolysis. Nature 212:508
58. Mueller-Eckhardt CH, Lüscher EF (1968) Immune reactions of human blood platelets. IV. Investigations on the problem of an immunologically-induced effect of endotoxin on human platelets. Thromb Diath Haemorrh 20:336–344
59. Hawiger J, Hawiger A, Steckley S, Timmons S, Cheng C (1977) Membrane changes in human platelets induced by lipopolysaccharide endotoxin. Br J Haematol 35: 285–299

60. Saba HI, Saba SR, Morelli G, Hartman RG (1984) Endotoxin-mediated inhibition of human platelet aggregation. Thromb Res 34:19–33
61. Lyons CR, Orloff GJ, Cunningham JM (1992) Molecular cloning and functional expression of an inducible nitric oxide synthase from a murine macrophage cell line. J Biol Chem 267:6470–6474
62. Radomski MW, Palmer RMJ, Moncada S (1990) Glucocorticoids inhibit the expression of an inducible but not the constitutive, nitric oxide synthase in vascular endothelial cells. Proc Natl Acad Sci USA 87:10043–10047
63. Rees DD, Cellek S, Palmer RMJ, Moncada S (1990) Dexamethasone prevents the induction by endotoxin of a nitric oxide synthase and the associated effects on vascular tone: An insight into endotoxin shock. Biochem Biophys Res Commun 173: 541–547
64. Mehta JL, Chen LY (1995) Identification of constitutive and inducible forms of nitric oxide synthase in human platelets. J Lab Clin Med 125:370–377
65. Lelchuk R, Radomski MW, Martin JF, Moncada S (1992) Constitutive and inducible nitric oxide synthases in human megakaryoblastic cells. J Pharmacol Exp Ther 262: 1220–1224
66. Radomski MW, Vallance P, Whitley G, Foxwell N, Moncada S (1993) Platelet adhesion to human vascular endothelium is modulated by constitutive and cytokine-induced nitric oxide. Cardiovasc Res 27:1380–1382
67. Schultz PJ, Raij L (1992) Endogenously synthesized nitric oxide prevents endotoxin-induced glomerular thrombosis. J Clin Invest 90:1718–1725
68. Palmer RMJ, Bridge L, Foxwell NA, Moncada S (1992) The role of nitric oxide in endothelial cell damage and its inhibition by glucocorticoids. Br J Pharmacol 105:11–12
69. Beckman JS, Beckman TW, Chen J, Marshall PA, Freeman BA (1990) Apparent hydroxyl radical production by peroxynitrite: Implications for endothelial injury from nitric oxide and superoxide. Proc Natl Acad Sci USA 87:1620–1624
70. Kooy NW, Royall JA (1994) Agonist-induced peroxynitrite production from endothelial cells. Arch Biochem Biophys 310:352–359
71. Moro MA, Darley-Usmar VM, Goodwin DA, et al (1994) Paradoxical fate and biological action of peroxynitrite on human platelets. Proc Natl Acad Sci USA 91: 6702–6706
72. Villa LM, Salas E, Darley-Usmar VM, Radomski MW, Moncada S (1994) Peroxynitrite induces both vasodilatation and impaired vascular relaxation in the isolated perfused rat heart. Proc Natl Acad Sci USA 91:12383–12387
73. Luscher TF, Tanner FC, Tschudi MR, Noll G (1993) Endothelial dysfunction in coronary artery disease Ann Rev Med 44:395–418
74. Flavahan NA (1992) Atherosclerosis or lipoprotein-induced endothelial dysfunction. Circulation 85:1927–1938
75. Cooke JP, Tsao P (1992) Cellular mechanisms of atherogenesis and the effects of nitric oxide. Curr Opin Cardiol 7:799–804
76. Chen L, Mehta JL (1994) High density lipoprotein antagonizes the stimulatory effect of low density lipoprotein on platelet function by affecting L-arginine-nitric oxide pathway. Circulation 90:I–30
77. Calver A, Collier J, Moncada S, Vallance P (1992) Effect of local infusion of N^G-monomethyl-L-arginine in patients with hypertension. The nitric oxide dilator mechanism appears abnormal. J Hypertension 10:1025–1031
78. Amado JA, Salas E, Botana MA, Poveda JJ, Berrazueta JR (1993) Low levels of intraplatelet cGMP in IDDM. Diabetes Care 16: 809–811
79. Yao SK, Ober JC, Golino P, Buja M, Willerson JT (1994) Both endogenous and exogenous nitric oxide protect against intracoronary thrombosis. Circulation 90:I–345
80. Komamura K, Node K, Kosaka H, Inoue M (1994) Endogenous nitric oxide inhibits microthromboembolism in the ischemic heart. Circulation 90:I–345
81. Myers PR, Wright TF, Tanner MA, Ostlund RE Jr (1994) The effects of native LDL and oxidized LDL on EDRF bioactivity and nitric oxide production in vascular endothelium. J Lab Clin Med 124:672–683

82. HogmanM, Frostell C, Arnberg H, Hedenstierna G (1993) Bleeding time prolongation and NO inhalation. Lancet 341:1664–1665
83. Malinski T, Taha Z, Grunfeld S, et al (1993) Diffusion of nitric oxide in the aorta wall monitored in situ by porphyrinic microsensors. Biochem Biophys Res Commun 193: 1076–1082
84. Feelisch M, Noack EA (1987) Correlation between nitric oxide formation during degradation of organic nitrates and activation of guanylyl cyclase. Eur J Pharmacol 139:19–30
85. Feelisch M (1991) The action and metabolism of organic nitrates and their similarity with endothelium-derived relaxing factor (EDRF). In: Moncada S, Higgs EA, Berrazueta JR (eds) Clinical relevance of nitric oxide in the cardiovascular system. Edicomplet, Madrid, pp 29–43
86. Gerzer R, Karrenbrock B, Siess W, Heim JM (1988) Direct comparison of the effects of nitroprusside, SIN-1 and various nitrates on platelet aggregation and soluble guanylyl cyclase activity. Thromb Res 52:11–21
87. Chirkov YY, Belushkina NN, Tyshchuk IS, Severina IS, Horowitz JD (1991) Increase in reactivity of human platelet guanylyl cyclase during aggregation potentiates the disaggregating capacity of sodium nitroprusside. Clin Exper Pharmacol Physiol 18: 517–524
88. Benjamin N, Dutton JAE, Ritter JM (1991) Human vascular smooth muscle cells inhibit platelet aggregation when incubated with glyceryl trinitrate: Evidence for generation of nitric oxide. Br J Pharmacol 102:847–850
89. De Caterina T, Giannesssi D, Crea F, et al (1984) Inhibition of platelet function by injectable isosorbide dinitrate. Am J Cardiol 53:1683–1687
90. Drummer C, Valta-Seufzer U, Karrenbrock B, Heim JM, Gerzer R (1991) Comparison of antiplatelet properties of molsidomine, isosorbide-5-mononitrate and placebo in healthy volunteers. Eur Heart J 12:541–549
91. Wallen NH, Larsson PT, Broijersen A, Andersson A, Hjemdahl P (1993) Effects of oral dose of isosorbide dinitrate on platelet function and fibrinolysis in healthy volunteers. Br J Clin Pharmacol 35:143–151
92. Chirkov YY, Naujalis JI, Sage RE, Horowitz JD (1993) Antiplatelet effects of nitroglycerin in healthy subjects and in patients with stable angina pectoris. J Cardiovasc Pharmacol 21:384–389
93. Karlberg KE, Ahlner J, Henriksson P, Torfgard K, Sylven C (1993) Effects of nitroglycerin on platelet aggregation beyond the effects of acetylsalicylic acid in healthy subjects. Am J Cardiol 71:361–364
94. Werns SW, Rote WE, Davis JH, Guevara T, Lucchesi BR (1994) Nitroglycerin inhibits experimental thrombosis and reocclusion after thrombolysis. Am Heart J 127: 727–737
95. Plotkine M, Allix M, Guillou J, Boulu R (1991) Oral administration of isosorbide dinitrate inhibits arterial thrombosis in rats. Eur J Pharmacol 201:115–116
96. Lam JYT, Chesebro JH, Fuster V (1988) Platelets, vasoconstriction, and nitroglycerin during arterial wall injury. A new antithrombotic role for an old drug. Circulation 78: 712–716
97. Gebalska J (1990) Platelet adhesion and aggregation in relation to clinical course of acute myocardial infarction. M.D. thesis, Warsaw, in Polish
98. Diodati J, Theroux P, Latour JG, et al (1990) Effects of nitroglycerin at therapeutic doses on platelet aggregation in unstable angina pectoris and acute myocardial infarction. Am J Cardiol 66:683–688
99. Sinzinger H, Virgolini I, O'Grady J, Rauscha F, Fitscha P (1992) Modification of platelet function by isosorbide dinitrate in patients with coronary artery disease. Thromb Res 65:323–335
100. De Caterina R, Giannessi A, Bernini W, Mazzone A (1988) Organic nitrates: Direct antiplatelet effects and synergism with prostacyclin. Antiplatelet effects of organic nitrates. Thromb Haemost 59:207–211

101. Sinzinger H, Fitscha P, O'Grady J, et al (1990) Synergistic effect of prostaglandin E_1 and isosorbide dinitrate in peripheral vascular disease. Lancet 335:627–628
102. Levin RL, Weksler BB, Jaffe EA (1982) The interaction of sodium nitroprusside with human endothelial cells and platelets: Nitroprusside and prostacyclin synergistically inhibit platelet function. Circulation 66:1299–1307
103. Hines R, Barash PG (1989) Infusion of sodium nitroprusside induces platelet dysfunction in vitro. Anesthesiology 71:805–806
104. Wautier JL, Weill D, Kadeva H, Maclouf J, Soria C (1989) Modulation of platelet function by SIN-1A. J Cardiovasc Pharmacol 14:S111–S114
105. Hogg N, Darley-Usmar VM, Wilson MT, Moncada S (1993) Oxidation of alpha-tocopherol in human low density lipoprotein by the simultaneous generation of superoxide and nitric oxide. FEBS Lett 326:199–203
106. Lefer DJ, Nakanishi K, Vinten-Johansen J (1993) Endothelial and myocardial cell protection by a cysteine-containing nitric oxide donor after myocardial ischemia and reperfusion. J Cardiovasc Pharmacol 22 (Suppl 7): S34–S43
107. Radomski MW, Rees DD, Dutra A, Moncada S (1992) S-nitrosoglutathione inhibits platelet activation in vitro and in vivo. Br J Pharmacol 107:745–749
108. de Belder AJ, MacAllister R, Radomski MW, Moncada S, Vallance PJ (1994) Effects of S-nitrosoglutathione in the human forearm circulation: Evidence for selective inhibition of platelet activation. Cardiovasc Res 28:691–694
109. Langford EJ, Brown AS, Wainwright RJ, et al (1994) Inhibition of platelet activity by S-nitrosoglutathione during coronary angioplasty. Lancet 344:1458–1460
110. ISIS-2 (Second International Study of Infarct Survival) (1988) Collaborative Group: Randomized trial of intravenous streptokinase, oral aspirin, both or neither among 17,187 cases of suspected acute myocardial infarction: ISIS-2. Lancet 2:349–360
111. ISIS-3 (1992) A randomized comparison of streptokinase vs tissue plasminogen activator vs anistreplase, and of aspirin plus heparin vs aspirin alone among 41,299 cases of suspected acute myocardial infarction. Lancet 339:753–770
112. Patrono C (1989) Aspirin and human platelets: From clinical trials to acetylation of cyclooxygenase and back. TIPS 10:453–458
113. Fernandez-Ortiz A, Jang IK, Fuster A (1994) Antiplatelet and antithrombin therapy. Coronary Artery Disease 5:297–305
114. Jensen BO, Holmsen H (1995) Nitric oxide (NO)-platelet interactions: Inhibition is independent of the prostanoid and ADP pathways. Platelets (In press)

Nitric Oxide as a Mediator of Hypotension and Inflammation in Sepsis

R. L. Danner, J. Cobb, and A. L. Van Dervort

Introduction

Septic shock is a severe systemic inflammatory response caused by infection with any of a wide variety of microorganisms. Clinically, it is characterized by fever, hypotension, and signs of organ hypoperfusion or dysfunction. The high mortality rate of septic shock, despite the administration of effective antimicrobial agents, fluids, and catecholamine vasopressors, has led to an extensive effort to identify the microbial products, host mediators, and cell-activation mechanisms that trigger and fuel the clinical manifestations of this syndrome. As a result of this effort, our understanding of septic shock has grown exponentially. However, although therapies targeting newly discovered pathogenic mechanisms have been developed, none have as yet demonstrated clinical efficacy [1]. One of the most interesting mediators recently implicated in septic shock is nitric oxide (NO).

NO is a reactive, membrane-permeable gas involved in intracellular and intercellular communication [2]. Unlike more familiar polypeptide, catecholamine, and steroid messengers, NO covalently interacts with its cellular target(s) to transduce its signal [3]. This volatile, simple molecule performs such vital homeostatic functions as neurotransmission [4] and the regulation of vascular tone [5, 6]. Under homeostatic conditions, endothelium-derived NO produces vasodilation by activating soluble guanylate cyclase in vascular smooth muscle, which causes a rise in intracellular cyclic GMP (cGMP) [3, 7]. In this situation NO is produced from L-arginine by a calcium- and calmodulin-dependent NO synthase. Tightly regulated signal transduction pathways that utilize calcium transients and are linked to receptors on the cell surface, control NO production, and therefore vascular tone, by this constitutive isoform of NOS (cNOS).

In septic shock, endotoxin, products of gram-positive bacteria and host cytokines such as TNF-α, IL-1β, and IFN-γ induce a calcium-independent isoform of NOS in certain cells and tissues [8–14]. This inducible NOS (iNOS) can produce large amounts of NO for long periods of time. In recent years, the concept that NO produced by iNOS is largely responsible for sepsis-related hypotension and contributes to the tissue injury and organ failures characteristics of septic shock has received widespread attention [15, 16]. As a consequence, L-arginine analogs that non-selectively inhibit NOS have been extensively studied in animal models of septic shock and have been taken quickly into clinical trials [17, 18].

However, since the inhibition of NO production was first proposed as a potentially useful therapeutic strategy, new information has become available that may necessitate a revision of the initial hypothesis. These still-evolving experimental data address salient issues such as the effects of NOS inhibitors in normal compared to septic animals, interspecies differences with regard to enzyme induction, mechanisms that control NO production by iNOS, and the role of NO as a modulator of inflammatory responses. The goal of this chapter is to review and revise some of the assumptions that led to the use of NOS inhibitors in septic shock.

Evidence Supporting the Use of NOS Inhibitors in Septic Shock

NO in Septic Shock

Hypotension is a predictor of organ injury and death during sepsis [19–21], and some patients with septic shock are relatively refractory to the vasopressors conventionally used to manage this complication. Approximately 30 to 50% of patients who die of septic shock do so with persistent high-cardiac output hypotension [20, 21]. It has been postulated that excess production of NO is the primary cause of this hyperdynamic shock syndrome and that NOS inhibition represents specific therapy that can normalize blood pressure and improve survival [22–25]. Although a corollary of the above hypothesis is that the overproduction of NO is due mainly to iNOS [15, 16], to date animal and patient studies have been limited to the use of non-selective inhibitors which block both iNOS and cNOS.

Several lines of evidence support the concept that NO is an important regulatory molecule during septic shock and is at least one of the determinants of vascular tone in this syndrome. Septic shock in animal models [26, 27] and in patients [28, 29] is associated with increased production of NO. Moreover, *in vitro* and *in vivo*, endotoxin and proinflammatory cytokines such as IL-1β, TNF-α, and IFN-γ induce iNOS in a variety of cells, including phagocytes [30, 31], hepatocytes [14], and endothelial [9] and vascular smooth muscle cells [10, 32]. Notably, the cell types capable of making a fully functional iNOS enzyme, and thus high-output NO production, appear to be species-specific [10, 33], a point that will be discussed more fully later.

Rationale for Using NOS Inhibitors in Septic Shock

Importantly, NOS inhibitors increase mean arterial pressure in animals with endotoxin- [24], TNF-α- [23], and IL-1β- [34] induced hypotension, and in patients with septic shock [17, 18]. At first, these inhibitors were reported to have a greater effect on blood pressure in septic than normal animals [23, 25]. This finding was taken as *prima facie* evidence that NO was the major endogenous vasodilator in septic shock. Therefore, the administration of NOS inhibitors would specifically treat the cause of this hypotension. Unlike conventional vasopressors, these agents would correct or normalize the vasodilatory state with little potential to cause hypertension or other untoward cardiovascular effects. This would have distinct advantages over catecholamine vasopressors, which are potentially dangerous and require sophisticated monitoring and precise titration. Further, NOS inhibitors appear to restore the depressed responsiveness of the vasculature in sepsis to catecholamines [35–37]. Early investigations suggested that these agents were nontoxic in normal animals with a therapeutic-to-toxic dose ratio >1/10 [25]. Lastly, the L-arginine analogs appeared to be competitive inhibitors and their effects could therefore be reversed, if necessary, by L-arginine [23–25].

In addition to reversing hypotension, NOS inhibitors may have other potentially beneficial effects in septic shock. An increasing body of literature has shown that the myocytes of a number of mammalian species can be induced to release NO [38]. Further, N^G-monomethyl-L-arginine (L-NMMA), a NOS inhibitor, has been shown to block the negative inotropic effects of TNF-α, IL-6, and IL-2 on isolated hamster papillary muscle preparations [39]. Another study, using N^ω-nitro-L-arginine methyl ester (L-NAME) and L-NMMA, found that these NOS inhibitors partially reversed the reduced contractility of ventricular myocytes harvested from endotoxin-challenged guinea pigs [40]. Collectively, these and other *in vitro* experiments suggest that NO may be the elusive myocardial depressant factor of septic shock. Further, NO has well-described cytotoxicity [41, 42] that could contribute to the tissue injury and multiple organ failure that, like vasodilatory shock, has been associated with death from sepsis [43]. Finally, NO has been shown *in vitro* to enhance cytokine production from phagocytes under certain conditions, a proinflammatory effect that could potentially worsen septic shock [44]. However, the overall effect on the inflammatory response to sepsis of blocking NO remains unknown. This issue will be discussed further in a later section.

Ultimately, proof that NO contributes to the pathogenesis of septic shock and that NOS inhibition has therapeutic efficacy awaits controlled, randomized clinical trials. However, an extensive number of preclinical animal studies have been conducted with non-selective NOS inhibitors. Most have used endotoxin or cytokine challenges rather than live bacteria and few have used mortality as an endpoint. Studies in rats [22] and dogs [23–25] indicated that NOS inhibitors could raise mean arterial pressure (MAP) in animals challenged with endotoxin, TNF-α, or IL-1β. The canine studies suggested that the effect of these agents was greater in septic than in normal animals. In a subsequent study, L-NMMA was reported to prevent the development of a premorbid state in dogs given high doses of IL-2 daily for 5 days [34]. Several well-designed studies of L-NAME administration to sheep challenged with a continuous infusion of low-dose endotoxin demonstrated that NOS inhibition raised MAP, normalized the endotoxin-induced hyperdynamic state, and reversed the increased organ blood flow without impairing oxygen consumption [45–47]. Mortality was not an endpoint in these investigations, so the impact of this intervention on survival could not be determined.

The only data showing that NOS inhibitors improve survival in septic shock are in rodents. Nava and colleagues [15] demonstrated that a moderate bolus dose of L-NMMA (30 mg/kg) given 90 min after endotoxin challenge in rats reversed hypotension and increased the number of animals surviving, but this latter effect did not reach statistical significance. In contrast, high doses of L-NMMA (300 mg/kg) precipitated hypotension, cardiovascular collapse, and death, effects that could be prevented by the co-administration of a NO donor [15]. These results suggested that complete blockade of NO production by both cNOS and iNOS with large doses of non-selective inhibitors could be harmful. Subsequent to these studies Teale and Atkinson [48] reported that inhibition of NOS improved survival in a mouse model of peritonitis [48]. This study had a number of important features. It employed an actual bacterial infection. Treatment with antibiotics and L-NMMA was begun late (7 h post-infection), somewhat analogous to what might occur clinically. Neither treatment alone improved outcome, but together they increased survival from 0 to 100% with increasing doses of L-NMMA up to 300 mg/kg. These experiments suggest that NOS inhibitors might be beneficial in septic shock and lend support to the ongoing clinical trials of these compounds. An attempt to reproduce these favorable results in mice challenged with either intraperitoneal or intravenous live *E. coli* was unsuccessful [49]. These authors suggested that the strain of mouse or differences in the bacteria used might have accounted for the different results.

NOS Inhibitors in Septic Shock: Areas of Controversy

Re-examining the NO Concept

Results from a number of investigations have challenged some of the assumptions inherent in the original hypothesis that NO is the major vasodilator in septic shock and that inhibition of its production will be therapeutically useful (Table 1) [8, 16, 26, 27, 37, 50–78]. Although NO production rises in sepsis, the concept that NO made by iNOS is the main endogenous vasodilator driving hypotension during an entire episode of septic shock deserves further scrutiny. The finding in several early studies that non-selective NOS inhibitors have greater effects in septic than in normal animals [23–25, 34] has not been confirmed by other investigators [27, 53–58]. This suggests that NO is not a more important determinant of vascular tone during sepsis than under homeostatic conditions. Further, other mediators with vasodilatory effects also rise in septic shock, such as eicosanoid products and platelet activating factor (PAF) [51]. A number of investigators have suggested a role

Table 1. Potential problems with the use of NOS inhibitors in septic shock

	Concern	References
1)	Vasodilators other than NO may be important.	[50–53]
2)	This therapy may not truly be specific as the effects of non-selective NO inhibitors are not greater in septic than in normal animals.	[27, 53–58]
3)	The therapeutic-to-toxic dose ratio is narrower than estimated from studies in normal animals. Septic compared to normal animals are harmed by lower doses of NOS inhibitors.	[27, 54]
4)	Both cNOS (early) and iNOS (late) may contribute to the regulation of vascular tone during septic shock.	[16, 37]
5)	Under certain conditions, NOS inhibitors are not competitive or reversible by L-arginine administration.	[54, 59, 60]
6)	NO production by iNOS in human vascular smooth muscle may be a regulated process. Control mechanisms include substrate (L-arginine) supply, cofactor (H_4 biopterin) production by the overlying endothelium, and NO inactivation of iNOS.	[8, 61–65]
7)	*In vivo*, NO may largely be cytoprotective and antiinflammatory.	[66–72]
8)	NO in septic shock may maintain microcirculatory blood flow by opposing the actions of sepsis-released vasoconstrictors such as thromboxane and endothelin-1 and by blocking platelet aggregation and leukocyte adhesion.	[55, 73–75]
9)	Non-selective NOS inhibitors may be harmful in sepsis.	[26, 27, 53–56, 74–78]

for vasodilators other than NO in this syndrome [52, 53], at least one of which, not yet chemically identified, appears to function via soluble guanylate cyclase [50].

It also appears likely that during the early stages of septic shock, cNOS activation by histamine, bradykinin, C5a, and other cell-signaling molecules plays a role in the vasodilatory response [16, 37]. Thus, iNOS *per se* may not be the major cause of hypotension throughout a septic event. Therefore, inhibitors that are highly selective for iNOS may not be effective in the early stages of septic shock.

Most animal studies employing NOS inhibitors to investigate the role of NO have examined time periods that range from immediately after the septic challenge to about 8 h. A few studies extend over one to two days. However, some patients with septic shock may have hypotension that requires vasopressor therapy over days. The role of iNOS and even NO itself in the late stages of septic shock is not completely clear.

In humans, septic mediators downregulate cNOS in the endothelium and upregulate tetrahydrobiopterin (H_4 biopterin) production, an essential cofactor required for NOS activity. This leads to an initial rise in endothelial NO production, followed by a decrease. In contrast to rodents, the calcium-independent enzyme, iNOS, is not induced [10, 65, 79]. Although iNOS is induced in the underlying vascular smooth muscle [10], resulting in an increase in total NO production, this enzyme is probably not fully activated and uncontrolled. Rather, NO production by vascular smooth muscle is regulated by substrate (L-arginine) supply [8, 64], and H_4 biopterin production by the endothelium [65, 79]. In support of this model, investigators have shown in septic animals [58, 77] and patients [17] that L-arginine infusion causes a further decrease in blood pressure, suggesting that substrate supply is a limiting factor in NO production. It has also been noted that H_4 biopterin production by endothelium can be upregulated 70-fold under experimental conditions [65]. In addition to these control mechanisms, both NO itself [61] and non-NO [65] mechanisms [62] have been shown to inactivate iNOS, thus representing potentially important negative feedback loops. Further, the synthesis of naturally-occurring inhibitors of NOS [63] and possibly development of tachyphylaxis after prolonged exposure to NO may control NO production or the effects of NO. Collectively, these data suggest that the pattern of NO production and its role in septic shock may change over time. Therefore, the pharmacologic manipulation of this pathway during septic shock may produce different consequences depending upon when the intervention takes place.

Effects of NO on the Inflammatory Response to Sepsis

NO has been shown to have many immunoregulatory functions (Table 2) [27, 44, 69, 70–72, 80–114]. However, the role of this messenger in modulating inflammatory responses during septic shock is poorly understood and very controversial. Many of the studies shown in Table 2, some of which have yielded conflicting results, infer their conclusions from the actions of NO donors and NOS inhibitors. As with other experiments which rely on chemical agonists and antagonists, unexpected non-specific effects and toxicities may confound the results [115, 117]. Despite an expanding foundation of basic research, it is still unclear in septic shock whether NO is primarily cytotoxic or cytoprotective, proinflammatory or antiinflammatory.

A large number of investigations into the immunobiology of NO have utilized rodents or rodent cells. As already noted, rodent models of sepsis have also been used to study NOS inhibitors and are the only experiments so far to show a definite survival benefit [15, 48]. Mouse and rat phagocytic cells can be induced by inflammatory mediators to make iNOS and, under these conditions, are capable of producing NO at very high rates [118, 119]. High-output NO production by rodent monocytes, macrophages, and neutrophils appears to be an essential antitumor [120] and antimicrobial defense system [121–124]. In mice and rats, NO is an important host defense, effector molecule in infections caused by intracellular pathogens such as Listeria monocytogenes [121, 122], malaria [123], and mycobacteria, and certain viral pathogens [124]. Although the NO pathway is activated in rodent phagocytes by common bacterial pathogens, the precise role of this product in rodent septic shock *per se* is less clear.

Table 2. Some effects of NO on the inflammatory response

Cell or Tissue	Effect	Species	Comment	Ref.
Neutrophil	NO inhibits chemotaxis, degranulation, O_2^- production and LTB4 release	Human	Non-specific effects of NO donors possible	[80]
	NO inhibits O_2^- production	Human	Demonstrated in whole cells and a cell-free system	[81]
	L-NMMA inhibits chemotaxis suggesting that NO modulates this function. Microbial killing of *S. aureus* was unaffected.	Human	L-arginine or dibutyryl cGMP reversed inhibition	[82, 83]
	NO activates the hexose monophosphate shunt (HMPS)	Human	NO nitrosylates glutathione. The HMPS turns on to regenerate it.	[84]
	NO augments degranulation	Human	Calcium ionophore and FMLP stimulated degranulation was increased by L-arginine and inhibited by LY-83583, a soluble guanylate cyclase blocking agent	[85]
	NO stimulates ADP-ribosylation of actin	Human	May inhibit F-actin assembly	[86]
	NO inhibits leukocyte (neutrophil) adhesion	Cat Human Rat	Mechanism appears to involve inhibition of O_2^- mediated adhesion	[87–89]
	NO scavenges O_2^-	Rat	See text	[90]
	NO increases TNF-α and IL-8 release from neutrophil preparations	Human	(See Fig. 3.) Mechanism may involve an interaction with a small number of contaminating monocytes (<2.0%).	[44, 91]

Table 2. (continued)

Cell or Tissue	Effect	Species	Comment	Ref.
Mast cells	NO inhibits degranulation, histamine release and H_2O_2 formation	Rat	May relate to mechanism by which NO blocks leukocyte adhesion to the endothelium	[72, 89, 92]
Monocytes and macrophages	NO increases cyclooxygenase activity and PGE_2 production	Rat (RAW264.7) (Kupffer cells)	In other systems, NO appears to depress eicosanoid production	[93, 94]
	NO decreases cyclooxygenase activity and decreases PGE_2 and TXB_2 production	Rat (Kupffer cells)	Effects inferred by the use of L-NMMA	[95]
	NO inhibits monocyte chemotaxis	Human	Used NO donors	[96]
	NO increases cytoplasmic motility	Rat (alveolar macrophages)	Inhibited by L-NMMA	[97]
	NO inhibits macrophage Ia expression	Mouse	Ia expression critical in T cell-macrophage interactions	[98]
	NO causes monocytic differentiation of HL-60 cells	Human (leukemia cell line)	Used NO donors	[99]
	NO induces or enhances TNF-α release from peripheral blood mononuclear cells	Human	Unclear whether target cells were monocytes or lymphocytes	[101, 102]
Endothelium	NO inhibits platelet aggregation and leukocyte adhesion	Human Cat Rat	May protect the endothelium in septic shock	[72, 87–90]
	NO scavenges and/or inhibits the formation of O_2^-	Human Rat	Reduces oxidative stress on the vasculature	[72, 88–90]
	NO decreases the permeability of endothelial cell monolayers	Bovine (aortic endothelial cells)	May reduce permeability by relaxing endothelial cells	[103]
	NO attenuates the biopathogenicity of homocysteine	Bovine (aortic endothelial cells)	Elevation of homocysteine is a risk factor for atherosclerosis	[71]

Microvasculature	NO decreases vasculature permeability	Cats Rats	L-NAME increased vascular permeability in mesenteric circulation of normal cats and in the coronary circulation of PAF or endothelin-1 challenged rats	[104, 105]
	NO increases vascular permeability	Guinea pig	L-NAME injected into skin sites challenged with bradykinin or histamine	[106]
Lymphocytes	NO inhibits lymphocyte proliferation	Rat Mouse	Under certain conditions, NO may depress cell mediated immunity	[107–109]
Hepatocyte	NO is cytoprotective	Dog Mouse	L-NMMA increases hepatic damage in LPS and *C. parvum* challenged animals	[27, 69, 70]
Myocardium	NO reduces myocardial reperfusion injury	Cat Dog	Mechanism may involve O_2^- scavenging and inhibition of leukocyte adhesion	[110, 111]
	NO contributes to reoxygenation injury	Piglet	L-NAME blocked injury	[112]
	NO production increases in cardiac allografts but L-NMMA does not affect graft rejection	Mouse	NO not essential for graft rejection	[113]
Brain	Brain NOS (bNOS) increases injury from cerebral ischemia but endothelial cNOS activity is protective	Mouse	Experiment used mutant mice deficient in bNOS	[114]

In contrast to rodents, the ability of human phagocytic cells to produce NO in appreciable quantities is controversial. Several investigators have reported that human macrophages can be induced to make NO when exposed to certain cytokines [125, 126] or viral antigens [127], or when infected with intracellular pathogens [128]. Other studies have detected iNOS and cNOS mRNA in unstimulated or stimulated human macrophages after amplification by reverse transcription polymerase chain reaction [129]. However, it appears that small quantities of iNOS mRNA can be found in apoptotic human monocytes/macrophages [129, 130], even though detectable quantities of functional NOS are not made. Overall, human monocytes/macrophages have not been demonstrated to produce NO using biochemical or enzymatic methods that easily detect NOS activity in human endothelial cells and activated rodent phagocytes [33].

A similar controversy surrounds NO production by human polymorphonuclear leukocytes (PMN). Many reports suggest that human PMN produce NO [82, 83, 85, 131–135]. However, much of this data differs significantly from the findings in rodents and many positive studies are inconsistent with each other. Unlike rodent neutrophils, human PMN NOS appears to be a constitutive enzyme [131, 132], but, unlike any other cNOS described, it appears to be calcium- and calmodulin-independent [132, 133]. Adding to the confusion, experiments from other laboratories demonstrate that human PMN make NO only after exposure to calcium ionophore [85] or chemotactic peptides [134, 135], which would indicate that the enzyme is calcium-dependent. As shown in Table 2 some studies have suggested that in human PMN NO functions as a second messenger in degranulation, chemotaxis, and regulation of superoxide (O_2^-) production [82, 83, 85].

In our laboratory, using cGMP assays, nitrite and nitrate determinations, and a radiolabelled L-arginine to L-citrulline conversion method to detect NOS activity enzymatically, we have been unable to demonstrate that human PMN have the ability to produce NO [44, 136]. Although it appears that human PMN contain a nitric oxide-sensitive guanylate cyclase, exposure to inflammatory mediators or a variety of phagocyte activators did not increase cGMP levels [44]. However, incubation of PMN with NO donors causes a prompt rise in intracellular cGMP, indicating that these cells can respond in a paracrine manner to a NO signal (Fig. 1). Neither the cytosol nor the membrane fraction of human PMN showed inhibitable NOS activity [136] using the L-arginine conversion assay (Fig. 2). Notably, NOS activity in human and bovine endothelial cells, which produce much less NO than induced rodent phagocytes, was easily detectable [136]. Further, *ex vivo* treatment with endotoxin, TNF-α, IL-1β, and IFN-γ, or *in vivo* exposure to endotoxin using normal volunteers (Table 3) failed to induce detectable NOS activity in human PMN.

Thus, human neutrophils could not be shown to contain measurable cNOS activity or a high output, inducible NO-generating system analogous to that seen in rodents. However, as noted above, human neutrophils can receive a NO signal [44]. During septic shock, human mononuclear cells and PMN are

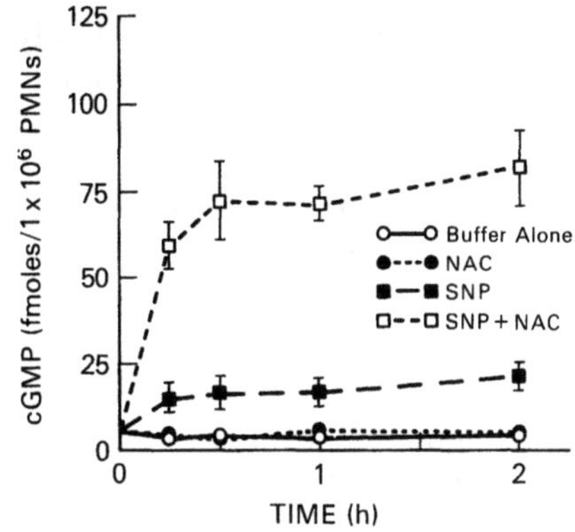

Fig. 1. Effect of sodium nitroprusside (SNP) and N-acetylcysteine (NAC) on human PMN cGMP. (Adapted from [44] with permission)

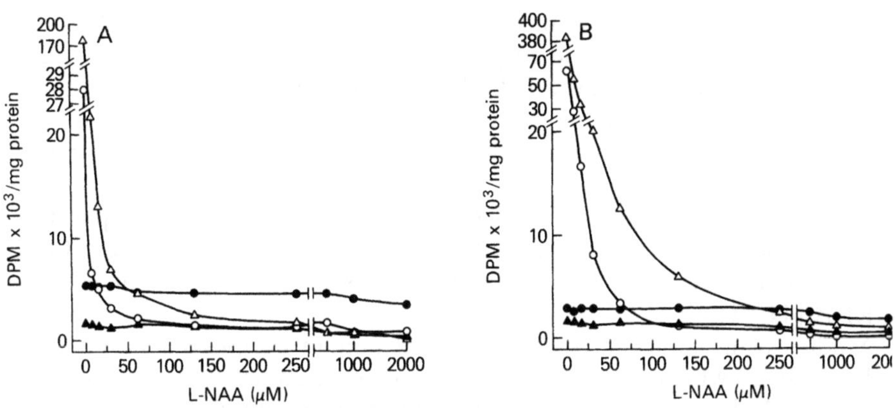

Fig. 2. Effect of N$^{\omega}$-amino-L-arginine (L-NAA), a NOS inhibitor, on the conversion of radiolabelled L-arginine to L-citrulline in (**A**) cytosol and (**B**) membrane fractions of human PMN (●), human aortic endothelial cells (○), and bovine pulmonary artery endothelial cells (△). Bovine serum albumin (▲) in the absence of cellular material was used as a control. (From [136] with permission)

Table 3. L-arginine to L-citrulline conversion in human PMN isolated from volunteers challenged with I.V. LPS[a]

Treatment	Before I.V. LPS	6 h after I.V. LPS[b]
Medium alone	3.39 ± 0.47	2.63 ± 0.24
L-NAA (1 mM)[c]	2.29 ± 0.23	1.88 ± 0.19

a Results are expressed as pmol/h/10^6 cells of citrulline (mean ± SE, n = 4)
b LPS, 4 ng/kg body weight; I.V., intravenous
c N$^{\omega}$-amino-L-arginine (L-NAA) was added to whole cells *ex vivo* before the L-arginine to L-citrulline conversion

undoubtedly exposed to relatively high concentrations of NO as they contact and cross the vasculature. The nature of the message delivered by NO under these conditions is not entirely clear, but answering this question is important to understanding the potential effect of NOS inhibitors on the inflammatory response and tissue injury in septic shock.

In the laboratory, NO has been shown to have direct cytotoxic effects [41, 42] and may augment certain phagocytic cell responses [82, 83, 85]. When NO and (O_2^-) interact, peroxynitrite ($OONO^-$), an extremely toxic molecule, is formed. However, other evidence suggests that NO protects cells by scavenging O_2^- (Table 2) [87–90]. It has been suggested that NO may divert O_2^- toward detoxifying sulfhydryl groups, the preferred target of $OONO^-$, and away from the metal centers of metabolically important enzymes [137].

In contrast to the conflicting studies listed in Table 2 that suggest NO either inhibits or augments several basic functions of phagocytic cells [80, 82, 83, 85, 93–97], we have been unable to show that NO donors or NOS inhibitors alter PMN chemotactic responses [138] or stimulated PGE_2 release (unpublished data). However, LPS-dependent TNF-α production by human PMN preparations is increased by incubation with sodium nitroprusside, a NO donor, and further augmented by addition of N-acetyl-cysteine, an agent which increases the bioavailability of NO (Fig. 3) [44]. In addition, NO has a similar effect on stimulated IL-8 release from human PMN preparations [91]. Although the mechanism of these effects are unknown, they do not appear to be mediated by cGMP or increased transcription of cytokine mRNA [44, 91].

Studies using adherent freshly isolated human peripheral mononuclear cells (PMBC) which are a mixture of lymphocytes, monocytes, and contaminating PMN, have also found that NO enhances or stimulates TNF-α secretion [101, 102]. Notably, this effect was not simulated by cGMP analogs. In contrast, using PMBC cultured for 14 days, which should primarily be differentiated macrophages, we found no effect of NO donors on LPS-elicited TNF-α production [44]. These studies suggest that NO, under specific conditions, does have the ability to augment the release of cytokines and there-

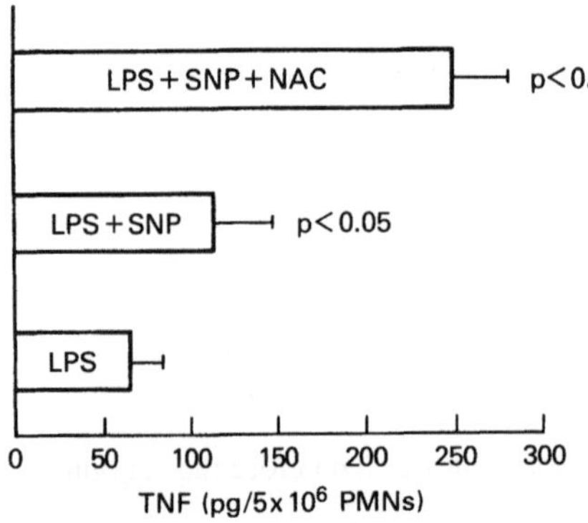

Fig. 3. Effect of sodium nitroprusside (SNP) and N-acetylcysteine (NAC) on LPS-induced TNF-α production by human PMN. (Adapted from [44] with permission)

by contribute to the systemic inflammatory response during septic shock. However, *in vivo* experiments have not shown that NOS inhibitors alter circulating cytokine levels in septic animals [27].

Although it has been demonstrated under experimental conditions that NO can be cytotoxic [41, 42] and that rodents use NO as an antitumor and antimicrobial effector molecule [120–124], in humans these actions of NO may be less relevant. As shown in Table 2, NO also has antiinflammatory and cytoprotective effects that may be important during septic shock in humans. NO has been shown to scavenge O_2^- [72, 88–90], block platelet aggregation [66], inhibit leukocyte adhesion to vascular endothelium [72, 87–90], stabilize mast cell membranes [72, 89, 92] and, in reperfusion and ischemia injury models, to decrease myocardial [110, 111] and brain damage [114]. Further, the cytoprotective properties of NO in septic shock are suggested by the finding that inhibition of hepatocyte iNOS using L-NMMA increases hepatic damage in septic animal models [27, 69, 70].

The ability of NO to dilate the microvasculature, block platelet aggregation, and inhibit leukocyte adhesion [66, 72, 87–90] suggests that increased NO production in septic shock might play a role in maintaining visceral and microvascular blood flow as well as protecting the endothelium from damage by activated leukocytes. In support of this concept, L-NAME administration to rabbits given PMN activating agents was shown to increase accumulation of platelets and neutrophils in the pulmonary circulation [139]. Further, *in vitro* and *in vivo* NO appears to decrease endothelial permeability and attenuate capillary leak [103–105]. However, NO increases extravascular edema in guinea pig skin after local injections of bradykinin and histamine (Table 2) [106].

Even though NO can be shown to have effects that are arguably proinflammatory, the net effect *in vivo* of blocking NO production is uncertain. By analogy, cyclooxygenase inhibitors in humans increase endotoxin-stimulated cytokine production *in vivo* by climinating an important negative feedback loop [140]. However, the net effect of these agents on the clinical manifestations and inflammatory response to endotoxin is antiinflammatory. The ability of NO to inhibit leukocyte adhesion and protect the endothelium is a potentially important antiinflammatory effect that may override its ability to augment or stimulate inflammation. This area of controversy has major implications for the clinical use of NOS inhibitors.

Preclinical Studies Demonstrating Harmful Effects of NOS Inhibition

Despite initial enthusiasm for using NOS inhibitors to test hypotension in septic shock, a number of studies have shown these agents to be harmful. In endotoxin- and PAF-challenged rats, L-NMMA and L-NAME, respectively, increase capillary leak and intestinal damage [67, 78]. Using videomicroscopy in a high cardiac output (CO) rat bacteremia model, Spain and colleagues [55] demonstrated that L-NAME normalized CO and increased MAP (+34%), but in the kidney, it exacerbated sepsis-induced interlobular artery and afferent arteriole constriction and further impaired blood flow (−90%) [55]. Several studies in endotoxin-challenged dogs have shown that NOS inhibitors in this hypodynamic shock model raised MAP, but further decreased cardiac index (CI), oxygen delivery index (DO_2I), and, in some cases, oxygen consumption index (VO_2I) [24, 27, 53, 54, 74, 76]. In a hyperdynamic pig model of endotoxin shock, L-NAME increased MAP, but depressed CI below control values in normal animals and exacerbated endotoxin-induced pulmonary hypertension [56]. Interestingly, L-NAME had no effect on mean pulmonary arterial pressure (MPAP) in normal animals, suggesting that NO production in the pulmonary vasculature may be inducible and partially compensates for the release of endotoxin-induced vasoconstrictors [56]. Five studies examining survival in endotoxin-challenged rabbits [75, 77], mice [26], and dogs [27, 54] found that 4 different non-selective NOS inhibitors all increased mortality rates.

The potential for NOS inhibitors to cause harm is illustrated by 2 studies in awake dogs conducted in our laboratory [27, 54]. In the first study, N^{ω}-amino-L-arginine (L-NAA) was infused for 22 h into normal and endotoxemic dogs (n = 23) at 1 or 10 mg/kg/h. L-NAA increased systemic vascular resistance index (SVRI), but decreased CI, DO_2I, and VO_2I and worsened lactic acidosis (Fig. 4). Unexpectedly, L-NAA decreased survival time in this model even at doses that did not significantly increase MAP. Surprisingly, this harmful effect on survival was not reversed by the administration of L-arginine. Further, normal dogs given the higher dose of L-NAA (10 mg/kg/h) developed muscular rigidity and seizure-like activity. It was therefore unclear whether the decreased survival time seen in endotoxemic animals was a direct effect of NOS inhibition or due to an idiosyncratic toxicity limited to L-NAA [54].

In an attempt to resolve this issue, a larger study (n = 66) was done using L-NMMA, which appeared to lack the neurotoxicity of L-NAA [27]. A similar design was used and a range of L-NMA doses was employed (1, 2, 4, and 10 mg/kg/h). The effects of L-NMMA on SVRI, CI, and DO_2I were similar to those in the L-NAA study, but VO_2I was not depressed. These cardiovascular effects of both L-NAA and L-NMA are similar to those reported with NOS inhibitors in other animal models [23, 24, 45, 46, 53, 56, 76] and in patients [17, 18] with sepsis. Comparing endotoxemic to normal dogs, L-NMMA had an increased effect in the venous and pulmonary vascular beds 6 h after endotoxin, suggesting that NO production was induced in these low resistance vessels (Fig. 5), a result which has been reported by others [16, 56, 141]. In contrast, the effect of L-NMMA on MAP was similar comparing endotoxemic and normal animals [27]. This differs from initial reports, also in endotoxemic dogs, that NOS inhibitors have a greater effect in septic compared to normal animals, which was taken as evidence that NO was a major cause of vasodilation in septic shock [23, 24]. However, the L-NAA study described above [54] and five other studies in rats [55], dogs [53], pigs [56], and sheep

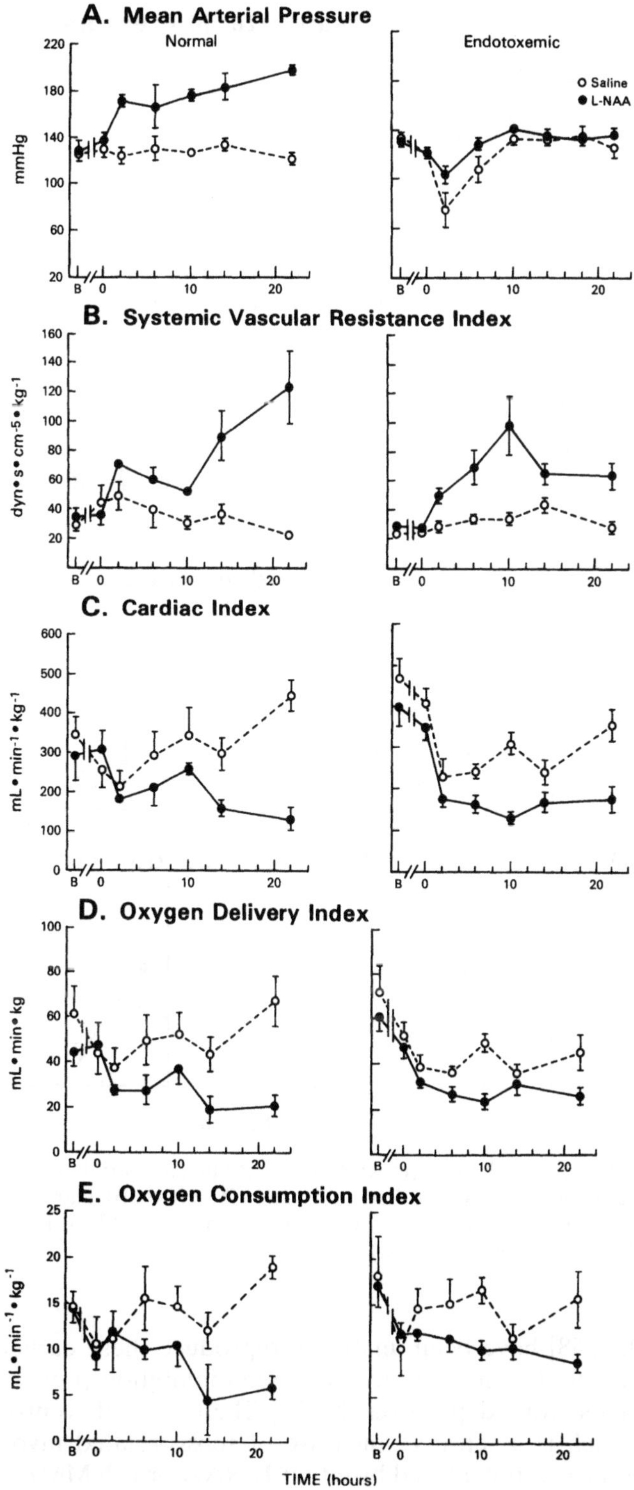

Fig. 4. Serial changes (mean ± SE) in (**A**) mean arterial pressure, (**B**) systemic vascular resistance index, (**C**) cardiac index, (**D**) oxygen delivery index, and (**E**) oxygen consumption index in normal and endotoxemic dogs treated with either saline (○- - -○) or N^{ω}-amino-L-arginine (L-NAA; ●- - -●). (Adapted from [54] with permission)

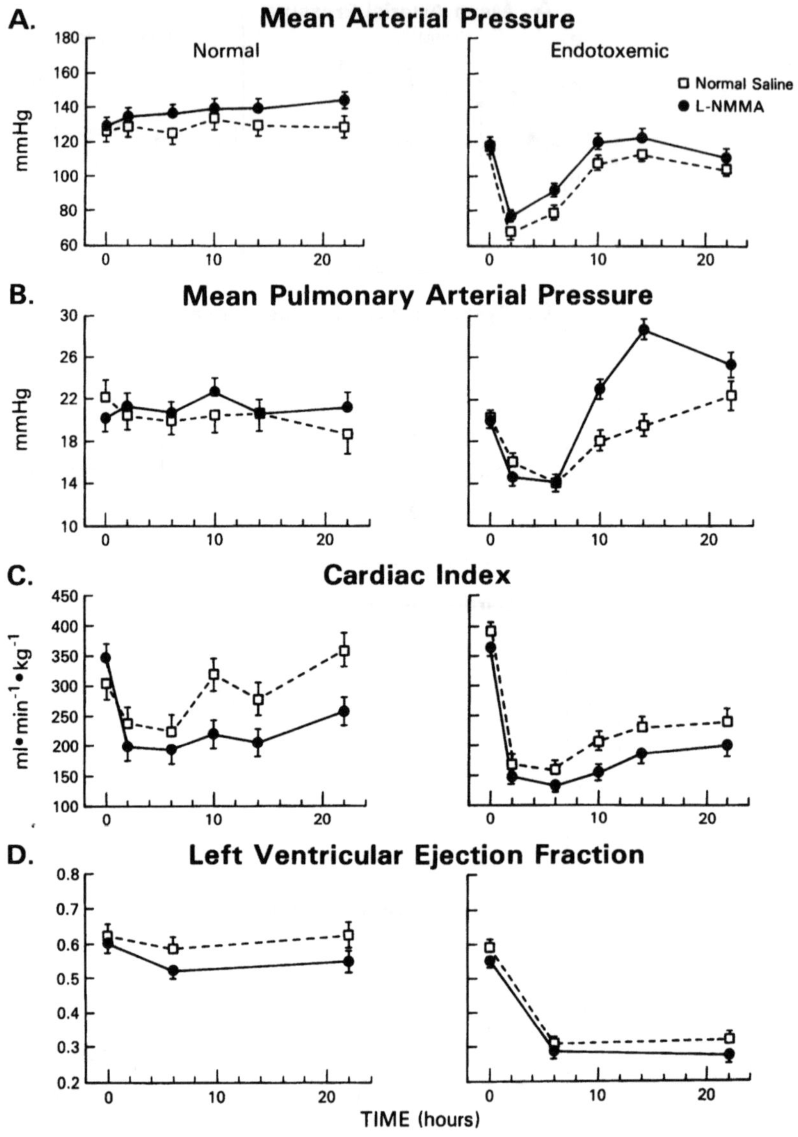

Fig. 5. Serial changes (mean ± SE) in (**A**) mean arterial pressure, (**B**) mean pulmonary arterial pressure, (**C**) cardiac index, and (**D**) left ventricular ejection fraction in normal and endotoxemic dogs treated with either saline (□---□) or L-NMMA (●---●). (Adapted from [54] with permission)

[57, 58] have been unable to reproduce these earlier results. We were also unable to show *in vivo* that NOS inhibition prevented endotoxin-induced myocardial depression [27, 54]. The fall in left ventricular ejection fraction, which is a sensitive measure of sepsis-related myocardial depression [21, 142], was not altered by either L-NAA or L-NMMA.

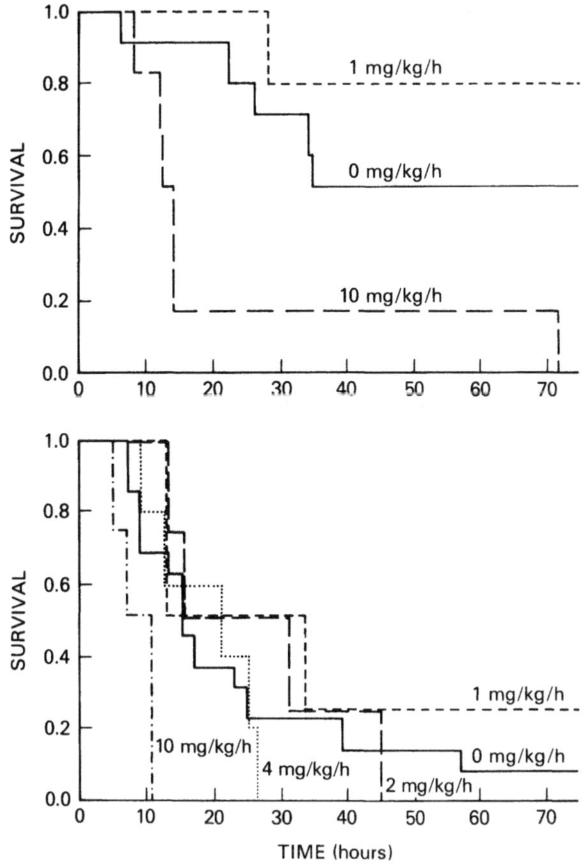

Fig. 6. Survival vs. time in dogs challenged with 8 mg/kg (top) and 16 mg/kg (bottom) of endotoxin and treated with saline (——), or L-NMMA at 1, 2, 4, or 10 mg/kg/h. (Adapted from [27] with permission)

Despite the absence of neurotoxicity in our L-NMMA study, this agent did not significantly improve outcome at any of the doses examined, and at the highest dose (10 mg/kg/h) survival time was decreased ($p<0.05$; Fig. 6). The reason why L-NMMA worsened outcome was not clear, but it was not explained by higher circulating levels of endotoxin or TNF-α [27]. Endotoxin-induced liver injury was worsened by L-NMMA, as has been reported by other groups [69, 70], but this did not seem severe enough to decrease survival times. Interestingly, survival rates and times were numerically, but not significantly, higher in animals treated with 1 mg/kg/h of L-NMMA, a dose that had minimal effects on MAP and on endotoxin-induced increases in serum nitrite/nitrate concentrations [27]. Further investigation of low-dose NOS inhibition in septic shock is warranted.

As reviewed earlier in this chapter, some preclinical results suggested that non-selective NOS inhibitors may be therapeutically useful in septic shock

[22–25]. Further, a more recent study in mice infected with live *E. coli* showed improved survival in animals treated relatively late (7 h) with both antibiotics and L-NMMA [48]. However, this favorable result could not be reproduced [49] and many other investigations in septic animal models have shown a disturbing tendency of NOS inhibitors in multiple species to cause harm [26, 27, 53–56, 74–78]. Although these data are compelling, it is important to point out some deficiencies.

First, virtually all of these studies, like many of those showing beneficial effects, used endotoxin challenge models of sepsis. These models vary in their ability to simulate the pathophysiology of human septic shock and, since an actual infection is not present, they may not accurately reflect the impact of NOS inhibition on host defenses. Second, most studies have used general anesthesia and positive pressure ventilation both of which have profound effects on the cardiovascular system. Notably, a study in anesthetized sheep demonstrated that halothane depressed the ability of L-NAME to raise MAP, but potentiated its myocardial depressant effects [143]. Conversely, urethane/α-chloralose- and pentobarbital-anesthetized rats, compared to conscious rats, had greater increases in blood pressure when given L-NMMA [144]. Interestingly, the pressor response to phenylephrine in this study was not different between groups, suggesting the presence of a specific interaction. Next, it is possible that the timing of administration or dose of NOS- inhibitor used in these studies precluded the demonstration of benefit. As described above, we found a non-significant trend toward benefit in endotoxin-challenged dogs treated with a very low dose of L-NMMA that had minimal effects on MAP [27]. Finally, all of these studies have used non-selective NOS inhibitors that block both cNOS and iNOS. Overall, the soundest conclusion from these preclinical studies is that additional work is necessary.

Conclusion

The importance and precise role of NO in regulating vascular tone and modulating inflammatory responses in septic shock is unclear. Depending on the experimental conditions and species, NO can demonstrate antiinflammatory or proinflammatory properties and can be cytoprotective or cytotoxic. Our data in endotoxin-challenged dogs, together with work from many other laboratories, suggest that non-selective NOS inhibitors may be deleterious in septic shock, particularly when cardiac output is low, as it is in up to 20% of patients with this syndrome. Even if non-selective NOS inhibitors are therapeutically useful in sepsis, it seems apparent that the proper dose and timing of administration are not well understood. Future efforts should be directed at better defining the effects of NO on host defenses and the study of low dose NOS inhibitors in animal models which use an actual nidus of infection. It would also seem prudent to further examine the effects of augmenting the NO pathway in septic shock by administering substrate (L-arginine), cofactor(s) (H_4biopterin), or sulfhydryl groups (N-acetylcysteine). The development of isoform selective NOS inhibitors will help define the relative roles of cNOS and iNOS in septic shock and improve the likelihood that we will be able to pharmacologically control the NO pathway in a clinically beneficial manner.

References

1. Natanson C, Hoffman WD, Suffredini AF, Eichacker PQ, Danner RL (1994) Selected treatment strategies for septic shock based on proposed mechanisms of pathogenesis. Ann Intern Med 120:771–783
2. Moncada S, Palmer RM, Higgs EA (1991) Nitric oxide: Physiology, pathophysiology, and pharmacology. Pharmacol Rev 43:109–142
3. Nathan C (1992) Nitric oxide as a secretory product of mammalian cells. FASEB J 6: 3051–3064
4. Garthwaite J, Charles SL, Chess-Williams R (1988) Endothelium-derived relaxing factor release on activation of NMDA receptors suggests role as intercellular messenger in the brain. Nature 336:385–388
5. Palmer RM, Ferrige AG, Moncada S (1987) Nitric oxide release accounts for the biological activity of endothelium-derived relaxing factor. Nature 327:524–526
6. Vallance P, Collier J, Moncada S (1989) Effects of endothelium-derived nitric oxide on peripheral arteriolar tone in man. Lancet 2:997–1000
7. Forsterman U, Schmidt HH, Pollock JS, et al (1991) Isoforms of nitric oxide synthase: Characterization and purification from different cell types. Biochem Pharmacol 42: 1849–1857
8. Morris SM Jr, Billiar TR (1994) New insights into the regulation of inducible nitric oxide synthase. Am J Physiol 266 (Endocrinol Metab 29):E829–E839
9. Radomski MW, Palmer RMJ, Moncada S (1990) Glucocorticoids inhibit the expression of an inducible, but not the constitutive, nitric oxide synthase in vascular endothelial cells. Proc Natl Acad Sci USA 87:10043–10047
10. MacNaul KL, Hutchinson NI (1993) Differential expression of iNOS and cNOS mRNA in human vascular smooth muscle cells and endothelial cells under normal and inflammatory conditions. Biochem Biophys Res Commun 196:1330–1334
11. Lonchampt MO, Auguet M, Delaflotte S, Goulin-Schulz J, Chabrier PE, Braquet P (1992) Lipoteichoic acid: A new inducer of nitric oxide synthase. J Cardiovas Pharmacol 20 (Suppl 12):S145–S147
12. Goodrum KJ, McCormick LL, Schneider B (1994) Group B streptococcus-induced nitric oxide production in murine macrophages is CR3 (CD11b/CD18) dependent. Infect Immun 62:3102–3107
13. Zembowicz A, Vane JR (1992) Induction of nitric oxide synthase activity by toxic shock syndrome toxin 1 in a macrophage-monocyte cell line. Proc Natl Acad Sci USA 89:2051–2055
14. Geller DA, Nussler AK, Di Silvio M, et al (1992) Cytokines, endotoxin, and glucocorticoids regulate the expression of inducible nitric oxide synthase in hepatocytes. Cell Biol 90:522–526
15. Nava E, Palmer RM, Moncada S (1992) The role of nitric oxide in endotoxin shock: Effects of N^G-monomethyl-L-arginine. J Cardiovasc Pharmacol 20 (Suppl 12): S132–S134
16. Thiemermann C (1994) The role of the L-arginine: Nitric oxide pathway in circulatory shock. Advances in Pharmacology 28:45–79
17. Lorente JA, Landín L, De Pablo R, Renes E, Liste D (1993) L-arginine pathway in the sepsis syndrome. Crit Care Med 21:1287–1295
18. Petros A, Lamb G, Leone A, Moncada S, Bennett D, Vallance P (1994) Effects of a nitric oxide synthase inhibitor in humans with septic shock. Cardiovasc Res 28:34–39

19. Danner RL, Elin RJ, Hosseini JM, Wesley RA, Reilly JM, Parrillo JE (1991) Endotoxemia in human septic shock. Chest 99:169–175
20. Parker MM, Shelhamer JH, Natanson C, Alling DW, Parrillo JE (1987) Serial cardiovascular variables in survivors and nonsurvivors of human septic shock: Heart rate as an early predictor of prognosis. Crit Care Med 15:923–929
21. Parker MM, Shelhamer JH, Bacharach SL, et al (1984) Profound but reversible myocardial depression in patients with septic shock. Ann Intern Med 100:483–490
22. Thiemermann C, Vane J (1990) Inhibition of nitric oxide synthesis reduces the hypotension induced by bacterial lipopolysaccharides in the rat in vivo. Eur J Pharmacol 182:591–595
23. Kilbourn RG, Gross SS, Jubran A, et al (1990) N^G-methyl-L-arginine inhibits tumor necrosis factor-induced hypotension: Implications for the involvement of nitric oxide. Proc Natl Acad Sci USA 87:3629–3632
24. Kilbourn RG, Jubran A, Gross SS, et al (1990) Reversal of endotoxin-mediated shock by N^G-methyl-L-arginine, an inhibitor of nitric oxide synthesis. Biochem Biophys Res Commun 172:1132–1138
25. Kilbourn RG, Griffith OW (1992) Overproduction of nitric oxide in cytokine-mediated and septic shock. J Natl Cancer Inst 84:827–831
26. Minnard EA, Shou J, Naama H, Cech A, Gallagher H, Daly JM (1994) Inhibition of nitric oxide synthesis is detrimental during endotoxemia. Arch Surg 129:142–148
27. Cobb JP, Natanson C, Quezado ZMN, et al (1995) The differential hemodynamic effects of N^ω-monomethyl-L-arginine in endotoxemic and normal canines. Am J Physiol 268 (Heart Circ Physiol 37):H1634–H1642
28. Ochoa JB, Udekwu AO, Billiar TR, et al (1991) Nitrogen oxide levels in patients after trauma and during sepsis. Ann Surg 214:621–626
29. Shi Y, Li H, Shen C, et al (1993) Plasma nitric oxide levels in newborn infants with sepsis. J Pediatr 123:435–438
30. Stuehr DJ, Cho HJ, Kwon NS, Weise MF, Nathan CF (1991) Purification and characterization of the cytokine-induced macrophage nitric oxide synthase: An FAD- and FMN-containing flavoprotein. Proc Natl Acad Sci USA 88:7773–7777
31. Adler H, Peterhans E, Nicolet J, Jungi TW (1994) Inducible L-arginine-dependent nitric oxide synthase activity in bovine bone marrow-derived macrophages. Biochem Biophys Res Commun 198:510–515
32. Beasley D, Schwartz JH, Brenner BM (1991) Interleukin 1 induces prolonged L-arginine-dependent cyclic guanosine monophosphate and nitrite production in rat vascular smooth muscle cells. J Clin Invest 87:602–608
33. Schneeman M, Schoedon G, Hofer S, Blau N, Guerrero L, Schaffner A (1993) Nitric oxide synthase is not a constituent of the antimicrobial armature of human mononuclear phagocytes. J Infect Dis 167:1358–1363
34. Kilbourn RG, Gross SS, Lodato RF, et al (1992) Inhibition of interleukin-1-alpha-induced nitric oxide synthase in vascular smooth muscle and full reversal of interleukin-1-alpha-induced hypotension by N^ω-amino-L-arginine. J Natl Cancer Inst 84:1008–1016
35. Hollenberg SM, Cunnion RE, Zimmerberg J (1993) Nitric oxide synthase inhibition reverses arteriolar hyporesponsiveness to catecholamines in septic rats. Am J Physiol 264 (Heart Circ Physiol 33):H660–H663
36. Gray GA, Schott C, Julou-Schaeffer G, Fleming I, Parratt JR, Stoclet JC (1991) The effect of inhibitors of the L-arginine/nitric oxide pathways on endotoxin-induced loss of vascular responsiveness in anaesthetized rats. Br J Pharmacol 103:1218–1224
37. Szabó C, Mitchell JA, Thiemermann C, Vane JR (1993) Nitric oxide-mediated hyporeactivity to noradrenaline precedes the induction of nitric oxide synthase in endotoxin shock. Br J Pharmacol 108:786–792
38. Balligand JL, Ungureanu D, Kelly RA, et al (1993) Abnormal contractile function due to induction of nitric oxide synthesis in rat cardiac myocytes follows exposure to activated macrophage-conditioned medium. J Clin Invest 91:2314–2319

39. Finkel MS, Oddis CV, Jacob TD, Watkins SC, Hattler BG, Simmons RL (1992) Negative inotropic effects of cytokines on the heart mediated by nitric oxide. Science 257: 387–389
40. Brady AJB, Poole-Wilson PA, Harding SE, Warren JB (1992) Nitric oxide production within cardiac myocytes reduces their contractility in endotoxemia. Am J Physiol 263 (Heart Circ Physiol 32): H1963–H1966
41. Estrada C, Gómez C, Martin C, Moncada S, González C (1992) Nitric oxide mediates tumor necrosis factor-α cytotoxicity in endothelial cells. Biochem Biophys Res Commun 186: 475–482
42. Nguyen T, Brunson D, Crespi CL, Penman BW, Wishnok JS, Tannenbaum SR (1992) DNA damage and mutation in human cells exposed to nitric oxide in vitro. Biochemistry 89: 3030–3034
43. Parrillo JE, Parker MM, Natanson C, et al (1990) Septic shock in humans: Advances in the understanding of pathogenesis, cardiovascular dysfunction, and therapy. Ann Intern Med 113: 227–242
44. Van Dervort AL, Yan L, Madara PJ, et al (1994) Nitric oxide regulates endotoxin-induced TNF-α production by human neutrophils. J Immunol 152: 4102–4109
45. Meyer J, Traber LD, Nelson S, et al (1992) Reversal of hyperdynamic response to continuous endotoxin administration by inhibition of NO synthesis. J Appl Physiol 73: 324–328
46. Meyer J, Lentz CW, Stothert JC, Traber LD, Herndon DN, Traber DL (1994) Effects of nitric oxide synthesis inhibition in hyperdynamic endotoxemia. Crit Care Med 22: 306–312
47. Meyer J, Hinder F, Stothert J Jr, et al (1994) Increased organ blood flow in chronic endotoxemia is reversed by nitric oxide synthase inhibition. J Appl Physiol 76: 2785–2793
48. Teale DM, Atkinson AM (1992) Inhibition of nitric oxide synthesis improves survival in a murine peritonitis model of sepsis that is not cured by antibiotics alone. J Antimicrob Chemother 30: 839–842
49. Evans T, Carpenter A, Silva A, Cohen J (1994) Inhibition of nitric oxide synthase in experimental gram-negative sepsis. J Infect Dis 169: 343–349
50. Wu CC, Szabó C, Chen SJ, Thiemermann C, Vane JR (1994) Activation of soluble guanylyl cylase by a factor other than nitric oxide or carbon monoxide contributes to the vascular hyporeactivity to vasoconstrictor agents in the aorta of rats treated with endotoxin. Biochem Biophys Res Commun 201: 436–442
51. Yamanaka S, Iwao H, Yukimura T, Kim S, Miura K (1993) Effect of the platelet-activating factor antagonist, TCV-309, and the cyclo-oxygenase inhibitor, ibuprofen, on the haemodynamic changes in canine experimental endotoxic shock. Br J Pharmacol 110: 1501–1507
52. McCormack DG, Paterson NAM (1993) Loss of hypoxic pulmonary vasoconstriction in chronic pneumonia is not mediated by nitric oxide. Am J Physiol 265 (Heart Circ Physiol 34): H1523–H1528
53. Klabunde RE, Helgren MC (1992) Cardiovascular actions of N^G-methyl-L-arginine are abolished in a canine shock model using high-dose endotoxin. Biochem Biophys Res Commun 78: 57–68
54. Cobb JP, Natanson C, Hoffman WD, et al (1992) N^ω-amino-L-arginine, an inhibitor of nitric oxide synthase, raises vascular resistance but increases mortality rates in awake canines challenged with endotoxin. J Exp Med 176: 1175–1182
55. Spain DA, Wilson MA, Garrison RN (1994) Nitric oxide synthase inhibition exacerbates sepsis-induced renal hypoperfusion. Surgery 116: 322–331
56. Robertson FM, Offner PJ, Ciceri DP, Becker WK, Pruitt BA (1994) Detrimental hemodynamic effects of nitric oxide synthase inhibition in septic shock. Arch Surg 129: 149–156
57. Landín L, Lorente JA, Renes E, Cañas P, Jorge P, Liste D (1994) Inhibition of nitric oxide synthesis improves the vasoconstrictive effect of noradrenaline in sepsis. Chest 106: 250–256

58. Lorente JA, Landín L, Renes E, et al (1993) Role of nitric oxide in the hemodynamic changes of sepsis. Crit Care Med 21:759–767
59. Olken NM, Rusche KM, Richards MK, Marletta MA (1991) Inactivation of macrophage nitric oxide synthase activity by N^G-methyl-L-arginine. Biochem Biophys Res Commun 177:828–833
60. Dwyer MA, Bredt DS, Snyder SH (1991) Nitric oxide synthase: Irreversible inhibition by L-N^G-nitroarginine in brain in vitro and in vivo. Biochem Biophys Res Commun 176:1136–1141
61. Assreuy J, Cunha FQ, Liew FY, Moncada S (1993) Feedback inhibition of nitric oxide synthase activity by nitric oxide. Br J Pharmacol 108:833–837
62. Vodovotz Y, Kwon NS, Pospischil M, Manning J, Paik J, Nathan C (1994) Inactivation of nitric oxide synthase after prolonged incubation of mouse macrophages with IFN-γ and bacterial lipopolysaccharide. J Immunol 152:4110–4118
63. Kimoto M, Tsuji H, Ogawa T, Sasaoka K (1993) Detection of N^G,N^G-dimethylarginine dimethylaminohydrolase in the nitric oxide-generating systems of rats using monoclonal antibody. Arch Biochem Biophys 300:657–662
64. Ghigo D, Alessio P, Foco A, et al (1993) Nitric oxide synthesis is impaired in glutathione-depleted human umbilical vein endothelial cells. Am J Physiol 265 (Cell Physiol 34):C728–C732
65. Schoedon G, Schneemann M, Blau N, Edgell CJS, Schaffner A (1993) Modulation of human endothelial cell tetrahydrobiopterin synthesis by activating and deactivating cytokines: New perspectives on endothelium-derived relaxing factor. Biochem Biophys Res Commun 196:1343–1348
66. Cooke JP, Tsao PS (1993) Cytoprotective effects of nitric oxide. Circulation 88:2451–2454
67. MacKendrick W, Caplan M, Hsueh W (1993) Endogenous nitric oxide protects against platelet-activating factor-induced bowel injury in the rat. Pediatr Res 34:222–228
68. Weyrich AS, Ma X, Lefer AM (1992) The role of L-arginine in ameliorating reperfusion injury after myocardial ischemia in the cat. Circulation 86:279–288
69. Harbrecht BG, Billiar TR, Stadler J, et al (1992) Nitric oxide synthesis serves to reduce hepatic damage during acute murine endotoxemia. Crit Care Med 20:1568–1574
70. Harbrecht BG, Stadler J, Demetrius AJ, Simmons RL, Billiar TR (1994) Nitric oxide and prostaglandins interact to prevent hepatic damage during murine endotoxemia. Am J Physiol 266 (Gastrointest Liver Physiol 29):G1004–G1010
71. Stamler JS, Osborne JA, Jaraki O, et al (1993) Adverse vascular effects of homocysteine are modulated by endothelium-derived relaxing factor and related oxides of nitrogen. J Clin Invest 91:308–318
72. Kubes P, Kanwar S, Niu XF, Gaboury JP (1993) Nitric oxide synthesis inhibition induces leukocyte adhesion via superoxide and mast cells. FASEB J 7:1293–1299
73. Parker JL, Adams HR (1993) Selective inhibition of endothelium-dependent vasodilator capacity by *Escherichia coli* endotoxemia. Circ Res 72:539–551
74. Statman R, Cheng W, Cunningham JN, et al (1994) Nitric oxide inhibition in the treatment of the sepsis syndrome is detrimental to tissue oxygenation. J Surg Res 57:93–98
75. Wright CE, Rees DD, Moncada S (1992) Protective and pathological roles of nitric oxide in endotoxin shock. Cardiovasc Res 26:48–57
76. Klabunde RE, Ritger RC (1991) N^G-monomethyl-L-arginine (NMA) restores arterial blood pressure but reduces cardiac output in a canine model of endotoxic shock. Biochem Biophys Res Commun 178:1135–1140
77. Pastor C, Teisseire B, Vicaut E, Payen D (1994) Effects of L-arginine and L-nitroarginine treatment on blood pressure and cardiac output in a rabbit endotoxin shock model. Crit Care Med 22:465–469
78. Hutcheson IR, Whittle BJR, Boughton-Smith NK (1990) Role of nitric oxide in maintaining vascular integrity in endotoxin-induced acute intestinal damage in the rat. Br J Pharmacol 101:815–820

79. Rosenkranz-Weiss P, Sessa WC, Milstien S, Kaufman S, Watson CA, Pober JS (1994) Regulation of nitric oxide synthesis by proinflammatory cytokines in human umbilical vein endothelial cells. J Clin Invest 93:2236–2243
80. Moilanen E, Vuorinen P, Kankaanranta H, Metsä-Ketelä, Vapaatalo H (1993) Inhibition by nitric oxide-donors of human polymorphonuclear leukocyte functions. Br J Pharmacol 109:852–858
81. Clancy RM, Leszczynska-Piziak J, Abramson SB (1992) Nitric oxide, an endothelial cell relaxation factor, inhibits neutrophil superoxide anion production via a direct action on the NADPH oxidase. J Clin Invest 90:1116–1121
82. Kaplan SS, Billiar T, Curran RD, Zdziarski UE, Simmons RL, Basford RE (1989) Inhibition of chemotaxis with N^G-monomethyl-L-arginine: A role for cyclic GMP. Blood 74:1885-1887
83. Belenky SN, Robbins RA, Rennard SI, Gossman GL, Nelson KJ, Rubinstein I (1993) Inhibitors of nitric oxide synthase attenuate human neutrophil chemotaxis in vitro. J Lab Clin Med 122:388–394
84. Clancy RM, Levartovsky D, Leszczynska-Piziak J, Yegudin J, Abramson SB (1994) Nitric oxide reacts with intracellular glutathione and activates the hexose monophosphate shunt in human neutrophils: Evidence for *S*-nitrosoglutathione as a bioactive intermediary. Proc Natl Acad Sci USA 91:3680–3684
85. Wyatt TA, Lincoln TM, Pryzwansky KB (1993) Regulation of human neutrophil degranulation by LY-83583 and L-arginine: Role of cGMP-dependent protein kinase. Am J Physiol 265 (Cell Physiol 34): C201–C211
86. Clancy RM, Leszczynska-Piziak J, Abramson SB (1993) Nitric oxide stimulates the ADP-ribosylation of actin in human neutrophils. Biochem Biophys Res Commun 191:847–852
87. Kubes P, Suzuki M, Granger DN (1991) Nitric oxide: An endogenous modulator of leukocyte adhesion. Proc Natl Acad Sci USA 88:4651–4654
88. Niu X, Smith W, Kubes P (1994) Intracellular oxidative stress induced by nitric oxide synthesis inhibition increases endothelial cell adhesion to neutrophils. Circ Res 74:1133–1140
89. Suematsu M, Tamatani T, Delano FA, et al (1994) Microvascular oxidative stress preceding leukocyte activation elicited by in vivo nitric oxide suppression. Am J Physiol 266 (Heart Circ Physiol 35): H2410–H2415
90. Gaboury J, Woodman RC, Granger DN, Reinhardt P, Kubes P (1993) Nitric oxide prevents leukocyte adherence: Role of superoxide. Am J Physiol 265 (Heart Circ Physiol 34): H862–H867
91. Van Dervort AL, Yan L, Madara PJ, et al (1993) Nitric oxide (NO) increases endotoxin (LPS)-induced cytokine production by human neutrophils (PMNs). Clin Res 41:281A (Abst)
92. Masini E, Di Bello MG, Pistelli A, et al (1994) Generation of nitric oxide from nitrovasodilators modulates the release of histamine from mast cells. J Physiol Pharmacol 45:41–53
93. Salvemini D, Misko TP, Masferrer JL, Seibert K, Currie MG, Needleman P (1993) Nitric oxide activates cyclooxygenase enzymes. Proc Natl Acad Sci USA 90:7240–7244
94. Roland CR, Goss JA, Mangino MJ, Hafenrichter D, Flye MW (1994) Autoregulation by eicosanoids of human Kupffer cell secretory products. Ann Surg 219:389–399
95. Stadler J, Harbrecht BG, Di Silvio M, et al (1993) Endogenous nitric oxide inhibits the synthesis of cyclooxygenase products and interleukin-6 by rat Kupffer cells. J Leukoc Biol 53:165–172
96. Bath PMW (1993) The effect of nitric oxide-donating vasodilators on monocyte chemotaxis and intracellular cGMP concentrations in vitro. Eur J Clin Pharmacol 45:53–58
97. Fukushima T, Sekizawa K, Jin Y, Sasaki H (1994) Interferon-γ increases cytoplasmic motility of alveolar macrophages via nitric oxide-dependent signaling pathways. Am J Respir Cell Mol Biol 10:65–71

98. Sicher SC, Vazquez MA, Lu CY (1994) Inhibition of macrophage Ia expression by nitric oxide. J Immunol 153:1293–1300
99. Magrinat G, Mason SN, Shami PJ, Weinberg JB (1992) Nitric oxide modulation of human leukemia cell differentiation and gene expression. Blood 80:1880–1884
100. Albina JE, Cui S, Mateo RB, Reichner JS (1993) Nitric oxide-mediated apoptosis in murine peritoneal macrophages. J Immunol 150:5080–5085
101. Lander HM, Sehajpal P, Levine DM, Novogrodsky A (1993) Activation of human peripheral blood mononuclear cells by nitric oxide-generating compounds. J Immunol 150:1509–1516
102. Eigler A, Sinha B, Endres S (1993) Nitric oxide-releasing agents enhance cytokine-induced tumor necrosis factor synthesis in human mononuclear cells. Biochem Biophys Res Commun 196:494–501
103. Oliver JA (1992) Endothelium-derived relaxing factor contributes to the regulation of endothelial permeability. J Cell Physiol 151:506–511
104. Kubes P, Granger DN (1992) Nitric oxide modulates microvascular permeability. Am J Physiol 262 (Heart Circ Physiol 31):H611–H615
105. Filep JG, Földes-Filep É, Sirois P (1993) Nitric oxide modulates vascular permeability in the rat coronary circulation. Br J Pharmacol 108:323–326
106. Paul W, Douglas GJ, Lawrence L, et al (1994) Cutaneous permeability responses to bradykinin and histamine in the guinea pig: Possible differences in their mechanism of action. Br J Pharmacol 111:159–164
107. Lejeune P, Lagadec P, Onier N, Pinard D, Ohshima H, Jeannin JF (1994) Nitric oxide involvement in tumor-induced immunosuppression. J Immunol 152:5077–5083
108. Schleifer KW, Mansfield JM (1993) Suppressor macrophages in African trypanosomiasis inhibit T cell proliferative responses by nitric oxide and prostaglandins. J Immunol 151:5492–5503
109. Hoffman RA, Langrehr JM, Wren SM, et al (1993) Characterization of the immunosuppressive effects of nitric oxide in graft vs host disease. J Immunol 151:1508–1518
110. Lefer DJ, Nakanishi K, Johnston WE, Vinten-Johansen J (1993) Antineutrophil and myocardial protecting actions of a novel nitric oxide donor after acute myocardial ischemia and reperfusion in dogs. Circulation 88:2337–2350
111. Ma X, Weyrich AS, Lefer DJ, Lefer AM (1993) Diminished basal nitric oxide release after myocardial ischemia and reperfusion promotes neutrophil adherence to coronary endothelium. Circ Res 72:403–412
112. Matheis G, Sherman MP, Buckberg GD, et al (1992) Role of L-arginine-nitric oxide pathway in myocardial reoxygenation injury. Am J Physiol 262 (Heart Circ Physiol 31):H616–H620
113. Bastian NR, Xu S, Shao XL, Shelby J, Granger DL, Hibbs JB Jr (1994) N^{ω}-monomethyl-L-arginine inhibits nitric oxide production in murine cardiac allografts but does not affect graft rejection. Biochem Biophys Acta 1226:225–231
114. Huang Z, Huang PL, Panahian N, Dalkara T, Fishman MC, Moskowitz MA (1994) Effects of cerebral ischemia in mice deficient in neuronal nitric oxide synthase. Science 265:1883–1885
115. McDonald LJ, Moss J (1993) Stimulation by nitric oxide of a NAD linkage to glyceraldehyde-3-phosphate dehydrogenase. Proc Natl Acad Sci USA 90:6238–6241
116. Peterson DA, Peterson DC, Archer S, Weir EK (1992) The non-specificity of specific nitric oxide synthase inhibitors. Biochem Biophys Res Commun 187:797–801
117. Gregory SH, Sagnimeni AJ, Wing EJ (1994) Arginine analogues suppress antigen-specific and -nonspecific T lymphocyte proliferation. Cell Immunol 153:527–532
118. Yui Y, Hattori R, Kosuga K, et al (1991) Calmodulin-independent nitric oxide synthase from rat polymorphonuclear neutrophils. J Biol Chem 266:3369–3371
119. Stuehr DJ, Gross SS, Sakuma I, Levi R, Nathan CF (1989) Activated murine macrophages secrete a metabolite of arginine with the bioactivity of endothelium-derived relaxing factor and the chemical reactivity of nitric oxide. J Exp Med 169:1011–1020

120. Takema M, Inaba K, Uno K, Kakihara KI, Tawara K, Muramatsu S (1991) Effect of L-arginine on the retention of macrophage tumoricidal activity. J Immunol 146: 1928–1933
121. Boockvar KS, Granger DL, Poston RM, et al (1994) Nitric oxide produced during murine listeriosis is protective. Infect Immun 62:1089–1100
122. Beckerman KP, Rogers HW, Corbett JA, Schreiber RD, McDaniel ML, Unanue ER (1993) Release of nitric oxide during the T cell-independent pathway of macrophage activation. Its role in resistance to *Listeria monocytogenes*. J Immunol 150:888–895
123. Taylor-Robinson AW, Phillips RS, Severn A, Moncada S, Liew FY (1993) The role of T_H1 and T_H2 cells in a rodent malaria infection. Science 260:1931–1934
124. Karupiah G, Xie Q, Buller ML, Nathan C, Duarte C, MacMicking JD (1993) Inhibition of viral replication by interferon-γ-induced nitric oxide synthase. Science 261: 1445–1448
125. Kolb JP, Paul-Eugene N, Damais C, Yamaoka K, Drapier JC, Dugas B (1994) Interleukin-4 stimulates cGMP production by IFN-γ-activated human monocytes. Involvement of the nitric oxide synthase pathway. J Biol Chem 269:9811–9816
126. Mautino G, Paul-Eugene N, Chanez P, et al (1994) Heterogeneous spontaneous and interleukin-4-induced nitric oxide production by human monocytes. J Leukoc Biol 56:15–20
127. Pietraforte D, Tritarelli E, Testa U, Minetti M (1994) gp120 HIV envelope glycoprotein increases the production of nitric oxide in human monocyte-derived macrophages. J Leukoc Biol 55:175–182
128. Denis M (1991) Tumor necrosis factor and granulocyte macrophage-colony stimulating factor stimulate human macrophages to restrict growth of virulent *Mycobacterium avium* and to kill *M. avium*: Killing effector mechanism depends on the generation of reactive nitrogen intermediates. J Leukoc Biol 49:380–387
129. Reiling N, Ulmer AJ, Duchrow M, et al (1994) Nitric oxide synthase: mRNA expression of different isoforms in human monocytes/macrophages. Eur J Immunol 24: 1941–1944
130. Kolb-Bachofen V, Alsdorf K, Fehsel K (1994) Expression of inducible NO synthase mRNA in human monocytes. Proceedings of the Eighth Annual Conference of the European Macrophage Study Group. p 57 [Abst]
131. Wright CD, Mülsch A, Busse R, Osswald H (1989) Generation of nitric oxide by human neutrophils. Biochem Biophys Res Commun 160:813–819
132. Bryant JL, Mehta P, Von der Porten A, Mehta JL (1992) Co-purification of 130 kD nitric oxide synthase and a 22 kD link protein from human neutrophils. Biochem Biophys Res Commun 189:558–564
133. Rao KMK, Padmanabhan J, Kilby DL, Cohen HJ, Currie MS, Weinberg JB (1992) Flow cytometric analysis of nitric oxide production in human neutrophils using dichlorofluorescein diacetate in the presence of a calmodulin inhibitor. J Leukoc Biol 51:496–500
134. Schmidt HHHW, Seifert R, Böhme E (1989) Formation and release of nitric oxide from human neutrophils and HL-60 cells induced by a chemotactic peptide, platelet-activating factor and leukotriene B_4. FEBS Lett 244:357–360
135. Riesco A, Caramelo C, Blum G, et al (1993) Nitric oxide-generating system as an autocrine mechanism in human polymorphonuclear leukocytes. Biochem J 292: 791–796
136. Yan L, Vandivier RW, Suffredini AF, Danner RL (1994) Human polymorphonuclear leukocytes lack detectable nitric oxide synthase activity. J Immunol 153:1825–1834
137. Stamler JS (1994) Redox signaling: Nitrosylation and related target interactions of nitric oxide. Cell 78:931–936
138. Corriveau CC, Tropea MM, Yan L, et al (1993) Effect of nitric oxide on neutrophil chemotaxis and PGE_2 production. Sixth Pediatric Critical Care Colloquium p 39
139. May GR, Crook P, Moore PK, Page CP (1991) The role of nitric oxide as an endogenous regulator of platelet and neutrophil activation within the pulmonary circulation of the rabbit. Br J Pharmacol 102:759–763

140. Martich GD, Danner RL, Ceska M, Suffredini AF (1991) Detection of interleukin-8 and tumor necrosis factor in normal humans after intravenous endotoxin: The effect of antiinflammatory agents. J Exp Med 173:1021–1024
141. Vallance P, Palmer RMJ, Moncada S (1992) The role of induction of nitric oxide synthesis in the altered responses of jugular veins from endotoxaemic rabbits. Br J Pharmacol 106:459–463
142. Natanson C, Fink MP, Ballantyne HK, MacVittie TJ, Conklin JJ, Parrillo JE (1986) Gram-negative bacteremia produces both severe systolic and diastolic dysfunction in a canine model that simulates human septic shock. J Clin Invest 78:259–270
143. Meyer J, Lentz CW, Herndon DN, Nelson S, Traber LD, Traber DL (1993) Effects of halothane anesthesia on vasoconstrictor response to N^G-nitro-L-arginine methyl ester, an inhibitor of nitric oxide synthesis, in sheep. Anesth Analg 77:1215–1221
144. Aisaka K, Mitani A, Kitajima Y, Ohno T, Ishihara T (1991) Difference in pressor responses to N^G-monomethyl-L-arginine between conscious and anesthetized rats. Japan J Pharmacol 56:245–248

Effects of Nitric Oxide on Intestinal Epithelial Structure and Function

M. P. Fink, N. Unno, and A. L. Salzman

Introduction

Nitric oxide (NO) has been identified as an important regulator and/or effector of many phenomena in the cardiovascular, nervous, and immune systems [1]. Recently, it has become apparent that NO also plays a versatile role in the physiology and pathophysiology of the gastrointestinal (GI) tract [2]. For example, NO appears to be one of several neurotransmitters responsible for the relaxation of non-vascular smooth muscle cells in the wall of the gut under the influence of non-adrenergic non-cholinergic (NANC) inhibitory enteric neurons [3]. NO-mediated inhibition of non-vascular smooth muscle contraction has been demonstrated at a number of sites in the GI tract, including the esophagus [4], stomach [5], small intestine [3, 6], colon [7], and sphincter of Oddi [8, 9]. NO is also an important chemical messenger serving to regulate the contraction of vascular smooth muscle cells, and hence vasomotor tone, in the mesenteric circulation [10–12]. While the effects of NO on non-vascular and vascular smooth muscle cells in the gut are potentially of considerable interest to clinicians interested in critical care, the focus of the present chapter will be on the actions of NO on intestinal enterocytes and closely related cells in the mucosa and submucosa of the GI tract.

Sources of NO in the GI Tract

A large variety of cell types in the gut are potential sources of NO. These cell types include myenteric neurons [13–16], vascular endothelial cells [16, 17], and interstitial cells of Cajal [18]. Inflammatory cells in the submucosa, including mast cells [19], macrophages [20, 21], and polymorphonuclear leukocytes [22, 23], are capable of producing NO. Luminal bacteria represent another potential source of NO in the GI tract [24].

Intestinal epithelial cells (i.e. enterocytes) also appear to be capable of producing NO, at least under certain conditions. Boughton-Smith and colleagues [25] reported that calcium-independent NO synthase (NOS) activity is markedly elevated in biopsy specimens of inflamed colonic mucosa from patients with ulcerative colitis. In studies using rats, this group also showed that administration of lipopolysaccharide (LPS) leads to the induction of calcium-independent NOS activity in homogenates of jejunal or colonic tissue [26]. Although these studies suggest that various inflammatory states (i.e. ulcerative colitis and acute endotoxicosis) are associated with induction of NOS activity in gut epithelial cells, the observed results also could be explained by NOS induction in resident or infiltrating inflammatory cells in the mucosa or lamina propria. In an effort to address this issue, Tepperman et al. [27] used enzymatic digestion to isolate enterocytes from the colons of control rats or rats injected 4 h earlier with LPS. Enterocytes from endotoxic rats evidenced markedly elevated levels of calcium-independent NOS activity. Moreover, prior administration of an anti-neutrophil antiserum prevented LPS-induced leukosequestration in the colonic mucosa (as assessed by measurements of myeloperoxidase activity), but had no effect on measured levels of NOS activity, a finding which argues against the notion that infiltrating neutrophils are the source of the inducible NOS in enterocyte preparations obtained from endotoxic rats. Similar preliminary data were obtained by Wilson et al. [28] who showed that levels of inducible NOS (iNOS) messenger ribonucleic acid (mRNA) are increased in mucosal scrapings from rats treated 60 min to 6 h earlier with LPS.

The idea that enterocytes can be induced to produce NO is further supported by preliminary findings obtained by several laboratories. Grisham et al. [29] showed that exposure of cultured intestinal epithelial (IEC-18) cells to LPS, interleukin-1 (IL-1), or tumor necrosis factor alpha (TNF-α) plus interferon gamma (IFN-γ), increases the production of nitrite plus nitrate (markers of NO production). IEC-18 is a cell line derived from normal rat intestinal crypt cells. Similarly, Dignass et al. [30] showed that incubation of cultures of a different rat enterocytic cell line (IEC-6) with a number of different cytokines (e.g. IL-2, TNF-α) results in modest increases in the production of nitrogen oxides. Recently, our group has shown that incubation of Caco-2BBe cells with the proinflammatory cytokine, IFN-γ, increases release of nitrite plus nitrate into the media (unpublished observations). Caco-2BBe is an immortal human enterocytic cell line that reproduces many of the features of normal small intestinal absorptive epithelium.

Effect of NO on Mucosal Secretion of Salt and Water

It is increasingly apparent that NO has profound affects on intestinal mucosal handling of salt and water. In 1993, Wilson et al. [31] showed that two NO donors, sodium nitroprusside (SNP) and S-nitroso-N-acetylpenicillamine (SNAP), stimulate an increase in short-circuit currents (I_{sc}) across sheets of rat colon mucosa mounted in Ussing chambers. I_{sc} is a measure of electrogenic chloride secretion. In the same study, Wilson et al. also showed that incubation of mucosal sheets with SNP inhibits non-electrogenic Na^+ and Cl^- absorption; i.e. by blocking salt absorption and stimulating anion secretion, NO favors net movement of electrolyte-rich fluid into the lumen of the gut. Similar results were obtained by MacNaughton [32] who employed sheets of guinea pig ileal mucosa mounted in Ussing chambers. Both Wilson et al. and MacNaughton found that the effects of NO-donors on mucosal electrolyte secretion are attenuated by inhibitors of cyclooxygenase, suggesting that NO-mediated modulation of electrolyte transport is, at least partly, prostaglandin-dependent [31, 32].

The hypothesis that NO promotes net secretion by the intestine is supported by findings obtained in a recent *in vivo* study in rats, which showed that two potent laxatives, bisacodyl and phenophthalein, increase intestinal NOS activity [33]. In this study, the increases in water and electrolyte secretion induced by the laxatives were inhibited by treatment with the NOS inhibitor, N^G-nitro-L-arginine methyl ester (L-NAME) (but, not the inactive enantiomer, D-NAME) [33]. These findings suggest not only that these laxatives work via an NO-dependent mechanism, but that endogenously generated NO promotes secretion of fluid and electrolytes into the lumen of the gut.

In marked contrast to the above results are data obtained by Barry et al. [34] who used vascularly and lumenally perfused segments of rabbit ileium, and showed that SNP and L-arginine (substrate for NOS) cause net absorption of salt and water. These investigators also showed that inhibition of endogenous NO production by infusion of L-NAME causes net secretion of ions and water [34], suggesting that endogenous NO fosters absorption across the ileal mucosa. Similar results were obtained by Rao et al. [35] who employed full-thickness sheets of mouse ileum mounted in Ussing chambers.

The conflicting results obtained in various studies regarding the effects of NO on intestinal electrolyte transport remain to be reconciled. The discrepancies in the findings among studies may be explained by species differences as well as the potential for differential effects of NO on the various cell types (e.g. mast cells, enterocytes) present in various *in vivo* or *ex vivo* intestinal preparations. As of this writing, data are unavailable regarding the effects of NO on electrolyte transport in cultured monolayers of an enterocytic cell line, such as T84, which is capable of agonist-induced chloride secretion.

Effect of NO on the Barrier Function of the Intestinal Epithelium

The intestinal epithelium normally functions as a selective barrier, permitting the absorption of nutrients, electrolytes, and water, but restricting the passage from the lumen of larger, potentially toxic compounds. This characteristic of the intestinal mucosa, which has been referred to as "selective permeability" [36], is thought to be mediated by the tight junctions ("zonula occludens") surrounding each cell in the epithelial sheet [37]. Previous studies have shown that intestinal epithelial permeability can be modulated by a number of factors, including changes in intracellular adenosine 3′-5′-cyclic monophosphate (cAMP) concentration [38], insulin [39], insulin-like growth factors [40], activators of protein kinase C [41], and cytokines [42, 43].

NO appears to be another signaling molecule capable of modulating intestinal epithelial permeability. Interestingly, depending upon the experimental system employed, NO has been shown to either diminish or increase intestinal permeability to various water-soluble molecules.

NO Maintains Gut Barrier Function

The notion that NO serves to decrease (or at least preserve normal) intestinal mucosal permeability derives largely from the work of Paul Kubes and his collaborators [44]. In 1992, Kubes reported that infusion of L-NAME into the mesenteric artery of cats leads to a sixfold increase in the blood-to-lumen clearance of ^{51}Cr-EDTA, a commonly used measure of intestinal epithelial permeability. This effect apparently is not due to the vasoconstricting effects of blocking NO synthesis with L-NAME, since infusion of phenylephrine causes a similar degree of vasoconstriction, but has no effect on epithelial permeability. By showing that L-NAME-induced mucosal hyperpermeability is not blocked by pretreatment with a monoclonal antibody to the CD 11b/CD18 integrin, Kubes also seemed to rule out a neutrophil-mediated mechanism [44].

In a subsequent study, Payne and Kubes [45] documented that infusion of a NO donor (SNP, SIN-1, or CAS-754) or exogenous arginine (NOS substrate) ameliorates gut mucosal hyperpermeability induced by ischemia-reperfusion, apparently via a guanosine 3′-5′-cyclic monophosphate (cGMP)-independent mechanism. Using a similar design, Boughton-Smith and colleagues [46] showed that infusion of SNAP ameliorates mucosal damage induced by intravenous injection of LPS in rats. Conversely, treatment with NOS inhibitors has been shown to exacerbate gut epithelial hyperpermeability induced by ischemia-reperfusion [47] and macroscopic mucosal damage induced by LPS [48].

In summary, therefore, data from a number of *in vivo* studies support the following conclusions: 1) under basal conditions, endogenous production of NO serves to maintain normal permeability of the intestinal mucosa; 2) under basal conditions, blocking NO synthesis increases gut mucosal permeability; 3) in certain pathologic conditions associated with derangements in gut mucosal barrier function and/or integrity (e.g. ischemia/reperfusion injury or endotoxicosis), provision of exogenous NO tends to maintain normal intestinal epithelial structure and function; and 4) in these same pathologic conditions, blockade of NO biosynthesis tends to exacerbate derangements in the structural and functional integrity of the mucosa.

NO Impairs Gut Barrier Function

In contrast to the findings just cited, results from other studies suggest that NO is capable of causing derangements in gut muscosal structure and function. For example, Lopez-Belmonte et al. [49] showed that close arterial infusion of NO donors (SNP or SNAP) can cause gastric mucosal damage in the rat, but only when the compounds are delivered at relatively high concentrations. Similarly, Tepperman et al. [27] recently reported that incubation of freshly harvest rat colonic epithelial cells with NO donors (SNP, SNAP, or S-nitrosoglutathione) causes a concentration-dependent decrease in viability. Thus, exogenous sources of NO appear to be capable of damaging gastric or colonic epithelial cells.

Using a rat model of acute endotoxicosis, Tepperman et al. [27, 50] have reported that *in vivo* administration of LPS decreases the viability of gut epithelial cells. Since this effect of LPS can be blocked by pretreatment of the rats with L-NAME (NOS inhibitor) or dexamethasone (inhibitor of iNOS induction), the effect of LPS on enterocyte viability appears to be a NO-dependent phenomenon.

Recently, our group conducted an extensive series of experiments to further investigate the effects of NO on intestinal epithelial permeability. In order to avoid the potential for confounding by other NO-mediated phenomenona, such as alterations in mucosal perfusion or modulation of the function of neighboring non-enterocyte cell populations, we performed our studies *in vitro*, using Caco-2BBe cells grown as monolayers on semipermeable supports as a reductionist model of the intestinal epithelium. The Caco-2 cell line has been used previously for studying intestinal permeability *in vitro* [41, 51]. In our initial studies, Caco-2BBe monolayers growing on Snapwell (permeable polycarbonate) inserts were mounted in modified Ussing chambers. The apical and basolateral compartments of the Ussing chambers were filled with complete growth medium and transepithelial resistance (TER) across the monolayer was measured using standard methods. Control Caco-2BBe monolayers incubated in the absence of the NO donor, SNP, exhibited a gradual, but eventually significant, decrease in TER over a 6 h period of observation [52]. In contrast, incubation of monolayers with SNP resulted in a much more rapid and profound decrease in TER, the extent of which was both time- and dose-dependent.

Although changes in TER are generally regarded as primarily reflecting alterations in the movement of ions across the monolayer via the paracellular pathway, TER also is somewhat sensitive to perturbations of the transcellular ionic conductance. In order to more specifically investigate the effects of NO on the paracellular pathway, we conducted a series of experiments, wherein we assessed the permeability of monolayers to a series of water-soluble macromolecular tracers. We used fluorescent or radioactive tracers which are cell impermeant by virtue of their size or charge [53, 54]. Permeability was determined by measuring the transepithelial passage of the tracers from the apical to the basolateral compartments of bicameral ("Transwell") tissue-culture chambers, wherein the two chambers are separated by Caco-2BBe monolayers grown on a support with 0.4 μM pores. Permeability was expressed as a clearance ($nL \cdot cm^{-2} \cdot h^{-1}$), calculated as the apical to basolateral flux of the probe divided by the concentration of the probe in the apical compartment.

In a preliminary study, we assessed the effect of SNP on permeability to fluorescein sulfonic acid (FS; mw=478 D), a molecule which is highly negatively charged at physiologic pH and accordingly is cell impermeant [53]. Permeability to FS increased in a concentration-dependent manner after monolayers were incubated for 6 h in the presence of SNP. After 6 h of incubation, 5.0 mM was the lowest concentration of SNP which significantly increased permeability. With a longer period of incubation (24 h), the threshold concentration of SNP required to induce changes in permeability was reduced to 1.25 mM (Fig. 1). When monolayers were incubated for 7 days in the presence of SNP (fresh SNP being added every 24 h), a significant increase in permeability to FS was observed when the concentration of the NO donor was as low as 100 μM (Fig. 2).

The effect of SNP on permeability in the Transwell system was not an immediate phenomenon. Indeed, as is shown in Fig. 3, there was no evidence of altered paracellular permeability to either of two different probes (^{3}H-mannitol or FS) after 6 h of exposure to 2.5 mM. However, after 12 h of incubation with SNP, there was a marked increase in paracellular permeability to both probes. Note also that the permeation of (control) monolayers incubated in medium without SNP was substantially greater for ^{3}H-mannitol (mw=176 D) than for the much larger molecule, FS (mw=478 D). Thus, the control monolayers exhibited a characteristic feature of normal epithelia, namely permselectivity based on molecular size [55].

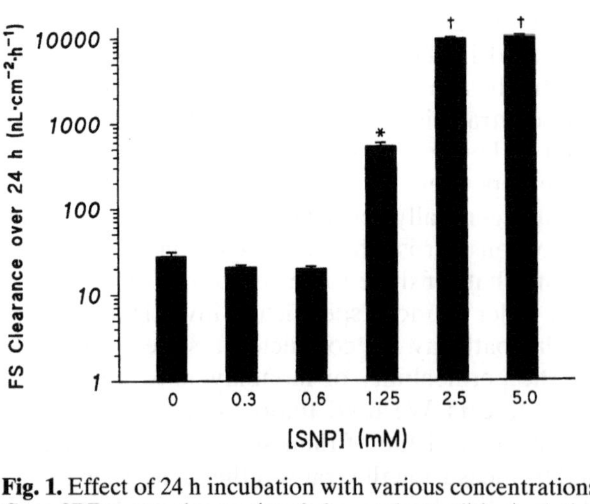

Fig. 1. Effect of 24 h incubation with various concentrations of SNP on the permeability of Caco-2BBe monolayers (n=3 for each condition). Apical chambers were loaded with fluorescein sulfonic acid (FS) and monolayers were incubated with control media or media containing the indicated concentrations of SNP. Permeability of the monolayers was expressed as a clearance, calculated as the apical to basolateral flux of the probe divided by the concentration of the probe in the apical compartment. Clearance values represent means ± SE. * $p<0.05$ $^{+}p<0.01$, respectively, for comparisons with (control) monolayers incubated in medium without SNP. (From [52] with permission)

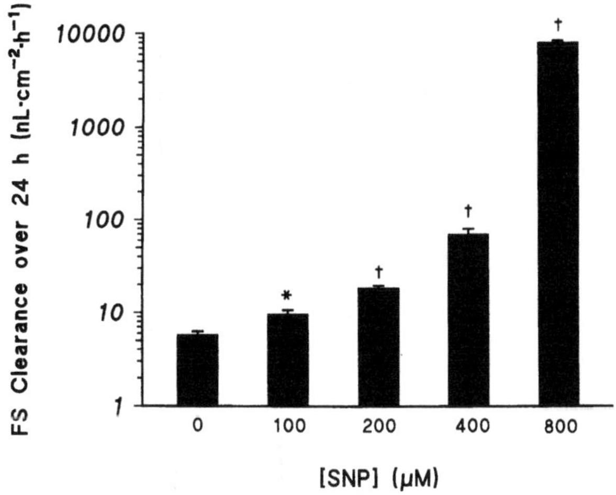

Fig. 2. Effect of incubation for 7 days with various concentrations of SNP on the permeability of Caco-2BBe monolayers (n = 4 for each condition). Media were aspirated from the incubation chambers every 24 h and replaced with freshly prepared media (with or without added SNP, as appropriate). FS was added for the last 24 h of incubation for the determination of permeability. * $p<0.05$ $^{+}p<0.01$, respectively, for comparisons with (control) monolayers incubated in medium without SNP. (From [52] with permission

In order to exclude the possibility that the effect of SNP on the permeability of Caco-2BBe cell monolayers was due to a toxic effect of the drug unrelated to the release of NO, additional experiments were performed using ferrocyanide and ferricyanide salts, compounds which are structurally identical to SNP except for the substitution of a CN^- group for the NO moiety present in SNP. Whereas incubation with 2.5 mM SNP markedly increased permeability, incubations with similar concentrations of ferricyanide or ferrocyanide resulted in no or only minimal changes in permeability. In order to further exclude a toxic effect of CN^- liberated from SNP, Caco-2BBe monolayers were incubated with 5.0 mM SNP or 5.0 mM NaCN for 6 h. Whereas the former compound increased clearance to FS, NaCN at this concentration did not affect permeability.

Additional confirmation that the release of NO was responsible for the effects of SNP on permeability of Caco-2BBe monolayers was provided by experiments using a structurally unrelated NO donor, SNAP. Incubation of monolayers with 5.0 mM SNAP for 24 h increased permeability by 8.9 fold compared to cells incubated in medium without SNAP (Fig. 4). Lower concentrations of SNAP did not affect permeability.

Additional experiments were performed using authentic NO gas. This highly reactive gas was mixed with N_2 rather than air, in order to prolong the half-life of NO and prevent its oxidation to toxic higher oxides of nitrogen. Accordingly, the controls for these experiments consisted of cells incubated under an atmosphere of 100% N_2. Despite these anoxic conditions, the con-

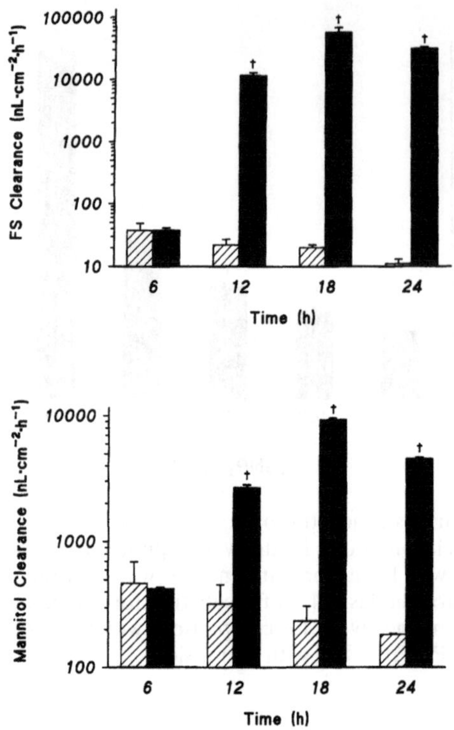

Fig. 3. Time course of the effect of sodium nitroprusside (SNP) on the permeability of Caco-2BBe monolayers (n = 6 for each condition). Apical chambers were loaded with FS (*top panel*) and ^{3}H-mannitol (*bottom panel*) and monolayers were incubated for 24 h with control medium (n = 6; *hatched bars*) or medium containing 2.5 mM SNP (n = 6; *solid bars*). Inserts were moved to new basolateral wells containing fresh medium at 6 h intervals. Thus, a series of four 6 h permeability studies were performed using the same monolayers over a 24 h period. Permeability of the monolayers was expressed as a clearance, calculated as the apical to basolateral flux of the probe divided by the concentration of the probe in the apical compartment. $^+$ $p < 0.01$ versus (control) monolayers incubated in medium without SNP. (From [52] with permission)

trol monolayers remained viable and maintained near normal permeability. As shown in Fig. 5, incubation under an atmosphere containing 1% NO, resulted in a significant increase in permeability to FS. The effect of NO was time-dependent, and increased progressively as the period of exposure was lengthened from 4 to 24 h.

Many of the actions of NO are mediated via activation of soluble guanylate cyclase leading to elevated intracellular concentrations of the second messenger, cGMP [56–58]. In order to determine whether NO-induced hyperpermeability of Caco-2BBe monolayers is due to activation of guanylate cyclase, we measured cGMP concentrations in Caco-2BBe cells incubated in control medium or medium containing 2.5 mM SNP. As a positive control for the assay, we also measured cGMP levels in Caco-2BBe monolayers incubated with *E. coli* heat-stable toxin, a known activator of the par-

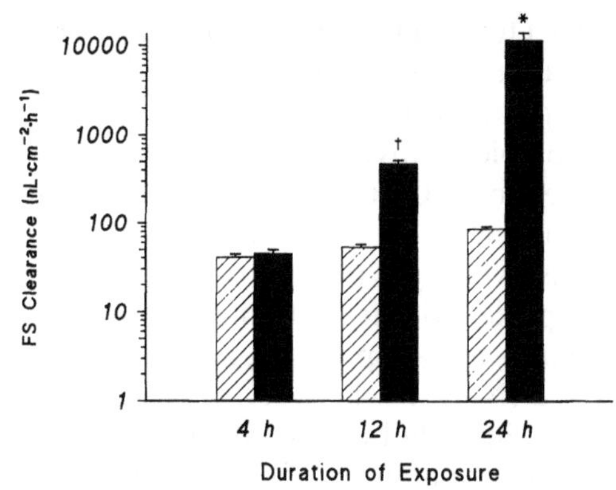

Fig. 4. Time course of the effect of authentic gaseous NO on the permeability of Caco-2BBe monolayers to FS. Apical chambers were loaded with FS and monolayers were incubated under an atmosphere of 100% N_2 (hatched bars) or 1%NO/ 99%N_2 (solid bars) for 4 h (n=12), 12 h (n=16), and 24 h (n=3). Permeability was calculated as in Fig. 2. * $p<0.05$, + $p<0.01$, respectively, for comparisons with (control) monolayers incubated under 100% N_2. (From [52] with permission)

ticulate form of guanylate cyclase in intestinal epithelial cells [59]. Incubation of Caco-2BBe cells with *E. coli* heat-stable toxin (50 nM for 1 h) increased intracellular cGMP levels from <5 to >80 picomoles/monolayer ($p<0.01$). In contrast, incubation of monolayers with SNP (2.5 mM for 1 h) had no effect on cGMP levels. Whereas incubation with SNP (2.5 mM for 6 h) significantly increased permeability to FS, incubation with *E. coli* heat-stable

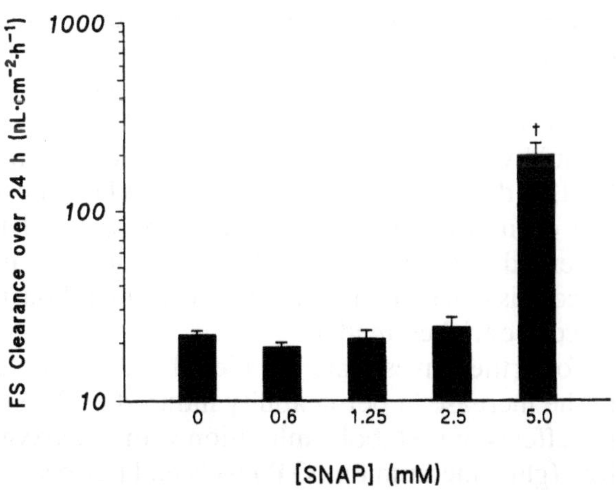

Fig. 5. Effect of SNAP on the permeability of Caco-2BBe monolayers to FS. Apical chambers were loaded with FS and monolayers were incubated for 24 h with control media, or media containing SNAP at a concentration of 0.6–5.0 mM. Methods are the same as for Fig. 1. + $p<0.01$ versus (control) monolayers incubated in medium without SNAP. (From [52] with permission)

toxin (100 nM for 6 h) had no effect on permeability to FS. Thus, the effect of NO on the permeability of Caco-2BBe monolayers appears to be a cGMP-independent phenomenon.

Soluble guanylate cyclase is by no means the only intracellular molecular target for NO [60]. Indeed, as a reactive, electrophilic radical, NO forms adducts with a variety of nucleophilic organic substituents (e.g. amines and thiols) which are present on numerous proteins. NO also forms coordination complexes with transition metal ions (e.g. Fe^{+2} and Fe^{+3}) which are present in many metalloproteins, and ternary complexes with the general structure X_n-Fe-$(NO)_m$, where X represents a nucleophilic group like -SH or $-NH_2$. Possibly because of these kinds of reactions, NO has been shown to inactivate a number of enzymes with Fe-S active centers, which are involved in mitochondrial respiration (namely, aconitase, NADH-ubiquinone oxido-reductase, and succinate-ubiquinone oxidoreductase) [21, 61–65]. NO also inactivates another important protein involved in the biosynthesis of adenosine triphosphate (ATP), namely the glycolytic enzyme, glyceraldehyde-3-phosphate dehydrogenase (G3PDH). Inactivation of G3PDH by NO apparently occurs via two (possibly inter-related) mechanisms: S-nitrosylation of thiol groups, which are essential for enzymatic activity [66]; and mono-ADP-ribosylation [66–69]. NO also activates a nuclear poly(ADP-ribosyl) synthase, an action which can lead to ATP and NAD depletion and cell death [70].

Since NO can interfere with cellular energy metabolism, we conducted additional experiments to determine the effect of NO or NO donors on ATP levels in cultured Caco-2BBe monolayers. The concentration of ATP in monolayers incubated under control conditions was 60 ± 1 nmoles/mg protein, which is in good agreement with previously reported values for cultured epithelial cell monolayers [71]. Exposure of monolayers to SNP or SNAP (24 h) or 1%NO/99%N_2 (12 h) decreased intracellular levels of ATP (Fig. 6). Decreased cellular levels of ATP were observed only with concentrations of NO donors (SNP or SNAP) that were observed to increase permeability to FS in other experiments. Control monolayers exposed to anoxia for 12 h did not manifest a decline in ATP concentration. In time course studies examining the effect of anoxia on Caco-2BBe monolayers, we have not observed a reduction in ATP levels until 24 h, suggesting that anaerobic glycolysis can support cellular energy metabolism of monolayers cultured in glucose-enriched media.

To further investigate the idea that ATP depletion may play a causal role in the increase in permeability induced by NO or NO donors, we measured the effects of metabolic inhibition with 2-deoxyglucose (2-DOG) in glucose-free (glu-) medium on ATP levels and permeability. Compared to cells incubated for 24 h in control medium, monolayers incubated with 2-DOG/glu- medium manifested significantly lower concentrations of ATP (11.7 ± 0.9 versus 5.1 ± 0.2 nmoles/monolayer, $P < 0.01$). Depletion of ATP with 2-DOG/glu- also was associated with an increase in permeability to FS (10.7 ± 0.3 for control, 633 ± 229 $nL \cdot cm^{-2} \cdot h^{-1}$ for DOG/glu-, $P < 0.01$). Interestingly,

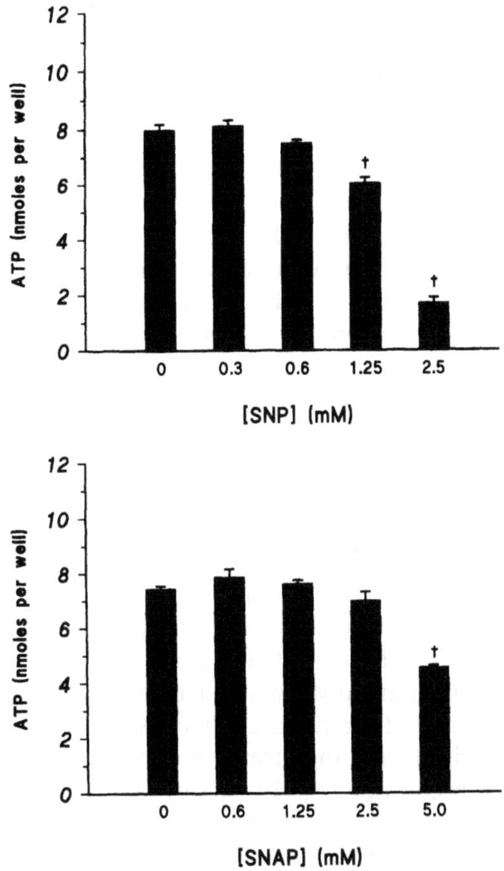

Fig. 6. Effect of SNP (*top panel*), SNAP (*middle panel*), or authentic NO gas (*bottom panel*) on the ATP content of Caco-2BBe monolayers. For the studies using SNP or SNAP, monolayers were incubated for 24 h with control medium, media containing various concentrations of SNP, or media containing various concentrations of SNAP (n = 4 per condition). For the studies using authentic NO gas, monolayers were incubated for 12 h under an atmosphere of 100% N_2 (*hatched bar*; n = 4) or 1% NO/99% N_2 (*solid bar*; n = 5). Cells were extracted with TCA/EDTA and ATP levels determined using a luciferin/luciferase assay. $^+$ $p<0.01$ versus (control) monolayers incubated in medium without SNP or SNAP. (From [52] with permission)

permeability bears a strong linear relationship to ATP level ($R^2 = 0.88$, $P < 0.00001$) when FS clearance (normalized to FS clearance in simultaneously studied controls) is plotted on a log-linear graph as a function of ATP concentration (normalized to ATP concentration in simultaneously studied controls) for monolayers incubated with authentic NO, NO donors, or 2-DOG/glu-. Thus, our results support (but do not prove) the notion NO-mediated hyperpermeability and ATP depletion are causally linked phenomena.

In order to address the question of whether NO-induced hyperpermeability was simply a manifestation of the loss of cellular viability, we measured the release of the cytosolic enzyme, lactate dehydrogenase (LDH), into the medium. Incubation of monolayers with 0.3–2.5 mM SNP for 24 h did not result in a significant increase in LDH release into the apical compartment of the bicameral Transwell chambers. Release of LDH into the basolateral compartment was very slightly increased by incubation with 2.5 mM SNP. Incubation with 1.25 mM SNP, which was sufficient to induce hyperpermeability in other studies (see above), had no effect on LDH release into the basolateral compartment. Incubation of monolayers with 1%NO/99%N_2 for 12 h, which markedly increased permeability in other experiments (see above), did not increase the release of LDH. Because SNAP interfered with the LDH assay, measurements of the effect of this NO donor on LDH release were not performed.

In an effort to further assess the effect of SNP on cell viability, we employed a commercially available cell viability assay system employing ethidium homodimer, a high-affinity fluorescent DNA stain, which is excluded from the nucleus by viable cells. For the assay, Caco-2BBe monolayers were incubated for 24 h in normal growth medium (pH 7.4) to which was added various concentrations of SNP (0.6–2.5 mM). As a negative control, cells were incubated in normal growth medium devoid of SNP. As a positive control, cells were incubated in normal growth medium titrated to pH 3.0 with HCl. At the end of the incubation period, ethidium homodimer (final concentration 2 μM) was added to the apical and basolateral compartments and incubated for 20 min. The cells were washed, and then the membranes were excised from the inserts and mounted for fluorescent microscopy. All of the cells exposed to pH 3.0 showed evidence of staining (i.e. cell death), whereas no staining was evident in the cells incubated at normal pH, irrespective of whether SNP was added to the medium.

Seeking still additional confirmation that the hyperpermeability induced by NO was not simply a consequence of cell death, we asked whether the derangement in tight junction function induced by SNP was a reversible phenomenon. Accordingly, we again documented that exposure of Caco-2BBe monolayers to 1.25 mM SNP for 24 h (i.e. t=0–24 h) increased permeability to FS (Fig. 7). The monolayers were then permitted to "recover" for 48 h (t=24–72 h) in media devoid of SNP. Finally, when permeability to FS was assessed again over a 24 h interval (t=72–96 h), the permeability defect induced by SNP was completely reversed. During the second clearance period, permeability was comparable for control monolayers, which had been incubated only with SNP-free media, and for monolayers, which had been rendered (temporarily) hyperpermeable by incubation in the presence of SNP. When this experiment was repeated using higher doses of SNP, the resulting derangement in permeability was irreversible, suggesting that a threshold of injury exists beyond which normal functioning of tight junctions cannot be restored.

The results of morphological studies suggest that the effects of NO on permeability of Caco-2BBe monolayers are related to alterations in the cytoskeleton and/or the tight junctions between adjacent enterocytes. Examina-

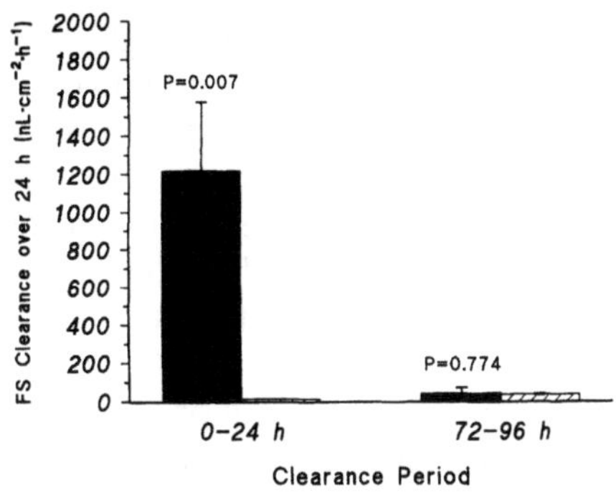

Fig. 7. Reversibility of SNP-induced hyperpermeability to FS in Caco-2BBe monolayers. Apical chambers were loaded with FS and monolayers were incubated with control medium or medium containing 1.25 mM SNP (n = 6 for each condition). Aliquots from the basolateral chambers were removed after 24 h for determination of fluorescence. Monolayers were washed twice with fresh medium without SNP and incubated for an additional 48 h. Monolayers were then washed twice in fresh medium and the apical chambers were loaded with FS. After 24 h of incubation, aliquots from the basolateral chambers were removed for determination of fluorescence. Permeability was calculated as described in Fig. 1. The p values represent comparisons at the two time points between monolayers exposed to SNP from 0–24 h (*solid bars*) to those incubated in the absence of SNP (*hatched bars*).(From [52] with permission)

tion of rhodamine-phalloidin stained control Caco-2BBe monolayers by confocal microscopy revealed localization of F-actin in the microvilli in the apical surface and in a contiguous beaded pattern at cell-cell contacts. In contrast, in SNP-treated cells, the association of F-actin with the plasma membrane was disrupted in some areas of cell-cell contact. The ultrastructure of the junctional complex in the apical region of the Caco-2BBe monolayer was investigated using electron microscopy. In cells incubated under control conditions, the tight junction, adherens junction, and desmosomes were intact. In SNP-treated cells, these junctions were separated by a wider space and contained less electron-dense material.

Reconciling the Discordant Results Regarding the Effects of NO on Gut Epithelial Permeability

At first glance, our results [52], along with those of Tepperman et al. [27, 50] and Lopez-Belmonte et al. [49], which suggest that NO impairs epithelial viability and/or function, seem to be at odds with those of Kubes and colleagues [44, 45], which suggest just the opposite. The discordance in findings, however, probably reflects differences in the model systems employed (or in some cases, the doses of NO donors used). Recently, Kanwar et al. [72] reported that inhibition of NO biosynthesis increases intestinal permeability as result of the loss of inhibition of release of other mediators, particularly histamine and platelet activating factor, from mast cells residing in proximity to the epithelium. Accordingly, it is probable that much of the data postulating a beneficial effect of NO on the maintenance of gut permeability may reflect effects of this signaling molecule on mast cells or other non-enterocytic cells residing in close proximity to the mucosa. Obviously, such effects would not be observed in experiments such as ours, which utilize a homogeneous population of epithelial cells. Concentration-dependent phenomena also may be important, particularly in the *in vivo* environment. Whereas low concentrations of NO or NO donors may have protective actions, perhaps by promoting vasodilating or inhibiting aggregation of neutrophils, higher concentrations may exert cytotoxic effects [49].

Role of NO as a Mediator of Cytokine-induced Intestinal Epithelial Hyperpermeability

In 1989, Madara and Stafford [42] reported that incubation of monolayers of cultured intestinal epithelial (T84) cells with the cytokine, IFN-γ, leads to marked increases in permeability to water-soluble molecules. This observation was subsequently confirmed by Adams et al. [43]. In view of our findings suggesting that NO is capable of modulating the permeability of enterocytic monolayers, we hypothesized that the effect of IFN-γ on permeability involved an NO-dependent mechanism. Accordingly, we incubated monolayers of Caco-2BBe cells with this cytokine and replicated the findings of the previous studies just cited [73]. Interestingly, we also found that incubation with IFN-γ increases nitrate plus nitrite concentrations in culture supernatants [73]. Additionally, we found that co-incubation with several different inhibitors of NO biosynthesis significantly blunted the induction of hyperpermeability induced by IFN-γ and blunted the release of nitrogen oxides into the culture media [73]. These findings support the notion that cytokine-induced intestinal epithelial hyperpermeability is mediated, at least in part, by a NO-dependent pathway.

Conclusion

It is becoming abundantly clear that NO has myriad effects on gut epithelial integrity and function. These effects include modulation of secretion, permeability, energy metabolism, and viability. Depending upon multiple factors, including the concentration of NO and/or the model system employed, the effects of NO can be widely disparate, ranging from cellular protection, on the one hand, to cytotoxicity on the other. In any case, however, because the intestinal epithelium appears to play such a central role in the pathophysiology of critical illness, modulation of the NO system as it relates to the gut is likely to become an increasingly active area of investigation for clinicians and scientists interested in problems relating to intensive care medicine.

References

1. Nathan C (1992) Nitric oxide as a secretory product of mammalian cells. FASEB J 6: 3051–3064
2. Stark ME, Szursewski JH (1992) Role of nitric oxide in gastrointestinal and hepatic function and disease. Gastroenterology 103:1928–1949
3. Osthaus LE, Galligan JJ (1992) Antagonists of nitric oxide synthesis inhibit nerve-mediated relaxations of longitudinal muscle in guinea pig ileum. J Pharmacol Exp Ther 260:140–145
4. Yamato S, Spechler SJ, Goyal RK (1992) Role of nitric oxide in esophageal peristalsis in the oppossum. Gastroenterology 103:197–204
5. Orihata M, Sarna SK (1994) Inhibition of nitric oxide synthase delays gastric emptying of solid meals. J Pharmacol Exp Ther 271:660–670
6. Hryhorenko LM, Woskowska Z, Fox-Threlkeld JET (1994) Nitric oxide (NO) inhibits release of acetylcholine from nerves of isolated circular muscle of the canine ileum: Relationship to motility and release of nitric oxide. J Pharmacol Exp Ther 271:918–926
7. Hata F, Ishii T, Kanada A, Yamano N (1990) Essential role of nitric oxide in descending inhibition in the rat proximal colon. Biochem Biophys Res Commun 172: 1400–1406
8. Kaufman HS, Shermak MA, May CA, Pitt HA, Lillemoe KD (1993) Nitric oxide inhibits resting sphincter of Oddi activity. Am J Surg 165:74–80
9. Slivkia A, Chuttani R, Carr-Locke DL, et al (1994) Inhibition of sphincter of Oddi function by the nitric oxide carrier S-nitroso-N-acetylcysteine in rabbits and humans. J Clin Invest 94:1792–1798
10. Diederich D, Yang Z, Bühler FR, Lüscher TF (1990) Impaired endothelium-dependent relaxations in hypertensive resistance arteries involve cyclooxygenase pathway. Am J Physiol 258:H445–H451
11. Bohlen HG, Lash JM (1992) Intestinal lymphatic vessels release endothelial-dependent vasodilators. Am J Physiol 262:H813–H818
12. Spain DA, Wilson MA, Bar-Natan MF, Garrison RN (1994) Role of nitric oxide in the small intestinal microcirculation during bacteremia. Shock 2:41–46
13. Bredt DS, Snyder SH (1990) Localization of nitric oxide synthase indicating a neural role for nitric oxide. Nature 347:768–770
14. Llewellyn-Smith IJ, Song ZM, Costa M, Bredt DS, Snyder SH (1992) Ultrastructural localization of nitric oxide synthase immunoreactivity in guinea-pig enteric neurons. Brain Res 577:337–342
15. Ward SM, Xue C, Shuttleworth CW, Bredt DS, Snyder SH (1992) NADPH diaphorase and nitric oxide synthase colocalization in enteric neurons of canine proximal colon. Am J Physiol 263:G277–G284
16. Nichols K, Staines W, Krantis A (1993) Nitric oxide synthase distribution in the rat intestine: A histochemical analysis. Gastroenterology 105:1651–1661
17. Pique JM, Whittle BJR, Esplugues JV (1989) The vasodilator role of endogenous nitric oxide in the rat gastric microcirculation. Eur J Pharmacol 174:293–296
18. Publicover NG, Hammond EM, Sanders KM (1993) Amplification of nitric oxide signaling by interstitial cells isolated from canine colon. Proc Natl Acad Sci 90:2087–2091
19. Hogaboam CM, Befus AD, Wallace JL (1993) Modulation of rat mast cell reactivity by IL-1β: Divergent effects on nitric oxide and platelet activating factor release. J Immunol 151:1367–3774
20. Granger DL, Hibbs JB, Perfect JR, Durack DT (1990) Metabolic fate of L-arginine in relation to microbiostatic capability of murine macrophages. J Clin Invest 85:264–273
21. Stuehr DJ, Nathan CF (1989) Nitric oxide: A macrophage produce responsible for cytostasis and respiratory inhibition in tumor target cells. J Exp Med 169:1543–1555
22. Grisham MB, Ware K, Gilleland HE Jr, Gilleland LB, Abell CL, Yamada T (1992) Neutrophil-mediated nitrosamine formation: Role of nitric oxide in rats. Gastroenterology 103:1260–1266

23. Malzwista SE, Montgomery RR, van Blaricom G (1992) Evidence for reactive nitrogen intermediates in killing of staphylococci by human neutrophil cytoplasts. A new microbicidal pathway for polymorphonuclear leukocytes. J Clin Invest 90:631–635
24. Goretski J, Zafiriou OC, Hollocher TC (1990) Steady-state nitric oxide concentrations during denitrification. J Biol Chem 265:11535–11538
25. Boughton-Smith NK, Evans SM, Hawkey CJ, et al (1993) Nitric oxide synthase activity in ulcerative colitis and Crohn's disease. Lancet 342:338–340
26. Boughton-Smith NK, Evans SM, Laszlo F, Whittle BJ, Moncada S (1993) The induction of nitric oxide synthase and intestinal vascular permeability by endotoxin in the rat. Br J Pharmacol 110:1189–1195
27. Tepperman BL, Brown JF, Korolkiewicz R, Whittle BJR (1994) Nitric oxide synthase activity, viability and cyclic GMP levels in rat colonic epithelial cells: Effect of endotoxin challenge. J Pharmacol Exp Ther 271:1477–1482
28. Wilson KT, Ciancio MJ, Chang EB (1994) Inducible nitric oxide synthase mRNA expression is increased in intestinal mucosa of endotoxemic rats and is inhibited by dexamethasone. Gastroenterology 106:A793 (Abst)
29. Grisham MB (1993) Nitric oxide production by intestinal epithelial cells. Gastroenterology 104:A710 (Abst)
30. Dignass A, Podolsky D, Rachmilewitz D (1994) Nitric oxide generation by intestinal epithelial cells is stimulated by cytokines and bacterial endotoxin. Gastroenterology 106:A673 (Abst)
31. Wilson KT, Xie Y, Musch MW, Chang EB (1993) Sodium nitroprusside stimulates anion secretion and inhibits sodium chloride absorption in rat colon. J Pharmacol Exp Ther 266:224–230
32. MacNaughton WK (1993) Nitric oxide-donating compounds stimulate electrolyte transport in the guinea pig intestine *in vitro*. Life Sciences 53:585–593
33. Gaginella TS, Mascolo N, Izzo AA, Autore G, Capasso F (1994) Nitric oxide as a mediator of bisacodyl and phenolphtalein laxative action: Induction of nitric oxide synthase. J Pharmacol Exp Ther 270:1239–1245
34. Barry MK, Aloisi JD, Pickering SP, Yeo CJ (1994) Nitric oxide modulates water and electrolyte transport in the ileum. Ann Surg 219:382–388
35. Rao PK, Revierc JM, Pascaud X, Junien JL, Porreca F (1994) Tonic regulation of mouse ileal ion transport by nitric oxide. J Pharmacol Exp Ther 269:626–631
36. Krugliak P, Hollander D, Le K, Ma T, Dadufalza VD, Katz KD (1990) Regulation of polyethylene glycol 400 intestinal permeability by endogenous and exogenous prostanoids. Influence on non-steroidal anti-inflammatory drugs. Gut 31:417–421
37. Madara JL (1989) Loosening tight junctions: Lessons from the intestine. J Clin Invest 83:1089–1094
38. Duffey ME, Hainau B, Ho S, Bentzel CJ (1981) Regulation of epithelial tight junction permeability by cyclic AMP. Nature 294:451–453
39. McRoberts JA, Aranda R, Riley N, Kang H (1990) Insulin regulates the paracellular permeability of cultured intestinal epithelial cell monolayers. J Clin Invest 85: 1127–1134
40. McRoberts JA, Riley NE (1992) Regulation of T84 cell monolayer permeability by insulin-like growth factors. Am J Physiol 262:C207–C213
41. Stenson WF, Easom RA, Riehl TE, Turk J (1993) Regulation of paracellular permeability in Caco-2 cell monolayers by protein kinase C. Am J Physiol 265:G955–G962
42. Madara JL, Stafford J (1989) Interferon-γ directly affects barrier function of cultured intestinal epithelial monolayers. J Clin Invest 83:724–727
43. Adams RB, Planchon SM, Roche JK (1993) IFN-γ modulation of epithelial barrier function: Time course, reversibility, and site of cytokine binding. J Immunol 150: 2356–2363
44. Kubes P (1992) Nitric oxide modulates epithelial permeability in the feline small intestine. Am J Physiol 262:G1138–G1142
45. PayneD, Kubes P (1993) Nitric oxide donors reduce the rise in reperfusion-induced intestinal mucosal permeability. Am J Physiol 265:G189–G195

46. Boughton-Smith NK, Hutcheson IR, Deakin AM, Whittle BJR, Moncada S (1990) Protective effect of S-nitroso-N-acetyl-penicillamine in endotoxin-induced acute intestinal damage in the rat. Eur J Pharmacol 191:485–488
47. Kubes P (1993) Ischemia-reperfusion in the feline small intestine: A role for nitric oxide. Am J Physiol 264:G143–G149
48. Hutcheson IR, Whittle BJ, Boughton-Smith NK (1990) Role of nitric oxide in maintaining vascular integrity in endotoxin-induced intestinal damage in the rat. Br J Pharmacol 101:815–820
49. Lopez-Belmonte J, Whittle BJ, Moncada S (1993) The actions of nitric oxide donors in the prevention or induction of injury to the rat gastric mucosa. Br J Pharmacol 108: 73–78
50. Tepperman BL, Brown JF, Whittle BJR (1993) Nitric oxide synthase induction and intestinal cell viability in rats. Am J Physiol 265:G214–G218
51. Hidalgo IJ, Raub TJ, Borchardt RT (1989) Characterization of human colonic carcinoma cell line (Caco-2) as a model system of intestinal epithelial permeability. Gastroenterology 96:736–749
52. Salzman AL, Menconi MJ, Unno N, et al (1995) Nitric oxide dilates tight junctions and depletes ATP in cultured Caco-2BBe intestinal epithelial monolayers. Am J Physiol 268:G361–G373
53. Kuwahara M, Berry CA, Verkman AS (1988) Rapid development of vasopressin-induced hydroosmosis in kidney collecting tubules measured by a new fluorescent technique. Biophys J 54:595–602
54. Tagesson C, Sjodahl R, Thoren B (1978) Passage of molecules through the wall of the gastrointestinal tract. I. A simple experimental model. Scand J Gastroenterol 13: 519–524
55. Pantzar N, Westrom BR, Luts A, Lundin S (1993) Regional small-intestinal permeability in vitro to different-sized dextrans and proteins in the rat. Scand J Gastroenterol 28:205–211
56. Geiger J, Nolte C, Butt E, Sage SO, Walter U (1992) Role of cGMP and cGMP-dependent protein kinase in nitrovasodilator inhibition of agonist-evoked calcium elevation in human platelets. Proc Natl Acad Sci 89:1031–1035
57. Garg UC, Hassid A (1989) Nitric oxide-generating vasodilators and 8-bromo-cyclic guanosine monophosphate inhibit mitogenesis and proliferation of cultured rat vascular smooth muscle cells. J Clin Invest 83:1774–1777
58. Waldman SA, Murad F (1988) Biochemical mechanisms underlying vascular smooth muscle relaxation: The guanylate cyclase-cyclic GMP system. J Cardiovasc Pharmacol 12 (Suppl 5):S115–S118
59. Guarino A, Cohen M, Thompson M, Dharmsathaphorn D, Gianella R (1987) T84 cell receptor binding and guanyl cyclase activation by Escherichia coli heat-stable toxin. Am J Physiol 253:G775–G780
60. Henry Y, Lepoivre M, Drapier JC, Ducrocq C, Boucher JL, Guissani A (1993) EPR characterization of molecular targets for NO in mammalian cells and organelles. FASEB J 7:1124–1134
61. Drapier JC, Hibbs JB Jr (1986) Murine cytotoxic activated macrophages inhibit aconitase in tumor cells. J Clin Invest 78:790–797
62. Hibbs JB Jr, Taintor RR, Vavrin Z, Rachlin EM (1988) Nitric oxide: A cytotoxic activated macrophage effector molecule. Biochem Biophys Res Commun 157:87–94
63. Drapier JC, Hibbs JB Jr (1988) Differentiation of murine macrophages to express nonspecific cytotoxicity for tumor results in L-arginine-dependent inhibition of mitochondrial iron-sulfur enzymes in the macrophage effector cells. J Immunol 140: 2829–2838
64. Stadler J, Billiar TR, Curran RD, Stuehr DJ, Ochoa JB, Simmons RL (1991) Effect of exogenous and endogenous nitric oxide on mitochondrial respiration rat hepatocytes. Am J Physiol 260:C910–C916
65. Stadler J, Curran RD, Ochoa JB, et al (1991) Effect of endogenous nitric oxide on mitochondrial respiration of rat hepatocytes in vitro and in vivo. Arch Surg 126:186–191

66. Molina y Vedia L, McDonald B, Reep B, et al (1992) Nitric oxide-induced S-nitrosylation of glyceraldehyde-3-phosphate dehydrogenase inhibits enzymatic activity and increases endogenous ADP-ribosylation. J Biol Chem 267:24929–24932
67. Dimmeler S, Lottspeich F, Brune B (1992) Nitric oxide causes ADP-ribosylation and inhibition of glyceraldehyde-3-phosphate dehydrogenase. J Biol Chem 267: 16771–16774
68. Zhang J, Snyder SH (1992) Nitric oxide stimulates auto-ADP-ribosylation of glyceraldehyde-3-phosphate dehydrogenase. Proc Natl Acad Sci 89:9382–9385
69. Dimmeler S, Ankarcrona M, Nicotera P, Brune B (1993) Exogenous nitric oxide (NO) generation or IL-1β-induced intracellular NO production stimulates inhibitory auto-ADP-ribosylation of glyceraldehyde-3-phosphate dehydrogenase in RINm5F cells. J Immunol 150:2964–2971
70. Zhang J, Dawson VL, Dawson TM, Snyder SH (1994) Nitric oxide activation of poly(ADP-ribose) synthetase in neurotoxicity. Science 263:687–689
71. Winter M, Wilson JS, Bedell K, Shasby DM (1990) The conductance of cultured epithelial cell monolayers: Oxidants, adenosine triphosphate, and phorbol dibutyrate. Am J Respir Cell Mol Biol 2:355–364
72. Kanwar S, Wallace JL, Befus D, Kubes P (1994) Nitric oxide synthesis inhibition increases epithelial permeability via mast cells. Am J Physiol 266:G222–G229
73. Unno N, Menconi MJ, Smith M, Fink MP (1995) Nitric oxide mediates interferon-gamma-induced hyperpermeability in cultured human intestinal epithelial monolayers. Crit Care Med (in press)

Nitric Oxide

Inhibition of Nitric Oxide Synthase Activity in Circulatory Shock: Friend or Foe?

C. Thiemermann

Introduction

Nitric oxide (NO) is a recently discovered autacoid, which is produced by a hitherto unrecognized enzymatic pathway in many mammalian cells. Thus, it is not surprising that NO has many diverse biological functions in the cardiovascular, nervous and immune system [1]. The biological activity of NO was first described by Furchgott and Zawadski in 1980 [2] as a labile, vasodilator substance released by endothelial cells in response to acetylcholine (and other endothelium-dependent vasodilators), and called "endothelium-derived relaxing factor (EDRF)". We know today that the formation of NO by the vascular endothelium accounts for the biological activity of EDRF [3, 4]. NO is produced from the guanidino nitrogen group of L-arginine by NO synthase (NOS). Once produced, NO diffuses to adjacent cells and activates soluble guanylyl cyclase by binding to the iron on its heme component. The subsequent rise in intracellular cyclic GMP mediates many, but not all, of the biological properties of NO [1].

Three different isoforms of NOS have been isolated, cloned, sequenced and expressed. The NOS in endothelial cells (eNOS) and neuronal cells (bNOS) are expressed constitutively, and both enzymes require an increase in intracellular calcium for activation. Under physiological conditions, the continuous release of NO following activation of eNOS, presumably by shear-stress, dilates blood vessels, and, in concert with vasoconstrictor catecholamines, regulates blood vessel diameter, organ blood flow and blood pressure. In addition, NO inhibits the adhesion of platelets and polymorphonuclear cells (PMN) to the surface of the endothelium [1]

Activation of macrophages and other cells with cytokines and endotoxin results in the expression of an "inducible" isoform of NOS (iNOS), the activity of which is independent of changes in intracellular calcium. Once expressed, iNOS produces large amounts of NO. This enhanced formation of NO by e.g. cytokine-activated macrophages accounts for the ability of these cells to kill bacteria or tumor cells and, hence, plays an important role in host defense. In contrast to the constitutive isoforms of NOS (eNOS or bNOS), the availability of L-arginine can become rate-limiting for a maximal generation of NO by iNOS [1, 5].

This chapter is intended to serve the following 3 functions:

1) to outline the role of an enhanced formation of NO in the pathophysiology of endotoxic and other forms of circulatory shock;
2) to discuss the mechanisms of iNOS induction and its regulation; and
3) to discuss novel therapeutic approaches designed to modulare the biosynthesis of NO with the aim to improve hemodynamics and survival in animal models and patients with septic shock.

Role of NO in the Pathophysiology of Septic Shock

The circulatory failure associated with shock of various etiologies is characterized by severe hypotension (or peripheral vasodilatation), hyporeactivity of the vasculature to endogenous and exogenous vasoconstrictor agents, myocardial dysfunction, maldistribution of organ blood flow and reduced tissue oxygen extraction, which ultimately lead to multiple organ failure (MOF) and death. There is now good evidence that an enhanced formation of NO contributes to several of these pathophysiological features of septic shock. For instance, an enhanced formation of NO due to activation of eNOS (acute phase of shock) and particularly following induction of iNOS in the vascular wall (late phase of shock) importantly contributes to the hypotension in various animal models (rat, dog, pig, sheep) of septic shock [5].

The peripheral vascular failure in animals and man with septic shock also results in a progressive attenuation of the pressor effects afforded by norepinephrine (Fig. 1) and other vasoconstrictor agents which contributes to the therapy-resistant hypotension. An enhanced formation of NO following induction of iNOS within the vascular smooth muscle also contributes to the hyporeactivity of vessels from animals with septic [6] (Fig. 1) or hemorrhagic shock [7] to endogenous or exogenous catecholamines (and other vasoconstrictors). In animals with septic shock, a vascular hyporeactivity (which is mediated by NO) has been demonstrated in conductance, resistance as well as venous vessels.

Prolonged periods of circulatory shock are associated with a reduction in cardiac contractility (myocardial dysfunction) and ultimately cardiac output. Cytokines including tumor necrosis factor (TNF) and interleukin-1 (IL-1) cause induction of iNOS and a reduction in the contractility of rat cardiac myocytes [8] and hamster papillary muscle [9]; and inhibition of NO synthesis causes positive inotropic effects in cardiac myocytes either treated with cytokines [9] or obtained from animals with septic shock [10]. Thus, it is conceivable that an enhanced formation of NO by iNOS also contributes to the myocardial dysfunction in septic shock. Further studies are, however, necessary to document that the systemic administration of NOS inhibitors also results in an improved contractility and cardiac output in animal models of septic shock.

Circulatory shock also results in a marked defect in tissue oxygen extraction resulting in an increased venous oxygen concentration. It is now well

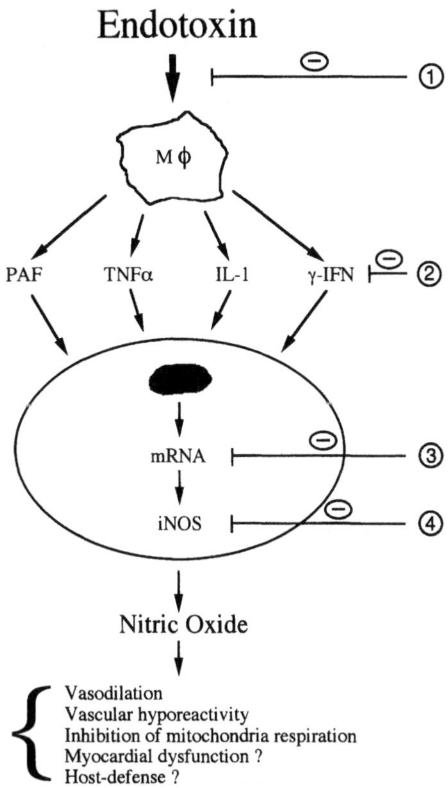

Fig. 1. Schematic diagram of the mechanisms leading to the induction of iNOS in endotoxic shock as well as of the therapeutic strategies (1–4) which inhibit either the process of induction (1–3) or the activity of iNOS. Endotoxin causes the release of TNF, IL-1, IFN-γ or PAF from macrophages (or other cells). These cytokines and PAF, either alone or in concert, cause the expression of iNOS in vascular smooth muscle (and other) cells, resulting in an enhanced formation of NO. In septic shock, the formation of NO by iNOS contributes to vasodilatation, vascular hyporeactivity to endogenous and exogenous vasoconstrictor agents (catecholamines), inhibition of mitochondrial respiration and possibly myocardial dysfunction, and may well be important for the host defense. The induction of iNOS can be prevented with therapeutic strategies which ① inhibit the activation of cells caused by endotoxin (e.g. monoclonal antibodies to endotoxin, BPI etc.); ② inhibit the release or the effects of cytokines or PAF (e.g. monoclonal antibodies to TNF, IL-1ra, PAFra etc.); or ③ inhibit transcription and/or translation of the iNOS gene (e.g. glucocorticoids, lipocortin-1, etc.); ④ The activity of iNOS can be inhibited with NOS inhibitors (e.g. L-NMMA, aminoguanidine, S-methyl-ITU etc.). It is suggested that therapies which prevent the induction of iNOS (1–3) have to be administered relatively early in septic shock (e.g. prior to iNOS induction and possibly prior to severe hypotension); while selective inhibitors of iNOS activity may exert beneficial hemodynamic effects even when administered at later stages of shock

established that the local generation of large amounts of NO causes autoinhibition of mitochondrial respiration (e.g. in hepatocytes or vascular smooth muscle cells), for NO inhibits several enzymes which are part of the mitochondrial respiratory chain (aconitase, NADH-ubiquinone reductase, succinate-ubiquinone oxidoreductase) [5]. These effects of NO are associated with an increased formation of lactate, indicating a shift in glucose metabolism from aerobic to anaerobic pathways. In addition, an enhanced formation of NO by iNOS within the vascular wall may well act as an "oxygen sink" (formation of nitrite and nitrate from oxygen and NO) for molecular oxygen diffusing from erythrocytes to the mitochondria of adjacent cells. This would result a shift in the "Krogh-cylinder" of the distribution of local tissue oxygen [11]. Thus, the generation of large quantities of NO may cause

1) maldistribution of regional blood flow (reduced oxygen supply),
2) formation of a diffusion barrier for oxygen (reduced oxygen extraction), and
3) inhibition of mitochondrial respiration.

In concert with the severe hypotension (reduced perfusion pressure), these effects of a local overproduction of NO may importantly contribute to the severe organ/tissue ischemia which ultimately facilitates the development of the multiple organ dysfunction syndrome associated with shock of various etiologies.

Prolonged periods of septic and hemorrhagic shock also result in a progressive endothelial dysfunction indicated by an impairment of endothelium-dependent vasodilatations. There is some evidence that this endothelial dysfunction is, in part, due to a reduced formation of NO by eNOS due to

1) downregulation of eNOS by TNF or other cytokines,
2) endothelial cell damage due to cytotoxic effects of NO itself, or
3) inactivation of NO by oxygen radicals.

The relative importance of this endothelial dysfunction in the pathogenesis of the alterations in organ blood flow distribution or MOF associated with circulatory shock are unclear.

In addition to endotoxin shock, an enhanced formation of NO (particularly following induction of iNOS) also plays an important role in the circulatory failure associated with hemorrhagic shock [11]. For instance, an enhanced formation of NO following induction of iNOS contributes to the delayed vascular decompensation and the vascular hyporeactivity to vasoconstrictor agents in rats with hemorrhagic hypotension [7]. Moreover, the dose-limiting hypotension associated with the chemotherapeutic use of IL-2 in patients with renal cancer or melanoma is due to an enhanced formation of NO by iNOS [12]. It has been suggested that the hyperdynamic circulatory failure in patients with liver cirrhosis, which is characterized by hypotension, low peripheral vascular resistance, increased cardiac output, and a vascular hyporeactivity to vasoconstrictor agents, is in part due to an augmented formation of NO within the vascular wall [13]. Although the circulatory failure associated with anaphylactic shock, traumatic shock or thermal (burns) injury also results in enhanced plasma levels of nitrite (metabolite of NO), it is yet unclear whether this enhanced formation of NO does, indeed, contribute to the hemodynamic alterations in these disorders [11].

Regulation and Expression of iNOS Induction

The induction of iNOS afforded by endotoxin *in vivo* is secondary to the release of the cytokines TNF, IL-1 and interferon-gamma (IFN-γ). TNF causes the expression of iNOS in cultured cells [14], and when injected into animals causes the formation of NO [15] as well as (NO-mediated) hypotension [16] and vascular hyporeactivity to vasoconstrictor agents [17]. Administration of TNF alone, or in combination with low doses of endotoxin causes hypotension, peripheral vasodilatation and organ injury; and septic shock is associated with elevated plasma levels of TNF. Moreover, pretreatment of rats with a monoclonal antibody to TNF prevents 1) the circulatory failure and the induction of iNOS afforded by endotoxin [14], and 2) the increase in plasma nitrite caused by injection of *Pseudomonas aeruginosa* [19]. Thus, the release by endotoxin of endogenous TNF importantly contributes to the induction of iNOS *in vivo*; and it is likely that some of the beneficial hemodynamic effects of either antibodies to TNF or of soluble TNF receptors in animal models of shock are due to the inhibition of iNOS induction.

Similarly, IL-1 causes the expression of iNOS in various cultured cells *in vitro* [20–22] and enhances the formation of NO *in vivo* [15]. In addition, incubation of isolated blood vessels with IL-1 results in vascular hyporeactivity to vasoconstrictor agents, which is due to the formation of NO [23]. The hypothesis that the formation of endogenous IL-1 in response to endotoxin contributes to the induction of iNOS *in vivo* is supported by the findings that 1) pretreatment of rats with the endogenous IL-1 receptor antagonist (IL-1ra) prevents the hemodynamic effects and the induction of iNOS caused by endotoxin in the rat [24]; and 2) that a polyclonal antibody to IL-1 attenuates the induction of iNOS in a murine model of endotoxemia [25].

The notion that endogenous IFN-γ contributes to the induction of iNOS in shock is supported by the findings that monoclonal antibodies to IFN-γ attenuate the expression of iNOS 1) in the liver of mice with gram-negative sepsis [26] and 2) in peritoneal macrophages from mice treated with M. bovis BCG [27]. Similarly, peritoneal macrophages obtained from either mice lacking the IFN-γ receptor or from mice with disrupted IFN-γ genes do not produce NO upon stimulation with endotoxin [28, 29].

Injection of platelet activating factor (PAF) itself causes expression of iNOS *in vivo* (rat), and the PAF receptor antagonist, WEB 2086, prevents the induction of iNOS caused by endotoxin *in vitro* and *in vivo* [30]. Thus, an enhanced formation of endogenous PAF either alone, or in concert with cytokines contributes to the expression of iNOS in endotoxemia (Fig. 2).

Interestingly, there is some evidence that anti-inflammatory cytokines including IL-4, IL-10 or IL-13 inhibit the induction of iNOS caused by endotoxin or pro-inflammatory cytokines. For instance, IL-4 and IL-10 inhibit the induction of iNOS in cultured murine macrophages [31–33]. The inhibition of iNOS induction caused by IL-4 or IL-10 is likely to be due to the inhibition

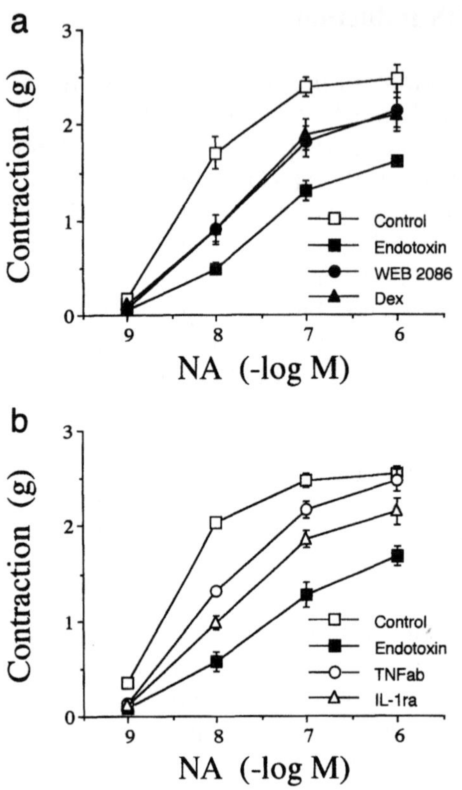

Fig. 2. Prolonged periods of endotoxemia cause a vascular hyporeactivity to vasoconstrictor agents. The figure depicts the dose-dependent contractions elicited by norepinephrine (10^{-9} to 10^{-6} M) in aortic rings prepared from rats treated with either vehicle (*panel a and b: open squares*), endotoxin (10 mg/kg I.V.; *panel a and b: black squares*), endotoxin plus the PAF-receptor antagonist WEB 2086 (5 mg/kg I.V. 20 min prior to endotoxin; *panel a: black circles*); endotoxin plus dexamethasone (Dex; 3 mg/kg I.V. 30 min prior to endotoxin; *panel a: black triangles*); endotoxin plus TNF antibody (*panel b:* TNFab; *open circles*) or endotoxin plus the endogenous IL-1ra (*panel b:* IL-1ra, 15 mg/kg at 15 min prior to endotoxin followed by an infusion of 2.4 mg/kg/h). Please note that all of the above therapies inhibit the expression of iNOS caused by endotoxin in the lung (not shown) and partially, but not completely, prevent the development of the vascular hyporeactivity to norepinephrine in rat aortic rings

by these cytokines of the formation of pro-inflammatory cytokines (TNF, IL-1, IFN-γ) which are known to cause the expression of iNOS *in vitro* and *in vivo*. For instance, IL-4 inhibits the release of TNF and IL-1 caused by endotoxin in human monocytes [34] and stimulates the formation of the endogenous IL-1ra [35]. Similarly, IL-10 attenuates the formation of IL-1, TNF and IL-8 by human monocytes or PMN activated with endotoxin [36]. Exogenous IL-10 reduces mortality in murine models of septic shock, while antibodies to IL-10 enhance the mortality caused by endotoxin in this species [37, 38]. It is likely that these effects of IL-10 are, at least in part, due to the inhibition of iNOS induction.

In addition, endogenous glucocorticoids exert a tonic suppression of the induction of iNOS, for 1) injection of small amounts of endotoxin to adrenalectomized animals results in a substantial degree of iNOS expression, while the same amount of endotoxin has no such effect in normal animals; and 2) this "supersensitivity" to the induction of iNOS in adrenalectomized animals is prevented by the replacement therapy with dexamethasone [39]. Moreover, a rise in the endogenous (plasma) levels of corticosterone contributes to the development of endotoxin-tolerance. For instance, repetitive exposure of rats to small amounts of endotoxin results in "endotoxin-tolerance", which is associated with an attenuation of hemodynamic effects (hypotension and vascular hyporeactivity) and iNOS induction caused by a latter injection of a large amount of endotoxin in this species [40]. Interestingly, the repetitive administration of endotoxin results in a substantial rise in endogenous corticosterone levels, and pretreatment of "endotoxin-tolerant" rats with the glucocorticoid receptor antagonist RU 486 restores the ability of endotoxin to cause a severe hypotension as well as induction of iNOS in these animals [40]. Thus, the cardiovascular tolerance to endotoxin *in vivo* is accompanied and explained by reduced induction of iNOS due to the elevation of endogenous glucocorticoid levels [40].

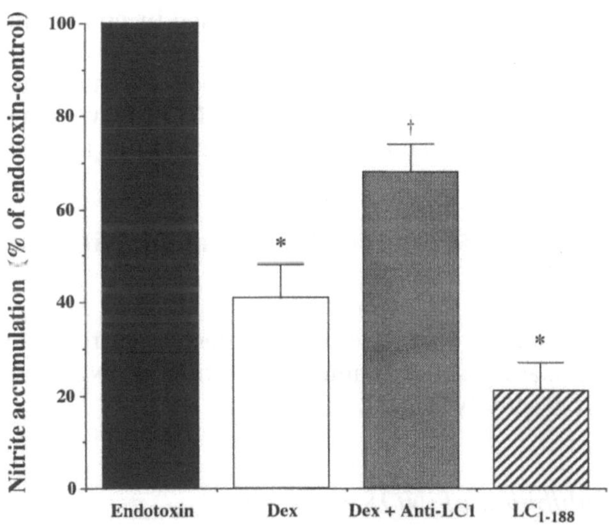

Fig. 3. Endogenous lipocortin-1 mediates the inhibition by dexamethasone of the induction of iNOS activity (measured as nitrite accumulation by the Griess reaction) in cultured macrophages treated with endotoxin (1 μg/mL for 24 h). Please note that exposure of J774.2 macrophages to endotoxin results in an enhanced formation of nitrite which is prevented by pretreatment of these cells with dexamethasone (Dex) or the lipocortin-1 fragment 1–188 (LC_{1-188}). The inhibition by dexamethasone of the induction of nitrite formation caused by endotoxin is attenuated by pretreatment of cells with a monoclonal antibody to lipocortin-1 (Dex + Anti-LC1)

The inhibition of the induction of iNOS by dexamethasone in cultured murine macrophages activated with lipopolysaccharide (LPS) is due to an enhanced expression of lipocortin-1, a member of the annexin superfamily of peptides, for

1) pretreatment of these cells with a monoclonal antibody against lipocortin-1 attenuates the inhibition by dexamethasone of the expression of iNOS protein and activity;
2) the lipocortin-1 fragment 1–188 also prevents the expression by endotoxin of iNOS protein and activity; and
3) pretreatment of murine macrophages with dexamethasone results in the expression of lipocortin-1 protein (cell surface pool) in these cells (Fig. 3) [41].

Similarly, pretreatment of rats with a monoclonal antibody to lipocortin-1 prevents the beneficial hemodynamic effects as well as the inhibition of the expression of iNOS caused by dexamethasone in the anesthetized rat [41]. Thus, dexamethasone causes the expression of lipocortin-1, which prevents the induction of iNOS caused by endotoxin *in vitro* and *in vivo*.

In addition to glucocorticoids, endogenous *polyamines* may also attenuate the degree of iNOS expression afforded by endotoxin or cytokines *in vivo* for spermine (and other polyamines) is metabolized by polyamine oxidase (present in plasma) to a yet unknown aldehyde metabolite(s), which in turn inhibits the expression of iNOS afforded by endotoxin in cultured macrophages [42]. As limitations of space do not allow an extensive review of all of the factors and cellular mechanisms which are involved in the expression and/or regulation of the expression of iNOS *in vitro* and *in vivo*, the interested reader is referred to a recent review of this topic [11].

Therapeutic Interventions to Inhibit Nitric Oxide Formation in Shock

In principle, there are two approaches for reducing the formation of NO in circulatory shock, namely inhibition of iNOS expression or inhibition of NOS activity (Fig. 2).

Inhibition of iNOS Expression

The induction of iNOS can be prevented by pretreatment (prior to the injection of endotoxin) of animals with dexamethasone [43], PAF receptor antagonists [30], the endogenous IL-1ra [24], antibodies to TNF [18], dihydropyridine-type calcium channel antagonists [44] or glibenclamide [45]; and the beneficial hemodynamic effects of these agents/drugs in animal models of septic shock are, at least in part, due to inhibition of iNOS induction (Fig. 2). It should, however, be pointed out that most of the above therapeutics have to be administered prior to the application of endotoxin to prevent the delayed circulatory failure as well as the induction of iNOS caused by endotoxin in animals *in vivo*. Thus, any therapy aimed to interfere with the effects of cytokines or other mediators which are released within the first 60 to 90 min of endotoxemia, are unlikely (at least in animal models) to achieve a maximal therapeutic benefit when given after the onset of hypotension and, hence, presumably iNOS induction.

Inhibition of NOS Activity

Following the observations that infusion of the NOS inhibitor N^G-methyl-L-arginine (L-NMMA) attenuates the hypotension afforded by endotoxin in rats [46] and enhances blood pressure and peripheral vascular resistance in dogs with endotoxemia [47], numerous experimental studies have documented that (lower doses) of this, and other non-selective NOS inhibitors, exert beneficial hemodynamic effects in animals (e.g. rats, dogs, pigs, sheep) and man with septic shock (see [5] for review). The use of non-selective inhibitors of NOS in animal models of circulatory shock is, however, hampered by the fact that these agents also inhibit eNOS activity and, hence, the release of NO from the vascular endothelium. As inhibition of eNOS activity leads to excessive vasoconstriction and potentially to an enhanced adhesion of platelets and PMN to the vascular endothelium, the use of non-selective NOS inhibitors may well lead to side effects which ultimately will limit the clinical usefulness of such agents.

Reported adverse effects of non-selective inhibition of NOS activity in animals with septic shock include an increased incidence of microvascular thrombosis as well as a reduced blood flow to critical organs including heart (resulting in a fall in cardiac output), liver (enhanced liver cell damage, thrombosis), kidney (thrombosis) and particularly the intestine [5]. Thus, it is not entirely surprising that high doses of L-NMMA increase mortality in a rabbit model of endotoxic shock [43]. Although numerous side effects of non-selective NOS inhibitors (particularly when used in high doses) have been reported in experimental endotoxemia, it should be noted that the use of lower doses of these agents in patients with circulatory shock caused a rise in blood pressure and peripheral vascular resistance without exacerbating the thrombocytopenia afforded by septic shock. Infusion of L-NMMA did, however, cause an increase in pulmonary vascular resistance as well as a fall in cardiac output [48].

To maximize any potential therapeutic effects of NOS inhibitors in patients with septic shock, several novel approaches have been suggested. These include 1) the combination of non-selective NOS inhibitors with NO-gas inhalation therapy and/or infusion of NO-donors; and 2) the use of agents which are relatively selective inhibitors of iNOS activity.

NO-Gas Inhalation Therapy

There are now several recent reports demonstrating that NO-gas inhalation causes beneficial hemodynamic effects in patients with pulmonary hypertension or adult respiratory distress syndrome (ARDS) [49]. Inhalation with NO-gas leads to a selective vasodilatation of blood vessels adjacent to ventilated areas, for NO is rapidly inactivated by hemoglobin and, hence, has no systemic hemodynamic effects. This selective dilatation of ventilated areas leads to a shift in blood flow from non-ventilated to ventilated areas of the lung (positive steal phenomenon), which in turn causes an improved ventilation to perfusion matching. Thus, NO inhalation therapy improves the efficacy of the Euler-Liljestrand reflex. It is conceivable that NO-inhalation therapy may be useful to either improve pulmonary hemodynamics and gas exchange in patients with septic shock or alternatively attenuate the rise in pulmonary vascular resistance afforded by non-selective NOS inhibitors when given together with these agents to patients with circulatory shock. Interestingly, treatment of pigs with NO-gas inhalation improves pulmonary hemodynamics, attenuates the fall in cardiac output and prevents the alterations in pulmonary gas exchanges caused by prolonged periods of endotoxemia. In addition, the combination therapy of NO-gas inhalation together with systemic administration of L-NMMA was superior to either treatment alone in preventing the alterations in hemodynamics and gas exchange as well as the mortality associated with septic shock in this species [50].

Selective Inhibition of iNOS Activity

Inhibition of NO formation by aminoguanidine was first reported by Corbett and coworkers in 1992 [51, 52]. The potency of aminoguanidine and L-arginine analogs as inhibitors of iNOS activity (*in vitro*) is similar. However, aminoguanidine is less potent as an inhibitor of eNOS in cultured cells, isolated blood vessels and enzyme preparations *in vitro* [52, 53]. Moreover, aminoguanidine attenuates the delayed fall in blood pressure and the vascular hyporeactivity to norepinephrine elicited by prolonged periods of endotoxemia in anesthetized rats. The hypothesis that aminoguanidine is a relatively selective inhibitor of iNOS *in vivo* is supported by the finding that a dose of aminoguanidine, which does not affect blood pressure (inhibition of eNOS) in normal rats, reverses the severe hypotension and inhibits iNOS activity when given at the same dose to rats at 180 min after injection of endotoxin (Fig. 4). Thus, low doses of aminoguanidine inhibit iNOS activity *in vivo* without causing a significant inhibition of eNOS activity. Interestingly, both the rise in blood pressure afforded by aminoguanidine *in vivo* as well as the inhibition of iNOS activity (*in vitro*) seen with aminoguanidine were slow in onset and only reached a maximum effect after 60 to 90 min [54]. In addition to causing beneficial hemodynamic effects, aminoguanidine improves survival in a murine model of severe endotoxemia [54]. Although it is tempting to

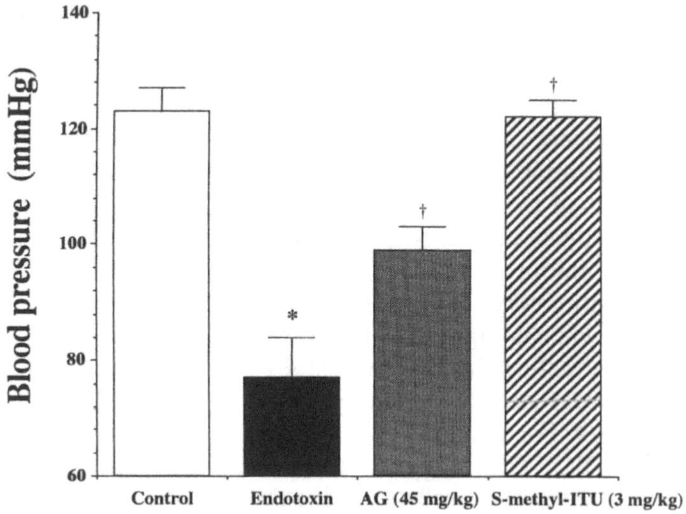

Fig. 4. The fall in mean arterial blood pressure caused by endotoxemia (10 mg/kg I.V.) for 180 min in the anesthetized rat is reversed by aminoguanidine (AG, 45 mg/kg I.V.) or S-methyl-isothiourea (S-methyl-ITU; 3 mg/kg I.V.). Please note that substantially larger dose of AG than S-methyl-ITU was needed to achieve a significant rise in blood pressure suggesting that AG is a weaker inhibitor of iNOS activity than S-methyl-ITU *in vivo*. The doses of AG and S-methyl-ITU used here cause significantly smaller increases in blood pressure (approximately 10 mm Hg) in rats without septic shock suggesting that these agents are relatively selective inhibitors of iNOS *in vivo*

speculate that the improvement in survival seen with aminoguanidine is due to (a relatively selective) inhibition of iNOS activity, aminoguanidine has many other pharmacological properties including

1) inhibition of the formation of advanced glycosylation endproducts,
2) inhibition of histamine metabolism,
3) inhibition of polyamine catabolism, and
4) inhibition of catalase activity (see [54] for references).

Thus, it is difficult to interpret, whether, and to what extent, the beneficial effects of aminoguanidine on hemodynamics or survival in rodent models of septic shock are due to inhibition of iNOS activity.

We have recently discovered that some S-substituted isothioureas are potent inhibitors of NOS activity with variable isoform selectivity [55]. In particular, *S-methylisothiourea* S-methyl-ITU), S-(2-aminoethyl)isothiourea (S-aminoethyl-ITU), S-ethylisothiourea (S-ethyl-ITU) and S-isopropyl-isothiourea (S-isopropyl-ITU) potently inhibit iNOS activity (murine J774.2 macrophages activated with LPS) with EC_{50} values 8–24 times lower than that of L-NMMA and 200 times lower than N^{w}-nitro-L-arginine (L-NAME). The inhibition of iNOS activity by these S-substituted isothioureas is dose-dependently prevented by an excess of L-arginine suggesting that these agents are competitive inhibitors of iNOS at the L-arginine binding site. Interestingly, S-ethyl-ITU and S-isopropyl-ITU are 4–6 times more potent than L-NMMA in inhibiting eNOS activity (bovine aortic endothelial cells) *in vitro* and are also more potent pressor agents than L-NMMA in the anesthetized rat. S-methyl-ITU and L-NMMA are approximately equipotent as inhibitors of eNOS activity, while, S-aminoethyl-ITU is 6 times less potent than L-NMMA in inhibiting eNOS activity *in vitro* and exerts only small increases in blood pressure (at higher doses) in the anesthetized rat. Thus, isothioureas represent a novel class of NOS inhibitors. S-methyl-ITU and S-aminoethyl-ITU are potent and relatively selective inhibitors of iNOS activity when compared to L-NMMA [55].

While writing this review, Garvey and colleagues [56] have reported that isothioureas are potent inhibitors of human NOS with variable isoform selectivity. For instance, S-ethyl-ITU is a potent competitive inhibitor of human iNOS (K_i value of 17 nM), eNOS (K_i value of 36 nM) and bNOS (K_i value of 29 nM) as well as murine iNOS (K_i value of 5.2 nM). It is suggested that

1) S-ethyl-ITU binds near the heme of NOS,
2) the sulfor atom plays a critical role in binding, and
3) the small S-ethyl or S-isopropyl substituents of S-ethyl- or S-isopropyl-ITU may well fit into a small hydrophobic pocket on the enzyme [56].

In addition, 2-NH_2-thiazoline and 2-NH_2-thiazole (the cyclized mimics of S-ethyl-ITU) are also competitive inhibitors of human NOS. Although bisisothioureas (such as S,S′-(1,3-phenylenebis(1,2-ethanediyl))bisisothiourea) appear to be the most selective inhibitors of purified human iNOS, the poor cellular penetration as well as the toxicity (rats, mice) of these doubly charged molecules may well limit their therapeutic potential [56]. Nevertheless, the above studies clearly demonstrate that selective inhibitors of iNOS can be designed.

These findings raise the question, whether isothioureas which are relatively selective inhibitors of iNOS activity, exert beneficial effects in animal models of septic shock. We have recently demonstrated that S-methyl-ITU dose-dependently (0.01–3 mg/kg I.V.) reverses the hypotension and the vascular hyporeactivity to norepinephrine in rats with endotoxic shock (Fig. 4). When given to conscious rats with endotoxemia, S-methyl-ITU (5 mg/kg i.p. given 2 h after endotoxin) attenuates the rise in plasma alanine and aspartate aminotransferases and bilirubin, suggesting that S-methyl-ITU protects against the liver injury afforded by endotoxemia. In the same model, endotoxemia for 6 h also causes an impairment of renal function, as indicated by rises in creatinine. While non-selective NOS inhibitors are known to exacerbate the renal injury afforded by endotoxemia by facilitating excessive vasoconstriction and glomerular thrombosis [5], S-methyl-ITU causes a significant improvement in renal function, possibly by raising blood pressure and, hence, perfusion pressure to the renal vascular bed. Most notably, S-methyl-ITU improves the 24-h survival in mice treated with a high dose of endotoxin

(60 mg/kg) from 8 (control) to 50% [57]. Similarly, aminoethyl-ITU reduces the degree of liver and renal injury and improves survival in rodent models of septic shock (unpublished data).

It is hoped that S-methyl-ITU, S-aminoethyl-ITU or other isothioureas, which are relatively selective inhibitors of iNOS activity, may be useful as tools in designing experiments which will help to gain a better insight into the physiological (host-defense) and pathophysiological role(s) of the generation of NO from iNOS. The outcome of such studies will also help in evaluating the potential risks and benefits of the use of NOS inhibitors in patients with circulatory shock.

Conclusion

There is now good evidence that an enhanced formation of NO, particularly due to induction of iNOS, contributes to several of the pathophysiological events leading to hemodynamic disturbances and MOF in animal models of septic and other forms of circulatory shock. Although non-selective NOS inhibitors exert beneficial hemodynamic effects in animals and man with septic shock, these agents (particularly when used in higher doses) also exert adverse effects which are due to the inhibition of endothelial eNOS. The clinical data regarding effects and side-effects of non-selective NOS inhibitors in patients with septic shock are limited. Animal studies suggest that the combination therapy of NOS inhibitors with either NO-gas inhalation and/or NO-donors or the use of iNOS-selective NOS inhibitors may result in fewer side-effects than the administration of non-selective NOS inhibitors alone. The finding that some S-substituted isothiourea derivatives are relatively selective inhibitors of iNOS clearly indicates that the development of highly selective inhibitors of iNOS may be possible. These agents will be a useful tool to gain a better understanding of the role of iNOS in host defense, and the side effects of iNOS inhibition in animal models of circulatory shock.

References

1. Moncada S, Higgs A (1993) The L-arginine-nitric oxide pathway. N Engl J Med 329: 2002–2012
2. Furchgott RF, Zawadski JV (1980) The obligatory role of endothelial cells in the relaxation of arterial smooth muscle by acetylcholine. Nature 288:373–376
3. Palmer RMJ, Ferridge AG, Moncada S (1987) Nitric oxide release accounts for the biological activity of endothelium-derived relaxing factor. Nature 327:524–526
4. Ignarro LJ, Buja GM, Wood KS, Byrns RE, Chaudhuri G (1987) Endothelium-derived relaxing factor produced and released from artery and vein is nitric oxide. Proc Natl Acad Sci USA 84:9265–9269
5. Thiemermann C (1994) The role of the L-arginine: Nitric oxide pathway in circulatory shock. Adv Pharmacol 28:45–79
6. Julou-Schaeffer G, Gray GA, Fleming I, Schott C, Parratt JR, Stoclet JC (1990). Loss of vascular responsiveness induced by endotoxin involves the L-arginine pathway. Am J Physiol 259:H1038–H1043

7. Thiemermann C, Szabo C, Mitchell JA, Vane JR (1993) Vascular hyporeactivity to vascoconstrictor agents and hemodynamic decompensation in hemorrhagic shock is mediated by nitric oxide. Proc Natl Acad Sci USA 90:267–271
8. Schulz R, Nava E, Moncada S (1992) Induction and potential biological relevance of a calcium-independent nitric oxide synthase in the myocardium. Br J Pharmacol 105: 575–580
9. Finkel MS, Oddis CV, Jacob TD, Watkins SC, Hattler BG, Simmons RL (1992) Negative inotropic effects of cytokines on the heart mediated by nitric oxide. Science 257: 387–389
10. Brady AJB, Poole-Wilson PA, Harding SE, Warren JB (1992) Nitric oxide production within cardiac myocytes reduces their contractivity in endotoxemia. Am J Physiol 263: H1963–H1966
11. Szabo C, Thiemermann C (1994) Invited opinion: Role of nitric oxide in hemorrhagic, traumatic and anaphylactic shock and thermal injury. Shock 2:145–155
12. Kilbourn RG, Griffith OW (1992) Overproduction of nitric oxide in cytokine-mediated and septic shock. J Natl Cancer Inst 84:827–877
13. Vallance P, Moncada S (1991) Hyperdynamic circulation in cirrhosis: A role for nitric oxide? Lancet 337:776–777
14. Drapier J, Wietxerbin D, Hibbs JB (1988) Interferon gamma and tumor necrosis factor induce the L-arginine dependent cytotoxic effector mechanism in murine macrophages. Eur J Immunol 18:1587–1591
15. Kosaka H, Harada N, Watanabe M, Yoshihara H, Katsuki Y, Shiga T (1992). Synergistic stimulation of nitric oxide hemoglobin production in rats by recombinant interleukin-1 and tumor necrosis factor. Biochem Biophys Res Commun 189:392–397
16. Kilbourn RG, Gross SS, Jubran A, et al (1990) N^G-methyl-L-arginine inhibits tumor necrosis factor-induced hypotension: Implications for the involvement of nitric oxide. Proc Natl Acad Sci USA 87:3629–3632
17. Vicaut E, Baudry N (1992) Nitric oxide and tumour necrosis factor in terminal arterioles of rat skeletal muscle. In: Moncada S, Marletta MA, Hibbs JB, Higgs EA (eds) Biology of nitric oxide, Vol 1. Portland Press, London & Chapel Hill, pp 216–218
18. Thiemermann C, Wu CC, Szabo C, Perretti M, Vane JR (1993) Tumour necrosis factor is an endogenous mediator of the induction of nitric oxide synthase in endotoxin shock in the rat. Br J Pharmacol 110:177–182
19. Shi Y, Li HQ, Shen CK, et al (1993) Association between protective efficacy of antibodies to tumour necrosis factor and suppression of nitric oxide production in neonatal rats with fetal injection. Pediatric Res 34:345–348
20. Mulsch A, Bassenge E, Busse R (1990) Nitric oxide synthesis in endothelial cytosol: Evidence for a calcium-dependent and calcium-independent mechanism. Naunyn-Schmiedeberg's Arch Pharmacol 340:767–770
21. Kilbourn RG, Belloni P (1990) Endothelial cell production of nitrogen oxides in response to interferon-gamma in combination with tumor necrosis factor, interleukin-1 or endotoxin. J Natl Cancer Inst 82:772–776
22. Curran RD, Billiar TR, Stuehr DJ, et al (1990) Multiple cytokines are required to induce hepatocyte nitric oxide production and inhibit total protein biosynthesis. Ann Surg 212:462–489
23. Robert R, Chapelain B, Neliat G (1993) Different effects of interleukin-1 on reactivity of arterial vessels isolated from various vascular beds in the rabbit. Circ Shock 40: 139–143
24. Szabo C, Wu CC, Gross SS, Thiemermann C, Vane JR (1993) Interleukin-1 contributes to the induction of nitric oxide synthase by endotoxin in vivo. Eur J Pharmacol 250:157–160
25. Cunha FQ, Assreuy J, Moss DW, et al (1994) Differential induction of nitric oxide synthase in various organs of the mouse during endotoxaemia: Role of TNF-alpha and IL-1 beta. Immunology 81:211–215
26. Evans T, Carpenter A, Silva A, Cohen J (1 992) Differential effects of monoclonal antibodies to tumor necrosis factor alpha and gamma interferon on induction of hepatic

nitric oxide synthase in experimental gram-negative sepsis. Infect Immun 60: 4133–4139
27. Green SJ, Nacy CA, Schreiber RD, et al (1993) Neutralization of gamma interferon and tumor necrosis factor alpha blocks in vivo synthesis of nitrogen oxides from L-arginine and protects against Francisella tularensis injection in mycobacterium bovis BCG-treated mice. Infect Immun 61:689–698
28. Huang S, Hendriks W, Althage A, et al (1993) Immune response in mice that lack the interferon-gamma receptor. Science 259:1742–1745
29. Dalton DK, Pitts-Meek S, Keshaw S, Figari IS, Bradley A, Stewart TA (1993) Multiple defects in immune function in mice with disrupted interferon-gamma genes. Science 259:1739–1741
30. Szabo C, Wu CC, Mitchell JA, Gross SS, Thiemermann C, Vane JR (1993) Platelet-activating factor contributes to the induction of nitric oxide synthase by bacterial lipopolysaccharide. Circ Res 73:991–999
31. Cunha FQ, Moncada S, Liew FY (1992) Interleukin-10 inhibits the induction of nitric oxide synthase by interferon-gamma in murine macrophages. Biochem Biophys Res Commun 182:1155–1159
32. Schneemann M, Schoeden G, Frei K, Schaffner A (1993) Immunovascular communication: Activation and deactivation of murine endothelial cells nitric oxide synthase by cytokines. Immunol Lett 35:159–162
33. Bogdan C, Vodovots Y, Paik J, Xie QW, Nathan C (1994) Mechanism of suppression of nitric oxide synthase expression by interleukin-4 in primary mouse macrophages. J Leukoc Biol 55:227–233
34. Wong HL, Costa GL, Lotze MT, Wahl SM (1993) Interleukin-4 differentially regulates monocyte IL-1 family gene expression and synthesis in vitro and in vivo. J Exp Med 177:775–781
35. Vannier E, Miller LC, Dinarello CA (1992) Coordinated anti-inflammatory effects of interleukin-4: Interleukin-4 suppresses interleukin-1 production but upregulates gene expression and synthesis of interleukin-1 receptor antagonists. Proc Natl Acad Sci USA 89:4076–4080
36. DeWaal Melefyt R, Abrams J, Bennet B, Figdor CG, De Vries JE (1991) Interleukin-10 inhibits cytokine synthesis by human monocytes: An autoregulatory role of IL-10 produced by monocytes. J Exp Med 174:1209–1220
37. Gerard C, Bruyns C, Marchant A, et al (1993) Interleukin-10 reduces the release of tumour necrosis factor and prevents lethality in experimental endotoxaemia. J Exp Med 177:547–550
38. Marchant A, Bruyns C, Vandenabeele P, et al (1994) Interleukin-10 controls interferon-gamma and tumour necrosis factor production during experimental endotoxaemia. Eur J Immunol 24:1167–1171
39. Szabo C, Thiemermann C, Vane JR (1993) Inhibition of the production of vasodilator prostaglandins attenuates the cardiovascular response to bacterial endotoxin in adrenalectomized rats. Proc Royal Soc London B 253:233–238
40. Szabo C, Thiemermann C, Wu CC, Perretti M, Vane JR (1994) Attenuation of the induction of nitric oxide synthase by endogenous glucocorticoids accounts for endotoxin tolerance in vivo. Proc Natl Acad Sci USA 91:271–275
41. Wu CC, Croxtall JD, Perretti M, et al (1995) Lipocortin-1 mediates the inhibition by dexamethasone of the induction by endotoxin of nitric oxide synthase in the rat. Proc Natl Acad Sci USA (In press)
42. Szabo C, Southan GJ, Thiemermann C, Vane JR (1994) The mechanism of the inhibitory effect of polyamines on the induction of nitric oxide synthase Role of aldehyde metabolites. Br J Pharmacol 113:757–766
43. Wright CE, Rees DD, Moncada S (1992) Protective and pathological roles of nitric oxide in endotoxin shock. Cardiovasc Res 26:48–57
44. Szabo C, Mitchell JA, Gross SS, Thiemermann C, Vane JR (1993) Nifedipine inhibits the induction of nitric oxide synthase by bacterial lipopolysaccharide. J Pharmacol Exp Ther 265:674–680

45. Wu CC, Thiemermann C, Vane JR (1995) Glibenclamide inhibits the expression of inducible nitric oxide synthase in cultured macrophages and in the anaesthetized rat. Br J Pharmacol (In press)
46. Thiemermann C, Vane JR (1990) Inhibition of nitric oxide synthesis reduces the hypotension induced by bacterial lipopolysaccharide in the rat. Eur J Pharmacol 182: 591–595
47. Kilbourn RG, Jubran A, Gross SS, et al (1993) Reversal of endotoxin-mediated shock by N^G-monomethyl-L-arginine, an inhibitor of nitric oxide synthesis. Biochem Biophys Res Commun 172:1132–1138
48. Petros A, Lamb G, Leone A, Moncada S, Bennett D, Vallance P (1994) Effects of a nitric oxide synthase inhibitor in humans with septic shock. Cardiovasc Res 28:34–39
49. Rossaint R, Pison U, Gerlach H, Falke KJ (1993) Inhaled nitric oxide: Its effects on pulmonary circulation and airway smooth muscle cells. Eur Heart Journal 14:133–140
50. Klemm P, Thiemermann C, Winklmaier G, Martorana P, Henning R (1995) Effects of nitric oxide synthase inhibition combined with nitric oxide inhalation in a porcine model of endotoxin shock. Br J Pharmacol (In press)
51. Corbett JA, Tilton RG, Chang K, et al (1992) Aminoguanidine, a novel inhibitor of nitric oxide formation, prevents diabetic vascular dysfunction. Diabetes 41:552–556
52. Misko TP, Moore WM, Kasten TP, et al (1993) Selective inhibition of the inducible nitric oxide synthase by aminoguanidine. Eur J Pharmacol 233:119–225
53. Griffith MJD, Messent M, MacAllister RJ, Evans TW (1993) Aminoguanidine selectively inhibits inducible nitric oxide synthase. Br J Pharmacol 110:963–968
54. Wu CC, Chen SJ, Szabo C, Thiemermann C, Vane JR (1995). Aminoguanidine attenuates the delayed circulatory failure and improves survival in rodent models of endotoxic shock. Br J Pharmacol (In press)
55. Southan GJ, Szabo C, Thiemermann C (1995) Isothioureas: Potent inhibitors of nitric oxide synthases with variable isoform selectivity. Br J Pharmacol (In press)
56. Garvey PE, Oplinger JA, Tanoury GJ, et al (1994) Potent and selective inhibition of human nitric oxide synthases. Inhibition by non-amino acid isothioureas. J Biol Chem 269:26669–26676
57. Szabo C, Southan GJ, Thiemermann C (1995) Beneficial effects and improved survival in rodent models of septic shock with S-methyl-isothiourea sufate, a novel, potent and selective inhibitor of inducible nitric oxide synthase. Proc Natl Acad Sci USA (In press)

Use of Nitric Oxide Synthase Inhibitors to reverse Hyperdynamic Circulation and MOF

M. Booke, L. D. Traber, and D. L. Traber

Introduction

Sepsis is the leading cause of morbidity and mortality in critically ill patients [1]. The number of cases of this lethal syndrome is increasing as more patients are immunosuppressed, treated with invasive techniques, and bacteria have become refractory to antibiotics [2, 3]. In the US, more than 400000 cases of sepsis are reported each year, and 50% of these patients die during their course of illness [1]. Death occurs mostly within 24 h after diagnosis and is the result of persistent hypotension.

Deaths occurring after the successful treatment of hypotension within the first 24 h can usually be attributed to multiple organ failure (MOF). MOF is most likely the result of reperfusion injury or irreversible tissue or organ damage sustained during the initial period of hypotension [4]. Sepsis may not only be caused by the endotoxin of gram-negative bacteria. Toxins released by gram-positive bacteria as well as viral and fungal products may contribute to the severity of this complication [5–8]. In fact, sepsis or SIRS (systemic inflammatory response syndrome), as defined by an international committee [9], is related to bacteria only 45% of the time, and less than 25% of SIRS cases are related to infections with gram-negative bacteria [10, 11].

Basic Pathophysiology of Sepsis

Sepsis, as a general inflammatory response, affects the entire organism. Intense vasodilation results in reduced mean arterial and perfusion pressure leading to tissue and organ damage [3]. Although regional tissue malperfusion during sepsis may even occur while the mean arterial pressure is within normal range [12], it is generally accepted that reduced mean arterial pressure correlates with mortality in sepsis [2]. Cardiopulmonary responses countering this vasodilation include an increase in cardiac output, but generally this cannot prevent the fall in arterial pressure. Further, local vasodilation may cause precapillary shunting, which deprives metabolically active tissues from adequate nutritive blood flow. The therapeutic goal in treating sepsis must be the restoration of blood and perfusion pressure as well as the redistribution of blood flow to metabolically active tissues. To achieve this goal, treating the *cause* of septic vasodilation is preferable to the use of vasopressors.

Nitric oxide (NO) is considered by some to be the final common mediator for vasodilation seen in sepsis [13, 14]. NO causes vasodilation by stimulating the formation of cyclic 3′,5′-guanosine monophosphate (cGMP) [15, 16]. cGMP, which is elevated during the course of sepsis [15, 17], relaxes the vascular smooth muscle by pumping calcium from the cell. cGMP is produced by guanylate cyclase, which occurs in a soluble and particulate form [18]. The latter enzyme can be stimulated by NO. In addition, enhanced formation of cGMP in the cardiac muscle reduces myocardial contractility and performance, increasing the inability to combat the hypotension [19, 20]. Vasodilation in the pulmonary microcirculation may cause an impaired ventilation/perfusion relationship, resulting in elevated intrapulmonary shunting and reduced oxygen transport. Renal function is reduced by the presence of NO, mainly because it causes intense vasodilation in the efferent rather than in the afferent vessels [21], thereby reducing the filtration pressure. NO is a main mediator in septic vasodilation as well as in MOF.

NO can be synthesized from its natural precursor L-arginine by two different enzymes, a constitutive and an inducible form of nitric oxide synthase (NOS) [22]. The constitutive form (cNOS) is calcium dependent and is mainly responsible for the basal release of NO, which regulates the vascular tone under physiological conditions [22]. The inducible form (iNOS) is calcium-independent and can be activated by various agents, including endo- and exotoxins and cytokines [23].

These enzymes can be blocked by analogs of L-arginine, such as N^G-monomethyl-L-arginine (L-NMMA) or N^ω-nitro-L-arginine methyl ester (L-NAME). With NO as the mediator causing vasodilation, NOS inhibition leads to vasoconstriction and restoration of blood pressure with the possibility of improvement to impaired organ functions. We tested the compounds in an established model of hyperdynamic ovine sepsis. This model closely recreates the multiple organ dysfunction seen in septic humans, including severe vasodilation accompanied by impaired myocardial, pulmonary, and renal function.

The Ovine Sepsis Model

There are many advantages to using sheep as experimental animals in the field of sepsis. Sheep can be monitored and resuscitated just like human intensive care patients, and its gentle nature allows it to be studied in an unrestrained, non-medicated state. The sheep is superior to other animals because of its high sensitivity to endotoxin, which is comparable to the level of sensitivity in humans [24, 25]. When small boluses of endotoxin were given, the sheep showed a transient hyperdynamic response [24]. To achieve a sustained hyperdynamic cardiopulmonary response, a continuous infusion of either bacteria [26] or endotoxin [27] is required. Under absolutely aseptic conditions, the sheep are surgically prepared for chronic study. After a recovery period of 5 days, a continuous infusion of either endotoxin (10 ng/kg/min) or live *Pseudomonas aeruginosa* ($2.5 \cdot 10^6$ CFU/min) is begun and maintained for 48 h. These infusions result in significant changes in systemic and pulmonary circulation and in organ function.

Systemic Circulation

Sepsis induced by either bacteria or endotoxin induces an intense vasodilation in sheep. Reduction in mean arterial pressure is accompanied by a significant increase in cardiac output, resulting in a very intense fall in systemic vascular resistance (Fig. 1; control) [24]. This hyperdynamic state of sepsis in sheep superbly mimics the hemodynamics of septic humans. Sepsis or septic syndrome is typically characterized by an elevated cardiac output and a concomitant reduced arterial pressure, resulting in a dramatically reduced systemic vascular resistance. This vasodilation leads to non-metabolically active tissues being perfused in preference to those which have metabolic needs. Consequently, the oxygen extraction falls during the course of sepsis, causing oxygen consumption to be dependent on oxygen delivery. Besides precapillary shunting of blood flow, microvascular and cellular abnormalities are responsible for the impaired capability of tissues to extract oxygen. An increase in systemic permeability leads to interstitial and intracellular edema formation, making it harder for oxygen to diffuse from the blood into the cell [28]. This overall reduced or impaired capability of tissue to extract oxygen from the bypassing blood extends the dependency of oxygen consumption on oxygen delivery. In healthy subjects, oxygen consumption stays stable over a wide range of oxygen delivery. Only when the oxygen delivery drops below 8 mL/kg/min, does the oxygen consumption fall [29]. Because of this flow dependency on oxygen consumption, the outcome in septic patients is highly dependent on the oxygen delivery [30]. This reduced capability of tissues to extract oxygen may end in local hypoxia and multiple organ dysfunction or even failure. This all may occur despite an increase in organ blood flow. In hyperdynamic ovine sepsis, most organs show an increase in blood flow when measured with radioactive or colored microspheres [31].

Pulmonary Circulation and Lungs

The pulmonary circulation behaves differently from that of the systemic circulation during the course of sepsis. Sepsis leads to a sustained increase in pulmonary arterial pressure. Pulmonary vascular resistance shows a marked elevation during the first 2 to 3 h after beginning the infusion of endotoxin or bacteria, but after 6 h pulmonary vascular resistance declines slightly. The early increase (0.5 to 1.5 h) in pulmonary pressure and resistance is closely

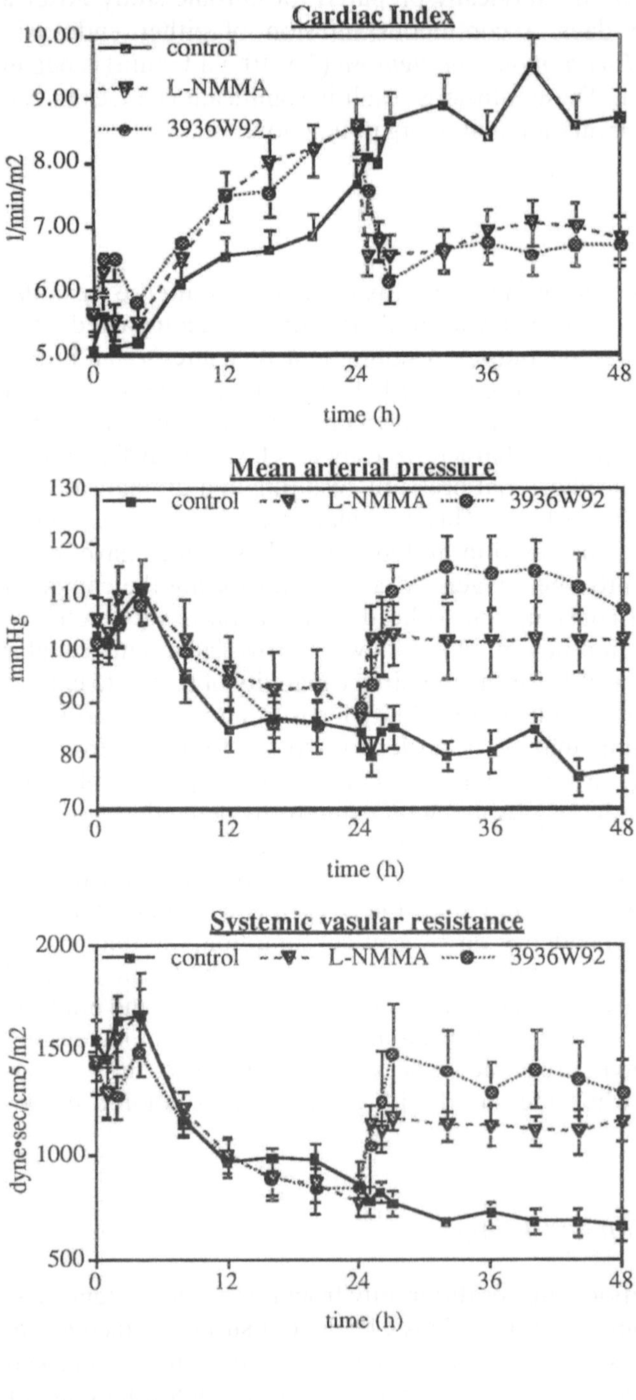

Fig. 1. Cardiac index, mean arterial pressure, and systemic vascular resistance index in sheep treated with a continuous infusion of live *Pseudomonas aeruginosa* ($2.5 \cdot 10^6$ CFU/min). After 24 h of sepsis, the sheep were randomly assigned to either saline (control, n = 9), L-NMMA (n = 7), or 3936W92 (n = 7). L-NMMA was given as a continuous infusion of 7 mg/kg/h without an initial loading dose, and 3936W92 was given in an equivalent dose of 9 mg/kg/h. Treatment with L-NMMA and 3936W92 led to a significant fall in cardiac index and simultaneously to a significant increase in mean arterial pressure and systemic vascular resistance

related to the release of thromboxane A_2. This release of eicosanoids appears to be related to the dose of endotoxin administered, because it is not released with lower dosages of lipopolysaccharide (LPS) [32]. The later increases in resistance (1.5 to 5 h) occur with both high and low dosages and may be related to the loss of constitutive NO from the pulmonary vessels [33]. The increased cardiac output is responsible for the elevation in pulmonary artery pressure during the later course of sepsis. The fall in pulmonary resistance may be the result of the formation of inducible NO synthase in the pulmonary vein rather than the artery. The pulmonary artery removed from animals 48 h after sepsis did not show a reduced contractile response to norepinephrine or KCl [34]. Further, the cGMP content of isolated pulmonary arteries was not elevated during sepsis [35]. Analysis of the pulmonary veins from these animals, on the other hand, shows a marked reduction in contractile response and an elevation in cGMP content. These data may also explain why we are unable to demonstrate an elevation in pulmonary microvascular pressure in sheep during the 12- to 48-h period of sustained sepsis [27].

The lungs are frequently involved in MOF in conjunction with sepsis. An increase in permeability to small molecules leads to interstitial edema followed by reduced diffusion capacity and often, alveolar flooding and collapse [36]. Besides edema formation, reduced pulmonary hypoxic vasoconstriction [37] contributes to a rise in the intracapillary shunt fraction that occurs during sepsis [38]. This is also mediated by vasodilation, in this case, of the pulmonary vasculature. This dilation abolishes hypoxic pulmonary vasoconstriction. Thus, blood that normally flows only to ventilated alveoli now perfuses unventilated alveoli, leading to a ventilation/perfusion mismatch and a fall in arterial oxygen content. These pulmonary events complicate the issue of oxygen delivery by reducing oxygen content of the arterial blood [37].

Heart

Myocardial depression is a common finding in septic humans [39, 40]. Myocardial performance during sepsis has been analyzed extensively [41, 42]. Left ventricular contractility is markedly reduced during sepsis, and an increase in cardiac output is achieved, mainly by an elevation in heart rate. There is evidence that there is a similar reduction in right ventricular contractility as well [42, 43].

The mediation of myocardial depression which is seen in sepsis is still controversial. A reduced influx of calcium ions [44] as well as a reduction in cardiac glutathione concentration [45] have been associated with this depression. The reduction in glutathione has been linked to the presence of oxygen free radicals. Concurrently, sepsis leads to a marked coronary dilation [45]. Both phenomena could be explained by an increase in myocardial NO synthesis that results in elevated cGMP levels, as the negative inotropic effect of cGMP is well known [19, 20]. The heart produces a significant amount of NO and the plasma levels of cGMP are elevated in sepsis. A reduction in myo-

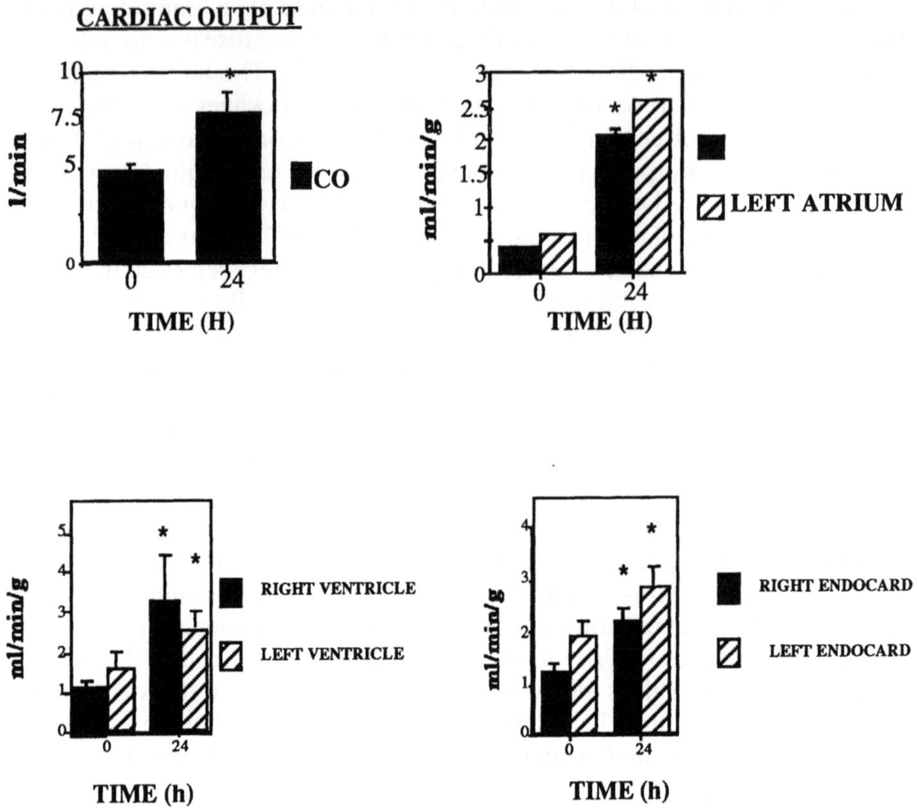

Fig. 2. Cardiac output and myocardial blood flow in sheep (n = 20) treated with a continuous infusion of live *Pseudomonas aeruginosa* ($2.5 \cdot 10^6$ CFU/min). Colored microspheres were injected into the left atrium at baseline and after 24 h of bacteria infusion. During the infusion, the sheep developed a hyperdynamic circulatory stage of sepsis. The increase in cardiac output is accompanied by similar increases in blood flow to the left and right atria, and ventricles. * $p \leq 0.05$ *versus* 0 h

cardial blood flow might be responsible for the impaired cardiac performance if, as a result of vasodilation, blood is diverted from metabolically-active areas to areas of low oxygen need. Even transient episodes of myocardial ischemia or acidosis can produce prolonged myocardial dysfunction [47, 48]. During hyperdynamic ovine sepsis, myocardial blood flow is elevated throughout the myocardium, even in the subendocardium (Fig. 2).

Kidney

Renal failure is a common occurrence in MOF associated with sepsis [49] and is highly correlated to outcome [50]. The measurement of creatinine clearance in sheep is used as an indicator for the glomerular filtration rate because sheep neither excrete nor reabsorb creatinine [51]. Sheep treated with a bolus of endotoxin showed no change in glomerular filtration rate, but a polyuric natriuresis occurred [52]. Despite increased renal blood flow and polyuria, the glomerular filtration rate drops during sustained ovine sepsis if induced by bacteremia or endotoxemia [53–55]. Two reasons may be responsible for the reduced renal function:

1) renal blood flow measured with an ultrasonic flow probe on the renal artery shows a significantly greater increase during sepsis than renal blood flow measured with colored microspheres, suggesting a precapillary shunting of blood (and microspheres) [56]; and
2) a strong correlation between glomerular filtration rate and mean arterial pressure or systemic vascular resistance has been reported during sepsis [57, 58].

The fall in renal vascular resistance may be profound because the renal microvasculature is known to be very sensitive to NO [59], and NO is known to be markedly elevated during sepsis. This reduced renal function is not only reflected by a reduction in creatinine clearance, but also by a fall in urine output, indicating a positive fluid balance [54].

The model of ovine sepsis accurately resembles the hemodynamic changes and the typical changes in organ dysfunction seen in the course of human sepsis. This makes the sheep an ideal experimental animal to investigate sepsis, its pathophysiology, and to look for new therapeutic approaches. In controlled studies new compounds can be tested. The administration of NOS inhibitors in sepsis, for example, was initially tested in our laboratory, and further tests are still ongoing.

Effects of NOS inhibition on Organ Function

Considering that NO is a major mediator of the vasodilation that occurs with sepsis [13, 14], the inhibition of NOS seems to be a logical therapeutic approach. Especially because NO is not only responsible for vasodilation, but most likely plays a major role in the development of MOF during sepsis. Consequently, the inhibition of NOS may not only restore blood pressure, but also improve organ functions that have become impaired during the course of sepsis.

Systemic Circulation

The inhibition of NOS, 24 h after the beginning of sepsis produced by the continuous infusion of endotoxin, leads to an immediate restoration of blood pressure [38]. Simultaneously, the cardiac output drops to baseline values, and the systemic vascular resistance index is restored to its pre-sepsis value. These effects can be achieved in our laboratory with L-NAME and L-NMMA as well as a non-amino acid inhibitor of NOS, 3936W92 (Fig. 2). The difference between these compounds is that L-NAME has a much longer half-life and therefore can be given as a single bolus, whereas L-NMMA and 3936W92 must be given as a continuous infusion because of their relatively short half-lives. One advantage of using L-NMMA or 3936W92 is that either drug can be discontinued if negative side-effects are observed. Further, the L-NMMA and 3936W92 have a higher affinity for iNOS than L-NAME [60, 61]. In contrast to L-NMMA, 3936W92 has the additional advantage of not being metabolized to L-arginine and further utilized by NOS [62, 63] because it is not an amino acid [61].

Although the fall in cardiac output occurs with a significant drop in oxygen delivery, the oxygen consumption remains unaffected. This is due to the peripheral tissue regaining the ability to extract enough oxygen from the blood. The precapillary shunting of blood flow with concomitant reduction in nutritive blood flow is reversed by NOS inhibition, leading to a redistribution of blood flow to metabolically active tissues.

The administration of NOS inhibitors, such as L-NAME or L-NMMA, brings the elevated organ blood flows back to baseline levels [64]. Organ blood flow is not reduced below baseline values. The only organ which shows a tendency to fall below pre-septic baseline values after treatment with L-NAME and is significantly lowered during the administration of L-NMMA, is the pancreas. Potentially, this may be harmful; however, in our experiments, there was no indication of pancreatitis, either clinically during the experiment or *post mortem*. A significant overall reduction in nutritive blood flow after NOS inhibition can definitely be excluded because this would have led to a fall in oxygen consumption and an increase in plasma lactate levels, neither occurred. The redistribution of blood flow to metabolically active tissues potentially improves tissue nourishment and prevents the further development of hypoxic tissue damage. This is a major step in preventing or minimizing MOF thereby improving the possible outcome in subjects suffering from severe sepsis.

Pulmonary Circulation and Lungs

As described above, NO reduces the efficacy of the hypoxic pulmonary vasoconstriction and leads to a ventilation/perfusion mismatch during sepsis. Inhibition of NOS causes a further increase in pulmonary hypertension but it restores the hypoxic pulmonary vasoconstriction, thereby reducing the fraction of intrapulmonary shunted blood flow [38]. However, the shunt fraction could not be brought back to baseline levels, suggesting that factors other than NO may reduce the pulmonary capability for oxygen uptake. One factor is definitely the interstitial pulmonary edema that develops during the course of sepsis [27]. The administration of NOS inhibitors cannot reverse this edema formation in the short term. However, long-term experiments are still to be performed.

There is a pulmonary hypertension in combination with the administration of the NOS inhibitors. This is, in part, responsible for the further but insignificant increase in pulmonary transvascular fluid flux after administration of NOS inhibitors [65, 66]. The reflection coefficient was not altered and the increase in lymph flow was due, mainly, to an increase in the filtration coefficient.

Heart

Assuming that NO is one of the mediators causing myocardial depression, the inhibition of NOS should lead to an improved cardiac performance. Indeed, there have been several studies that have found the inhibition of NOS to reverse the depression of myocytes produced by cytokines or endotoxin [67–69]. In our ovine model of sepsis, there is a tendency for reduced stroke work and elevated left atrial pressures to be reversed by administration of the NOS inhibitors. Experiments are presently being performed in which we are looking for myocardial contractility in septic sheep treated by NOS inhibition using sophisticated crystallographic techniques. However, an increase in pulmonary arterial and thus right ventricular pressure may produce geometrical changes in the left ventricle that will compromise most of the conventional indices of contractility [70].

As mentioned above, a reduction in myocardial perfusion can be excluded as a reason for myocardial depression; however, the increase in myocardial blood flow is caused by NO-mediated coronary vasodilation. Consequently, the inhibition of NOS could induce hypoperfusion, causing cardiac performance to deteriorate. In experiments carried out with L-NAME, L-NMMA, and 3936W92, there is no evidence that this occurs. The reduction in myocardial perfusion observed in our experiments is most likely the result of a reduction in heart rate [31]. The cardiac blood flow did not fall below pre-septic baseline levels in any of the septic sheep given NOS inhibitors, the blood flow in the subendocardial layer also stayed above pre-septic baseline levels.

The fall in heart rate that is commonly observed after or during the administration of NOS inhibitors [71,72] is most likely caused by a reduction in sympathetic nerve activity. It may contribute to the myocardial perfusion that takes place in the prolonged diastolic phase.

Kidney

As stated above, the glomerular filtration rate is reduced during sepsis, perhaps as a result of NO-induced dilatation of the efferent arteriole with consecutive reduced glomerular filtration pressure. Thus, the inhibition of NOS should improve renal function. On the other hand, several studies sup-

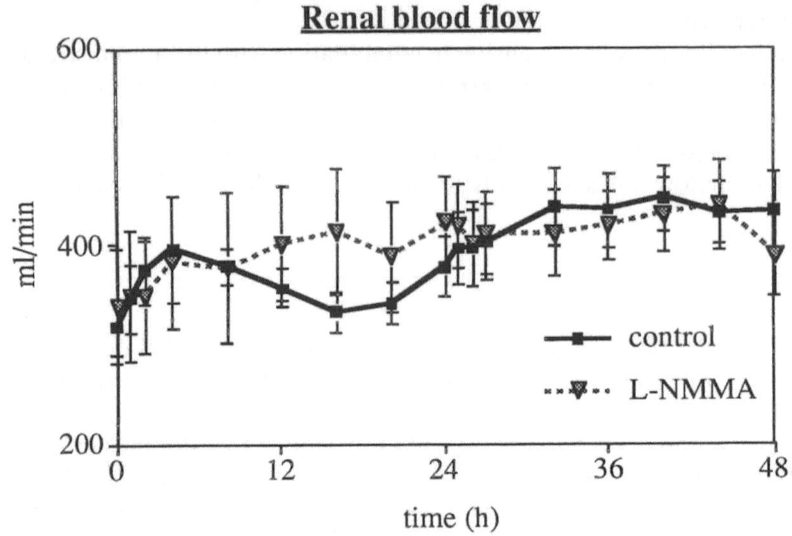

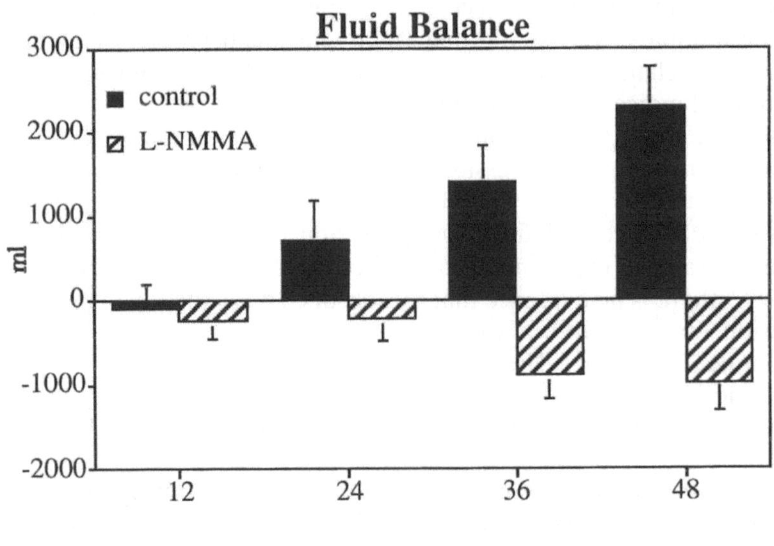

Fig. 3. Renal blood flow and fluid balance during the course of ovine sepsis (continuous infusion of live *Pseudomonas aeruginosa*, $2.5 \cdot 10^6$ CFU/min). After 24 h of sepsis, the sheep were randomly assigned to either saline (control, n = 9) or L-NMMA (n = 7). Sepsis led to a significant increase in renal blood flow, which was not altered by the administration of L-NMMA; although, L-NMMA significantly lowered cardiac output (Fig. 2). The administration of the NOS inhibitor (Fig. 2) led to a negative fluid balance. This is in contrast to control animals, which retained fluid especially during the second 24 h of sepsis

port the theory that the activity of NO in the kidney is beneficial, by guaranteeing a minimal renal perfusion by local vasodilation, making the inhibition of NOS harmful [73].

In the ovine model of sepsis, the administration of NOS inhibitors proved to be dramatically beneficial. Not only did the creatinine clearance increased significantly [53], but even more impressively, in some animals that had become anuric in the course of sepsis, the administration of NOS inhibitors resulted in significant urine production. Most likely, NOS inhibition causes constriction of the efferent vessel thereby increasing glomerular filtration pressure. Of course, the rise in mean arterial pressure contributes to this phenomenon. Further, when given in hyperdynamic ovine sepsis, NOS inhibition does not cause a reduction in renal blood flow below pre-septic baseline values (Fig. 3). Therefore, the needed blood flow for an adequate urine output is still provided, and, in fact, treatment with L-NMMA causes a negative fluid balance in ovine sepsis, whereas untreated sheep developed a tremendous positive fluid balance. This may be different when NOS inhibitors are given in hypodynamic sepsis because under these circumstances, the reduced blood flow is the main determinant for reduced renal function. NOS inhibition at this stage would cause further reduction in renal blood flow by reducing cardiac output and causing renal vasoconstriction. This, of course, would be very harmful and could trigger the development of MOF in sepsis.

Conclusion

The present data support the theory that NO is one of the main mediators for MOF associated with sepsis. The reversal of the hyperdynamic state of sepsis and thereby the restoration of blood pressure is the most obvious effect of NOS inhibitors. Additionally, NOS inhibition improves pulmonary oxygen uptake, cardiac performance, and renal function. Organs that are most often involved in MOF associated with sepsis show improved functionality as a result of NOS inhibition in hyperdynamic sepsis.

References

1. (1990) Increase in National Hospital Discharge Survey rates for septicemia-United States, 1979–1987. MMWR Morb Mortal Wkly Rep39:31–34
2. Martich GD, Boujoukos AJ, Suffredini AF (1993) Response of man to endotoxin. Immunobiology 187:403–416
3. Jeremy JY, Nystrom ML, Barradas MA, Mikhailidis DP (1994) Eicasonoids and septicemia. Prostaglandins Leukot Essent Fatty Acids 50:287–297
4. Cerra FB (1990) The multiple organ failure syndrome. Hosp Pract (Off) 25:169–176
5. Deutschman CS, Konstantinides FN, Tsai M, Simmons RL, Cerra FB (1987) Physiology and metabolism in isolated viral septicemia. Further evidence of an organism-independent, host-dependent response. Arch Surg 122:21–25

6. Wiles JB, Cerra FB, Siegel JH, Border JR (1980) The systemic septic response: Does the organism matter? Crit Care Med 8:55–60
7. Thijs LG, Schneider AJ, Groeneveld AB (1990) The haemodynamics of septic shock. Intensive Care Med 16 (Suppl 3): S182–S186
8. Redl H, Schlag G, Thurnher M, Traber LD, Traber DL (1989) Cardiovascular reaction pattern during endotoxin or peptidoglycan application in awake sheep. Circ Shock 28: 101–108
9. Bone RC, Balk RA, Cerra FB, et al (1992) Definitions for sepsis and organ failure and guidelines for the use of innovative therapies in sepsis. The ACCP/SCCM Consensus Conference Committee. American College of Chest Physicians/Society of Critical Care Medicine. Chest 101:1644–1655
10. Bone RC, Fisher CJ, Clemmer TP, Slotman GJ, Metz CA (1987) Early methylprednisolone treatment for septic syndrome and the adult respiratory distress syndrome. Chest 92:1032–1036
11. Knaus WA, Harrell FE, Fisher CJ, et al (1993) The clinical evaluation of new drugs for sepsis. A prospective study design based on survival analysis. JAMA 270:1233–1241
12. Vincent JL, Bihari D (1992) Sepsis, severe sepsis or sepsis syndrome: Need for clarification. Intensive Care Med 18:255–257
13. Nava E, Palmer RM, Moncada S (1991) Inhibition of nitric oxide synthesis in septic shock: How much is beneficial? Lancet 338:1555–1557
14. Geller DA, Nussler AK, Di Silvio M, et al (1993) Cytokines, endotoxin, and glucocorticoids regulate the expression of inducible nitric oxide synthase in hepatocytes. Proc Natl Acad Sci USA 90:522–526
15. Fleming I, Julou-Schaeffer G, Gray GA, Parratt JR, Stoclet JC (1991) Evidence that an L-arginine/nitric oxide-dependent elevation of tissue cyclic GMP content is involved in depression of vascular reactivity by endotoxin. Br J Pharmacol 103:1047–1052
16. Spiro JG, Scott S, MacMillan J, et al (1985) Treatment of uremic pruritus with blue light. Photodermatol 2:319–321
17. Schuller F, Fleming I, Stoclet JC, Gray GA (1992) Effect of endotoxin on circulating cyclic GMP in the rat. Eur J Pharmacol 212:93–96
18. Martin W, White DG, Henderson AH (1988) Endothelium-derived relaxing factor and atriopeptin II elevate cyclic GMP levels in pig aortic endothelial cells. Br J Pharmacol 93:229–239
19. Smith JA, Shah AM, Lewis MJ (1991) Factors released from endocardium of the ferret and pig modulate myocardial contraction. J Physiol (Lond) 439:1–14
20. Shah AM, Spurgeon HA, Sollott SJ, Talo A, Lakatta EG (1994) 8-bromo-cGMP reduces the myofilament response to Ca^{2+} in intact cardiac myocytes. Circ Res 74: 970–978
21. Hoffend J, Cavarape A, Endlich K, Steinhausen M (1993) Influence of endothelium-derived relaxing factor on renal microvessels and pressure-dependent vasodilation. Am J Physiol 265:F285–F292
22. Moncada S, Higgs A (1993) The L-arginine-nitric oxide pathway. N Engl J Med 329: 2002–2012
23. Nicolson AG, Haites NE, McKay NG, Wilson HM, MacLeod AM, Benjamin N (1993) Induction of nitric oxide synthase in human mesangial cells. Biochem Biophys Res Commun 193:1269–1274
24. Traber DL, Redl H, Schlag G, et al (1988) Cardiopulmonary responses to continuous administration of endotoxin. Am J Physiol 254:H833–H839
25. Suffredini AF, Fromm RE, Parker MM, et al (1989) The cardiovascular response of normal humans to the administration of endotoxin. N Engl J Med 321:280–287
26. Lingnau W, McGuire R, Traber LD, Matsumoto N, Traber DL (1994) Changes in organ blood flows after nitric oxide inhibition during ovine bacteremia. Crit Care Med 22:A198(Abst)
27. Nakazawa H, Noda H, Noshima S, et al (1993) Pulmonary transvascular fluid flux and cardiovascular function in sheep with chronic sepsis. J Appl Physiol 75:2521–2528

28. Hersch M, Gnidec AA, Bersten AD, Troster M, Rutledge FS, Sibbald WJ (1991) Histologic and ultrastructural changes in non-pulmonary organs during early hyperdynamic sepsis. Surgery 107:397–410
29. Schuhmacker P, Cain S (1994) The concept of a critical oxygen delivery. Intensive Care Med 13:223–229
30. Tuchschmidt J, Fried J, Astiz M, Rackow E (1992) Elevation of cardiac output and oxygen delivery improves outcome in septic shock. Chest 102:216–220
31. Meyer J, Hinder F, Stothert J, et al (1994) Increased organ blood flow in chronic endotoxemia is reversed by nitric oxide synthase inhibition. J Appl Physiol 76:2785–2793
32. Traber DL, Flynn JT, Herndon DN, Redl H, Schlag G, Traber LD (1989) Comparison of the cardiopulmonary responses to single bolus and continuous infusion of endotoxin in an ovine model. Circ Shock 27:123–138
33. Spath JA, Sloane PJ, Gee MH, Albertine KH (1994) Loss of endothelium-dependent vasodilation in the pulmonary vessels of sheep after prolonged endotoxin. J Appl Physiol 76:361–369
34. Nelson S, Stewart RH, Traber L, Traber D (1991) Endotoxin-induced alterations in contractility of isolated blood vessels from sheep. Am J Physiol 260:H1790–H1794
35. Nelson SH, Ehardt JS, Lingnau W, Traber L, Traber D (1994) Influence of sepsis and nitric oxide synthase (NOS) inhibition in pulmonary artery and vein. Shock 1:60(Abst)
36. Anderson RR, Holliday RL, Driedger AA, Lefcoe M, Reid B, Sibbald WJ (1979) Documentation of pulmonary capillary permeability in the adult respiratory distress syndrome accompanying human sepsis. Am Rev Respir Dis 119:869–877
37. Theissen JL, Loick HM, Curry BB, Traber LD, Herndon DN, Traber DL (1991) Time course of hypoxic pulmonary vasoconstriction after endotoxin infusion in unanesthetized sheep. J Appl Physiol 70:2120–2125
38. Meyer J, Lentz CW, Stothert JC, Traber LD, Herndon DN, Traber DL (1994) Effects of nitric oxide synthesis inhibition in hyperdynamic endotoxemia. Crit Care Med 22: 306–312
39. Cunnion RE, Parrillo JE (1989) Myocardial dysfunction in sepsis. Recent insights. Chest 95:941–945
40. Cunnion RE, Parrillo JE (1989) Myocardial dysfunction in sepsis. Crit Care Clin 5: 99–118
41. Noshima S, Noda H, Herndon DN, Traber LD, Traber DL (1993) Left ventricular performance during continuous endotoxin-induced hyperdynamic endotoxemia in sheep. J Appl Physiol 74:1528–1533
42. Redl G, Newald J, Schlag G, Traber LD, Traber DL (1991) Cardiac function in an ovine model of endotoxemia. Circ Shock 35:31–36
43. Schroeder D, Quinn M, Leonova E, Raymond RM (1993) Right ventricular contractility during acute endotoxin shock in the dog. FASEB J 7:A322 (Abst)
44. McDonough KH, Lang CH, Spitzer JJ (1985) Effect of cardiotropic agents on the myocardial dysfunction of hyperdynamic sepsis. Circ Shock 17:1–19
45. Goldfarb RD, Lee KJ, Andrejuk T, Dziuban SW (1990) End–systolic elastance as an evaluation of myocardial function in shock. Circ Shock 30:15–26
46. Shah AM (1993) Nitric oxide, synthase activities in human myocardium. Lancet 341:448
47. Lew WY (1993) Mechanisms of volume-induced increase in left ventricular contractility. Am J Physiol 265:H1778–H1786
48. Hung J, Lew WY (1993) Temporal sequence of endotoxin-induced systolic and diastolic myocardial depression in rabbits. Am J Physiol 265:H810–H819
49. Baur M, Schmid TO, Landauer B (1989) Role of phospholipase A in multiorgan failure with special reference to ARDS and acute renal failure. Klin Wochenschr 67:196–202
50. Rayner BL, Willcox PA, Pascoe MD (1990) Acute renal failure in community-acquired bacteraemia. Nephron 54:32–35
51. Nawaz M, Shah BH (1984) Renal clearance of endogenous creatinine and urea in sheep during summer and winter. Res Vet Sci 36:220–224

52. Lübbesmeyer HJ, Woodson L, Traber LD, Flynn JT, Herndon DN, Traber DL (1988) Immunoreactive atrial natriuretic factor is increased in ovine model of endotoxemia. Am J Physiol 254: R567–R571
53. Lingnau W, McGuire R, Traber LD, Traber DL (1994) Renal function after nitric oxide inhibition in ovine bacterial sepsis. Crit Care Med 22: A122 (Abst)
54. Hinder F, Matsumoto N, Traber LD, Traber DL (1994) Nitric oxide synthase inhibition during experimental sepsis increases urine output in the presence of chronically elevated atrial natriuretic peptide. Crit Care Med 22: A123 (Abst)
55. Weber A, Schwieger IM, Poinsot O, Klohn M, Gaumann DM, Morel DR (1992) Sequential changes in renal oxygen consumption and sodium transport during hyperdynamic sepsis in sheep. Am J Physiol 262: F965–F971
56. Booke M, Traber LD, Traber DL (1995) Vasodilators in sepsis. In: Schlag G, Redl H, Traber DL (eds) Shock, Sepsis, and Organ Failure. Springer, Berlin Heidelberg New York (In press)
57. Weber A, Schwieger IM, Poinsot O, Morel DR (1994) Time course of systemic and renal plasma prostanoid concentrations and renal function in ovine hyperdynamic sepsis. Clin Sci 86: 599–610
58. Cumming AD (1994) Sepsis and acute renal failure. Ren Fail 16: 169–178
59. Lahera V, Salom MG, Miranda Guardiola F, Moncada S, Romero JC (1991) Effects of N^G-nitro-L-arginine methyl ester on renal function and blood pressure. Am J Physiol 261: F1033–F1037
60. Gross SS, Jaffe EA, Levi R, Kilbourn RG (1991) Cytokine-activated endothelial cells express an isotype of nitric oxide synthase which is tetrahydrobiopterin-dependent, calmodulin-independent and inhibited by arginine analogs with a rank-order of potency characteristic of activated macrophages. Biochem Biophys Res Commun 178: 823–829
61. Garvey EP, Oplinger JA, Tanoury GJ, et al (1994) Potent selective inhibition of human nitric oxide synthases – Inhibition by non-amino acid isothioureas. J Biol Chem 269: 2669–2676
62. Hecker M, Sessa WC, Harris HJ, Anggard EE, Vane JR (1990) The metabolism of L-arginine and its significance for the biosynthesis of endothelium-derived relaxing factor: Cultured endothelial cells recycle L-citrulline to L-arginine. Proc Natl Acad Sci USA 87: 8612–8616
63. Hecker M, Mitchell JA, Harris HJ, Katsura M, Thiemermann C, Vane JR (1990) Endothelial cells metabolize N^G-monomethyl-L-arginine to L-citrulline and subsequently to L-arginine. Biochem Biophys Res Commun 167: 1037–1043
64. Meyer J, Stothert JC, Pollard V, et al (1993) Nitric oxide synthesis inhibition in ovine endotoxemia: Effects on brain blood flow. Anesth Analg 76: S264 (Abst)
65. Hinder F, Meyer J, Nakazawa H, Traber LD, Traber DL (1993) Inhibition of nitric oxide synthase during chronic endotoxemia does not increase lung microvascular permeability to protein. Anesthesiology 79: A220 (Abst)
66. Lingnau W, McGuire R, Traber LD, Traber DL (1994) Changes in pulmonary microvascular pressure and permeability in septic sheep after nitric oxide inhibition. Anesth Analg 78: S251 (Abst)
67. Balligand JL, Ungureanu D, Kelly RA, et al (1993) Abnormal contractile function due to induction of nitric oxide synthesis in rat cardiac myocytes follows exposure to activated macrophage-conditioned medium. J Clin Invest 91: 2314–2319
68. Kumar A, Kosuri R, Kandula P, et al (1994) Cardiomyocyte depression induced by human septic serum is mediated by nitric oxide and cyclic GMP generation. Crit Care Med 22: 119 (Abst)
69. Kuman A, Parillo JE (1995) Myocardial depression in septic shock: Role of nitric oxide and other mediators. In: Schlag G, Redl H, Traber DL (eds) Shock, Sepsis, and Organ Failure. Springer, Berlin Heidelberg New York (In press)
70. Sugi K, Newald J, Traber LD, et al (1991) Cardiac dysfunction after acute endotoxin administration in conscious sheep. Am J Physiol 260: H1474–H1481

71. Gardiner SM, Compton AM, Kemp PA, Bennett T (1990) Regional and cardiac haemodynamic effects of N^G-nitro-L-arginine methyl ester in conscious, Long Evans rats. Br J Pharmacol 101:625–631
72. Tresham JJ, Dusting GJ, Coghlan JP, Whitworth JA (1991) Haemodynamic and hormonal effects of N^ω-nitro-L-arginine, an inhibitor of nitric oxide biosynthesis, in sheep. Clin Exp Pharmacol Physiol 18:327–330
73. Yukimura T, Yamashita Y, Miura K, Okumura M, Yamanaka S, Yamamoto K (1992) Renal effects of the nitric oxide synthase inhibitor, L-N^G-nitroarginine, in dogs. Am J Hypertens 5:484–487

Cardiovascular Alterations associated with Interleukin-2 Therapy

G. A. Fonseca and R. G. Kilbourn

Introduction

Renal cell carcinoma affects approximately 28000 patients per year in the United States. The prognosis of patients with this malignancy is poor with a 5-year survival of approximately 15% in patients with metastatic disease. Two major characteristics of the disease contribute to the low survival. First, patients usually present with advanced disease because of the lack of early clinical symptoms and the lack of cost effective screening tests. Second, this neoplasia possesses innate resistance to conventional chemotherapeutic agents. Interleukin-2 (IL-2) is among the most active agents for the treatment of disseminated renal cell carcinoma. The antitumor activity against renal cell carcinoma was first reported during early, phase I studies at the National Institute of Health [1]. Subsequent clinical trials confirmed the activity of IL-2 which lead to approval by the U.S. Food and Drug Administration for the treatment of patients with metastatic kidney cancer. Approval was based on the finding that significant, durable responses were observed (17–33%) and that some patients were cured of their disease (approximately 5%).

IL-2 Therapy as a Model of Sepsis

Attempts to increase the response rate to IL-2 have involved combinations with other cytokines, such as tumor necrosis factor (TNF) or α-interferon, or high doses of IL-2. Both approaches have been limited by the development of severe, sometimes life threatening, side effects. Treatment with high doses of IL-2 induces a syndrome similar to the systemic inflammatory response syndrome associated with sepsis [2–5]. The similarities between IL-2 and sepsis are likely to be the result of the endogenous production of cytokines such as TNF, γ-interferon (IFN-γ), and IL-1 [6–8]; all of which are implicated in the pathogenesis of sepsis. The side effects associated with IL-2 administration can be summarized into:

1. Hypotension
2. Vasopressor insensitivity

3. Major organ dysfunction
4. Vascular leak
5. Lactic acidosis
6. Coagulopathies and platelet abnormalities.

Infusion of high dose IL-2 can induce hemodynamic compromise characterized by severe hypotension, high cardiac output, and low systemic vascular resistance [9–15]. This hyperdynamic state has also been described in patients with septic shock [3, 5], or humans who receive endotoxin [5, 16].

Nitric Oxide is a Mediator of Cytokine-Induced Hypotension

Control of vascular tone is a balance between the constrictive actions of α-adrenergic agonists and the vasorelaxant influence of endothelium-derived relaxing factors (EDRF). EDRF was first described in 1980 as a labile factor released from vascular endothelial cells in response to acetylcholine. Subsequent studies [17] showed that EDRF was chemically indistinguishable from nitric oxide (NO), a diffusible free radical that decomposed spontaneously to nitrite and nitrate. The biological effects of NO appears to be mediated by the activation of guanylyl cyclase, which results in vascular smooth muscle relaxation.

NO is formed as a result of the action of NO synthase (NOS) which catalyzes the tetrahydrobiopterin (H_4 biopterin-) and NADPH-dependent oxidation of L-arginine to citrulline and NO [18]. The best characterized inhibitors of NOS activity are guanidino-substituted arginine analogs for which N^G-monomethyl-L-arginine (L-NMMA), an arginine antagonist that acts as a mechanism-based inhibitor of NOS [19], is the prototype. While three different NOS isoforms have been identified, only the inducible isoform of NOS (iNOS) appears to be controlled by immune stimulation. Initially, it was thought that only macrophages could make NO in response to cytokine stimulation. The finding that vascular endothelial cells could also be induced to make large amounts of nitrite as a result of exposure to cytokines [20] first implied a role for NO in cytokine-induced shock states such as IL-2-induced hypotension and septic shock. Figure 1 shows the time course of NOS activation following cytokine exposure. Messenger RNA (mRNA) for the cytokine-inducible iNOS in endothelial cells is first detectable within 2 h, and appearance of NOS by western blot analysis can be observed within 4 h. Following the appearance of NOS protein, enzyme activity increases and nitrite accumulation in the culture medium is detectable. The induction of NO formation has also been demonstrated in patients receiving IL-2. Elevated levels of nitrates have been detected in the serum and urine of these patients [21, 22]. Figure 2 shows the accumulation of nitrate in a patient who developed severe, refractory hypotension after 24 h of IL-2 therapy requiring the early discontinuation of IL-2. Additional studies with isotopically labelled arginine have confirmed that nitrate is derived from the guanidino nitrogen; the same

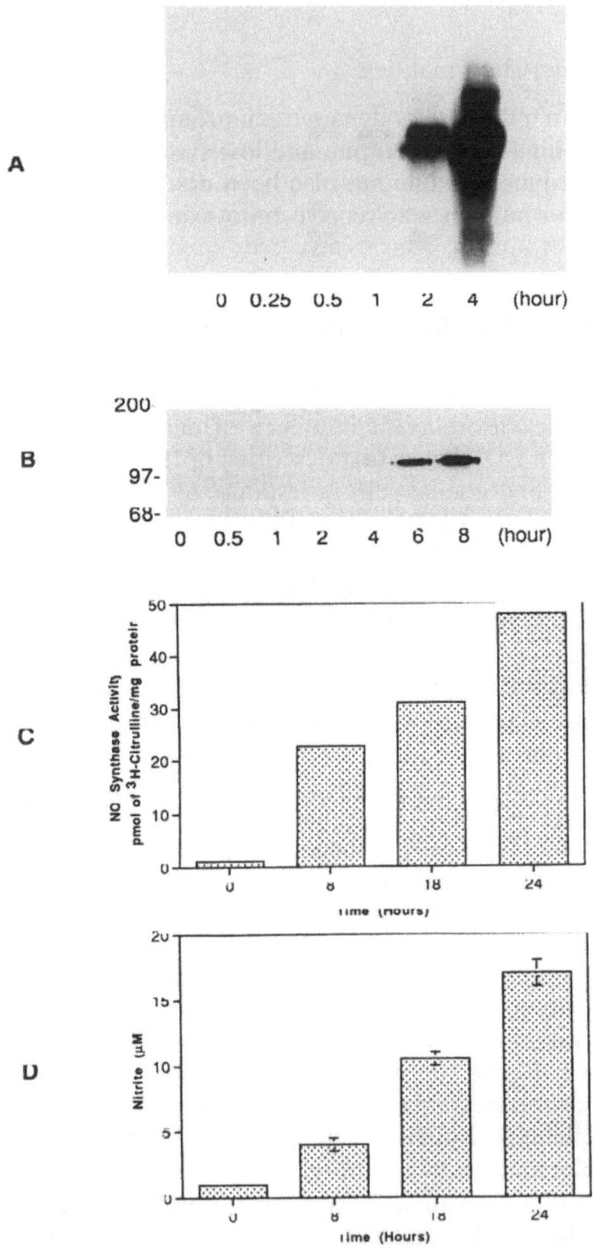

Fig. 1. Time course of NOS activation after cytokine exposure. Southern and western blot analysis reveal mRNA for the cytokine-inducible iNOS in endothelial cells first detectable within 2 h (*panel A*), while protein is observed at 4 h (*panel B*). Following the appearance of iNOS protein, enzyme activity increases (*panel C*) and nitrites accumulate in the culture medium (*panel D*)

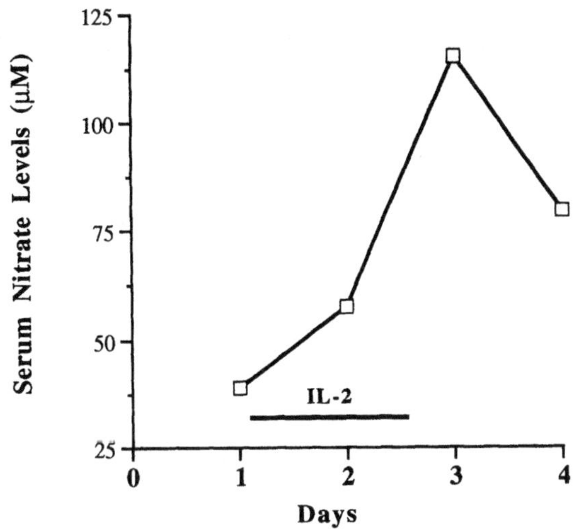

Fig. 2. Accumulation of nitrates and NO in a patient who developed severe, refractory hypotension after 24 h of IL-2 therapy requiring the early discontinuation of IL-2. Due to effects of IL-2 therapy the patient had become anorexic when serum samples were obtained. NO levels were measured by a Sievers® chemiluminescence instrument that reduces all nitrogen radicals in a vanadium chloride solution to NO

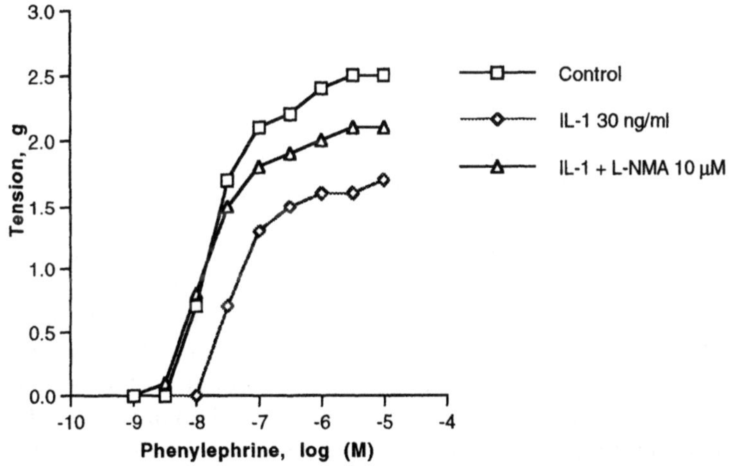

Fig. 3. In isolated vascular rings, preparation equimolar quantities of phenylephrine fail to induce vasoconstriction after cytokine exposure. This hyporesponsivity is reversed towards normal by the addition of NO inhibitors

site previously demonstrated to derive NO. Cardiovascular studies in dogs have provided additional experimental evidence for the role of NO in cytokine-induced vasodilation [23, 24]. These studies have shown that inhibitors of NOS rapidly and completely reverse this hypotension and restore vascular tone.

In addition to hypotension, IL-2 also induces relative insensitivity to vasopressors [14, 25]. Studies using isolated vascular rings in our laboratory provide an example of how incubation of vessels with cytokines leads to this hyporesponsiveness. Figure 3 shows how equimolar quantities of phenylephrine fail to induce vasoconstriction after cytokine exposure and the reversal of this hyporesponsivity by the addition of NO inhibitors. This vascular insensitivity has also been described after endotoxin exposure [26]. In the course of septic shock, even multiple vasopressors may be insufficient to reverse the severe hypotension. Similarly, the use of multiple vasopressors during IL-2-induced hypotension can fail to fully reverse the relaxation of the vascular smooth muscle which leads to vasodilation.

Role of NO in Major Organ Dysfunction

Hypotension may also account for some of the other side effects of IL-2 such as major organ dysfunction and lactic acidosis because of disordered vascular perfusion, of organs resulting in decreased oxygen delivery. Reversible renal insufficiency, intrahepatic cholestasis, myocardial depression, skin toxicity, and central nervous system dysfunction have all been reported during cytokine therapy and commonly become dose-limiting toxicities [27–29]. Renal insufficiency is of major importance in the care of kidney cancer patients that have undergone unilateral nephrectomy to remove the primary tumor mass. Anecdotal evaluation of patients who develop renal insufficiency during IL-2 therapy has failed to reveal any structural or functional abnormalities by renal ultrasound with Doppler flow analysis (unpublished observation). Sonographic changes in the renal parenchyma suggest that intramedullary edema may contribute to the decrease in renal function.

Additionally, cardiac depression has been noted. The cardiac depression is similar to that observed in sepsis: a biventricular decrease in cardiac efficiency [30]. The cause has been postulated to be due to circulating myocardial depressant factors [31]. More recently, NO induced by inflammatory cytokines has been implicated [32]. The mechanism of this NO-induced alteration is unclear, however it does not appear to be explained by alteration in coronary blood flow. It may be related to the inhibition of oxidative metabolism that has been previously described in tumor cells exposed to NO generating macrophages (discussed below). This occurs because of a direct nitrosylation of crucial iron-sulfur containing enzymes in the electron transport system of the mitochondria by NO [33]. It would be anticipated that the reversal of this NO-mediated lesion would require new enzyme synthesis requiring a prolonged period for normalization of respiratory function compared to the immediate effects observed in vascular tone discussed above. It is also possible that this inhibition of mitochondrial metabolism could account for the increased lactic acidosis observed in patients receiving cytokines or septic patients [34]. NO potently and reversibly deenergizes mitochondria at low oxygen tension [35]. Effects on the mitochondrial oxidative metabolism [36, 37] may be also responsible for the intrahepatic cholestasis and decreased liver synthetic capability reported in patients receiving IL-2 [29, 38].

Another important side effect of IL-2 involves the central nervous system (CNS). The spectrum of CNS toxicity induced by cytokines is usually characterized by acute cognitive deficits and depression; however frank psychoses, seizures, and loss of consciousness have been described [9, 39]. Incubation of microglial cells with cytokines induces NO synthesis. Figure 4 shows the accumulation of nitrite, a stable breakdown product of NO, in the culture medium of brain astrocytes exposed to cytokines. The appearance of nitrites is delayed by several hours after the addition of cytokines, and nitrite formation is blocked by inhibitors of protein synthesis suggesting that this nitrite accumulation is due to iNOS similar to that observed in cytokine-treated endothelial cells [40]. Additional studies of the rank-order potency of various arginine analog inhibitors (Fig. 5) confirm that the enzyme induced in brain astrocytes is similar to that described in cytokine-treated endothelial cells [40], and macrophages [41]. While the cause of cytokine-induced neuronal toxicity is not well defined, an induced state of NO overproduction may account for some of the CNS alterations observed with bioimmunotherapy.

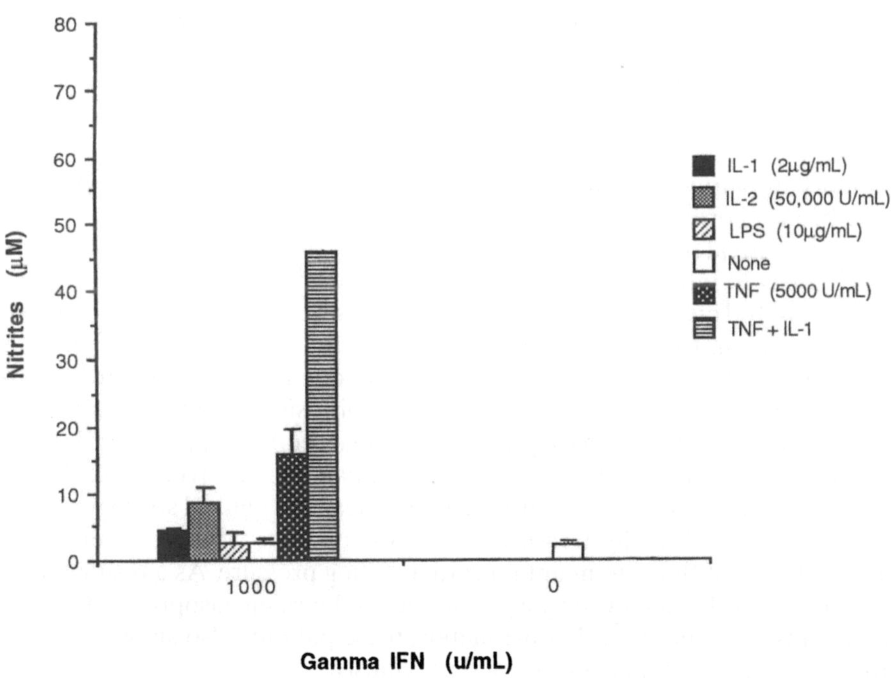

Fig. 4. Differential accumulation of nitrites in the culture medium of brain astrocytes exposed to cytokines. IL-1 = Interleukin-1; IL-2 = Interleukin-2; LPS = lipopolysaccaride; IFN = γ-interferon; TNF = tumor necrosis factor

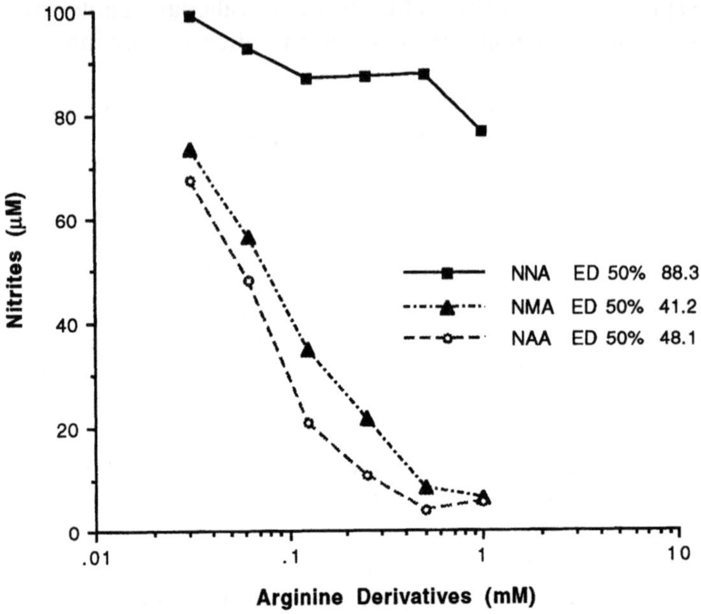

Fig. 5. Comparison of the potency of different arginine-analog inhibitors in cultures of brain astrocytes show the inferior effects of NNA and comparably greater potency of NMA and NAA. This pattern of rank-order potency is similar to that observed for the cytokine-induced isoform of NOS present in macrophages (rodent), cytokine-treated endothelial cells, and vascular smooth muscle cells
NNA = N^G-nitro-L-arginine; NMA = N^G-methyl-L-arginine; NAA = N^G-amino-L-arginine

Memory impairments which are commonly seen in patients receiving IL-2 have been reported to be modulated by NO [42].

Of even greater concern is the development of capillary leak syndrome which leads to intravascular depletion and contributes to major organ dysfunction [43]. This phenomenon appears to involve vascular endothelial cell damage leading to leakage of intravascular fluid into tissues [44] which is most noticeable in the lung. This leads to a unique form of non-cardiac pulmonary edema resulting in a decrease in the diffusion capacity. In some patients, this results in decreased oxygen exchange requiring mechanical ventilation. Because of this known complication, patients receiving IL-2 should not be given large amounts of intravenous fluids. This clinical strategy is contrary to that used in the treatment of septic patients who are often given large volumes of fluids to maintain cardiac filling pressure. As a result, blood pressure control is achieved by a greater dependence on vasopressor therapy in patients receiving IL-2. Unfortunately, these patients also show an insensitivity to pressor agents as discussed previously.

Therapy with IL-2 has also been associated with a variety of reversible hematological toxicities, in particular anemia, leukopenia and thrombocytopenia [45]. These effects appear directly related to the anti-proliferative effects of IL-2 and IL-2-induced endogenous cytokines on bone marrow progenitor cells [45, 46]. Additionally, NO has strong antiplatelet effects [47] inhibiting platelet aggregation induced by neurohumoral factors, or by increased vessel wall shear-stresses [48, 49]. This in turn, may explain the observed increased bleeding time in patients with high NO levels [50, 51]. Immunotherapy with IL-2 leads to a decreased synthesis of coagulation factors [38], intravascular activation of the coagulation cascade, and fibrinolysis [52]. This disregulated state of platelet function and coagulation can potentially induce further organ dysfunction by microvascular thrombosis as characteristically occurs in septic shock [53].

Various approaches have been used to counteract the severe side effects of IL-2. The use of concurrent steroids, pentoxifylline, and anti-cytokine therapies have been attempted in order to ameliorate the side effects of endogenous cytokine generation [11, 12, 54–56]. However, the presence of endogenous cytokines, such as TNF of IL-1, may be important in the anti-cancer response to IL-2 [57, 58]. In contrast, studies of the NO inhibitor, L-NMMA, show no effect on human lymphokine activated killer cell activity, IL-2-induced lymphocytosis, or generation of natural killer cell activity [59, 60]. Studies of alloantigen-induced activation of rat splenocytes suggest that L-NMMA further stimulates the IL-2-mediated generation of cytotoxic lymphocytes [61]. As a result of preclinical results that implicated NO in the toxicity of IL-2 administration, we performed a phase I, dose-ranging safety study. We report here our observations of notable biological activity of L-NMMA in hypotensive patients.

Phase I Study of L-NMMAin Humans

From the spring of 1992, patients at MD Anderson Cancer Center (MDACC) with biopsy-proven metastatic renal cell carcinoma with good performance status have been eligible for inclusion into a phase I, dose-ranging study of L-NMMA. All patients had either a nephrectomy or embolization of the primary tumor prior to enrollment. Patients with a prior history of atherosclerotic or cardiovascular disease, angina, myocardial infarction, or hypertension were excluded. No concurrent nitrovasodilators, anti-anginals, anti-hypertensive, steroids, or vasopressor drugs were allowed during the study. Patients receiving arginine-containing total parenteral or enteral solutions were excluded. Informed, written consent was obtained as per the guidelines of the University of Texas, MDACC Surveillance Committee.

Dosage and Administration of IL-2

IL-2 was administered at 18×10^6 IU/m^2/day by continuous infusion for 5 days according to the regimen defined by West et al. [28]. Patients received

2 treatments with IL-2 in the first month (induction) and then treatment was repeated every 28 days (maintenance) until tumor progression was documented. For the treatment of IL-2-related side effects patients received acetaminophen, prochlorperazine, clemastine fumarate, misoprostol, and ketorolac tromethamine. Due to the high rate of central line infections with gram-positive organisms previously observed in patients receiving IL-2 [62], patients also received intravenous vancomycin with dosages adjusted according to serum level determinations. IL-2 was discontinued in patients who developed thrombocytopenia, renal insufficiency, oliguria, respiratory distress, or hypotension refractory to vasopressors. The latter was defined as a systolic blood pressure <100 which was unresponsive to dopamine dosages >10 μg/kg/min and neosynephrine dosages >150 μg/min. Patients whose blood systolic pressure declined <90 mm Hg during the initial courses of IL-2 therapy became eligible to receive L-NMMA during their next scheduled IL-2 cycle.

Dosage and Administration of L-NMMA

Administration of L-NMMA in humans was approved by the U.S. Food and Drug Administration and the Hospital Surveillance Committee at MDACC. Sterile L-NMMA acetate was synthesized according to the method of Corbin & Reporter using Dowex-1 to convert the flavianate intermediate to the acetate salt at Cornell Medical College by Dr. Owen Griffith. Sterile, pyrogen-free L-NMMA acetate was formulated by the Pharmaceutical Services, College of Pharmacy at The University of Iowa.

The selection of the starting dosage of L-NMMA was guided by preclinical toxicology studies performed at MDACC. L-NMMA was found to be non-toxic in mice when given in single doses as high as 1250 mg/kg and in rats when given at a dose of 900 mg/kg per day for 14 consecutive days. The LD_{10} dose in mice was found to be 1750 mg/kg; a LD_{50} dose could not be defined because of the low level of toxicity observed. In dogs, 600 mg/kg of L-NMMA was non-toxic when administered in 3 divided doses over 1 h. Given chronically over a 14-day period, L-NMMA caused no adverse side effects in dogs at a dose of 50 mg/kg per day. Dogs receiving 150 or 300 mg/kg daily for 14 consecutive days experienced weight loss, gastrointestinal disturbances, uncoordinated behavior, and hypoactivity. Based on these results, the starting dose in humans was chosen to be 3 mg/kg given as a loading dose over 1 h in 3 divided doses followed by a maintenance dose of 1 mg/kg every 4 h as needed to maintain blood pressure for a maximum duration of treatment of 5 days.

L-NMMA dose escalation ranged from 3 to 24 mg/kg for the initial loading dose and 1 to 8 mg/kg for maintenance doses. At least 3 patients were treated at each dose level. Dose-level escalation within a patient group was performed when the duration of the antihypotensive effect was shorter than the predetermined maintenance schedule of 4 h. If the systolic blood pressure remained <100 mm Hg for 2 successive measurements spaced 15 min apart, despite a one-level dose escalation, no further L-NMMA was administered and therapy with standard vasopressors was instituted. In the initial protocol design, patients also received a test dose of L-NMMA prior to the start of IL-2 therapy. This dose of L-NMMA was then repeated when hypotension developed during IL-2 therapy to observe its antihypotensive effects. Test-doses were limited to 3, 6 and 12 mg/kg levels as discussed below.

Patient Studies and Analysis

All patients had an arterial and right heart catheterization. Hemodynamic measurements were taken immediately prior to and 15 min after administration of each dose of L-NMMA. Cardiac output was obtained by a flow-directed catheter using the thermodilution technique (Swan-Ganz® catheter – Abbott Laboratories, North Chicago, IL) [63]. Cardiac index, systemic vascular resistance, and pulmonary vascular resistance were calculated by standard methods [14].

The hemodynamic changes that resulted from administration of the initial loading dose of L-NMMA in each patient group were analyzed separately from changes loading that resulted from the administration of maintenance doses of L-NMMA. This allowed the separate evaluation of the initial effects of L-NMMA without overrepresenting data from patients responding to IL-2 (who received a greater number of IL-2 cycles combined with L-NMMA than patients with progression of their disease who were removed from the study earlier). Differences noted prior to and after administration of the loading and maintenance doses of L-NMMA were analyzed using the paired two-tailed *t*-test. A p value ≤ 0.05 was considered significant.

Toxicity data was collected for all clinical and laboratory abnormalities, and categorized according to standardized National Cancer Institute (NCI) guidelines for cancer patients.

Antihypotensive Actions of L-NMMA in Cancer Patients Receiving IL-2

Twelve [12] patients received a total of 39 courses of IL-2 therapy; 27 courses combined with L-NMMA. The dose levels of L-NMMA ranged from 3 to 24 mg/kg for the initial loading doses and from 1 to 8 mg/kg for the maintenance doses. Test-doses of 3 and 6 mg/kg did not significantly alter the blood pressure at baseline (122 ± 8 to 124 ± 12 mm Hg systolic blood pressure) (Fig. 6). However, transient hypertension was encountered at the 12 mg/kg dose level (128 ± 14 to 165 ± 24 mm Hg systolic blood pressure) (Fig. 6). No clinical symptoms developed as a consequence of the development of transient hypertension, nevertheless; further test doses were discontinued.

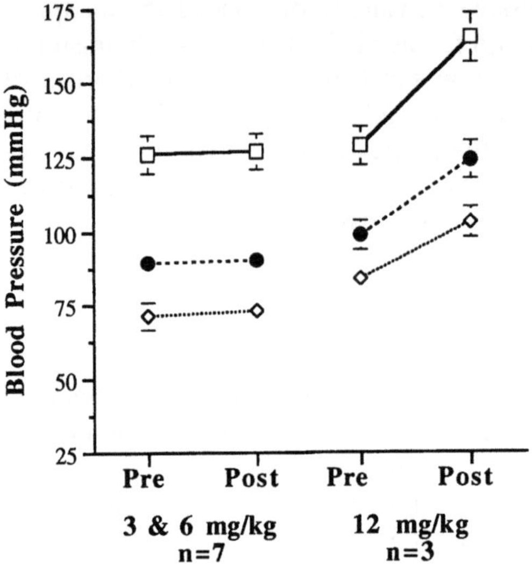

Fig. 6. Resting blood pressure response in humans to L-NMMA. Patients received a test dose of L-NMMA at baseline (normotensive state) prior to the initiation of IL-2 therapy, and then after the development of the cytokine-induced hypotension. Means ± SEM

At dosages of either 3 or 6 mg/kg of L-NMMA, IL-2 induced hypotension was reversed (84±3 to 110±6 mm Hg systolic blood pressure) (Fig. 7). Initial L-NMMA loading doses reversed the hypotension at all dose-levels (Tables 1–3). The magnitude of the increase in blood pressure was greater in hypotensive patients receiving either 3 or 6 mg/kg of L-NMMA when compared with the change in the resting systolic blood pressure after administration of the same doses of L-NMMA administered as a test doses (2 vs 26 mm Hg).

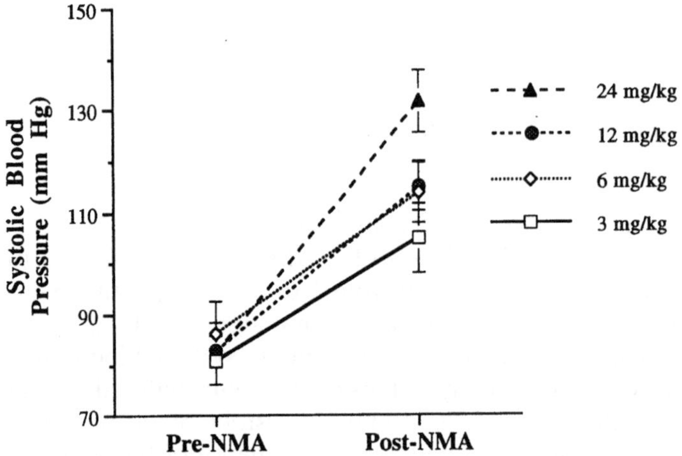

Fig. 7. Systolic blood pressure response to the initial loading dose of L-NMMA in patients with IL-2-induced hypotension. Means ± SEM (n = 20)

Table 1. Blood pressure effects induced by the initial administration of L-NMMA to normotensive patients at the dose levels of 3 and 6 mg/kg. Means ± SEM

mm Hg	Before	After	p Value
3 mg/kg			
SBP	81	105.0	0.01
DBP	51	59	NS
MBP	61	74.4	<0.05
6 mg/kg			
SBP	86.3	114	0.01
DBP	47.6	56.6	<0.05
MBP	60.5	75.8	<0.05

SBP = systolic blood pressure; DBP = diastolic blood pressure; MBP = mean blood pressure

Table 2. Hemodynamic effects induced by the initial administration of L-NMMA at the dose level of 12 mg/kg. Means ± SEM

	Before	After	p Value
SBP, mm Hg	83.5	117.6	<0.01
DBP, mm Hg	48.6	67.8	<0.05
MBP, mm Hg	60.2	84.4	<0.01
MPA, mm Hg	16.5	21.4	NS
CI, L/min/m^2	4.57	3.85	<0.05
SVR, dyne · sec/cm^{-5}	507.6	644.2	<0.01
PVR, dyne · sec/cm^{-5}	80.1	155.4	NS
PCWP, mm Hg	7.6	8.4	NS
CVP, mm Hg	4	5	NS
Heart Rate, min	121.4	108	NS

SBP = systolic blood pressure; DBP = diastolic blood pressure; MBP = mean blood pressure; MPA = mean pulmonary artery; CI = cardiac index; SVR = systemic vascular resistance; PVR = pulmonary vascular resistance; PCWP = pulmonary capillary wedge pressure; CVP = central venous pressure

The hemodynamic effects were readily observed within minutes of the administration of L-NMMA and were dose dependent. The IL-2-induced hyperdynamic state of high cardiac output and low systemic vascular resistance was also reversed by L-NMMA. The predominant effect of L-NMMA appeared to be due to the restoration of normal vascular tone based on the observed increase in the systemic vascular resistance (Fig. 8) (Tables 2, 3). Following infusion of L-NMMA, a decrease in the cardiac index below normal levels was observed. This effect, however, was reversed by the co-infusion of dobutamine as previously described in animal studies [64]. The

Table 3. Hemodynamic effects induced by the initial administration of L-NMMA at the dose level of 24 mg/kg. Means ± SEM

	Before	After	p Value
SBP, mm Hg	82.4	131.6	<0.001
DBP, mm Hg	40.6	78	0.001
MBP, mm Hg	54.5	95.87	<0.001
MPA, mm Hg	17.7	27.6	<0.05
CI, $L/min/m^2$	4.53	3.65	<0.05
SVR, dyne·sec/cm^{-5}	455.7	862.7	<0.01
PVR, dyne·sec/cm^{-5}	84.5	230.5	<0.01
PCWP, mm Hg	8.75	10	<0.01
CVP, mm Hg	5.5	8.75	NS
Heart Rate, min	110	112.4	NS

SBP = systolic blood pressure; DBP = diastolic blood pressure; MBP = mean blood pressure; MPA = mean pulmonary artery; CI = cardiac index; SVR = systemic vascular resistance; PVR = pulmonary vascular resistance; PCWP = pulmonary capillary wedge pressure; CVP = central venous pressure

co-infusion of dobutamine augmented the antihypotensive effect of L-NMMA and significantly prolonged the duration of action (data not shown). As a result of treatment with IL-2, most patients developed elevated pulmonary artery pressures, which were further increased by administration of L-NMMA (Fig. 8) (Tables 2, 3). Pulmonary vascular resistance increased after administration of L-NMMA to much the same degree as the systemic vascular resistance (+189 and +199%, respectively, at a L-NMMA dose of 24 mg/kg). No clinical evidence of right heart failure secondary to increased pulmonary artery pressures was observed. Pulmonary capillary wedge pressure increased moderately after L-NMMA administration, while heart rate and right atrial pressures did not change significantly.

The duration of action of L-NMMA increased in a dose-dependent manner (Fig. 9). This allowed the IL-2 therapy to be continued for a longer time. The average number of L-NMMA maintenance doses administered at dose-level 1 (1 mg/kg) and dose-level 2 (2 mg/kg) was 12 doses. At dose-levels 3 (4 mg/kg) and 4 (8 mg/kg), an average of 19 doses were administered.

The duration of action of L-NMMA decreased sequentially with continued IL-2 administration until the duration of effect was unacceptably short prompting the cessation of L-NMMA therapy and the start of vasopressors. This was especially noticeable at the lower dose levels. At the highest dose level tested, 66% of the L-NMMA doses achieved normalization of the blood pressure >3 h. After failing treatment with L-NMMA, most patients exhibited marked vasopressor insensitivity requiring rapid pressor dose escalation and subsequent discontinuation of IL-2 therapy.

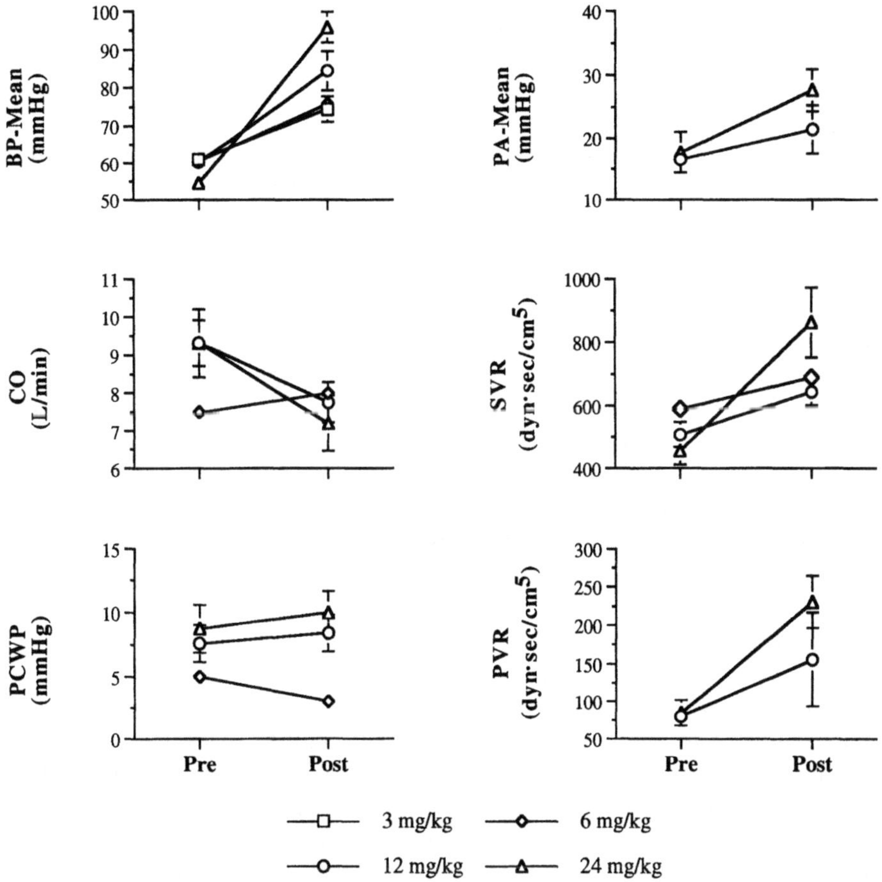

Fig. 8. Hemodynamic responses of cancer patients receiving IL-2 to L-NMMA at different dose-levels. Means ± SEM

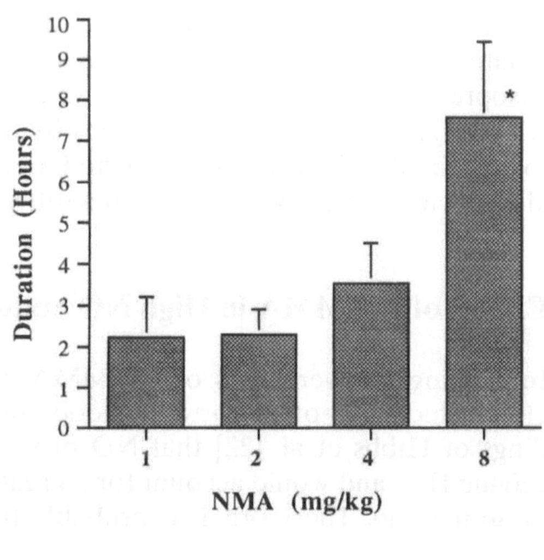

Fig. 9. Duration of the antihypotensive effect of the initial loading doses of L-NMMA in cancer patients with IL-2-induced hypotension. Means ± SEM (n = 20)

Toxicity of L-NMMA in Shock States

No toxicity was induced by L-NMMA administration as an IV bolus at doses as high as 24 mg/kg (8 mg/kg maintenance dose; cumulative dose range 30.3 to 163.2 mg/kg/24 h at dose-level 8 mg/kg). No evidence of cardiac ischemia, bleeding diathesis, bowel damage, or hepatic injury were observed to be induced by L-NMMA. Plasma lactate concentration stayed within the normal range (1 to 2.2 mmol/L). Oxygen delivery and consumption was not significantly affected and platelet function appeared unchanged (unpublished observation).

IL-2 therapy resulted in clinically severe toxicity (NCl grade >2) in the majority of patients. The major IL-2 toxicities encountered consisted of hypotension (82.1%), capillary leak syndrome defined as weight gain >4.5 kg (66.7%), pulmonary edema with hypoxemia (2.5%), fever ≥40°C (15.5%), creatinine elevation >3 times the normal value (5.1%), liver enzyme elevation ≥5 times the normal values (7.7%), anemia ≤7.9 g/dL hemoglobin) (10.3%), and CNS toxicity (12%).

NOS Inhibitors have a Unique Mechanism of Action

After the administration of an initial loading dose of L-NMMA, all patients in our study showed evidence of a clinically significant anti-hypotensive effect with the rapid reversal of the IL-2-induced hemodynamic abnormalities. This was illustrated by changes in cardiac output, systemic vascular resistance, pulmonary artery pressure, and pulmonary vascular resistance. Additional bolus doses were effective in maintaining vascular tone and blood pressure. The mechanism of action of L-NMMA is unique, compared with that of standard vasopressors, since it reverses an induced state of vasoplegia resulting in the normalization of blood pressure.

Previous studies in which lower doses of L-NMMA were administered to humans, showed an anti-hypotensive effect by L-NMMA when used in combination with vasopressors in patients with septic shock [65]. A decrease in vasopressor requirements was observed, consistent with studies showing that L-NMMA reverses the hyporesponsitivity to α-adrenergic agents in endotoxic animals [25]. The study reported here details the first use of L-NMMA alone (without vasopressors) as an antihypotensive agent.

Effects of L-NMMA in High NO States

Interestingly, lower doses of L-NMMA had a significant anti-hypotensive effect which did not alter *resting* blood pressure. This result supports the findings of Hibbs et al. [22] that NO production is increased in patients receiving IL-2 and would account for a greater net effect on blood pressure in these patients. Therefore, it is probable that the increased NO production overcomes adrenergic vasoconstrictive influences and decreases the sensitivity to adrenergic agonists, resulting in a state of vascular relaxation (vasoplegia) with an associated counter-regulatory response (hyperdynamic state). The duration of the anti-hypotensive response at these low doses, however, is short. Duration is increased at higher doses, but resting pressure is also increased which precludes prophylactic treatment of normotensive patients with these doses of L-NMMA.

The Need for Selective Inhibitors of NOS

NO is produced by the enzyme NOS, which exists in both constitute and inducible isoforms. The inducible isoenzyme thought to be responsible for NO overproduction following cytokine exposure has been demonstrated in the vascular endothelium [66] and smooth muscle cells [67, 68]. *In vitro* studies have shown that NOS inhibitors differ in potency and selectivity from the inducible iNOS [41], but L-NMMA has the lowest toxicity. It is possible that the effects on the constitutive enzyme can be reduced, resulting in a decreased effect on resting blood pressure by the use of more selective inhibitors of iNOS [69].

Another consideration for the design of selective inhibitors has been the possible protective effects of NO . Reports in rodent models using high doses of endotoxin have demonstrated bowel damage, abnormal liver function tests, and platelet dysfunction induced by L-NMMA [70–73]. We saw no clinical indication of such adverse effects in humans receiving L-NMMA. Notably, some animal models differ widely from the human cardiovascular response during septic shock [30, 74]. A dissimilar response to endotoxin is particularly evident in rodents which are insensitive to endotoxin and refractory to the development of shock [75]. In contrast to rodent studies, no bowel, liver, or hematologic toxicity was observed in dogs receiving IL-2 and L-NMMA [59]. The canine response to IL-2 administration was similar to that observed in humans and predicted the hemodynamic response to L-NMMA observed in this study.

Effects of L-NMMA on the Heart, Lung, and Vascular Endothelium

Elevation of blood pressure appeared to be due to a L-NMMA-mediated increase in systemic vascular resistance, even though cardiac output was decreased. The mechanism by which L-NMMA decreases the cardiac output may be related to the increase in afterload that results as a consequence of the sudden increase in systemic vascular resistance [14]. This conclusion is supported by animal experiments that also demonstrated a decreased cardiac output after administration of phenylephrine, an α-adrenergic agent [23]. The decreased cardiac output seen with L-NMMA administration can be acutely reversed, however, by administration of dobutamine, an inotropic agent [76]. Studies of myocardial function in animal models of sepsis also suggest that the cytokine-induced overproduction of NO may cause myocardial depression, an effect not acutely reversed by L-NMMA administration [77, 78].

The elevation in pulmonary pressure was also noted but was not associated with clinical symptoms. As seen in septic shock, IL-2 also induces pulmonary hypertension [79]. We observed that the administration of a NOS inhibitor further potentiates this effect. No adverse clinical effects were observed as a consequence of the increased pulmonary artery pressures. Local administration of NO by inhalation would be one possible way to decrease pulmonary pressures without systemic side effects [80].

In our study, coadministration of L-NMMA to patients receiving IL-2 did not reverse the vascular leak previously associated with IL-2 therapy since the patients continued to develop tissue edema and gained weight (data not shown). It is unlikely that the intermittent bolus schedule of L-NMMA administration used in the study would totally prevent vascular damage due to excess NO since patients were not treated prophylactically. Continuous (not intermittent) suppression of NO production may be necessary to judge the role of NO in IL-2-induced vascular leak syndrome. Alternatively, other mediators of vascular leak may be responsible for microvascular dysfunction as demonstrated by animal models [81, 82].

Conclusion

Therapy with high dose IL-2 is associated with severe side effects which are more intense during subsequent cycles of therapy. The addition of L-NMMA significantly and rapidly attenuated the hypotension, without adding toxicity. At the doses studied thus far, other important side-effects induced by IL-2, such as capillary leak syndrome, oliguria, liver dysfunction, or central nervous system toxicity are not modified by L-NMMA. Additional studies of larger doses of L-NMMA are in progress. In addition, further refinement of the schedule of administration of L-NMMA (e.g. bolus combined with continuous infusion) may be necessary to provide continuous suppression of NO production in order to assess its usefulness in modulating the other side effects of IL-2.

The antitumor response in this study was low (5.5% complete response rate), a finding consistent with the low response rate to IL-2 therapy previously reported in metastatic renal cell carcinoma [83, 84] and with the small number of patients treated in this study. It is possible that L-NMMA could favorably enhance the response rate to IL-2 by permitting an increase in the dose intensity of IL-2 treatments, either by prolonging treatment or by allowing an increase in the maximum tolerated dose. Additionally, L-NMMA treatment may permit a decrease in the need for intensive care monitoring for patients undergoing high dose IL-2 therapy, thereby decreasing overall treatment-related costs.

References

1. Lotze MT, Chang AE, Seipp CA, et al (1986) High-dose recombinant interleukin-2 in the treatment of patients with disseminated cancer. JAMA 256:3117–3124
2. Lee R, Lotze M, Skibber J, et al (1989) Cardiorespiratory effects of immunotherapy with interleukin-2. J Clin Oncol 7:7–20
3. Ognibene F, Rosenberg S, Skibber J, et al (1986) Interleukin-2 hemodynamics mimic septic shock. Clin Res 34:413A (Abst)
4. Rosenberg SA (1988) The development of new immunotherapies for the treatment of cancer using interleukin-2. A review. Ann Surg 208:121–135
5. Suffredini A, Fromm R, Parker M, et al (1989) The cardiovascular response of normal humans to the administration of endotoxin. N Engl J Med 321:280–287
6. Chong ASF, Scuderi P, Gremes WJ, et al (1989) Tumor targets stimulate IL-2 activated killer cells to produce interferon-γ and human tumor necrosis factor. J Immunol 142: 2133–2139
7. Gemlo B, Palladino M, Jaffe H, et al (1988) Circulating cytokines in patients with metastatic cancer treated with recombinant interleukin-2 and lymphokine-activated killer cells. Cancer Res 48:5864–5867
8. Mier JW, Vachino G, Van der Meer JWM, et al (1988) Induction of circulating tumor necrosis factor (TNFα) as the mechanism for the febrile response to interleukin-2 (IL-2) in cancer patients. J Clin Immunol 8:426-436
9. Rosenberg SA, Lotze MT, Yang JC, et al (1993) Prospective randomized trial of high-dose interleukin-2 alone or in conjunction with lymphokine-activated killer cells for the treatment of patients with advanced cancer. J Nat Cancer Inst 85:622–631
10. Parkinson DR (1990) Lessons from the clinical trials of interleukin-2. Nat Immun Cell Growth Regul 9:242–252
11. Thompson JA, Benyunes MC, Fefer A (1994) Reducing IL-2 toxicity: Efforts to improve the therapeutic index of IL-2. Hem/Onc Annals 2:351–355
12. Mier J, Vachino G, Klempner M, et al (1990) Inhibition of interleukin-2-induced tumor necrosis factor release by dexamethasone: Prevention of an acquired neutrophil chemotaxis defect and differential suppression of interleukin-2-associated side effects. Blood 76:1933–1940
13. Glauser FL, DeBlois GG, Bechard DE, et al (1988) A comparison of the cardiopulmonary effects of continuous versus bolus infusion of recombinant interleukin-2 in sheep. Cancer Res 48:2221–2225
14. Zeilender S, Davis D, Fairman RP, et al (1989) Inotropic and vasoactive drug treatment of interleukin-2-induced hypotension in sheep. Cancer Res 49:4423–4426
15. West WH, Tauer KW, Yannelli JR, et al (1987) Constant-infusion recombinant interleukin-2 in adoptive immunotherapy of advanced cancer. N Engl J Med 316:898–905
16. Taveira da Silva AM, Kaulbach HC, Chuidian FS, et al (1993) Brief report: Shock and multiple-organ dysfunction after self-administration of Salmonella endotoxin. N Engl J Med 328:1457–1460
17. Palmer R, Ferrige A, Moncada S (1987) Nitric oxide release accounts for the biological activity of endothelium-derived relaxing factor. Nature 327:524–526
18. Stuehr D, Griffith OW (1992) Mammalian nitric oxide synthases. Adv Enzymol Relat Areas Mol Biol 65:287–346
19. Hibbs J, Taintor R, Vavrin Z (1987) Macrophage cytotoxicity: Role for L-arginine deiminase andimino nitrogen oxidation to nitrite. Science 235:473–476
20. Kilbourn RG, Belloni P (1990) Endothelial cell production of nitrogen oxides in response to interferon gamma in combination with tumor necrosis factor, interleukin-1, or endotoxin. J Natl Cancer Inst 82:772–776
21. Ochoa JB, Udekwu AO, Billiar TR, et al (1991) Nitrogen oxide levels in patients after trauma and sepsis. Ann Surg 214:621–626
22. Hibbs J, Westenfelder C, Taintor R, et al (1992) Evidence for cytokine-inducible nitric oxide synthesis from L-arginine in patients receiving interleukin-2 therapy. J Clin Invest 89:867–877

23. Kilbourn RG, Jubran A, Gross SS, et al (1990) Reversal of endotoxin-mediated shock by N^G-methyl-L-arginine, an inhibitor of nitric oxide synthesis. Biochem Biophys Res Commun 172:1132–1138
24. Kilbourn RG, Gross SS, Lodato RF, et al (1992) N^{ω}-aminoarginine inhibits interleukin-1α-induced nitric oxide synthase in vascular smooth muscle and fully reverses interleukin-1α-induced hypotension. J Natl Cancer Inst 84:1008–1016
25. Fleming I, Gray G, Julou-Schaeffer G, et al (1989) Impaired vascular reactivity in the rat following endotoxin treatment can be endothelium independent, yet involves the L-arginine pathway. J Physiol 423:105P–107P
26. Julou-Schaeffer G, Gray G, Fleming I, et al (1990) Loss of vascular responsiveness induced by endotoxin involves L-arginine pathway. Am J Physiol 259:H1038–H1043
27. Rosenberg SA, Lotze MT, Muul LM, et al (1987) A progress report on the treatment of 157 patients with advanced cancer using lymphokine-activated killer cells and interleukin-2 or high-dose interleukin-2 alone. N Engl J Med 316:889–897
28. West WH, Tauer KW, Yannelli JR, et al (1987) Constant-infusion recombinant interleukin-2 in adopotive immunotherapy of advanced cancer. N Engl J Med 316: 898–905
29. Rosenberg SA, Lotze MT, Yang JC, et al (1989) Experience with the use of high-dose interleukin-2 in the treatment of 652 cancer patients. Ann Surg 210:474–484
30. Natanson C, Danner R, Fink M, et al (1988) Cardiovascular performance with E. coli challenges in a canine model of human sepsis. Am J Physiol 254:H558–H569
31. Parker M, Shelhamer J, Bacharach S, et al (1984) Profound but reversible myocardial depression in patients with septic shock. Ann Intern Med 100:483–490
32. Finkel M, Oddis C, Jacob T, et al (1992) Negative inotropic effects of cytokines on the heart mediated by nitric oxide. Science 257:387–389
33. Lancaster J, Hibbs J (1990) EPR demonstration of iron-nitrosyl complex formation by cytotoxic activated macrophages. Proc Natl Acad Sci USA 87:1223–1227
34. Gilbert EM, Haupt MT, Mandanas RY, et al (1986) The effect of fluid loading, blood transfusion, and catecholamine infusion on oxygen delivery and consumption in patients with sepsis. Am Rev Respir Dis 134:873–878
35. Schweizer M, Richter C (1994) Nitric oxide potently and reversibly deenergizes mitochrondia at low oxygen tension. Biochem Biophys Res Commun 204:169–175
36. Stadler J, Curran RD, Ochoa JB, et al (1991) Effect of endogenous nitric oxide on mitochondrial respiration of rat hepatocytes in vitro and in vivo. Arch Surg 126: 186–191
37. Billiar T, Curran R, Harbrecht B, et al (1990) Modulation of nitrogen oxide synthesis in vivo: N^G-monomethyl-L-arginine inhibits endotoxin-induced nitrate/nitrate biosynthesis while promoting hepatic damage. J Leukoc Biol 48:565–569
38. Richard V, Bernier M, Themelin L, et al (1991) Blood coagulation abnormalities during adoptive immunotherapy with interleukin-2 (r-Met Hu IL-2 [ala125]). Ann Oncol 2:67–68
39. Triozzi PL, Kinney P, Rinehart JJ (1990) Central nervous system toxicity of biological response modifiers. Ann NY Acad Sci (USA) 594:347–354
40. Gross SS, Jaffe EA, Levi R, et al (1991) Cytokine-activated endothelial cells express an isotype of nitric oxide synthase which is tetrahydrobiopterin-dependent, calmodulin-independent and inhibited by arginine analogs with a rank order of potency characteristic of activated macrophages. Biochem Biophys Res Commun 178:823–829
41. Gross S, Stuehr D, Aisaka K, et al (1990) Macrophage and endothelial cell nitric oxide synthesis. Cell-type selective inhibition by N^G-amino-arginine, N^G-nitro-arginine and N^G-methyl-arginine. Biochem Biophys Res Commun 170:96–103
42. Zhuo M, Small SA, Kandel ER, et al (1993) Nitric oxide and carbon monoxide produce activity-dependent long-term synaptic enhancement in hippocampus. Science 260: 1946–1950
43. Rosenberg SA, Lotze MT, Yang JC, et al (1993) Prospective randomized trial of high-dose interleukin-2 alone or in conjunction with lymphokine-activated killer cells for the treatment of patients with advanced cancer. J Natl Cancer Inst 85:622–632

44. Damle N, Doyle L, Bender J, et al (1987) Interleukin-2 activated lymphocytes exhibit enhanced adhesion to normal vascular endothelial cells. J Immunol 138: 1779–1785
45. Ettinghausen SE, Moore JG, White DE, et al (1987) Hematologic effects of immunotherapy with lymphokine-activated killer cells and recombinant interleukin-2. Blood 69:1654–1660
46. Raefsky E, Platanias L, Zoumbos N, et al (1985) Studies of interferon as a regulator of hematopoietic cell proliferation. J Immunol 135:2507–2512
47. Radomski MW, Palmer RM, Moncada S (1987) The anti-aggregating properties of vascular endothelium: Interactions between prostacyclin and nitric oxide. Br J Pharmacol 92:639–646
48. Durante W, Kroll MH, Vanhoutte PM, et al (1992) Endothelium-derived relaxing factor inhibits thrombin-induced platelet aggregation by inhibiting platelet phospholipase C. Blood 79:110–116
49. Durante W, Schafer AI, Hrbolich JK, et al (1993) Endothelium-derived relaxing factor inhibits shear stress-induced platelet aggregation. Platelets 4:135–140
50. Högman M, Frostell C, Arnberg H, et al (1993) Bleeding time prolongation and NO inhalation. Lancet (letter) 341:1664–1665
51. Bassenge E (1991) Antiplatelet effect of endothelium-derived relaxing factor and nitric oxide donors. Eur Heart J 12 (Suppl):12–15
52. Baars JW, De Boer JP, Wagstaff J, et al (1992) Interleukin-2 induces activation of coagulation and fibrinolysis: Resemblance to the changes seen during experimental endotoxemia. Brit J Haematol 82:295–301
53. Moalli R, Doyle JM, Tahhan R, et al (1989) Fibrinolysis in critically ill patients. Am Rev Respir Dis 140:287–293
54. Thompson JA, Benyunes MC, Bianco JA, et al (1993) Treatment with pentoxifylline and ciprofloxacin reduces the toxicity of high-dose interleukin-2 and lymphokine-activated killer cells. Semin Oncol 20:46–51
55. Thompson J, Nemunitis J, Vogelzang NJ, et al (1994) Phase 1 trial of CT1501R in cancer patients receiving high-dose interleukin-2 (IL-2). Proc ASCO 13:299 (Abst)
56. Edwards MJ, Heniford BT, Klar EA, et al (1992) Pentoxifylline inhibits interleukin-2-induced toxicity in C57BL/6 mice but preserves antitumor efficacy. J Clin Invest 90: 637–641
57. Boccoli G, Masciulli E, Ruggeri E, et al (1990) Adoptive immunotherapy of human cancer: The cytokine cascade and monocyte activation following high-dose interleukin-2 bolus treatment. Cancer Res 50:5795–5800
58. Boldt DH, Mills BJ, Gemlo BT, et al (1988) Laboratory correlates of adoptive immunotherapy with recombinant interleukin-2 and lymphokine-activated killer cells in humans. Cancer Res 48:4409–4416
59. Kilbourn R, Owen-Schaub L, Cromeens D, et al (1994) N^G-methyl-L-arginine, an inhibitor of nitric oxide formation, reverses IL-2 mediated hypotension in dogs. J Appl Physiol 76:1130–1137
60. Kilbourn R, Owen-Schaub L, Griffith O, et al (1992) Interleukin-2 mediated hypotension in dogs is reversed by N^G-monomethyl-L-arginine (NMA), an inhibitor of nitric oxide (NO) formation. Proc AACR 33:328 (Abst)
61. Hoffman RA, Langrehr JM, Billiar TR, et al (1990) Alloantigen-induced activation of rat splenocytes is regulated by the oxidative metabolism of L-arginine. J Immunol 145:2220–2226
62. Pockaj BA, Topalian SL, Steinberg SM, et al (1993) Infectious complications associated with interleukin-2 administration: A retrospective review of 935 treatment courses. J Clin Oncol 11:136–147
63. Ganz W, Donoso R, Marcus HS, et al (1971) A new technique for measurement of cardiac output by thermodilution in man. Am J Cardiol 27:392–396
64. Kilbourn RG, Cromeens DW, Chelly FD, et al (1994) N^G-methyl-L-arginine, an inhibitor of nitric oxide formation acts synergistically with dobutamine to improve cardiovascular performance in endotoxemic dogs. Crit Care Med 22:1835–1840

65. Petros A, Bennett D, Vallance P (1991) Effects of nitric oxide synthase inhibitors on hypotension in patients with septic shock. Lancet 338:1557–1558
66. Kilbourn RG, Belloni P (1990) Endothelial cell production of nitrogen oxides in response to interferon gamma in combination with tumor necrosis factor, interleukin-1, or endotoxin. J Natl Cancer Inst 82:772–776
67. Beasley D, Schwartz JH, Brenner BM (1991) Interleukin 1 induces prolonged L-arginine-dependent cyclic guanosine monophosphate and nitrite production in rat vascular smooth muscle cells. J Clin Invest 87:602–608
68. Kilbourn RG, Gross SS, Lodato RF, et al (1992) N^{ω}-amino-arginine inhibits interleukin-1α-induced nitric oxide synthase in vascular smooth muscle and fully reverses interleukin-1α-induced hypotension. J Natl Cancer Inst 84:1008–1016
69. Joly G, Ayres M, Chelly F, et al (1994) Effects of N^G-methyl-L-arginine, N^G-nitroarginine, and aminoguanidine on constitutive and inducible nitric oxide synthase in rat aorta. Biochem Biophys Res Commun 199:147–154
70. Nava E, Palmer R, Moncada S (1991) Inhibition of nitric oxide synthesis in septic shock: How much is beneficial? Lancet 338:1555–1557
71. Hutcheson IR, Whittle BJ, Boughton SNK (1990) Role of nitric oxide in maintaining vascular integrity in endotoxin-induced acute intestinal damage in the rat. Br J Pharmacol 101:815–820
72. Billiar TR, Curran RD, Harbrecht BG, et al (1990) Modulation of nitrogen oxide synthesis in vivo: N^G-monomethyl-L-arginine inhibits endotoxin-induced nitrate/nitrate biosynthesis while promoting hepatic damage. J Leukoc Biol 48:565–569
73. Stark ME, Szurszewski JH (1992) Role of nitric oxide in gastrointestinal and hepatic function and disease. Gastroenterology 103:1928–1949
74. Parrillo JE (1993) Pathogenic mechanisms of septic shock. N Engl J Med 328:1428 (letter)
75. Wang Q, Jacobs J, DeLeo J, et al (1991) Nitric oxide hemoglobin in mice and rats in endotoxic shock. Life Sciences 49:55–60
76. Kilbourn RG, Cromeens DM, Chelly FD, et al (1994) N^G-methyl-L-arginine, an inhibitor of nitric oxide formation, acts synergistically with dobutamine to improve cardiovascular performance in endotoxic dogs. Crit Care Med 22:1835–1840
77. Stein B, Frank P, Mulsch A (1993) Long-term treatment of cardiomyocytes with endotoxin or cytokines induce negative inotropic effects mediated by nitric oxide. Endothelium 1:s63 (Abst)
78. Grocott-Mason R, Lewis M, Shah A (1993) Nitric oxide has direct myocardial relaxant effects in the isolated ejecting heart. Endothelium 1:s62 (Abst)
79. Clowes GH, Farrington GH, Zuschneid W, et al (1970) Circulating factors in the etiology of pulmonary insufficiency and right heart failure accompanying severe sepsis (peritonitis). Ann Surg 171:663–678
80. Rossaint R, Falke KJ, Lopez F, et al (1993) Inhaled nitric oxide for the adult respiratory distress syndrome. N Engl J Med 328:399–405
81. Rabinovici R, Sofronski M, Renz J, et al (1992) Platelet activating factor mediates interleukin-2-induced lung injury in the rat. J Clin Invest 89:1669–1673
82. Leder GH, Oppenheim M, Rosenstein M, et al (1993) Aminoguanidine decreases IL-2-induced nitric oxide production but not the IL-2-induced capillary leak syndrome. Endothelium 1:s51 (Abst)
83. Abrams JS, Rayner AA, Wiernik PH, et al (1990) High-dose recombinant interleukin-2 alone: A regimen with limited activity in the treatment of advanced renal cell carcinoma. J Natl Cancer Inst 82:1202–1206
84. Parkinson DR (1990) Interleukin-2 further progress through greater understanding. J Natl Cancer Inst 82:1374–1376

Blocking the Effects of Nitric Oxide in Septic Shock

J. L. Vincent, J. C. Preiser, and H. Zhang

Introduction

That isolated vessels and organs and intact organisms markedly increase nitric oxide (NO) release during septic challenge has been amply demonstrated in both animal and human studies. Many experimental protocols using competitive antagonists of L-arginine, the natural precursor of NO, have been performed *in vitro*, *in vivo*, and in clinical pilot trials. NO production can be inhibited by several analogs of L-arginine, such as N^G-monomethyl-L-arginine (L-NMMA), N^w-nitro-L-arginine (L-NNA), N^G-nitro-L-arginine-methyl ester (L-NAME) and N^w-amino-L-arginine (L-NAA). Methylene blue can be also used to block the effects of NO mediated by the formation of cyclic guanosine monophosphate (cGMP) (Table 1). Trials with these agents have documented the role of NO in the development of vasodilation, hyporesponsiveness to catecholamines, and myocardial depression associated with severe sepsis.

The inducible NO synthase (iNOS) is activated under the influence of lipopolysaccharide (LPS), endotoxin, tumor necrosis factor-α (TNF), interleukin-1 (IL-1) and interferon gamma (IFN-γ), as well as supernatant from activated neutrophils or macrophages. In severe sepsis, the vascular smooth muscle, the macrophages and other cellular elements (primarily outside the

Table 1. Various mechanisms to reduce NO activity

1. anti-endotoxin antibodies
2. anti-TNF antibodies, IL-1ra
3. PAF antagonists
4. inhibitors of NO induction: corticosteroids, IL-10, IL-8, TGF, nifedipine
5. NOS inhibitors:
   a) L-arginine analogs:
      - N^G-monomethyl-L-arginine (L-NMMA)
      - N^w-amino-L-arginine (L-NAA)
      - N^w-nitro-L-arginine (L-NNA)
      - N^G-nitro-L-arginine methyl ester (L-NAME)
   b) specific inhibitors of iNOS:
      - aminoguanidine
      - isothioureas
6. methylene blue

endothelium) are involved in this NO release. Typically, mouse peritoneal macrophages exposed to endotoxin generate nitrites and nitrates [1]. L-NMMA has been shown to be most effective in reversing LPS or TNF hypotension [2] and endotoxin-induced loss of vascular responsiveness [3].

Patients with severe sepsis and especially septic shock have elevated nitrite/nitrate levels [4, 5]. Shi et al. [6] determined plasma levels of nitrite and nitrate in neonates with sepsis and found these levels to be elevated at the time of entry compared with those of control subjects; the levels were significantly higher in the patients with sepsis and shock than in those without shock. Elevations of nitrite/nitrate have also been correlated with TNF and severity of illness [6]. The important role of the cytokine response was emphasized by the observation that cancer patients treated with IL-2, a cytokine known to induce a strong inflammatory response, exhibited increased serum and urine concentration of L-arginine-derived nitrates [4, 7, 8].

Since cytokine levels are directly related to morbidity and mortality rates and are also involved in NO release, it would seem logical to suppose that the more severe the sepsis, the more severe the NO release. This is commensurate with the observations that the more severe the sepsis, the more severe the reduction in systemic vascular resistance [9] and that nitrites/nitrates levels correlate with systemic vascular resistance [4]. Furthermore, nitrites/nitrates levels parallel the disease severity, as reflected by the APACHE III score [10, 11]. High plasma levels of nitrates/nitrites were observed in septic patients in association with low systemic vascular resistance and high endotoxin levels [4].

There are, however, some controversial aspects surrounding the theory of immunologically-induced NO synthase. First, although iNOS would require several hours to be activated, LPS is able to induce the rapid release of NO from bovine endothelial cells [12]. Second, several studies demonstrated an impairment of the endothelium-dependent relaxation in endotoxemic animals [13, 14]. Many studies have been performed on aortic conductance vessels in the rat, but the decreased responsiveness to norepinephrine is less significant in small femoral arteries where the sensitivity to NO is reduced. In addition, NO activity in vascular endothelial cells [15] or in coronary arteries [13] is reduced following endotoxin administration. Thus, LPS may impair the agents-stimulated release of NO [16]. Graier et al. [17] recently showed that altered calcium mobilization in the endothelial cells is a fundamental mechanism for impaired agonist-mediated NO release.

Increased NO Release in Septic Shock: Is it Beneficial or Harmful?

With its potent vasodilating action and negative inotropic effects, NO has been incriminated in the hemodynamic disturbances related to severe sepsis and septic shock. NO can also exert direct toxic effects on the cells. Therefore blocking NO appears, on first inspection, to be an attractive option. Nevertheless, there are some detracting points as well. First, NO plays a fun-

Table 2. Arguments for and against NO blockade in septic shock

For:
- NO is largely implicated in the vascular hyporeactivity responsible for arterial hypotension
- NO is directly cytotoxic in high concentration

Against:
- NO may preserve capillary ultrastructure
- There is a lack of evidence that NO is involved in the alterations in oxygen extraction capabilities that characterize severe sepsis
- NO can be protective against infections
- NO blockade
  - can reduce cardiac output and oxygen delivery to the cells
  - has been shown to alter hepatic function in several models
  - may also alter renal function in endotoxic shock
  - may reduce myocardial oxygen supply by inducing coronary vasoconstriction
  - can worsen pulmonary hypertension
  - may promote platelet aggregation and thrombosis
  - has been shown to increase mortality rates in some experimental models

damental role in the maintenance of peripheral tissue perfusion (through vascular relaxation and inhibition of platelet aggregation). Second, NO plays an important role in intercellular signaling and endothelial cell function. Third, the macrophage-mediated response to some microorganisms requires NO. Hence, blocking NO production may be very hazardous. Indeed, some animal studies using competitive antagonists of L-arginine showed that such interventions may increase organ damage and mortality rates [18]. Therefore, the question of whether NOS blockade is beneficial therefore requires further debate. In patients with septic shock, as in animal models, administration of L-arginine analogs resulted in a transient increase in arterial pressure, but had no beneficial effect on oxygen supply or on mortality [19–21]. The administration of methylene blue had similar effects [22]. This chapter aims to summarize the arguments for and against NO blockade in septic shock (Table 2).

Arguments in Favor of NO Blockade

Increased NO Release can have Deleterious Cardiovascular Effects

Kilbourn and co-workers [23] observed that the profound hypotension occurring 2 h after endotoxin infusion, can be reversed by L-NMMA. They also found that L-NMMA antagonized the TNF-induced hypotension and that administration of excess L-arginine restored the hypotension [2]. In a recent study, Kilbourn et al. [24] reported that administration of L-NMMA decreased serum nitrate levels and increased the arterial pressure in dogs treated with IL-2. Meyer et al. [25] investigated the effects of L-NAME in awake sheep. 24 h after the start of the endotoxin infusion, the animals exhibited a hyperdynamic state with increased cardiac index, decreased systemic vascular resistance, impaired oxygenation, and increased pulmonary shunt fractions. Oxygen delivery (DO_2) was significantly increased, while oxygen consumption (VO_2) remained virtually unchanged, resulting in a decreased oxygen extraction ratio (O_2ER). In the untreated control group, the significant alterations in systemic hemodynamic, lung function and DO_2 persisted for the remainder of the study. Administration of L-NAME in the other group normalized cardiac index and systemic vascular resistance index, and decreased heart rate. Although DO_2 significantly decreased after administration of L-NAME, VO_2 did not change, as O_2ER increased. Despite a significant reduction of pulmonary shunt fraction, oxygenation did not improve. Pulmonary arterial pressure and pulmonary vascular resistance showed a peak 2 h after administration of L-NAME. In contrast, the effects of L-NAME on the systemic circulation persisted for the remainder of the study [25].

Thiemermann and Vane [26] showed that inhibition of NO with L-NMMA resulted in an increase in arterial pressure and markedly attenuated the LPS-induced hypotension in rats. These effects were significantly reversed with L-arginine. In a study of endotoxic shock in anesthetized rats, Nava et al. [27] found that administration of 10 mg/kg of L-NMMA prevented the fall in blood pressure and partially prevented the increase in nitrate and nitrite and alanine aminotransferase (ALT). None of the animals in this group died. In contrast, when 300 mg/kg of L-NMMA was administered, this caused an initial increase in blood pressure followed by a rapid fall and enhanced the increase in ALT whilst abolishing the elevation of nitrate and nitrite. All of these animals died before the end of the experiment. However, continuous infusion of S-nitroso-N-acetylpenicillamine (SNAP) in rats treated with high doses of L-NMMA resulted in maintenance of blood pressure at control levels. No mortality was observed in this group.

Gray et al. [28] observed that in anesthetized rats, L-NMMA (30 mg/kg), but not D-NMMA, caused an increase in arterial pressure of approximately 30 mm Hg and restored responses to norepinephrine in LPS-treated rats. This effect was reversed by L-arginine but not by D-arginine. Fleming et al. [29] found that L-NMMA (300 μM) restored the contractile response to norepinephrine of endotoxin-treated aortic rings to control group levels. The effect of L-NMMA was reversed by L-arginine (1 mM). These data indicate that endotoxin can act directly on vascular tissue to induce a hyporeactivity to norepinephrine which is secondary to the activation of the L-arginine pathway.

Elevated NO levels are associated with hypotension and also hyporeactivity to norepinephrine leading to vascular collapse. It is widely accepted that the higher the requirements for vasopressor agents, the poorer the prognosis. The severity of decrease in systemic vascular resistance has also been directly related to the severity of septic shock and the likelihood of death [9, 30]. L-arginine analogs that block NO release can improve tissue perfusion pressure either directly or indirectly by increasing responsiveness to norepinephrine [3]. NO blockade could have less deleterious effects on the distribution of blood flow than α-adrenergic stimulation, which may tend to limit renal and splanchnic blood flow. Using NO inhibitors instead of α-adrenergic agents to increase vascular tone in septic shock may offer advantages in terms of distribution of blood flow. In sheep, Meyer et al. [31] observed that arginine analogs reduce blood flow similarly in all organs. In rats with cecal ligation/perforation, Martin and Sibbald [32] observed no change in distribution of blood flow with L-NAME. Thus this advantage has not been entirely substantiated. Some studies even raised the possibility of a selective reduction in hepato-splanchnic blood flow with NO blockade.

NO can have Toxic Effects on Cells

NO is a highly reactive free radical. As an uncharged particle, it can pass freely through cell membranes and thereby exert profound local intracellular damage. With a half-life of only a few seconds, the direct effects of NO are localized. However, by reacting with superoxide, NO forms peroxynitrite and this toxic molecule has a relatively long persistence, so that NO can indirectly damage more distant targets. In the rat, NO can inhibit hepatic mitochondrial enzymes [33] and disrupt DNA synthesis by inhibiting ribonucleotide reductase, a rate limiting enzyme in DNA production [34] and also directly damage DNA by deamination [35]. Furthermore, NO can inhibit adenosine triphosphate (ATP) production at 3 levels:

1) inactivation of glycolytic enzyme glyceraldehyde-3-phosphate dehydrogenase;
2) disruption of the Krebs cycle by binding to the heme group of aconitase; and
3) inhibition of oxidative phosphorylation by binding to the heme group of the ubiquinone reductase.

NO can depress Myocardial Contractility

A rapid onset of myocardial depression has been documented on numerous occasions in septic shock, even with normal or increased cardiac output [30]. Endotoxin administration in healthy volunteers can result in an acute reduction in myocardial contractility [36]. Various mediators including TNF, platelet activating factor (PAF), thromboxane and oxygen free radicals have been incriminated in this sepsis-related myocardial depression. Since the degree of inflammatory response is proportional to fatality rate, one can expect a more severe myocardial depression to be associated with a worse outcome, and indeed this has been observed in a number of studies [30, 37]. Various forms of immunotherapy, including anti-TNF antibodies or antioxidants such as N-acetylcysteine (NAC) and 21 aminosteroids can result in an acute improvement of myocardial function [38–40]. TNF, IL-2, and IL-6 inhibited contractility of isolated hamster papillary muscles in a concentration-dependent, reversible manner. L-NMMA blocked these negative inotropic effects but L-arginine administration could, in turn, reverse the inhibition by L-NMMA [41].

The direct negative inotropic effect of the cytokines may be mediated through myocardial NO synthase resulting in an overproduction of NO [42]. In physiological conditions, the heart does not produce appreciable amounts of NO. However, iNOS can be activated within myocytes under the influence of mediators, resulting in a loss of myocardial force of contraction. NO production within cardiac myocytes reduces their contractility during endotoxemia [43], and abnormal contractile function due to induction of NO synthesis in rat cardiac myocytes follows exposure to activated macrophage-conditioned medium [44].

Brady et al. [43] investigated whether increased NO synthase activity within cardiac myocytes contributes to the depressed cardiac contractility observed in endotoxic shock. Isolated ventricular myocytes were studied to examine the effects of substances and inhibitors of NO synthase on myocyte contractility. When stimulated electrically, the resting length of myocytes from control animals shortened by 5.3±0.3%. Baseline contraction of myocytes from endotoxin-treated animals was reduced to 3.0±0.3%. L-NAME had no effect on myocytes from control animals, but it increased the contraction of myocytes from endotoxin-treated animals by 40%.

Arguments against NO Blockade

NO Blockade may reduce Cellular Oxygen Availability

The vasoconstricting effects of NO may be responsible for impairing capillary blood flow. This reduction in vascular tone is not by itself deleterious. On the contrary, the reduction in systemic vascular resistance facilitates an increase in cardiac output and thus helps to match the increase in oxygen demand of the cells. Vasodilation can also reduce the distance required for oxygen diffusion from the capillaries to the cells.

An almost universal finding is that an increase in arterial pressure is associated with NOS inhibitors [5, 22, 27, 31, 45–48]. In some studies, this was attributed to an increased response to catecholamines [49]. However, most studies also reported a reduction in cardiac output [31, 45, 47, 50], and thus DO_2. This was a subject of concern for some authors [47, 50], especially since VO_2 fell simultaneously in some instances [5, 50]. As mentioned before, Meyer et al. [25] noted a simultaneous increase in O_2ER that could compensate for a reduction in DO_2, so that VO_2 could remain constant. Kilbourn et al. [51] recently proposed a combination of L-NMMA with dobutamine to restore and maintain cardiac output. Evidently, hypotension is only one symptom of septic shock and the administration of therapeutic interventions to inhibit vasodilation may not be beneficial overall.

NO has important Antiplatelet Effects

NO can inhibit adhesion and, together with prostacyclin, can also inhibit platelet aggregation, thus maintaining the patency of the capillaries. NO blockers may increase platelet activation and induce thrombosis. This was shown by Schultz and Raij [52] who observed increased thrombosis in kidneys following L-NMMA in endotoxic shock.

NO inhibits PMN Adhesion to Endothelial Cells

This is a complicated relation. As a highly reactive radical, NO can be toxic by itself, directly damaging capillary structures and inducing edema. However, NO also reacts with superoxide radical, another potentially toxic substance and a strong activator of leukocytes [53]. On the one hand, this mechanism may inhibit the interactions between polymorphonuclear cells and endothelial cells, thus preserving vascular integrity. On the other hand, the reaction of NO with superoxide can result in the formation of peroxynitrites which are highly toxic. Hence, it is difficult to identify whether these interactions are, in balance, beneficial. Indeed, various studies have produced conflicting results. In some studies, NO blockers increased permeability, suggesting that NO may preserve capillary integrity [54, 55]. In rat mesenteric venules, NO inhibitors increased the albumin leakage but in other models, NOS inhibitors prevented the vascular permeability changes. NO blockade has been linked to intestinal damage [55], and L-NAME administration during endotoxic shock may result in intestinal ischemia related to the strong vasoconstriction [56].

Measurements of the leakage of ^{125}I-labelled human serum albumin revealed that L-NMMA (10-50 mg/kg intravenous) pretreatment augmented the macroscopic and histological intestinal damage and the increases in vascular permeability evident 15 min after endotoxin administration. The increased arterial pressure did not appear to be the mechanism by which endotoxin-induced intestinal damage was enhanced, since similar systemic pressor responses induced by phenylephrine had no such effect. It would thus seem that synthesis of NO from L-arginine has a role in maintaining the microvascular integrity of the intestinal mucosa following acute endotoxin challenge [55].

NO does not Significantly alter Oxygen Extraction Capabilities

A major argument in favor of NO blockade would be that NO alters the oxygen extraction capabilities by altering vascular tone. This has not been demonstrated so far and, indeed, Vallet et al. [57] showed that NO does not contribute to extraction abnormalities. Furthermore, Zhang et al. [58] showed that sodium nitroprusside does not alter oxygen extraction. On the contrary, NAC which acts in part as endothelium-derived relaxing factor (EDRF)-like substance, increases the oxygen extraction capabilities in endotoxic shock [39].

NO exerts Important Microbicidal Properties

NO has toxic effects on microorganisms, especially the intracellular organisms or those such as fungi or parasites which are too large to be phagocytosed. NO can have antimicrobial properties against some bacteria (e.g. *Mycobacterium*, *Listeria*, *Chlamydia*), fungi (e.g. *Cryptococcus*), viruses (e.g. *Herpes*) and parasites (e.g. *Leishmania*). This may be explained, perhaps in part, via inhibition of the superoxide ion [59].

NO can exert important immunological effects which have not been completely elucidated yet. NO regulates lymphocyte function and may inhibit some T helper cells. NO can also interfere with TNF-α release. Napolitano et al. [60] showed, in murine thermal injury, a reduction in IL-10 production, which was normalized by NO.

NO Inhibition may increase Pulmonary Artery Pressures and precipitate Right Heart Failure

Since NO exerts potent vasodilating effects on the pulmonary vasculature, NO blockade is expected to increase pulmonary artery pressures. Indeed, in many experimental studies [31, 46, 48], NO blockade resulted in a significant increase in pulmonary artery pressures, with an associated risk of right ventricular failure. Clinical experience with L-NMMA also indicates that pulmonary hypertension may develop in some patients. A combination of arginine analogs and inhaled NO may offer a means of compensating for this.

NO Inhibition may have Deleterious Effects on Hepato-Splanchnic Blood Flow and Hepatic Function

Although the vasodilating effects of NO are exhibited throughout the body, some authors found that NO blockade may, nevertheless, selectively decrease hepato-splanchnic blood flow. Wright et al. [18] reported that L-NMMA administration either before or after LPS in rabbits, resulted in intense vasoconstriction associated with a decrease in blood flows to the portal vein and hepatic artery leading to hepatic necrosis. In the study by Meyer et al. [25] on awake sheep, L-NAME administration reduced regional blood flow to the brain, splanchnic and renal beds following endotoxemia. Mulder et al. [61] observed that L-NNA reduced blood flow to the renal and gastrointestinal tract. Pastor et al. [47] also reported that L-NNA reduces blood flow to the splanchnic-hepatic regions.

Several other studies showed that NO blockers such as L-NMMA may alter liver function [62]. Harbrecht et al. [63, 64] reported that although L-NMMA by itself did not induce any liver injury, its administration with endotoxin augmented liver damage. These authors proposed that NO may be important in maintaining liver perfusion and a low portal pressure, perhaps by preventing platelet aggregation in the sinusoids. L-NNA effectively blocked the increase in serum nitrate/nitrite levels seen in endotoxin and also inhibited the increase in hepatic protein synthesis in endotoxin rats. This effect may be related to a reduction in hepatic blood flow [65]. *In vitro*, Billiar et al. [66] showed that NO downregulates protein synthesis in hepatocytes, but a feasible explanation for this is the fact that high doses were used. In an *in vivo* study to rats, Frederick et al. [65] reported that endotoxin increased NO release and increased hepatic protein synthesis. Possible explanatory mechanisms for these phenomena are: increased liver blood flow; scavenging superoxide ions; increased cellular levels of cGMP. Thus maintenance of NO may be necessary to maintain liver blood flow and may prevent the hepatic necrosis associated with enzyme release. However, Evans et al. [67] found that L-NMMA does not influence hepatic damage in a live *E. coli* model in mice.

NO Inhibition may have Deleterious Effects on Renal Function

Severe sepsis can induce vasoconstriction in the kidneys, resulting in a reduction in renal blood flow and reduced glomerular filtration. Many substances have been incriminated, including thromboxane, leukotrienes, endothelins and prostaglandin F2. In common with vasodilating prostaglandins, NO may be essential to counterbalance the effects of these vasoconstrictive substances and maintain renal blood flow. Some studies [68] showed that NO blockers can reduce renal blood flow and increase renal vascular resistance. Inhibition of LPS-stimulated NO release as studied by Schulz and Raij [52] resulted in a loss of renal function; there was also histologic evidence of fibrin deposition and glomerular thrombosis. Spain et al. [69] showed that NO inhibition exacerbates microvascular constriction as assessed by *in vivo* videomicroscopy. The same authors [70] studied NOS inhibition in rats that were subjected to *E. coli* bacteremia. Here, these authors observed an increase in arterial pressure but reductions in cardiac output and preglomerular constriction, the overall effect being a reduction in renal blood flow. These observations suggest that NO may play a key part in counteracting the vasoconstrictive effects of severe sepsis.

Not all authors have reported a deleterious effect of NO blockers on renal function. Papadimitriou et al. [71] demonstrated that endotoxin reduced urine output and creatinine clearance acutely and that L-NMMA restored these parameters.

NO Inhibition may reduce Coronary Blood Flow

The vasoconstricting effects of NO blockade may threaten the coronary blood supply, especially in patients with coronary artery disease. This may be a theoretical concern; there is no evidence that NO inhibitors alter coronary blood flow in normal subjects, although the possibility may exist for patients with coronary heart disease. Nevertheless, in septic shock, coronary blood flow is usually well-preserved, so that this should not have major consequences.

NO Inhibition has increased Organ Damage and worsened Mortality in some Animal Models

The effects of NOS inhibition on survival have been variable and it is difficult to draw conclusions from acute studies. There have been examples of both increased and decreased survival [47, 72, 73]. Teale et al. [74] documented an improvement in survival associated with NOS inhibition during intraperitoneal *E. coli* in mice. Cobb et al. [5] studied the effects of NOS inhibitors in awake endotoxically challenged dogs. L-NAA was chosen as the most specific inhibitor available and its effects were compared with those of a normal saline treatment. Compared with the control animals, the NOS-inhibited dogs exhibited higher systemic and pulmonary vascular resistance, and decreased heart rate, cardiac index, DO_2 and VO_2. Furthermore, mortality rates were higher in L-NAA-treated animals. These authors concluded that this was due either to NO inhibition in particular or to unique properties of L-NAA . In a study by Evans et al. [67], L-NMMA had no effect on survival in mice challenged with either intraperitoneal or intravenous live *E. coli*. It was proposed by these authors that differences in mouse strain or in bacteria samples may contribute to the contradicting results obtained with NOS inhibitors. Tiao et al. [75] also observed increased mortality in their study. They pretreated rats with either saline solution or L-NNA before subjecting them to either intravenous endotoxic challenge or a normal saline solution. A mortality rate of 33% was attributed to endotoxin alone but in those animals pretreated with the NOS inhibitor this rose to 74%. In this latter group, there were also increased levels of TNF and IL-6 which these authors cited as a possible cause of increased mortality. Similarly, in anesthetized rabbits with endotoxin-induced septic shock, Pastor et al. [47] found that L-NNA treatment was associated with a restoration in blood pressure close to normal levels but also with reduced flow. In these animals, mortality was again found to be increased.

How can these Discrepancies be reconciled?

The Agents administered are Usually not Specific for iNOS

Excessive doses may block the physiologic release of NO which is so essential for cellular function. Agents such as aminoguanidine may have more specific effects, however.

The Dose used may be of Critical Importance

Reversal of endotoxin-induced hypotension by NO synthase inhibitors is dose-dependent. Some investigators have found that limited doses may increase survival, but higher doses may worsen survival. This has been demonstrated by Nava et al. [76] in rats: a dose of 3 mg/kg of L-NMMA was ineffective; 30 mg/kg reversed hypotension; 300 mg/kg worsened hypotension and increased mortality rate. Similarly, Wright et al. [18] observed that 300 mg/kg of L-NMMA worsened hypotension and increased mortality rate in rabbits.

The Timing of Intervention may represent Another Crucial Factor

It takes several hours for iNOS to be expressed and yet many studies have examined the early effects of NO blockade with interventions started at the onset of sepsis [48, 50, 73]. The importance of the time course of events was recently illustrated by Wang et al. [77] who found that in rats submitted to cecal ligation and puncture, endothelium-dependent relaxation increased at 2 h, returned to baseline at 3.5 h but decreased at 10–20 h. Hence, a triphasic response was indicated.

It would seem, therefore, that the type of agent used, the dose and the timing are all significant influencing factors.

NO Inhibition in Humans

Although still quite limited in humans to date, the experience obtained with the use of NO blockers is consistent with animal studies. That is, NO blockade can increase arterial pressure but may also reduce cardiac output (or at least have no beneficial effect on blood flow). Lorente et al. [20] studied 8 patients with septic shock who were treated with a 20 mg/kg intravenous bolus of L-NNA. They observed profound vasoconstriction in the systemic and pulmonary circulation with hypertension and a reduction of cardiac output. These effects could be reversed by L-arginine. Blood pressure increased but cardiac index decreased from 3.51 to a minimum of 2.65 L/min $\cdot$ m^2. An increase in left and right sided filling pressures was also observed. Systemic vascular resistance increased as did pulmonary vascular resistance. On the other hand, arginine reduced blood pressure but increased cardiac index, DO_2 and VO_2.

Petros et al. [78] studied 12 patients with septic shock who received L-NMMA in doses from 0.3 to 1 mg/kg. They noted a dose-dependent increase in arterial pressure, systemic vascular resistance, and pulmonary vascular resistance but also a reduction in cardiac output and an increase in cardiac filling pressures.

An ongoing study (open multicentric) using escalating doses indicates that L-NMMA is generally well tolerated. This study indicates that L-NMMA is associated with an increase in arterial pressure but also a reduction in cardiac output. The requirements for vasopressor agents are simultaneously reduced. Some patients also showed an increase in pulmonary artery pressure.

NO Donor in Septic Shock

Some experimental studies also described the effects of NO donors in septic shock and sometimes demonstrated beneficial effects of these interventions. Animals pretreated with L-NAME and SNAP showed reduced hemodynamic changes when compared with control (LPS only) or LPS and L-NAME treated animals.

Administration of 100 μg/kg SNAP 10 min post-trauma followed by 10 μg/kg/h infusion prolonged survival time, attenuated the increase in plasma free aminonitrogen and tissue myeloperoxidase activities. Moreover, SNAP significantly preserved superior mesenteric artery endothelial function and the vasorelaxation to acetylcholine. SNAP maintained systemic blood pressure, preserved vascular endothelial integrity, and inhibited neutrophil-endothelial interaction and the resultant reduced microvascular leakiness.

Pastor et al. [47] observed that arginine increased arterial pressure during septic shock. We recently observed that SIN-1, a NO donor, increased blood flow but did not decrease arterial pressure during endotoxic shock in dogs [79].

Effects of Methylene Blue in Septic Shock

The vascular relaxation, myocardial depression and inhibition of platelet aggregation elicited by NO have been associated with high intracellular levels of cGMP, resulting from the activation of soluble guanylate cyclase (GC), whereas the microbiocidal properties and the cytotoxicity against tumor cells are mediated through other pathways [80, 81]. Hence, inhibition of GC should represent means of attenuating the hemodynamic effects of NO in septic shock, while maintaining the other effects.

Methylene blue (MB) could be used for this purpose, since it is an inhibitor of the GC activity which çounteracts the effects of NO and other nitrovasodilators in endothelium and vascular smooth muscle [82]. MB, or bis(dimethylamino)phenazathionium chloride trihydrate, has been used for many years as an hydrogen acceptor in numerous enzyme systems. Once MB enters the cell, it is reduced (by the enzyme NADPH reductase in the erythrocyte, for e.g.) and this reduced compound ("leucobleu") is reoxidized into MB, by reduction of the ferrous ion linked with an intracellular protein (contained either in a heme or in a non-heme group). For instance, the ability of MB to reverse toxic methemoglobinemia is due to the reduction of the heme group of methemoglobin. Notably, a high dose of MB can produce methemoglobinemia, probably when the reductor system of the erythrocyte is overwhelmed.

MB may have other effects than inhibition of enzyme GC. MB may in part inhibit NOS [83, 84]. A recent study [84] challenged the specificity of the action of MB on GC, and suggested an inhibitory effect of MB on the C^{2+}-dependent iNOS, which also contains tightly bound iron, and a heme group. This assumption was based on the observation that the formation of L-citrulline (the byproduct of the formation of NO from L-arginine) by porcine cerebellum NOS was completely inhibited by 30 μmol of MB, whereas GC stimulated by the NO donor S-nitrosoglutathione was far less sensitive to MB, even at higher concentrations. However, MB does not appear to inhibit NOS significantly, at least when compared with L-arginine competitive antagonists. Furthermore, different forms of GC have been characterized; whether MB is able to inhibit all of them remains to be determined [85]. MB has been demonstrated to prevent the endotoxin- or IL-1-induced vasorelaxation by abolishing cGMP production [86].

Other effects of MB include generation of oxygen free radicals [83, 87] and diminution of glutathione levels. In the rabbit vascular ring, MB enhanced relaxation in response to ultraviolet light, suggesting that the UV-generated free radicals could switch the action of MB from inhibition to activation of GC [87].

Experimental Effects of MB on Vascular Tone

The hemodynamic effects of MB were studied in different models of sepsis. Isolated vascular rings, cardiac myocytes, platelets or *ex vivo* tissues were exposed to various sepsis-inducing agents. The amplitude of contractions in response to phenylephrine, blunted by IL-1 and LPS, was reversed by MB [86]. In rat aortic rings incubated with LPS, exposure to MB restored the pre-LPS responsiveness to norepinephrine and calcium and the pre-LPS intracellular cGMP content [29, 88, 89]. In addition, the reactivity to norepinephrine, along with the rise in intracellular cGMP content were reversed by cycloheximide [90], indicating that the hyporesponsiveness to catecholamines involved the synthesis of iNOS. In a similar model, MB (10 μM) reversed the depression in vascular contractile response to phenylephrine, induced by LPS [91] or by lipoteichoic acid from *Staphylococcus aureus* [92]. In addition,

MB abolished the neutrophil-induced relaxation of rabbit aorta and dog femoral artery [93]. In a rat model, MB (3 mg/kg) restored the reactivity to norepinephrine which was blunted by LPS and reduced the intracellular cGMP content of small arteries [94]. Importantly, in a preparation of aorta harvested from rats previously given endotoxin, the contraction in response to norepinephrine was completely restored by MB, but only partially by a competitive analog of L-arginine, indicating that GC can also be activated by other factors than NO [95].

Other studies showed that MB reversed endotoxin- or cytokines-induced hypotension, hyporeactivity to vasoconstrictors, and myocardial depression [86, 94]. Following endotoxic shock in rabbits, MB at doses of 1.5 and 10 mg/kg produced a prompt dose-dependent increase in arterial pressure in association with normalization of plasma NO concentration [83].

Experimental Effects of MB on Myocardial Function

MB may also reverse the sepsis-related myocardial depression. In cardiac myocytes, exposure to LPS resulted in a decrease of the spontaneous fractional shortening, associated with an increase in myocardial cGMP, which was reversed by MB [96, 97]. MB could also improve myocardial function in endotoxic shock. This last observation is consistent with that of Brady et al. [96], who found that MB increased myocardial contractility *in vitro* in endotoxemia.

Administration of MB in Patients

The current clinical use of MB includes the treatment of toxic methemoglobinemia (by reduction of the heme group of methemoglobin), the detection of postoperative fistula or the delineation of organ limits [98, 99]. Except methemoglobinemia observed following very high doses of MB, the adverse effects observed associated with intravenous administration of MB are relatively minor [98]. Transient increase in blood pressure and decrease in renal blood flow were noted in patients immediately following intravenous infusion of MB [100]. Also, in patients with lactic acidosis, MB administration was followed by a brisk decrease in blood lactate levels [101].

In pilot studies in patients with septic shock, MB administration was shown to increase arterial pressure and systemic vascular resistance [102, 103]. We studied the effects of MB in 14 patients with profound septic shock requiring adrenergic support [103] (Fig. 1). Intravenous MB (2 mg/kg) administration was followed by an increase in mean arterial pressure (from 61.1 ± 7.6 to 71.7 ± 12.0 mm Hg at 60 min, $p < 0.01$). Pulmonary artery pressure, cardiac filling pressures, cardiac output, were not significantly affected. Left ventricular stroke work increased from 42.5 ± 17.9 to 48.9 ± 14.5 g·m after 60 min ($p < 0.05$). Arterial lactate decreased from 3.4 ± 1.4 to 2.7 ± 1.3

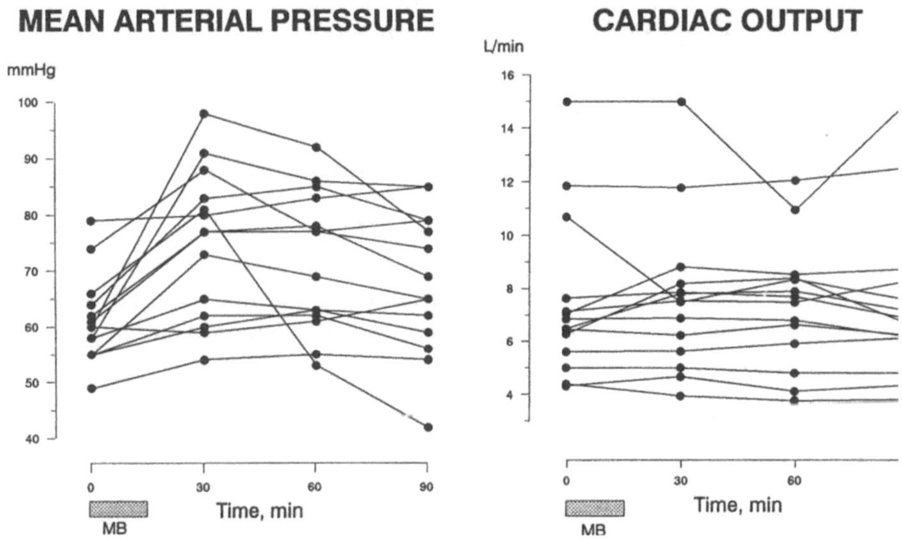

Fig. 1. Time course of mean arterial pressure (*left panel*) and cardiac output (*right panel*) following the intravenous administration of methylene blue (MB) (2 mg/kg) just after baseline (time 0) in 14 septic shock patients. Mean arterial pressure significantly increased from baseline, at 30 and 60 min ($p<0.01$). Reproduced with permission from [103]

mEq/L ($p<0.05$). Interestingly, these effects were again observed after the administration of a second dose in 6 patients. A decrease in blood lactate level was observed, in the absence of increase in oxygen supply or consumption. This could perhaps be attributed to a reduction of the lactate/pyruvate ratio resulting from the reoxidation of NADPH [101]. No methemoglobinemia higher than 1% was observed. Ultimately, 11 of the patients died. Therefore, a clear benefit of MB in the treatment of septic shock is not demonstrated.

A Relevant Question: Does MB alter Regional Blood Distribution?

We [104] studied the effects of MB in anesthetized dogs ventilated with room air and equipped with arterial and pulmonary artery catheters, and Doppler flow probes to monitor mesenteric femoral and renal blood flows. MB was infused at increasing doses (from 2.5 to 20 mg/kg over 30 min), either in control conditions ($n=7$) or following the intravenous administration of 2 mg/kg of LPS ($n=7$). In the non-septic group, MB administration did not significantly increase arterial pressure, and at the higher doses, arterial pressure and cardiac index decreased, and pulmonary artery pressure and systemic and pulmonary vascular resistance increased. Femoral blood flow rose, while renal blood flow declined and mesenteric blood flow was unaffected. In the septic group, MB administration was followed by progressive reversal of the hypotension induced by LPS. MB also improved ventricular function.

Mesenteric and femoral blood flows, both of which were reduced following LPS, increased after MB (up to 10 mg/kg). In both groups, arterial and mixed venous oxygen content dose-dependency declined in relation to an increased oxygen consumption.

Thus, these data indicate the dramatically different effects of MB in the absence and in the presence of sepsis and in differing doses. In particular, MB can reverse the abnormalities in vascular tone and myocardial contractility induced by LPS. Low and moderate doses of MB can significantly raise arterial pressure but not cardiac output; MB infusion may selectively increase mesenteric blood flow at low to moderate doses, but not renal nor femoral blood flows. High doses of MB can even worsen hypotension and myocardial depression [104].

Conclusion

It has been widely shown that NO is released in increased amounts during septic shock. As might be expected, there are both beneficial and deleterious consequences to this increased release but these have yet to be defined satisfactorily. Blocking NO release may yield benefits in some patients but this needs further study. Another option, that of blocking the effects of NO with MB, has increased mesenteric blood flow in experimental models. Such results suggest that MB treatment may be effective in septic shock patients but the full effects on morbidity and mortality have not yet been identified.

Although no harmful side effects have been reported after NO inhibition in human sepsis, animal data strongly suggest that excessive reduction of endogenous NO release may cause organ damage and death under certain circumstances. It is possible that low doses of NOS inhibitors, either as a single injection, or more probably, as an infusion, will be beneficial. Higher doses of NOS inhibitors to achieve greater inhibition of endogenous NOS may need to be accompanied by a NO generating vasodilator.

References

1. Stuehr DJ, Marletta MA (1985) Mammalian nitrate biosynthesis: Mouse macrophages produce nitrite and nitrate in response to Escherichia coli lipopolysaccharide. Proc Natl Acad Sci USA 82:7738–7742
2. Kilbourn RG, Gross SS, Jubran A, et al (1990) N^G-methyl-L-arginine inhibits tumor necrosis factor-induced hypotension: Implications for the involvement of nitric oxide. Proc Natl Acad Sci USA 87:3629–3632
3. Julou-Schaeffer G, Gray GA, Fleming I, Schott C, Parratt JR, Stoclet JC (1990) Loss of vascular responsiveness induced by endotoxin involves L-arginine pathway. Am J Physiol 259:H1038–H1043
4. Ochoa JB, Udekwu AO, Billiar TR, et al (1991) Nitrogen oxide levels in patients after trauma and during sepsis. Ann Surg 214:621–626
5. Cobb JP, Natanson C, Hoffman WD, et al (1992) N-amino-L-arginine, an inhibitor of nitric oxide synthase, raises vascular resistance but increases mortality rates in awake canines challenged with endotoxin. J Exp Med 176:1175–1182

6. Shi Y, Li HQ, Shen CK, et al (1993) Plasma nitric oxide levels in newborn infants with sepsis. J Pediatr 123:435–438
7. Evans T, Carpenter A, Kinderman H, Cohen J (1993) evidence of increased nitric oxide production in patients with the sepsis syndrome. Circ Shock 41:77–81
8. Hibbs J, Westenfelder C, Taintor R, et al (1992) Evidence for cytokine-Inducible nitric oxide synthesis from L-arginine in patients receiving interleukin-2 therapy. J Clin Invest 89:867–877
9. Groeneveld AB, Nauta JJ, Thijs LG (1988) Peripheral vascular resistance in septic shock: Its relation to outcome. Intensive Care Med 14:141–147
10. Baker C, Sutton ET (1993) Arteriolar endothelium-dependent vasodilation occurs during endotoxin shock. Am J Physiol 264: H1118–H1123
11. Bone RC, Fisher CJJ, Clemmer TP, Slotman GJ, Metz CA, Balk RA (1989) Sepsis syndrome: A valid clinical entity. Crit Care Med 17:389–393
12. Salvemini D, Korbut R, Änggard E, Vane J (1990) immediate release of a nitric oxide-like factor from bovine aortic endothelial cells by Escherichia-coli lipopolysaccharide. Proc Natl Acad Sci USA 87:2593–2597
13. Parker J, Keller RS, Defily D, Laughlin J, Novotny M, Adams H (1991) Coronary vascular smooth muscle function in E. coli endotoxemia in dogs. Am J Physiol 260: H832–H841
14. Wylam NE, Samsel RW, Umans JG, Mitchell RW, Leff AR, Schumacker PT (1990) Endotoxin in vivo impairs endothelium-dependent relaxation of canine arteries in vitro. Am Rev Respir Dis 142:1263–1267
15. Myers PR, Wright TF, Tanner MA, Adams HR (1992) Endothelial-derived relaxing factor and nitric oxide production in cultured endothelial cells: Direct inhibition by E. coli endotoxin. Am J Physiol 262:H710–H718
16. Umans JG, Wylam ME, Samsel RW, Edwards J, Schumacker PT (1993) Effects of endotoxin in vivo on endothelial and smooth-muscle function in rabbit and rat aorta. Am Rev Respir Dis 148:1638–1645
17. Graier WF, Myers PR, Rubin LJ, Adams HR, Parker JL (1994) Escherichia coli endotoxin inhibits agonist-mediated cytosolic Ca2+ mobilization and nitric oxide biosynthesis in cultured endothelial cells. Circ Res 75:659–668
18. Wright CH, Rees DD, Moncada S (1992) Protective and pathological roles of nitric oxide in endotoxin shock. Cardiovasc Res 26:48–57
19. Petros A, Bennett D, Vallance P (1991) Effect of nitric oxide synthase inhibitors on hypotension in patients with septic shock. Lancet 338:1557–1558
20. Lorente JA, Landin L, De Pablo R, Renes E, Liste D (1993) L-arginine pathway in the sepsis syndrome. Crit Care Med 21:1287–1295
21. McMahon TJ, Kadowitz PJ (1992) Methylene blue inhibits neurogenic cholinergic vasodilator responses in the pulmonary vascular bed of the cat. Am J Physiol 263: L575–L584
22. Preiser JC, Lejeune P, Roman A, Carlier E, Leeman M, Vincent JL (1993) Use of methylene blue in patients with refractory septic shock. Circ Shock S2:68 (Abst)
23. Kilbourn RG, Jubran A, Gross SS, et al (1990) Reversal of endotoxin-mediated shock by N^G-methyl-L-arginine, an inhibitor of nitric oxide synthesis. Biochem Biophys Res Commun 172:1132–1138
24. Kilbourn RG, Owen-Schaub L, Cromeens D, et al. (1994) N^G-methyl-L-arginine, an inhibitor of nitric oxide formation, reverses IL-2 mediated hypotension in dogs. J Appl Physiol 76:1130–1137
25. Meyer J, Lentz CW, Stothert JC, Traber LD, Herndon DN, Traber DL (1994) Effects of nitric oxide synthesis inhibition in hyperdynamic endotoxemia. Crit Care Med 22: 306–312
26. Thiemermann C, Vane J (1990) Inhibition of nitric oxide synthesis reduces the hypotension induced by bacterial lipopolysaccharides in the rat in vivo. Eur J Pharmacol 182:591–595
27. Nava E, Palmer RM, Moncada S (1992) The role of nitric oxide in endotoxic shock: Effects of N^G-monomethyl-L-arginine. J Cardiovasc Pharmacol 20: S132–S134

28. Gray GA, Schott C, Julou-Schaeffer G, Fleming I, Parratt JR, Stoclet JC (1991) The effect of inhibitors of the L-arginine/nitric oxide pathway on endotoxin-induced loss of vascular responsiveness in anaesthetized rats. Br J Pharmacol 103:1218–1224
29. Fleming I, Gray GA, Julou-Schaeffer G, Parratt JR, Stoclet JC (1990) Incubation with endotoxin activates the L-arginine pathway in vascular tissue. Biochem Biophys Res Commun 171:562–568
30. Vincent JL, Gris P, Coffernils M, et al (1992) Myocardial depression characterizes the fatal course of septic shock. Surgery 111: 660–667
31. Meyer J, Traber LD, Nelson S, et al (1992) Reversal of hyperdynamic response to continuous endotoxin administration by inhibition of NO synthesis. J Appl Physiol 73:324–328
32. Martin CM, Sibbald WJ (1994) Modulation of hemodynamics and organ blood flow by nitric oxide synthase inhibition is not altered in normotensive, septic rats. Am J Respir Crit Care Med 150:1539–1544
33. Stadler J, Billiar TR, Curran RD, Stuehr DJ, Ochoa JB, Simmons RL (1991) Effect of exogenous and endogenous nitric oxide on mitochondrial respiration of rat hepatocytes. Am J Physiol 260:C910–C916
34. Nguyen T, Brunson D, Crespi CL, Penman BW, Wishnok JS, Tannenbaum SR (1992) DNA damage and mutation in human cells exposed to nitric oxide in vitro. Proc Natl Acad Sci USA 89:3030–3034
35. Wink DA, Kasprzak KS, Maragos CM, Elespuru RK, Misra M, Dunams TM (1991) DNA determinating ability and genotoxicity of nitric oxide and its progenitors. Science 254:1001–1003
36. Suffredini AF, Shelhamer JH, Neumann RD, Brenner M, Baltaro RJ, Parrillo JE (1992) Pulmonary and oxygen transport effects of intravenously administered endotoxin in normal humans. Am Rev Respir Dis 145:1398–1403
37. D'Orio V, Mendes P, Carlier P, Fatemi M, Marcelle R (1991) Lung fluid dynamics and supply dependency of oxygen uptake during experimental endotoxic shock and volume resuscitation. Crit Care Med 19:955–961
38. Vincent JL, Bakker J, Marécaux G, Schandene L, Kahn RJ (1992) Administration of anti-TNF antibody improves left ventricular function in septic shock patients. Chest 101: 810–815
39. Zhang H, Spapen H, Nguyen DN, Rogiers P, Bakker J, Vincent JL (1995) Effects of N-acetyl-L-cysteine on regional blood flow during endotoxic shock. Eur Surg Res (In press)
40. Zhang H, Spapen H, Manikis P, et al (1995) Tirilazad mesylate (U74006F) inhibits the effets of endotoxin in dogs. Am J Physiol 268:H1847–H1855
41. Finkel MS, Oddis CV, Jacob TD, Watkins SC, Hattler BG, Simmons RL (1992) Negative inotropic effects of cytokines on the heart mediated by nitric oxide. Science 257: 378–389
42. Finkel MS, Oddis CV, Jacob TD, Watkins SC, Hattler BG, Simmons RL (1992) Negative inotropic effects of cytokines on the heart mediated by nitric oxide. Science 257:378–389
43. Brady AJ, Poole-Wilson PA, Harding SZ, Warren JB (1992) Nitric oxide production within cardiac myocytes reduces their contractility in endotoxemia. Am J Physiol 32: H1963–H1966
44. Balligand JL, Ungureanu D, Kelly RA, et al (1993) Abnormal contractile function due to induction of nitric oxide synthesis in rat cardiac myocytes follows exposure to activated macrophage-conditioned medium. J Clin Invest 91:2314–2319
45. Klabunde RE, Ritger RC (1991) N^G-monomethyl-L-arginine (NMA) restores arterial blood pressure but reduces cardiac output in a canine model of endotoxic shock. Biochem Biophys Res Commun 178:1135–1140
46. Lorente JA, Landin L, Renes E, et al (1993) Role of nitric oxide in the hemodynamic changes of sepsis. Crit Care Med 21:759–767
47. Pastor C, Teisseire B, Vicaut E, Payen D (1994) Effects of L-arginine and L-nitroarginine treatment on blood pressure and cardiac output in rabbit endotoxin shock model. Crit Care Med 22:465–469

48. Robertson FM, Offner PJ, Ciceri DP, Becker WK, Pruitt BA (1994) Detrimental hemodynamic effects of nitric oxide synthase inhibition in septic shock. Arch Surg 129:149–156
49. Landin L, Lorente JA, Renes E, Canas P, Jorge P, Liste D (1994) Inhibition of nitric oxide synthesis improves the vasoconstrictive effect of noradrenaline in sepsis. Chest 106:250–256
50. Statman R, Cheng W, Cunningham JN, et al (1994) Nitric oxide inhibition in the treatment of the sepsis syndrome is detrimental to tissue oxygenation. J Surg Res 57: 93–98
51. Kilbourn RG, Cromeens DM, Chelly FD, Griffith OW (1994) N^G-methyl-L-arginine, an inhibitor of nitric oxide formation, acts synergistically with dobutamine to improve cardiovascular performance in endotoxemic dogs. Crit Care Med 22:1835–1840
52. Shultz PJ, Raij L (1992) Endogenously synthesized nitric oxide prevents endotoxin-induced glomerular thrombosis. J Clin Invest 90:1718–1725
53. Rubanyi GM, Ho EH, Cantor EH, Lumma WC, Parker Botelho LH (1991) Cytoprotective function of nitric oxide: Inactivation of superoxide radicals produced by human leukocytes. Biochem Biophys Res Commun 181:1392–1397
54. Kubes P, Granger DN (1992) Nitric oxide modulates microvascular permeability. Am J Physiol 262: H611–H615
55. Hutcheson IR, Whittle BJR, Boughton-Smith NK (1990) Role of nitric oxide in maintaining vascular integrity in endotoxin-induced acute intestinal damage in the rat. Br J Pharmacol 101: 815–820
56. Mitchell JA, Kohlhaas KL, Sorrentino R, Warner TD, Murad F, Vane JR (1993) Induction by endotoxin of nitric oxide synthase in the rat mesentery: Lack of effect on action of vasoconstrictors. Br J Pharmacol 109:265–270
57. Vallet B, Curtis SE, Winn MJ, King CE, Chapler CK, Cain SM (1994) Hypoxic vasodilation does not require nitric oxide (EDRF/NO) synthesis. J Appl Physiol 76: 1256–1261
58. Zhang H, Nguyen DN, Spapen H, Moock M, Maciel F, Vincent JL (1995) Sodium nitroprusside does not influence tissue oxygen extraction capabilities during a critical reduction in oxygen delivery. Cardiovasc Res (In press)
59. Curnutte JT, Whitten DM, Babior BM (1974) Defective superoxide production by granulocytes from patients with chronic granulomatous disease. N Engl J Med 290: 593–597
60. Napolitano L, Campbell C (1994) Nitric oxide inhibition normilizes splenocyte interleukin-10 synthesis in murine thermal injury. Arch Surg 129:1276–1283
61. Mulder MF, Lambalgen AA, Huisman E, Visser JJ, Van den Bos GC, Thijs LG (1994) Protective role of NO in the regional hemodynamic changes during acute endotoxemia in rats. Am J Physiol 266:H1558–H1564
62. Billiar TR, Curran RD, Harbrecht BG, Stuehr DJ, Demetris AJ, Simmons RL (1990) Modulation of nitrogen oxide synthesis in vivo: N^G-monomethyl-L-arginine inhibits endotoxin-induced nitrite/nitrate biosynthesis while promoting hepatic damage. J Leukoc Biol 48:565–569
63. Harbrecht BG, Billiar TR, Stadler J, et al (1992) Nitric oxide synthesis serves to reduce hepatic damage during acute murine endotoxemia. Crit Care Med 20:1568–1574
64. Harbrecht BG, Di Silvio M, Demetris AJ, Simmons RL, Billiar TR (1994) Tumor necrosis factor-a regulates in vivo nitric oxide synthesis and induces liver injury during endotoxemia. Hepatology 20:1055–1060
65. Frederick JA, Hasselgren PO, Davis S, Higashiguchi T, Jacob TD, Fischer JE (1993) Nitric oxide may upregulate in vivo hepatic protein synthesis during endotoxemia. Arch Surg 128:152–157
66. Billiar TR, Curran RD, Stuehr DJ, West MA, Bentz BG, Simmons RL (1989) A L-arginine-dependent mechanism mediates Kupffer cell inhibition of hepatocyte protein synthesis in vitro. J Exp Med 169:1467–1472
67. Evans T, Carpenter A, Silva A, Cohen J (1994) Inhibition of nitric oxide synthase in experimental gram-negative sepsis. J Infect Dis 169:343–349

68. Walder CE, Thiemermann C, Vane JR (1991) The involvement of endothelium-derived relaxing factor in the regulation of renal cortical blood flow in the rat. Br J Pharmacol 102:967–971
69. Spain DA, Wilson MA, Bloom IT, Garrison RN (1994) Renal microvascular responses to sepsis are dependent on nitric oxide. J Surg Res 56:524–529
70. Spain DA, Wilson MA, Garrison RN (1994) Nitric oxide synthase inhibition exacerbates sepsis-induced renal hypoperfusion. Surgery 116:322–331
71. Papadimitriou M, Economidou D, Vakianis P, et al (1994) Effect of inhibition of nitric oxide synthase in acute renal failure due to endotoxin shock in dogs. Nephrol Dial Transplant 9:82–87
72. Cobb JP, Natanson C, Quezado ZMN, et al (1993) Lack of protection by-N-methyl-1-arginine (NMA), a nitric oxide synthase inhibitor, in endotoxin(LPS)-challenged canines. Crit Care Med 21:S280 (Abst)
73. Minnard EA, Shou J, Naama H, Cech A, Gallagher H, Daly JM (1994) Inhibition of nitric oxide synthesis is detrimental during endotoxemia. Arch Surg 129:142–148
74. Teale DM, Atkinson AM (1992) Inhibition of nitric oxide synthesis improves survival in a murine peritonitis model of sepsis that is not cured by antibiotics alone. J Antimicrob Chemother 30:839–842
75. Tiao G, Rafferty J, Ogle C, Fischer JE, Hasselgren PO (1994) Detrimental effect of nitric oxide synthase inhibition during endotoxemia may be caused by high levels of tumor necrosis factor and interleukin-6. Surgery 116:332–337
76. Nava E, Palmer RM, Moncada S (1991) Inhibition of nitric oxide synthesis in septic shock: How much is beneficial? Lancet 338:1555–1557
77. Wang P, Ba ZF, Chaudry IH (1994) Nitric oxide. To block or enhance its production during sepsis ? Arch Surg 129:1137–1143
78. Petros A, Lamb G, Leone A, Moncada S, Bennett D, Vallance P (1994) Effects of a nitric oxide synthase inhibitor in humans with septic shock. Cardiovasc Res 28: 34–39
79. Zhang H, Rogier P, Spapen H, Friedman G, Preiser JC, Vincent JL (1995) NO donor SIN-1 can increase splanchnic blood flow during endotoxic shock. Am J Respir Crit Care Med 151:A16 (Abst)
80. Henry Y, Lepoivre M, Drapier JC, Ducrocq C, Boucher JL, Guissani A (1993) EPR characterization of molecular targets for NO in mammalian cells and organelles. FASEB J 7:1124–1134
81. Schneider F, Lutun P, Couchot A, Bilbault P, Tempé JD (1993) Plasma cyclic guanosine 3′-5′ monophosphate concentrations and low vascular resistance in human septic shock. Intensive Care Med 19:99–104
82. Martin W, Villani GM, Jothianandan D, Furchgott RF (1985) Selective blockade of endothelium-dependent and glyceryl trinitrate-induced relaxation by hemoglobin and by methylene blue in the rabbit aorta. J Pharmacol Exp Ther 232:708–716
83. Keaney JFJ, Puyana JC, Francis S, Loscalzo JF, Stamler JS, Loscalzo J (1994) Methylene blue reverses endotoxin-induced hypotension. Circ Res 74:1121–1125
84. Mayer B, Brunner F, Schmidt K (1993) Inhibition of nitric oxide synthesis by methylene blue. Biochem Pharmacol 45:367–374
85. Wong SK, Garbers DL (1992) Receptor guanylyl cyclases. J Clin Invest 90:299–305
86. Beasley D (1990) Interleukin-1 and endotoxin activate soluble guanylate cyclase in vascular smooth muscle. Am J Physiol 259:R38–R44
87. Chen X, Gillis CN (1993) Methylene blue enhanced photorelaxation in aorta, pulmonary artery and corpus cavernosum. Biochem Biophys Res Commun 190:559–563
88. Gray GA, Julou-Schaeffer G, Oury K, Fleming I, Parratt JR, Stoclet JC (1990) An L-arginine-derived factor mediates endotoxin-induced vascular hyposensitivity to calcium. Eur J Pharmacol 191:89–92
89. Andrade SP, Hart IR, Piper PJ (1992) Inhibitors of nitric oxide synthase selectively reduce flow in tumor-associated neovasculature. Br J Pharmacol 107:1092–1095
90. Fleming I, Gray GA, Stoclet JC (1993) Influence of endothelium on induction of the L-arginine-nitric oxide pathway in rat aortas. Am J Physiol 264:H1200–H1207

91. Auguet M, Guillon JM, Delaflotte S, Etiemble E, Chabrier PE, Braquet P (1990) Endothelium independent protective effect of N^G-monomethyl-L-arginine on endotoxin-induced alterations of vascular reactivity. Life Sci 48:189–193
92. Auguet M, Lonchampt MO, Delaflotte S, Goulin-Schulz J, Chabrier PE (1992) Induction of nitric oxide synthase by lipoteichoic acid from Staphylococcus aureus in vascular smooth muscle cells. FEBS 297:183–185
93. Rimele TJ, Armstrong SJ, Grimes D, Sturm RJ (1991) Rat peritoneal neutrophils selectively relax vascular smooth muscle. J Pharmacol Exp Ther 258:963–971
94. Paya D, Gray GA, Stoclet JC (1993) Effects of methylene blue on blood pressure and reactivity to norepinephrine in endotoxemic rats. J Cardiovasc Pharmacol 21:926–930
95. Wu CC, Szabo C, Chen SJ, Thiemermann C, Vane JR (1994) Activation of soluble guanylyl cyclase by a factor other than nitric oxide or carbon monoxide contributes to the vascular hyporeactivity to vasoconstrictor agents in the aorta of rats treated with endotoxin. Biochem Biophys Res Commun 201:436–442
96. Brady AJ, Warren JB, Poole-Wilson PA, Williams TJ, Harding SE (1993) Nitric oxide attenuates cardiac myocyte contraction. Am J Physiol 265:H176–H182
97. Schulz R, Nava E, Moncada S (1992) Induction and potential biological relevance of a Ca^{2+}-independent nitric oxide synthase in the myocardium. Br J Pharmacol 105: 575–580
98. (1993) Martindale. The extra pharmacopoeia. The Pharmaceutical Press, London Ed. 13 pp 622–623
99. Kupfer A, Aeschlimann C, Wermuth B, Cerny T (1994) Prophylaxis and reversal of ifosfamide encephalopathy with methylene-blue. Lancet 343:763–764
100. Birch AA, Boyce WH (1976) Hypertension and decreased renal blood flow following methylene blue injection. Anesth Analg 55:674–676
101. Tranquada RE, Bernstein S, Grant WJ (1963) Methylene blue in the treatment of lactic acidosis. Clin Res 11:230–235
102. Schneider F, Lutun P, Hasselmann M, Stoclet JC (1992) Methylene blue increases systemic vascular resistance in human septic shock. Preliminary observations. Intensive Care Med 18:309–311
103. Preiser JC, Lejeune P, Roman A, et al (1995) Methylene blue administration in septic shock: A clinical trial. Crit Care Med 23:259–264
104. Zhang H, Rogiers P, Preiser JC, Spapen H, Manikis P, Vincent JL (1995) Effects of methylene blue on oxygen availability and regional blood flow during endotoxic shock. Crit Care Med (In press)

Inhibition of Nitric Oxide Production in Clinical Septic Shock

E. D. Bennett

Introduction

There is now an increasing body of evidence which strongly suggests that excess production of nitric oxide (NO) is responsible for the severe hypotension commonly seen in septic shock. Very recently, Gomez-Jimenez et al. [1] have shown that plasma nitrite and nitrate levels were not different between a control group of patients and those with sepsis, but the values in both these groups were significantly lower than in a group of patients with septic shock. In addition, they were able to show a correlation between nitrite and nitrate levels and cardiac output and plasma endotoxin levels in the septic group.

It seems that this excess production is due to the activity of the calcium-independent inducible enzyme, nitric oxide synthase (NOS). This enzyme appears to be present throughout the vessel wall and, when stimulated by various inflammatory mediators such as TNF, leads to the uncontrolled production of large amounts of NO. The resulting profound vasodilatation leads to severe hypotension and a doubling or even tripling of cardiac output. This is despite an overproduction of various endothelins which would normally cause vasoconstriction and hypertension.

It is clear therefore the resistance vessels are unable to respond to agents which would normally cause vasoconstriction and the clinical relevance of this is profound. The severe hypotension leads, despite the high cardiac output, to hypoperfusion of various organs particularly the renal and hepato-splanchnic beds. As the blood vessels are so unresponsive to the commonly used α-agonists such as norepinephrine and angiotensin, these have to be administered in extremely high concentrations to achieve small rises in blood pressure. This clinical situation of profound reduction in systemic resistance with low blood pressure, high cardiac output and organ hypoperfusion is associated with a mortality in excess of 60% and a very considerable morbidity. It has also been suggested that the excess production of NO in the tissues particularly the liver may, in addition to producing hypotension, be cytotoxic and thus worsen the local tissue damage which is such a feature of septic shock.

The ability pharmacologically to block the production of NO using agents such as N^G-monomethyl-L-arginine (L-NMMA) led to several groups undertaking clinical studies to determine whether the administration of these blocking agents might result in an increase in systemic vascular resistance as the excess NO production is inhibited. These studies have evoked a good deal of criticism [2] as it was felt that the administration of agents such as L-NMMA and N^G-nitro-L-arginine methyl ester (L-NAME) would likely lead to excess vasoconstriction because these agents block both the inducible and the constitutive forms of NOS. These critics felt that blocking all NO production could be potentially harmful as low levels of NO were physiological and exerted beneficial effects. There was indeed some experimental data to suggest that this is the case, and when the blocking agents were given to various animal models of sepsis an increase in mortality and morbidity was seen [3, 4]. They suggested that studies should not be carried out until such time that selected agents were available which only blocked the inducible form of NOS.

Clinical Implications

The first clinical administration of a NOS inhibitor was undertaken by Petros et al. [5] who gave bolus doses of L-NMMA to patients with profound hypotension associated with sepsis and who were also receiving high dose norepinephrine. They showed a dose-related, proportionally similar rise in systemic and pulmonary artery pressure which was maintained for up to 1 h. In a third patient, the longer acting L-NAME was administered and this resulted in a much more prolonged rise in blood pressure which persisted for up to 2 h. In addition, the rise in pulmonary artery pressure was disproportionally greater than that in the systemic circulation. This study was designed as a pilot study and invoked some criticism because of the potential uncontrolled vasoconstriction that might have occurred leading to further organ underperfusion and tissue damage.

In 1993, Lorente et al. [6] undertook a controlled study in a group of 15 patients with a diagnosis of septic syndrome. Eight patients received 20 mg/kg of N^{ω}-nitro-L-arginine (L-NNA) as an intravenous bolus. 45 min later, they then received 200 mg/kg of L-arginine also as an intravenous bolus. The remaining 7 patients received only the dose of L-arginine. The administration of L-NNA resulted in a significant rise of blood pressure from a mean of 89 ± 8 to 140 ± 12 mm Hg, and this was associated with significant fall in cardiac output from 3.51 ± 0.39 to 2.65 ± 0.21 L/min/m^2 with increases in both right atrial and pulmonary occlusion pressure. As a consequence, the systemic vascular resistance increased from 1871.1 ± 302.3 to 3825.6 ± 244.4 dyne$\cdot$sec/cm$^{-5} \cdot$m^2 and pulmonary vascular resistance from 533.2 ± 125.8 to 816 ± 117.3 dyne$\cdot$sec/cm$^{-5} \cdot$m^2.

The expected parallel significant changes occurred in oxygen delivery but there was no change in oxygen consumption. These changes were rapidly reversed following the administration of L-arginine. The 7 patients who received L-arginine only, showed a reduction in mean blood pressure and an increase in cardiac index from 3.57 ± 0.15 to 4.74 ± 0.54 L/min/m^2. The systemic and pulmonary vascular resistances decreased appropriately. The increase in cardiac output led to the expected increase in oxygen delivery which resulted in a significant increase in oxygen consumption from 127.0±19.0 to 182.5±37.3 mL/min/m^2. The authors concluded that the basal release of NO plays an important role in the regulation of systemic and pulmonary vascular tone in patients with the sepsis syndrome. In addition, L-arginine has systemic and pulmonary vasodilatory effects.

In 1994, Petros et al. [7] published a randomized, double blind, placebo controlled study investigation of the effects of L-NMMA in 12 patients with severe sepsis associated with hypotension. In the first part of the study, bolus doses of 0.3 and 1.0 mg/kg were administered to 5 patients separated by an interval of 15 min. At the end of the second 15-min period, an infusion of 1 mg/kg/h was commenced and maintained for 6 h. Routine hemodynamic data were collected throughout this period. In addition, blood was collected for the measurement of routine hematological, renal and liver function tests. The patients in the two groups were well matched in most respects except that those in the protocol group had a significantly higher cardiac output and blood pressure than the control group. Both groups were receiving norepinephrine at a mean dose of 1 μg/kg/min. Bolus injections of L-NMMA produced a dose-dependent increase in mean blood pressure representing a rise of approximately 20%. A rise of about 15% was maintained during the 6-h infusion period falling to the baseline value 2 h after the infusion was stopped. A dose-related fall in cardiac output occurred following bolus injection, and this reduction of almost 30% was maintained during the 6-h infusion, and the cardiac output only returned towards the baseline value 6 h after stopping the infusion. This represented a fall in cardiac output from 11.2±2.1 to 8.9±1.9 L/min.

The changes in blood pressure, and cardiac output and blood pressure resulted in a 50% increase in systemic vascular resistance which persisted for the duration of the infusion. This represents an increase in SVR from 547±92 to 889±143 dynes/sec/cm^{-5}. The central venous pressure and pulmonary artery occlusion pressure showed similar changes rising by about 15%. Pulmonary vascular resistance increased by about 50%. No significant changes in any of the variables were seen in the patients in the control group.

Indices of renal and hepatic function showed evidence of minor deterioration although none of the changes were significant with creatinine rising from 180±40 to 231±49 μmol/L and bilirubin rising from 25.5±9.2 to 51.0±8.3 μmol/L. Alanine transaminase increased from 53±21 to 77±40 IU/L. Oxygen delivery showed a significant 29% fall which did not result in any significant change in lactate. The plasma concentration of L-NMMA 5 min after 1 mg/kg bolus was 3.2±0.6 μmol/L. There were no significant changes in any of the hematological variables. Platelet numbers fell in both groups but there was no significant difference between the groups.

The authors concluded that the study demonstrated that overproduction of NO was likely to be an important causative factor in the profound hypotension seen in septic shock. Relatively low dose administration on L-

NMMA caused a widespread increase in vascular tone leading to a marked increase in blood pressure. This was associated in a reduction in tissue oxygen delivery, the implications of which were uncertain. They felt a large scale study of agents which inhibit NOS was required.

It should perhaps be pointed that the major difference between this study [7] and that of Lorente et al. [6] is that the patients in the Petros study had a mean cardiac output 11.2 L/min whereas the mean cardiac index in the latter study was only 3.5 L/min. It perhaps implies that NOS inhibitors, if they are to be used clinically, should only be administered to patients with sepsis showing evidence of classic hyperdynamic shock.

Within the last month, Preiser et al. [8] have published data showing that the administration of 2 mg/kg of methylene blue to 14 patients with septic shock produced a significant increase in blood pressure from 61.1 ± 7.6 to 71.7 ± 12.0 mm Hg 60 min after the administration of the agent. There was however no significant effect on pulmonary artery pressure, cardiac filling pressures, cardiac output, oxygen delivery and oxygen consumption. Arterial lactate decreased significantly from 3.4 ± 1.4 to 2.7 ± 1.3 mmol/L. These effects were transient lasting up to 1 h with a second dose of methylene blue producing a similar result. The authors argued that methylene blue is a powerful inhibitor of guanylate cyclase which is the target enzyme of NO. They suggest that this approach maintains the potential beneficial immunological effects of NO and avoids major toxicity as methylene blue is free from obvious side effects.

It is quite clear from the limited number of clinical studies where the NO pathways have been manipulated that significant increases in systemic vascular resistance have been obtained usually with concomitant reduction in cardiac output. None of the studies have been able to demonstrate any beneficial effect on outcome nor of course were they designed to do so. This is, of course, the crucial issue relating to the role of NO in sepsis. Is the inhibition of both the constitutive and inducible forms of NOS ever likely to lead to a reduction in morbidity and mortality, or is it just a rather expensive way of raising blood pressure in a group of very ill patients? Would the development of agents which selectively block only the inducible NOS confere any extra benefit? Indeed is the excess production of NO harmful in any other way other than probably being responsible for the hypotension seen in the septic syndrome? The correspondence columns of the February 1995 edition [9, 10, 11] of *Critical Care Medicine* nicely illustrate the controversy surrounding this subject.

The evidence of Gómez-Jiménez et al. [1] would perhaps suggest that, as plasma nitrate and nitrites levels are only raised in septic patients with hypotension, any other potential damaging effects are unimportant. On the other hand, the intriguing result obtained by Yu et al. [12] who showed that the administration of ketaconazole to a group of patients at high risk of developing ARDS led to significant reductions in the incidence of ARDS and mortality in the treatment group may be relevant. The authors suggested that their results might be related to reduction in thomboxane B_2 levels that occurred in the treatment group. However it is of some consideration that Bogle et al. [13] showed that the addition of several of the imidazole group of drugs to macrophages in cell culture completely inhibited the production of NO when stimulated by lipopolysaccharide or gamma interferon. Ketaconazole, of course, is a member of the imidazole group of drugs and it is fascinating to contemplate that this is indeed the explanation of this interesting study.

Conclusion

It is clear therefore that NO appears to play a pivotal role in the causation of the profound hypotension seen in hyperdynamic septic shock, and that the administration of agents which block both the constitutive and inducible forms of NOS will raise blood pressure and lower cardiac output. This pivotal role is confirmed by the demonstration of the excess production of nitrates and nitrites in patients with hypotensive septic shock. What is much less clear is whether the administration of such agents has any other effects either harmful of beneficial. The experiments undertaken in animal models of sepsis have given conflicting results depending on the particular model, and the dose and timing of the particular blocking agents used. This dilemma will of course only finally be answered when good, double-blind, clinical trials have been completed.

References

1. Gómez-Jiménez J, Salgado A, Mourelle M, et al (1995) L-arginine: Nitric oxide pathway in endotoxemia and human septic shock. Crit Care Med 23:253–258
2. Cobb JP, Cunnion RE, Danner RL (1993) Nitric oxide as a target for therapy in septic shock. Crit Care Med 21:1261–1263
3. Cobb JP, Natanson C, Hoffman WD, et al (1992) N-amino-L-arginine, an inhibitor of nitric oxide synthase, raises vascular resistance but increases mortality in canines challenged with endotoxin. J Exp Med 176:1175–1182
4. Biliar TR, Curran RD, Harbrecht BG, et al (1990) Modulation of nitrogen oxide synthesis in vivo: N mono-methyl-L-arginine inhibits endotoxin-induced nitrite/nitrate biosynthesis while promoting hepatic damage. J Leukoc Biol 48:565–569
5. Petros A, Bennett D, Vallance P (1991) Effect of nitric oxide synthase inhibition on hypotension in patients with septic shock. Lancet 338:1557–1558
6. Lorente JA, Landín L, De Pablo R, et al (1993) L-arginine pathway in the sepsis syndrome. Crit Care Med 21:1287–1285
7. Petros A, Lamb, Leone A, et al (1994) Effects of a nitric oxide synthase inhibitor in humans with septic shock. Cardiovasc Res 28:34–39
8. Preiser JC, Lejeune P, Roman A, et al (1995) Methylene blue administration in septic shock: A clinical trial. Crit Care Med 23:259–264
9. Brüssel T, Meyer J (1995) Nitric oxide synthesis inhibition in septic shock. Crit Care Med 23:412 (Letter)
10. Cobb JP, Cunnion RE, Danner RL (1995) Nitric oxide synthesis inhibition in septic shock. Crit Care Med 23:413–414 (Letter)

11. Lorente JA Landín, Renes E, Liste D (1995) Nitric oxide synthesis inhibition in septic shock. Crit Care Med 23:414–415 (Letter)
12. Yu M, Tomasa G (1993) A double blind, prospective, randomized trial ketoconazole, a thromboxane synthetase inhibitor, in the prophylaxis of adult respiratory distress syndrome. Crit Care Med 21:1635–1642
13. Bogle RG, Whitley GS, Soo SC, et al (1994) Effect of anti-fungal imidazoles on mRNA levels and enzyme activity of inducible nitric oxide synthase. Br J Pharmacol 111: 1257–1261

Nitric Oxide and the Lung

Role of Nitric Oxide in Pulmonary Vascular Dysfunction

M. J. D. Griffiths and T. W. Evans

Introduction

Early descriptions of patients with septic shock in which invasive monitoring was used emphasized the failure of a sustained response of the central venous pressure and cardiac output to volume replacement indicating pathological dilation of peripheral blood vessels [1]. Apart from the hemodynamic changes associated with shock, metabolic derangement and failure of multiple organ systems, including acute lung injury (ALI), were recorded in the majority of patients. Since early reports that a calcium-independent nitric oxide synthase (NOS) could be induced by endotoxin (LPS) and proinflammatory cytokines in extra-endothelial vascular tissue [2], considerable evidence (*vide supra*) has accumulated implicating nitric oxide (NO), derived from inducible iNOS, as a final mediator of sepsis-associated vasodilation and hyporesponsiveness to vasopressors. More recently, with the increasing awareness of the importance of oxygen delivery in determining the outcome of patients with sepsis, attention has focused on the regional distribution of blood flow both at an organ system and microcirculatory level. In this chapter, the evidence implicating the L-arginine/NO pathway as an important effector of sepsis-induced pulmonary vascular dysfunction is discussed.

Sepsis, SIRS, ALI and ARDS

The definitions of sepsis and related syndromes derived at a recent consensus conference [3] reflect the increased emphasis now placed on the response of the host by recognizing that some patients with a variety of conditions (severe trauma, burns, pancreatitis, etc.) may present with signs of sepsis in the absence of infection. Hence, the all-encompassing term "systemic inflammatory response syndrome (SIRS)" has gained popularity. The effect of the presence of an infective agent on the outcome of SIRS is uncertain, but the clinical course may be similar regardless of the initiating event activating the inflammatory cascade [4]. A sustained inflammatory response, as may be associated with extensive trauma, unresolved infection or inadequate resuscitation from shock, frequently progresses to the multiple organ dysfunction syndrome (MODS). ALI and the acute respiratory distress syndrome (ARDS) are the most common manifestations of this phenomenon affecting a third of patients with non-pulmonary sepsis [5].

It is now recognized that a spectrum of lung injury exists, the most severe manifestation of which is ARDS (Table 1) [6]. ALI results from inflammatory damage to the alveolar-capillary membrane causing increased permeability and non-hydrostatic pulmonary edema. The lung may be damaged directly or indirectly as a result of acute SIRS: the clinical correlates of these processes are described in Table 2. The spread of pulmonary inflammation outside areas directly involved in injury [7] and the perpetuation of pulmonary damage after the initial insult suggest that humoral inflammatory mediators play a major role in ALI regardless of its cause. Indeed, the majority of patients with established ARDS have clinical signs of SIRS regardless of the presence of infection (Fig. 1) [8].

Table 1. Recommended criteria for ALI and ARDS. (Adapted from [6])

	Onset	Oxygenation*	Chest X-ray (frontal)	Pulmonary artery wedge pressure
ALI	acute	$PaO_2/FiO_2 \leq 300$	Bilateral infiltrates	≤ 18 mm Hg or no clinical evidence of raised left atrial pressure
ARDS	acute	$PaO_2/FiO_2 \leq 200$	Bilateral infiltrates	≤ 18 mm Hg or no clinical evidence of raised left atrial pressure

* Oxygenation to be considered regardless of the level of positive end-expiratory pressure

Table 2. Clinical risk factors for ARDS

Respiratory	Non-respiratory
Aspiration of gastric contents	Sepsis syndrome
Pneumonia – bacterial, fungal, viral, etc.	Disseminated intravascular coagulation
Pulmonary contusion	Major trauma/Shock
Inhalation of smoke and toxins	Massive burns
Thoracic irradiation	Pregnancy associated – amniotic fluid embolism, pre-eclampsia, etc.
Near drowning	Multiple transfusions
Ischemia-reperfusion	Fat embolism
Oxygen toxicity	Pancreatitis
Pulmonary vasculitis (eg. Goodpasture's syndrome)	Head injury/raised intracranial pressure
Post-pneumonectomy	Drugs/toxins – paraquat, heroin, amiodarone
	Tumor lysis syndrome
	Post-cardiopulmonary bypass

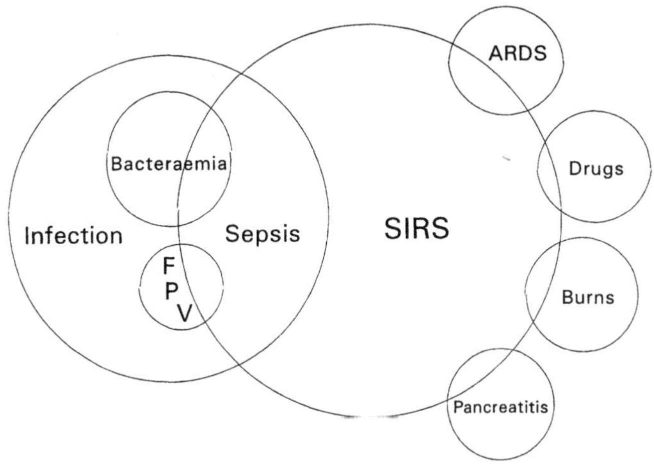

Fig. 1. Relationships between Sepsis, SIRS and ARDS. F: fungi; P: parasites; V: viruses; SIRS: systemic inflammatory response syndrome; ARDS: acute respiratory distress syndrome. (Adapted from [3])

Hence, the inflammatory processes that cause SIRS may cause ALI, and local pulmonary injury that presents as ALI is likely to be associated with SIRS during the course of the patient's illness. Experimental [9] and clinical [10] studies have shown that established ARDS is a multisystem vascular disorder, characterized not only by impaired pulmonary gas exchange, but also by failure of oxygen uptake in the periphery which can itself lead to MODS. Respiratory failure may account for only 16% of deaths in patients with ARDS, the majority being attributable to MODS [11]. Given the degree of overlap in causation and cardiovascular pathophysiology between SIRS and ALI, the role of NO in the pulmonary vascular dysfunction of both conditions is considered together.

Pulmonary Vascular Dysfunction in SIRS and ALI

Pulmonary vascular dysfunction, in particular pulmonary hypertension (PHT), is commonly associated with ARDS in critically ill patients. Pulmonary vascular resistance (PVR) and pulmonary artery pressure (Ppa) are elevated in patients with respiratory failure [12] and in animal models of ALI [13]. PVR is increased in early ARDS, even after correction for arterial hypoxemia. Early in the clinical course, functional factors causing increased vascular tone predominate, whilst structural factors become more important as the pulmonary vasculature remodels (Table 3) [14]. Remodeling reduces the capillary surface area for gas exchange which probably accounts for the exercise-induced increase in alveolar-arterial oxygen gradient (A-aDo_2) seen in the majority of survivors of ARDS [15].

Table 3., Mechanisms underlying increased pulmonary vascular resistance in ALI

Vasomotor	Hypoxic pulmonary vasoconstriction Vasoconstrictor mediators TXA_2, LTs, PAF, ET-1 Decreased vasodilation
Structural	Intraluminal obstruction - emboli - *in situ* thrombosis - cellular aggregates Wall thickening - endothelial swelling - intimal proliferation - medial hypertrophy External compression - edema/hemorrhage - PPV/PEEP

TX: thromboxane; LT: leukotrienes; PAF: platelet activating factor; ET: endothelin; PPV: positive pressure ventilation; PEEP: positive end-expiratory pressure

PHT contributes to impaired right ventricular performance [16] and has been associated with increased mortality in patients with ARDS [10]. The degree of PHT correlates with overall ALI severity, raising the possibility that PHT is a merely marker of vascular dysfunction or itself has an important influence on outcome. A correlation between Ppa and pulmonary edema formation has been demonstrated in patients with ARDS [17]. Interstitial edema may raise Ppa by causing vascular compression, which may itself promote edema formation if the site of vasoconstriction is distal to the leaking vessels. However, in canine oleic acid-induced lung injury, hypoxic ventilation of the left lower lobe was associated with reduced edema formation in that lobe [18]. This effect was attributed to hypoxic pulmonary vasoconstriction (HPV), since inhibition of the latter with minoxidil abolished hypoxia-induced protection against edema formation. The putative mechanism underlying this depends on HPV causing constriction proximal to the leaking vessels. Hence, the precise site of action of vasoactive drugs may influence their effects on the complex pathophysiology of ALI.

Equally important in the pathogenesis of ALI is the loss of microvascular function which underlies the observed refractory hypoxemia. For example, the loss of HPV in ALI [19] has an especially marked effect on gas exchange because alveolar edema and atelectasis cause enormous increases in physiological dead space. Studies using the multiple inert gas technique have demonstrated that the degree of intra-pulmonary shunting is large enough to account for the observed $A\text{-}aDo_2$, without invoking a reduction in diffusion capacity [20].

The use of vasoconstrictors in patients with ARDS, whilst being superficially unattractive because of their negative effects on Ppa and right ventricular function, may benefit gas exchange if HPV is restored. Preliminary results from work on patients with ARDS suggest that the peripheral chemoreceptor stimulant almitrine improves intrapulmonary shunting without decreasing cardiac output despite the rise in right ventricular afterload [21]. Conversely, infused vasodilators improve cardiac output but exacerbate intrapulmonary shunt. Inhaled vasodilators improve oxygenation by decreasing intrapulmonary shunting. Recent studies suggest that the use of inhaled NO in combination with intravenous almitrine may have additive beneficial effects on gas exchange [22].

NO in the Normal Pulmonary Circulation

The pulmonary circulation is a low pressure system with a great capacity to adapt to total and local changes in blood flow. Both pulmonary arterial and venous endothelia express NOS constitutively and release NO in response to a variety of chemical and mechanical stimuli. The contribution of endothelium-derived NOS (eNOS) to the maintenance of low pulmonary vascular tone has been the subject of much research. Different answers to this question have been derived from the use of a variety of preparations from several species and controversy remains. For example, using specific inhibitors of NOS and cyclooxygenase (COX) in an isolated lung preparation has produced evidence suggesting that the low vascular resistance and recruitment-distensibility properties of the pulmonary circulation in the dog depend on COX products, whilst in the rat NO production is of primary importance [23]. In normal humans, low dose infusion of the NOS inhibitor N^G-monomethyl-L-arginine (L-NMMA) increases systemic vascular resistance (SVR) by approximately 60% and PVR approximately 40% [24], suggesting an important role for eNOS in determining vascular tone.

Chand and Altura [25] in 1981 first demonstrated endothelium-dependent pulmonary artery (PA) dilatation in response to acetylcholine (ACh) and bradykinin (BK). This phenomenon has also been demonstrated in human PA [26]. The administration of NOS inhibitors in isolated lung preparations demonstrates the role of eNOS in modulating HPV [27] and the response of PA to various vasoconstrictors [28] and increases in flow [29]. Finally, NO mediates endothelium-dependent and independent non-adrenergic-non-cholinergic (NANC)-induced relaxation of guinea pig PA *in vivo* [30].

The effects of eNOS on pulmonary vascular function have been elucidated by studies using low concentrations of inhaled NO. At concentrations of 5–80 parts per million (ppm), NO produces selective and rapidly-reversible pulmonary vasodilation in animal models of PHT [31, 32]. In healthy humans, inhaling NO has no effect on pulmonary hemodynamics, but inhaled NO reverses the PHT and increase in PVR induced by breathing 12% oxygen without affecting systemic hemodynamics [33].

Mediators of Pulmonary Hypertension in SIRS and ARDS

SIRS and ARDS are associated with an uncontrolled release of vasoactive agents, both dilators and constrictors. The balance of the effects of these agents determines vascular tone at local, organ and whole circulation levels. Involvement of NO in acute PHT would require either: 1) decreased activity of constitutive eNOS, 2) insensitivity of pulmonary vascular smooth muscle to NO, or 3) failure of NOS induction in pulmonary vessels, any of which may tip the balance of pulmonary dilation in favor of vasoconstriction. The effects of mediators of SIRS on endothelial function (receptor-mediated endothelium-dependent dilation) in intact tissues are variable and several contradictory reports have been published. In the isolated rat lung, treatment with tumor necrosis factor (TNF) augments HPV and diminishes ACh-induced responses without affecting sodium nitroprusside (SNP)-induced endothelium-independent relaxation [34]. Relaxation to ACh and SNP is preserved in isolated rat main PA from LPS-treated rats, whilst in the isolated lungs of identically treated animals, SNP-mediated vasorelaxation is attenuated and the response to ACh is abolished (unpublished observation). The responsiveness to SNP is not restored by inhibition of the activity of inducible NOS, as has recently been demonstrated in rat aorta incubated for 24 h with LPS [35], and so is unlikely to result simply from overstimulation of vascular smooth muscle by NO. The discrepancy between effects on conduit and resistance PA may relate to the ability of activated neutrophils to adhere to and damage the endothelium of smaller vessels, although no data support this speculation. Similar "anti-EDRF" (endothelium-derived relaxing factor) effects have been elicited in isolated arteries from cat (carotid) [36], cow (coronary and PA) [37] and rat (PA and aorta) [38] following treatment with large doses of TNF. However, incubating ovine PA in a lower concentration of TNF that corresponded to the plasma level of that cytokine measured following LPS administration, failed to alter endothelium-dependent relaxation or the contractile response to norepinephrine [39]. Endothelium-dependent relaxation has similarly been demonstrated in guinea pig (aorta) [40], canine (coronary) [41], rabbit and rat (aorta) [42] endotoxemia. LPS infusion in the pithed rat preparation impaired vasodepressor responses to 5-hydroxytryptamine (5-HT), ACh and BK, as well as to SNP, endothelin and salbutamol [43]. 24 h after cecal ligation and puncture, resistance vessels from pulmonary and a variety of systemic beds showed impaired responsiveness to vasoconstrictors but all vessels were dilated normally by ACh [44]. In contrast, femoral, mesenteric and renal artery rings from LPS-treated dogs had uniformally impaired endothelium-dependent relaxation but normal contractile responses [45]. Finally, during an infusion of LPS into conscious rats, ACH and BK caused hypotension but there were marked regional variations in the vasodilator responses [46]. Notably, vasodilation to both agonists was decreased in the renal vascular bed. It seems likely that the dose of LPS or cytokine, the duration of incubation and the type of vessel tested influence their effects on endothelial-function and vascular smooth muscle sensitivity to NO.

Treatment of rats with a large dose of LPS leads in main PA to steroid-sensitive induction of mRNA for iNOS and functional evidence of increased extra-endothelial NO production [47]. Induction of NOS in the PA of patients with SIRS or ALI has not yet been demonstrated, although the amount of NO in exhaled breath increases in a tracheotomized septic rat model [48]. Post-translational control of iNOS activity by endogenous NO-induced negative feedback has recently been demonstrated in isolated rat aorta [49]. This phenomenon may explain some of the properties of inhaled NO in ARDS, for example the rebound PHT sometimes observed on cessation of therapy. The level of NO production by pulmonary vessels in patients has important implications for the use of NOS inhibitors in the treatment of septic shock. Inhibiting a significant production of NO would have a more marked adverse effect on PHT than if NOS isoforms were already relatively inactive. Data from patients with septic shock are scarce, but using unselective NOS inhibitors, the increase in PVR was approximately half that of the rise in SVR [50]. Hence, available evidence suggests that iNOS-derived NO actively attenuates the PHT associated with SIRS and ALI.

The acute pulmonary hypertensive response to LPS or live bacteria has been well characterized. This typically consists of an early peak (within minutes), a decline over a few hours, and then a later sustained response. The early PHT is associated with elevated thromboxane (TX) A_2 levels in serum and lung lymph and is inhibited by pretreatment with TX synthetase inhibitors [51]. Leukotriene (LT) levels are elevated in pulmonary edema fluid of patients with ARDS and in lung lymph from sheep challenged with LPS [51]. Pretreatment with a LTB_4-receptor antagonist attenuates the effects of LPS, including systemic hypotension and acute PHT in pigs [52]. Since LT antagonists block the rise in TXA_2 *in vivo*, but not *in vitro*, it has been proposed that PHT may involve the sequential release of LT followed by TXA_2. Similarly, platelet activating factor (PAF) infusion causes PHT in various animal models, and pretreatment of animals with PAF antagonists inhibits both the rise in TXA_2 and the pressor response to LPS [53]. Antagonism of neutrophil-derived proteases and reactive oxygen species, as well as tumor necrosis factor α (TNF), attenuates the rise in circulating TXA_2 metabolites and LPS-induced early and late phase PHT. These data suggest that LPS infusion stimulates the release of vasoactive arachidonic acid metabolites predominantly TXA_2, which mediates acute pulmonary vasoconstriction [51].

Pulmonary sequestration of platelets and depletion of platelet 5-hydroxytryptamine (5-HT) in patients with early ARDS suggests that the vasoconstrictor effects of 5-HT may contribute to PHT. Ppa was decreased in these patients by ketanserin, a 5-HT antagonist [54]. Increased circulating levels of endothelin-1 (ET-1) have been reported in patients with ARDS owing to both reduced clearance and net ET-1 release by the lungs [55]. ET-1 is not only a potent vasoconstrictor, but also promotes mitogenesis in vascular smooth muscle. In ARDS and experimental models, structural events such as endothelial swelling, interstitial edema, the formation of cellular/fibrin thrombi and eventually vascular remodeling become more important contributors to the hypertensive response as the condition progresses (Table 3). Nevertheless, vasomotor influences may persist beyond the first few hours: for example, PHT observed in dogs 24 h after intraperitoneal implantation of *E. coli*-containing thrombus may be reversed by COX inhibitors [56]. In chronic or established ARDS, the associated chronic hypoxia may affect endothelium-dependent vasodilation and thus contribute to PHT. Pulmonary vessels from 7 day hypoxic rats exhibit reduced vasodilation in response to ACh associated with decreased activity of soluble guanylyl cyclase [57]. Similarly, isolated pulmonary vessels from patients with chronic hypoxic lung disease demonstrate reduced endothelium-dependent vasodilation [58].

Mediators of the Impaired Control of Pulmonary Vascular Tone in SIRS and ARDS

In animal models of ALI induced by pneumonia, oxygen toxicity, bleomycin or sepsis, pulmonary vasoconstrictor responses, to hypoxia and vasopressors, are substantially attenuated [59]. The role of NO as an etiological agent in the acute phases of ALI has been little explored, but increased NO release from iNOS might be expected to contribute to vascular hyporesponsiveness and to the dampening of HPV. In studies using larger animal models of sepsis in which pulmonary artery catheters were used, NO inhibition predictably increased SVR causing a decrease in cardiac output and hence oxygen delivery. PVR was increased by LPS and by NOS inhibition, but was restored by inhaled NO which also improves oxygenation [60]. Meyer and coworkers [61] infused a low dose of LPS into sheep for 48 h, a model which perhaps most accurately reflects human SIRS. A non-selective NOS inhibitor given after 12 h decreased oxygen delivery, but improves oxygen extraction suggesting improved peripheral oxygen uptake. However, intrapulmonary shunting increased after NOS inhibition although not as markedly in the LPS-treated group as in the controls. COX inhibitors enhance HPV in animal models of ALI and prevent LPS-induced loss of HPV [14]. In dogs with oleic acid-induced pulmonary edema, the structurally dissimilar agents, indomethacin and aspirin, increase pulmonary vascular tone and improve intrapulmonary shunt by approximately 30% [62]. Conflicting evidence has been provided by small studies in patients with established sepsis or ARDS, on the one hand showing that indomethacin [63] and ibuprofen [64] improve gas exchange, but on the other that dazoxiben, a specific antagonist of TX synthetase, has no effect on pulmonary hemodynamics or gas exchange [65]. These data are consistent with the release during ALI of a vasodilating prostaglandin, which impairs gas exchange by counteracting HPV [14].

Non-Hemodynamic Effects of NO

NO acts on vascular smooth muscle by binding the heme moiety and directly activating soluble guanylyl cyclase to produce guanosine 3′,5′-cyclic guanosine monophosphate (cGMP) [66]. The mechanisms underlying cGMP-mediated smooth muscle relaxation are incompletely understood, but include the phosphorylation of poorly-defined substrates by a cGMP-dependent protein kinase. NOS may be induced in the pulmonary vasculature [47], in alveolar macrophages [67] and in alveolar epithelial cells [68]. Hence, regardless of the use of inhaled NO, the environment of the alveolar-capillary membrane of a patient with ARDS may be rich in NO.

The much greater amounts of NO produced continuously by iNOS lead to its interaction with a diverse array of intra- and extracellular targets through which the cytotoxic [69], tumoricidal [70] and anti-microbial [71] effects of NO are mediated independently of cGMP. For example, NO reacts with protein sulphydryl groups leading to S-nitrosylation and mono-adenosine diphosphate (ADP)-ribosylation [72]. S-nitrosylation of plasma proteins such as albumin provide a sink of NO-like activity following NOS induction [73]. In human neutrophils, NO-induced depletion of glutathione, by S-nitrosylation, is associated with concomitant activation of the hexose monophosphate shunt with inhibition of superoxide production [74]. However, ADP-ribosylation of glycolytic enzymes inhibits their activity and may contribute to NO-induced cytotoxicity of certain cell lines.

Both exogenous and endogenous NO cause dose-dependent and reversible inhibition of rat hepatic mitochondrial enzymes involved in the electron transport chain [75], but the mechanism of this action remains obscure. The cytostatic effect of NO on tumor cells is at least partially explained by inhibition of ribonucleotide reductase, a rate-limiting enzyme in DNA synthesis [70], demonstrating that NO may modulate the activity of several enzymes in all cellular compartments. In tissue culture, a product of L-arginine derived from LPS-treated alveolar macrophages inhibits epithelial sodium ion transport by inactivating amiloride-sensitive non-selective cation channels [67]. If this observation were translated into an intact animal, the implication is that endogenous or exogenous NO may interfere with the ability of epithelial cells to clear alveolar edema. Finally, increased NO synthesis may deprive the cell of oxygen by direct competition of other enzymes with NOS and direct reaction of NO with available oxygen.

NO, itself a reactive radical, can react with superoxide to form peroxynitrite [76]. Under acidic conditions, peroxynitrite degenerates forming hydroxyl radicals which damage cell membranes, proteins and DNA [77]. For example, peroxynitrite and its products inhibit pulmonary surfactant function by lipid peroxidation and damage to surfactant proteins [78]. Local formation of NO and superoxide by alveolar macrophages during ALI may be expected to lead to peroxynitrite formation. Contrasting conclusions have been drawn by others, suggesting that NO is a scavenger of the superoxide anion, and that NO may provide a chemical barrier to cytotoxic reactive oxygen species [79]. These largely biochemical observations indicate that NO can have profound effects on cellular metabolism and cell survival. Excessive local NO production may thus contribute to alveolar-capillary membrane damage thus exacerbating capillary leak and contributing to surfactant dysfunction leading to atelectasis and intra-pulmonary shunting.

Therapeutic Implications

Insights into the role of NO in the pathogenesis of SIRS and ALI must now be considered in the context of the therapeutic use of NOS inhibitors in systemic hypotension [80], and inhaled NO, which has become an established adjunct to ventilatory support in ARDS [81].

The NO donor SNP has long been known to decrease PVR in patients with ARDS, but, in common with other intravenously administered vasodilators, is associated with a prohibitive deterioration in gas exchange. The experience of Zapol and colleagues [81] in treating 10 young patients with severe ARDS with inhaled NO has been published recently. Used over short periods, 18 ppm NO caused small but significant decrements in Ppa, PVR and shunt fraction, without affecting cardiac output or systemic pressures. Oxygenation improves during NO inhalation, but deteriorated significantly, as did systemic pressures, during intravenous infusion of the vasodilator prostacyclin. NO was inhaled by 7 patients for 3 to 53 days without loss of beneficial hemodynamic effect or evidence of toxicity. Similarly, concern about the toxicity of low concentrations of inhaled NO has been allayed by experiments involving long-term exposure in mice [82]. Concentrations of inhaled NO in the ppb range, improves oxygenation without affecting PVR in 3 patients with ARDS [83]. Hence, given the toxicity of nitrogen oxides, using the minimum effective concentration of NO [84], monitoring of inspired and expired gases and assessment of methemoglobin levels are advisable precautions [85]. Despite the undeniable beneficial effects of inhaled NO on pulmonary hemodynamics and gas exchange, there are currently no data on its effects on the pathogenesis of ALI and the possibility remains that NO toxicity may emerge much in the same way as oxygen toxicity.

The use of NOS inhibitors in septic shock is more controversial. Whilst, the imperative of increasing the patient's systemic blood pressure is satisfied, the importance of deleterious effects on the pulmonary and splanchnic [86, 87] circulations are difficult to assess. Using selective inhibitors of iNOS or low doses of the unselective inhibitors have theoretical advantages. Currently, selective inhibitors for the various isoforms of NOS are being developed [88, 89]. Large scale clinical trials of these agents in patients with SIRS and septic shock are eagerly awaited.

Conclusions

The mechanisms underlying the opposite blood pressure response of the systemic and pulmonary circulations in SIRS remain obscure. NOS appears to be inducible in the vascular smooth muscle of both circulations, and there is evidence of increased synthesis of iNOS-derived NO in the pulmonary circulation in experimental lung injury. In both the systemic and pulmonary circulations, there is conflicting evidence concerning the effects of inflammatory mediators on vascular smooth muscle sensitivity to NO and on EDRF-dependent vasodilatation. In LPS-treated rats, endogenous NO attenuates increased HPV and PHT associated with SIRS and ALI. In contrast, patients with ARDS have diminished HPV, but exogenous inhaled NO decreases PHT. Despite the potential for NO to interact with reactive oxygen species to form highly damaging moieties, the toxic effects on the injured lung of increased endogenous NO synthesis and inhaled NO are largely unknown.

References

1. MacLean LD, Mulligan WG, McLean APH, Duff JH (1967) Patterns of septic shock in man – A detailed study of 56 patients. Annal Surg 166:543–562
2. Rees DD, Cellek S, Palmer RM, Moncada S (1990) Dexamethasone prevents the induction by endotoxin of a nitric oxide synthase and the associated effects on vascular tone: An insight into endotoxin shock. Biochem Biophys Res Commun 173:541–547
3. Bone RC, Balk RA, Cerra FB, et al (1992) Definitions for sepsis and organ failure and guidelines for the use of innovative therapies in sepsis. The ACCP/SCCM Consensus Conference Committee. American College of Chest Physicians/Society of Critical Care Medicine. Chest 101:1644–1655
4. Marshall J, Sweeney D (1990) Microbial infection and the septic response in critical surgical illness. Arch Surg 125:17–23
5. Rubin DB, Wiener-Kronish JP, Murray JF (1990) Elevated von Willebrand factor antigenis: An early predictor of impending acute lung injury in nonpulmonary sepsis syndrome. J Clin Invest 86:474–480
6. Bernard GR, Artigas A, Brigham KL, et al. (1994) The American-European consensus conference on ARDS: Definitions, mechanisms, relevant outcomes and clinical trial co-ordination. Am J Respir Crit Care Med 149:818–824
7. St John RC, Mizer LA, Kindt GC, et al (1993) Acid aspiration-induced acute lung injury causes leukocyte-dependent systemic organ injury. J Appl Physiol 74:1994–2003
8. Meduri GU, Belenchia JM, Estes RJ, et al (1991) Fibroproliferative phase of ARDS. Clinical findings and effects of corticosteroids. Chest 100:943–952
9. Mizer LA, Weisbrode SE, Dorinsky PM (1989) Neutrophil accumulation and structural changes in nonpulmonary organs after acute lung injury induced by phorbol myristate acetate. Am Rev Respir Dis 139:1017–1026
10. Bone RC, Balk R, Slotman G, et al (1992) Adult respiratory distress syndrome. Sequence and importance of development of multiple organ failure. The Prostaglandin E1 Study Group. Chest 101:320–326
11. Montgomery AB, Stager MA, Carrico CJ, Hudson LD (1985) Causes of mortality in patients with the adult respiratory distress syndrome. Am Rev Respir Dis 132:485–489
12. Zapol WM, Snider MT (1977) Pulmonary hypertension in severe acute respiratory failure. N Engl J Med 296:476–480

13. Jones R, Reid L (1992) Pulmonary vascular changes in adult respiratory distress syndrome. In: Artigas A, et al (eds) Adult respiratory distress syndrome. Churchill Livingstone, New York, pp 45–59
14. Leeman M (1991) The pulmonary circulation in acute lung injury: A review of some recent advances. Intensive Care Med 17:254–260
15. Buchser E, Leuenberger P, Chiolero R, Perret C, Freeman J (1985) Reduced pulmonary capillary blood volume as a long-term sequel of ARDS. Chest 87:608–611
16, Sibbald WJ, Driedger AA (1983) Right ventricular function in acute disease states: Pathophysiologic considerations. Crit Care Med 11:339–345
17. Prewitt RM, McCarthy J, Wood LDH (1981) Treatment of acute low pressure pulmonary edema in dogs. J Clin Invest 67:409–418
18. Cheney FW, Eisenstein BL, Bishop MJ (1989) Alveolar hypoxia, inhibition of hypoxic pulmonary vasoconstriction, and permeability edema. Respir Physiol 75:11–18
19. Weir EK, Milczoch J, Reeves JT, Grover RF (1976) Endotoxin and prevention of hypoxic pulmonary vasoconstriction. J Lab Clin Med 68:975–983
20. Dantzker DR, Brook CJ, Dehart P, Lynch JP, Weg JG (1979) Ventilation-perfusion distributions in the adult respiratory distress syndrome. Am Rev Respir Dis 120: 1039–1052
21. Reyes A, Roca J, Rodriguez-Roisin R, et al. (1988) Effect of almitrine on ventilation-perfusion distribution in adult respiratory distress syndrome. Am Rev Respir Dis 137: 1062–1067
22. Wysocki M, Delclaux C, Roupie E, et al (1994) Additive effect on gas exchange of inhaled nitric oxide and intravenous almitrine bismesylate in the adult respiratory distress syndrome. Intensive Care Med 20:254–259
23. Barnard JW, Wilson PS, Moore TM, Thompson WJ, Taylor AE (1993) Effect of nitric oxide and cyclooxygenase products on vascular resistance in dog and rat lungs. J Appl Physiol 74:2940–2948
24. Stamler JS, Loh E, Roddy M, Hoffman KE, Creager MA (1994) Nitric oxide regulates systemic and pulmonary vascular resistance in normal subjects. Circulation 89: 2035–2040
25. Chand N, Altura BM (1981) Acetylcholine and bradykinin relax intrapulmonary arteries by acting on endothelial cells: Role in lung vascular diseases. Science 213: 1376–1379
26. Greenberg B, Rhoden K, Barnes PJ (1987) Endothelium-dependent relaxation of human pulmonary arteries. Am J Physiol 252:H434–H438
27. Liu SF, Crawley DE, Barnes PJ, Evans TW (1991) Endothelium-derived relaxing factor inhibits hypoxic pulmonary vasoconstriction in rats. Am Rev Respir Dis 143: 32–37
28. Liu SF, Crawley DE, Evans TW, Barnes PJ (1991) Endogenous nitric oxide modulates adrenergic neural vasoconstriction in guinea-pig pulmonary artery. Br J Pharmacol 104:565–569
29. Cremona G, Higenbottam T, Wood AM, Dinh Xuan AT (1992) importance of endogenous nitric oxide in the adaptation to increased blood flow in isolated human and sheep lungs. Am Rev Respir Dis 145:A206 (Abst)
30. Liu SF, Crawley DE, Rohde JA, Evans TW, Barnes PJ (1992) Role of nitric oxide and guanosine 3′,5′-cyclic monophosphate in mediating nonadrenergic, noncholinergic relaxation in guinea-pig pulmonary arteries. Br J Pharmacol 107:861–866
31. Frostell C, Fratacci MD, Wain JC, Jones R, Zapol WM (1991) Inhaled nitric oxide. A selective pulmonary vasodilator reversing hypoxic pulmonary vasoconstriction. Circulation 83:2038–2047
32. Fratacci MD, Frostell CG, Chen TY, et al. (1991) Inhaled nitric oxide. A selective pulmonary vasodilator of heparin-protamine vasoconstriction in sheep. Anesthesiology 75:990–999
33. Frostell CG, Blomqvist H, Hedenstierna G, Lundberg J, Zapol WM (1993) Inhaled nitric oxide selectively reverses human hypoxic pulmonary vasoconstriction without causing systemic vasodilation. Anesthesiology 78:427–435

34. Liu SF, Dewar A, Crawley DE, Barnes PJ, Evans TW (1992) Effect of tumor necrosis factor on hypoxic pulmonary vasoconstriction. J Appl Physiol 72:1044–1049
35. Tsuchida S, Hiraoka M, Sudo M, Kigoshi S, Muramatsu I (1995) Attenuation of sodium nitroprusside responses after prolonged incubation of rat aorta with endotoxin. Am J Physiol 267:H2305–H2310
36. Aoki N, Siegfried M, Lefer AM (1989) Anti-EDRF effect of tumor necrosis factor in isolated, perfused cat carotid arteries. Am J Physiol 256:H1509–H1512
37. Xie J, Wang Y, Lippton H, et al. (1993) Tumor necrosis factor inhibits stimulated but not basal release of nitric oxide. Am Rev Respir Dis 148:627–636
38. Wang P, Ba ZF, Chaudry IH (1994) Administration of tumor necrosis factor alpha in vivo depresses endothelium-dependent relaxation. Am J Physiol 266:H2535–H2541
39. Spath JA, Sloane PJ, Gee MH, Albertine KH (1994) Loss of endothelium-dependent vasodilation in the pulmonary vessels of sheep after endotoxin. Am J Physiol 76: 361–369
40. Parker JL, Adams HR (1993) Selective inhibition of endothelium-dependent vasodilator capacity by Escherichia coli endotoxemia. Circ Res 72:539–551
41. Parker JL, Keller RS, DeFily DV, Laughlin MH, Adams HR (1991) Coronary vascular smooth muscle function in E. coli endotoxemia in dogs. Am J Physiol 260: H832–H841
42. Umans JG, Wylam ME, Samsel RW, Edwards J, Schumacker PT (1993) Effects of endotoxin in vivo on endothelial and smooth muscle function in rabbit and rat aorta. Am Rev Respir Dis 148:1638–1645
43. Guc MO, Gray GA, Furman BL, Parratt JR (1991) Endotoxin-induced impairment of vasodepressor responses in the pithed rat. Eur J Pharmacol 204:63–70
44. Martin CM, Yaghi A, Sibbald WJ, McCormack D, Paterson NA (1993) Differential impairment of vascular reactivity of small pulmonary and systemic arteries in hyperdynamic sepsis. Am Rev Respir Dis 148:164–172
45. Wylam ME, Samsel RW, Umans JG, et al. (1990) Endotoxin in vivo impairs endothelium-dependent relaxation of canine arteries in vitro. Am Rev Respir Dis 142: 1263–1267
46. Waller J, Gardiner SM, Bennett T (1994) Regional haemodynamic reponses to acetylcholine, methoxamine, salbutamol and bradykinin during lipopolysaccharide infusion in conscious rats. Br J Pharmacol 112:1057–1064
47. Griffiths MJD, Liu S, Curzen NP, Messent M, Evans TW (1995) In vivo treatment with endotoxin-induced nitric oxide synthase in rat main pulmonary artery. Am J Physiol 268:L509–L518
48. Stewart TE, Valenza F, Ribiero SP, et al (1995) Increased nitric oxide in exhaled gas as an early marker of lung inflammation in a model of sepsis. Am J Respir Crit Care Med 151:713–718
49. Kiff RJ, Moss DW, Moncada S (1994) Effect of nitric oxide gas on the generation of nitric oxide by isolated blood vessels: Implications for inhalational therapy. Br J Pharmacol 113:496–498
50. Lorente JA, Landin L, de Pablo R, Renes E, Liste D (1993) L-arginine pathway in the sepsis syndrome. Crit Care Med 21:1287–1295
51. Gossage JR, Christman BW (1994) Mediators of acute and chronic pulmonary hypertension. Sem Respir Crit Care Med 15:190–198
52. Fink MP, O'Sullivan BP, Menconi MJ, et al (1993) A novel leukotriene B4-receptor antagonist in endotoxin shock: A prospective, controlled trial in a porcine model. Crit Care Med 21:1825–1837
53. Christman BW, Lefferts PL, Blair IA, Snapper JR (1990) Effect of platelet-activating factor receptor antagonism on endotoxin-induced lung dysfunction in awake sheep. Am Rev Respir Dis 142:1272–1278
54. Huval NW, Lelcuk S, Shepro D, Hechtman HB (1984) Role of serotonin in patients with acute respiratory failure. Ann Surg 200:166–172
55. Druml W, Steltzer H, Waldhausl W, et al (1993) Endothelin-1 in adult respiratory distress syndrome. Am Rev Respir Dis 148:1169–1173

56. Fink MP, Macvittie TJ, Casey LC (1984) Inhibition of prostaglandin synthesis restores normal haemodynamics in canine hyperdynamic sepsis. Ann Surg 200:619–626
57. Crawley DE, Zhao L, Giembycz MA, et al (1992) Chronic hypoxia impairs soluble guanylyl cyclase-mediated pulmonary arterial relaxation in the rat. Am J Physiol 263: L325–L332
58. Dinh-Xuan AT, Higenbottam TW, Clelland CA, et al (1991) Impairment of endothelium-dependent pulmonary artery relaxation in chronic obstructive lung disease. N Engl J Med 324:1539–1547
59. Fox GA, McCormack DG (1992) The pulmonary physician and critical care. 4. A new look at the pulmonary circulation in acute lung injury. Thorax 47:743–747
60. Weitzberg E (1993) Circulatory responses to endothelin-1 and nitric oxide. Acta Physiol Scand 148:S611 (Abst)
61. Meyer J, Lentz CW, Stothert JC, et al (1994) Effects of nitric oxide synthesis inhibition in hyperdynamic endotoxemia. Crit Care Med 22:306–312
62. Hanly P, Sienko A, Light RB (1987) Effect of cyclooxygenase blockade on gas exchange and hemodynamics in Pseudomonas pneumonia. J Appl Physiol 63: 1829–1836
63. Steinberg SM, Rodriguez JL, Bitzer LG, et al (1990) Indomethacin treatment of human adult respiratory distress syndrome. Circ Shock 30:375–384
64. Bernard GR, Reines HD, Metz CA, et al (1988) Effects of a short course of ibuprofen in patients with severe sepsis. Am Rev Respir Dis 137:A138 (Abst)
65. Leeman M, Boeynaems JM, Degaute JP, Vincent JL, Kahn RJ (1985) Administration of dazoxiben, a selective thromboxane synthetase inhibitor, in the adult respiratory distress syndrome. Chest 87:726–730
66. Ignarro LJ, Burke TM, Wood KS, Wolin MS, Kadowitz PJ (1993) Association between cyclic GMP accumulation and acetylcholine-induced relaxation of bovine intrapulmonary artery. J Pharmacol Exp Ther 228:682–690
67. Compeau CG, Rotstein OD, Tohda H, et al (1994) Endotoxin-stimulated alveolar macrophages impair lung epithelial sodium ion transport by an L-arginine-dependent mechanism. Am J Physiol 266:C1330–C1341
68. Gutierrez HH, Pitt BR, Schwarz M, et al (1995) Pulmonary alveolar epithelial inducible NO synthase gene expression: Regulation by inflammatory mediators. Am J Physiol 268:L501–L508
69. Heiss LN, Lancaster JR, Corbett JA, Goldman WE (1994) Epithelial autotoxicity of nitric oxide: Role in the respiratory cytopathology of pertussis. Proc Natl Acad Sci USA 91:267–270
70. Kwon NS, Stuehr DJ, Nathan CF (1991) Inhibition of tumor cell ribonucleotide reductase by macrophage-derived nitric oxide. J Exp Med 174:761–767
71. Liew FY, Millott S, Parkinson C, Palmer RM, Moncada S (1990) Macrophage killing of Leishmania parasite in vivo is mediated by nitric oxide from L-arginine. J Immunol 144:4794–4797
72. Brune B, Dimmeler S, Vedia LM, Lapetina EG (1994) Nitric oxide: A signal for ADP-ribosylation of proteins. Life Sci 54:61–70
73. Stamler JS, Jaraki O, Osborne J, et al (1992) Nitric oxide circulates in mammalian plasma primarily as an S-nitroso adduct of serum albumin. Proc Natl Acad Sci USA 89:7674–7677
74. Clancy RM, Levartovsky D, Leszcznska-Piziak J, Yegudin J, Abramson SB (1994) Nitric oxide reacts with intracellular glutathione and activates the hexose monophosphate shunt in human neutrophils: Evidence for S-nitrosoglutathione as a bioactive intermediary. Proc Natl Acad Sci USA 91:3680–3684
75. Stadler J, Billiar TR, Curran RD, et al (1991) Effect of exogenous and endogenous nitric oxide on mitochondrial respiration of rat hepatocytes. Am J Physiol 260: C910–C916
76. Beckman JS, Beckman TW, Chen J, Marshall PA, Freeman BA (1990) Apparent hydroxyl radical production by peroxynitrite: Implications for endothelial injury from nitric oxide and superoxide. Proc Natl Acad Sci 87:1620–1624

77. Nguyen T, Brunson D, Crespi CL, et al (1992) DNA damage and mutation in human cells exposed to nitric oxide in vitro. Proc Natl Acad Sci 89:3030–3034
78. Haddad IY, Ischiropoulos H, Holm BA, et al (1993) Mechanisms of peroxynitrite-induced injury to pulmonary surfactants. Am J Physiol 265:L555–L564
79. Rubanyi GM, Ho E, Cantor EH, Lumma WC, Parker Botelho LH (1991) Cytoprotective function of nitric oxide: Inactivation of superoxide radicals produced by human leukocytes. Biochem Biophys Res Commun 181:1392–1397
80. Griffiths MJD, Curzen NP, Sair M, Evans TW (1994) Nitric oxide synthase inhibitors in septic shock: Theoretical considerations. Clin Intensive Care 5:29–36
81. Rossaint R, Falke KJ, Lopez F, et al (1993) Inhaled nitric oxide for the adult respiratory distress syndrome. N Engl J Med 328:399–405
82. Oda H, Nogami H, Kusomoto S, et al. (1976) Long-term exposure to nitric oxide in mice. J Jpn Soc Air Pollut 11:150–160
83. Gerlach H, Pappert D, Lewandowski K, Rossaint R, Falke KJ (1993) Long-term inhalation with evaluated low doses of nitric oxide for selective improvement in oxygenation in patients with adult respiratory distress syndrome. Intensive Care Med 19: 443–449
84. Zapol WM (1993) Minidose inhaled nitric oxide: Less is better. Intensive Care Med 19:433–434
85. Wessel DL, Adatia I, Thompson JE, Hickey PR (1994) Delivery and monitoring of inhaled nitric oxide in patients with pulmonary hypertension. Crit Care Med 22: 930–938
86. Walder CE, Thiemermann C, Vane JR (1991) The involvement of endothelium-derived relaxing factor in the regulation of renal cortical blood flow in the rat. Br J Pharmacol 102:967–973
87. Hutcheson IR, Whittle BJ, Boughton-Smith NK (1990) Role of nitric oxide in maintaining vascular integrity in endotoxin-induced acute intestinal damage in the rat. Br J Pharmacol 101:815–820
88. Umans JG, Samsel RW (1992) L-canavanine selectively augments contraction in aortas from endotoxemic rats. Eur J Pharmacol 210:343–346
89. Griffiths MJD, Messent M, MacAllister RJ, Evans TW (1993) Aminoguanidine selectively inhibits inducible nitric oxide synthase. Br J Pharmacol 110:963–968

Endogenous Nitric Oxide Formation in the Respiratory System

L. E. Gustafsson, P. A. Lönnqvist, and M. G. Persson

Introduction

Synthesis of nitric oxide (NO) derived from L-arginine is known to occur in a number of tissues. NO is believed to exert regulatory functions in the circulatory, nervous and immune systems [1,2]. In the respiratory system, NO is known to modulate vascular tone and has been suggested to modulate the pulmonary hypoxic pressor response *in vitro* and *in vivo* [3,4]. Administration of NO synthase inhibitors have been found to cause hypoxia in anesthetized rabbits [4] and to interfere with the normal vascular adaptation at birth in a sheep model [5]. Early studies showed that NO could be formed in rat lung homogenates, but also that the formation was markedly smaller than in other tissues, and mainly derived from calcium-independent (inducible) NO synthesis (iNOS) [6]. It was therefore important to elucidate whether NO was actually formed in lungs *in vivo*, and how NO formation might be affected by various challenges such as hypoxia or allergen challenge. The discovery of NO in exhaled air [7] has made it possible to directly monitor NO formation in the respiratory system, and we will presently focus mainly on knowledge obtained from this method.

Demonstration of Authentic NO in Exhaled Air

The first study on exhaled NO showed the presence of NO in three species, man, rabbit and guinea pig, and statistical analysis showed that rabbit had higher exhaled concentrations than man, in mixed exhaled air. The exhaled NO was clearly dependent on formation from L-arginine since it was inhibited by guanidino-substituted L-arginine analogs, the effect being reversed by L-arginine but not D-arginine [7]. The demonstration of exhaled NO was first made with chemiluminescence analysis, but the identity of NO was made likely also from chemical trapping experiments followed by nitrate analysis, and by Gc-MS analysis of NO trapped onto an organic compound [7]. The presence of NO in exhaled air has been proven also by direct mass spectrometric analysis on fractions of exhaled gas [8]. A simple test for the presence of NO as cause for measured chemiluminescence in exhaled air has been suggested, using differential freezing out of the NO signal by liquid nitrogen but not by dry ice [7,9].

Origin of Exhaled NO

In the initial study of exhaled NO [7], it was demonstrated that NO in exhaled air persisted for several hours after circulation was stopped. This strongly indicated that NO derived from the respiratory system, and was not a passive excretion product from other organs. Definite proof of origin in the respiratory system has been obtained by demonstration of NO in the exhaled gas from perfused lungs [9]. An interesting aspect of these studies is that NO increased upon injection of helium intravenously in order to halt the circulation [7], or when blood perfusion was stopped by introducing saline perfusion [9]. The most likely explanation for this is that pulmonary NO is formed to a large extent by cells in the air-carrying parts of the lungs and is eliminated by uptake and reaction with hemoglobin. This notion has strong support from single-breath data demonstrating lower NO concentrations in alveolar air, than in air exhaled from the airways [10]. The extent of elimination by uptake onto hemoglobin is probably quite large, since exhaled NO more than doubles upon switching to saline perfusion of rabbit lungs (Fig. 1) [9].

The initial single-breath experiments clearly indicated an airway origin of exhaled NO [10] . This does not allow any conclusion on the cell of origin of exhaled NO, and does not exclude NO formation in the alveoli, although very little alveolar NO should be expected to be found in exhaled air under

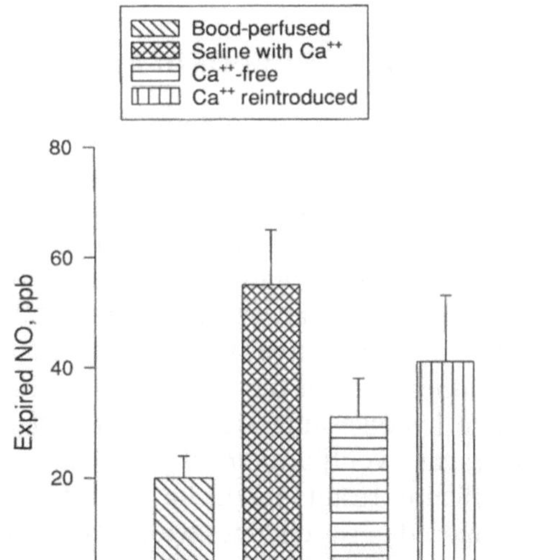

Fig. 1. Exhaled NO from rabbit lungs. Rabbits were anesthetized with sodium pentobarbital. The trachea was cannulated and the animals were artificially ventilated. NO in mixed exhaled air was determined by chemiluminescence during normal blood perfusion (▧) or after perfusion of the lungs with saline buffer had been commenced (▩). Shown is also the effect of switching from perfusion with normal calcium (▤) to a solution without calcium, and the switching back to calcium-containing perfusate (▥)

normal conditions due to the uptake by red blood cells. NOS has been found in nerves in the lungs [11–15]. Airway epithelial cells must also be considered as possible sources of exhaled NO [10]. Accordingly, NOS in epithelial cells has been demonstrated immunohistochemically using antibodies or probes both for constitutive (cNOS) [12, 16] and inducible (iNOS) [14, 16, 17] NOS. An interesting possibility is that iNOS might be present in healthy airway mucosa of humans [14].

We have examined the effects of various inhibitors on pulmonary NO formation as determined from exhaled NO [9]. The obtained pattern, together with that from earlier studies [7] does not lend support for any specific known type of enzyme involved. Instead, it might be that exhaled NO derives from both calcium-dependent and calcium-independent NOS. In agreement, we found that calcium depletion decreased but did not abolish exhaled NO. The effect was partially reversed upon reintroduction of calcium in the perfusion medium (Fig. 1). Another clue to the enzyme involved might be the demonstration of a dependence on bacterial presence. Among healthy rats, germfree animals had similar exhaled NO concentrations as did control animals [9], making a marked bacterial influence in the lower airways of these animals less likely in the normal situation. In the human, we have been able to demonstrate high upper airway NO concentrations already at birth in babies delivered by cesarean section [18]. Thus, we may firmly state that if iNOS is involved in respiratory system NO synthesis, this enzyme(s) is present constitutively, and is not critically dependent on the presence of bacteria.

Regulation and Effects of Pulmonary NO Formation

Since NOS inhibitors can enhance hypoxic pulmonary vasoconstriction, it is of importance to consider how NO formation is affected by hypoxia. Direct measurement of exhaled NO has demonstrated that exhaled NO decreases during inhalation of hypoxic gas [7]. This lends support to the suggestion that pulmonary NO counterbalances hypoxic vasoconstriction, acting to improve pulmonary ventilation/perfusion matching [4]. In agreement, inhalation of NO counteracts hypoxic vasoconstriction in animals and man [19–21]. On the other hand, it has been suggested that pulmonary NO formation might dilate hypoxic lung areas [22], an effect which could at least theoretically aggravate ventilation/perfusion disturbances. A reason for the seemingly conflicting results might be that, in the latter study, NO mechanisms were compared between lungs which were subjected to anoxia or to 100% oxygen. We have found that the vasoconstricting effect of NOS inhibition is more or less abolished by inhalation of oxygen [21]. In consequence, to understand the physiological role of endogenous NO, we need to study the pulmonary circulation under normoxic or hypoxic conditions. Our finding also suggests that NO closely interacts with the hypoxic pulmonary vasoconstrictor mechanism. One possible pathway for the interaction is that of NO and oxygen function on a common target molecule, for example a guanylate cyclase or another molecule with a regulatory heme unit. Such a target molecule might be localized directly in the smooth musculature, with the task of directly regulating vascular muscle tone. Another possibility is that oxygen availability controls the release of a vasoconstrictor principle from cells e.g. in the airway mucosa [23], and that endogenously formed NO modulates the release or the action of such a principle. In support for such a notion, NO has been found to regulate endothelin, a known pulmonary vasoconstrictor. Thus, NO inhibited the formation of endothelin in vascular endothelial cells, and inhibited hypoxia-induced increased expression of endothelin [24].

It was mentioned above that NO might be involved in the vascular adaptation at birth [5]. A prominent feature of this period is the sudden distension of the lungs, and the increased pulmonary oxygen concentration. Our data suggest that increase in oxygen from a state with relative hypoxia will lead to increased NO formation, and that hyperoxia will not further increase NO, tentatively supporting the possibility of NO involvement in neonatal vascular adaptation. Furthermore, we have been able to demonstrate that stretching the lungs by application of positive end-expiratory pressure (PEEP) will lead to increased pulmonary NO formation *in vivo* [25]. In addition, stretching the lungs by applying negative end-tidal extrathoracic pressure (NEEP) will also enhance pulmonary NO formation (Fig. 2) (unpublished results). A neuronal component might be involved in the stretch-effect of PEEP, since increased exhaled NO was blunted by severing of vagus and depressor nerves [25]. Further studies are needed to establish the role of NO during neonatal adaptation, but inhalation of 5–20 ppm NO in fetal lambs can fully substitute for the vasodilating effect of an FiO_2 of 1 [26]. The obser-

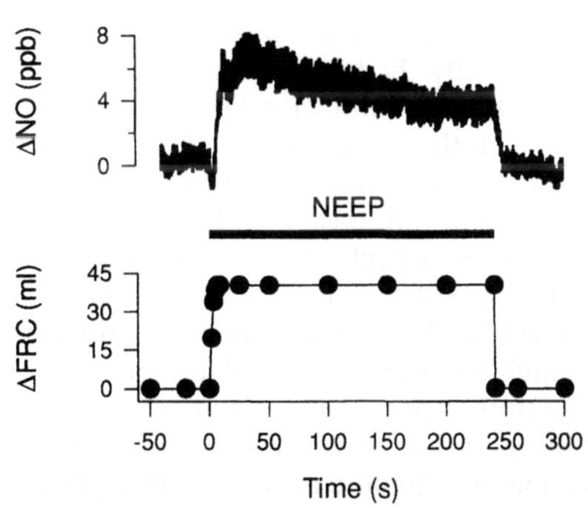

Fig. 2. Anesthetized rabbit on artificial ventilation by means of intermittent positive pressure ventilation. The animal was enclosed in an acrylic chamber, and during the period indicated by the horizontal bar, the animal was subjected to a negative extrathoracic end-expiratory pressure (NEEP). This led to a distension of the lungs as indicated by an increase in functional residual capacity (FRC) determined by a pneumotachygraph (*bottom trace*) and to an increase in exhaled NO (ppb) determined in the tracheal gas (*top trace*)

vation of PEEP-induced effects on pulmonary NO formation might also have bearing on the critical care situation where PEEP is often used to improve oxygenation. In the rabbit model, we were able to demonstrate a beneficial effect of PEEP on oxygenation, and this effect remained in the presence of a maximal dose of an NOS inhibitor [25]. Thus, increased NO formation is not a critical mediator for the acute effect of PEEP on oxygenation.

We have discussed above the origin of exhaled pulmonary NO, showing that it might derive both from calcium-dependent and calcium-independent sources, but that pulmonary NO is normally not critically dependent on induction by bacteria. However, administration of bacterial toxins can cause increased exhaled NO in rats [27]. This was suggested to be an early marker of lung inflammation, since it was evident already at 100 min after *Salmonella typhosa* lipopolysaccaride (LPS) infusion, and since hypotension alone did not cause increase in exhaled NO. It is quite possible that increased pulmonary NO formation upon stimulation with endotoxins [4, 27] is triggered by the action of cytokines, since cytokines can increase the expression of iNOS and formation of NO in lung cells [16, 17].

Pulmonary NO and Asthma

Reaction to allergen such as in asthma is another mechanism involving inflammation, and might thus be expected to cause changes in NO formation. Asthma is of further interest since inhibitory nerves of the airways might utilize NO [28, 29] and since inhalation of NO can cause bronchodilation [30, 31]. In sensitized guinea pigs, we found that inhalation of antigen (ovalbumin) caused a rapid and marked increase in exhaled NO [32]. Most importantly, the bronchial obstruction caused by the allergen was enhanced by NOS inhibition, clearly indicating that endogenous NO was acting as a brake on the acute experimental asthma reaction [33, 34]. Further studies in humans [35] showed that asthmatic patients had increased exhaled NO concentrations, findings simultaneously reported by other groups [36, 37]. An interesting aspect is that asthmatics treated with steroids were found to exhale less NO than untreated patients, possibly suggesting an involvement of iNOS [37]. Steroid treatment was also found to counteract cytokine-induced stimulation of iNOS in isolated bronchial epithelial cells [17].

Another possibility is that steroids depressed the mediator release reactions and thus depressed NO originating from formation subsequent to the action of asthmatic mediators. In support for this is an animal study showing that histamine and leukotriene antagonists can abolish both the bronchial obstruction response and the increased exhaled NO upon allergen inhalation [38]. The initial studies on allergen challenge in animals [32] showed that after the first rapid increase in exhaled NO followed a period of decreased exhaled NO. It is thus possible that a complex pattern might occur during asthma, with periods of both increased and decreased NO formation, and involvement of both constitutive and inducible enzymes. Further studies in humans are necessary to clarify the exact role of NO in asthma. The weak bronchodilator effect of inhaled NO in humans [39] might either be taken to indicate that NO has little effect on asthmatic bronchial obstruction, or that endogenous NO is already maximally active. Studies with NOS inhibitors followed by addition of inhaled NO might be one way to elucidate this, although care should be taken since animal studies indicate that marked enhancement of bronchial obstruction might be seen upon maximal inhibition of NOS.

Upper versus Lower Airway Formation, and Autoinhalation of NO

The initial studies on single-breath measurements of exhaled NO pioneered the work on respiratory system NO formation in that they made it clear that exhaled NO derived mainly from the airways [10]. The instrumentation used employed carbon dioxide analysis and the procedure involved breath-holding. During such measures, it is not unlikely that carbon dioxide will be added to dead space from various sources, and the technique will not allow an exact localization of NO formation within the airways. It soon became evident that NO can be added to pulmonary NO not only from stomach gas [40], but also from gas in the pharynx or the nose [36, 41–43]. It is quite possible that the nasal NO can reach concentrations high enough for evoking pulmonary vasodilation when inhaled into the lungs. Intermittent concentrations in the pharynx above 0.1 ppm have thus been demonstrated [18, 41]. Of even further importance is that concentrations of 2–4 ppm can be demonstrated in the nose already at birth [18]. This has two major implications. First, the toxicology of NO will have to be re-evaluated, taking these high endogenous concentrations into account. It might implicate that NO at these concentrations is not toxic in itself, but only after conversion to higher oxides, or other derivatives. Second, it means that autoinhalation occurs already at birth, with possible effects on vascular adaptation, and that this autoinhalation primarily occurs independent of bacterial colonization [18], which otherwise has been suggested for the adult situation [41]. The observation does not exclude subsequent bacterial stimulation, since nasal NO production increased during the first 24 h of extrauterine life [18]. That microbial or other challenges may be important in stimulating NO formation in proximal airways is evident from the finding of higher tracheal NO concentrations in chronically tracheostomized patients than in normal controls [43].

Non-arginine Dependent NO Exhalation

Exhaled NO can be used for monitoring NO formation from infused NO donating vasodilator drugs [44–46]. In rabbits, it has been shown that the potency of vasodilators in depressing blood pressure and in inhibiting smooth muscle contractility *in vitro* correlates with the capacity to stimulate exhaled NO in the presence of inhibitors of endogenous formation [45]. Tachyphylaxis to nitrovasodilators is a well known phenomenon. We have been able to demonstrate in rabbits that tachyphylaxis *in vivo* is directly correlated to a decreased NO excretion from the infused nitrovasodilator [44]. This type of measurement also offers a way of monitoring tachyphylaxis mechanisms in humans [46].

Measurement of exhaled NO can monitor effects of inhibitors of NO formation. An indication of this is the finely graded effect of substituted L-arginines on exhaled NO in rabbits and guinea pigs [7, 9]. That the exhaled NO concentrations are affected by endogenous inhibitors is suggested from the observation that administration of L-arginine can cause increments in exhaled NO [9]. Also non-arginine compounds might be of interest to study possible inhibitors, since ethanol has been found to decrease exhaled NO in animals and man [47, 48].

Exhaled NO in Special Situations

It has been shown that endogenous NO concentration drops in exhaled air during hyperventilation and exercise [10], but that the total amount of exhaled NO increases in exercise [10, 49, 50]. One explanation for this might be an increased blood flow or diffusional area in the lungs, since exhaled NO in humans is correlated to pulmonary diffusion capacity [51]. However, if endogenous NO is rapidly taken up in the alveoli, as discussed above, increased blood flow or diffusional capacity in alveoli seem unlikely explanations. This does not rule out an increased diffusion from increasingly exposed non-alveolar areas. An effect of blood flow seems less probable also from experimental data on the lack of blood flow effects [25, 27]. The increased stretching of the lung parenchyma during increased breathing should be expected to increase pulmonary NO formation, as judged from the effect of PEEP in animals [25].

Decreased exhalation of NO has been observed in smokers [32] and in a group of patients with pulmonary hypertension [51]. Abolition of nasal NO production has been found in patients with Kartagener's syndrome [42], a disease where ciliated cells are abnormal and exhibit depressed ciliary activity. This suggests that endogenous NO has a role in the control of ciliary function, since NO has been found to stimulate ciliary activity in airway epithelium [52, 53].

Conclusion

Measurement of exhaled NO constitues the first experimental situation where authentic NO has been unequivocally demonstrated to be formed *in vivo* in humans. It offers a possibility to monitor, by on-line biochemistry, the activity in an important messenger and host defense system, and also the bioactivation and effect of drugs. It has already proven to be of value in the study of asthma, and we hypothesize that in the future it may become a standard clinical procedure. Exhaled NO from the lower airways and lungs is a result of NO synthesis which is regulated by availability of oxygen, by stretching of the lungs, and by cytokines. Exhaled NO is a reflection of a larger NO synthesis, but where a significant amount of this NO is rapidly taken up by lung blood flow. We propose that regulation of lung blood flow is perhaps the most important role of endogenous pulmonary and airway NO formation, but that NO clearly has a role also in respiratory system host-defense mechanisms, as originally suggested.

Acknowledgments: We would like to thank Drs Ulla Schedin and Claes Frostell for valuable discussions. Supported by the Swedish National Environment Protection Board, the Swedish Heart-Lung Foundation, The AGA Medical Research Fund, the Swedish Medical Research Council (proj 7919), Tore Nilsson, Magnus Bergvall and Fraenkel Foundations, and the Karolinska Institute.

References

1. Moncada S, Palmer RMJ, Higgs EA (1991) Nitric oxide: Physiology, pathophysiology and pharmacology. Pharmacol Rev 43:109–142
2. Moncada S, Higgs A (1993) The L-arginine-nitric oxide pahway. N Engl J Med 329: 2002–2012
3. Archer SL, Tolins JP, Raij L, Weir EK (1989) Hypoxic pulmonary vasoconstriction is enhanced by inhibition of the synthesis of an endothelium-derived relaxing factor. Biochem Biophys Res Commun 164:1198–1205
4. Persson MG, Gustafsson LE, Wiklund NP, Moncada S, Hedqvist P (1990) Endogenous nitric oxide as a probable modulator of pulmonary circulation and hypoxic pressor response in vivo. Acta Physiol Scand 140:449–457
5. Abman SH, Chatfield BA, Hall SL, McMurtry IF (1990) Role of endothelium-derived relaxing factor during transition of pulmonary circulation at birth. Am J Physiol 259: H1921–H1927
6. Palacios M, Knowles RG, Moncada S (1992) Enhancers of nonspecific immunity induce nitric oxide synthase: Induction does not correlate with toxicity or adjuvancy. Eur J Immunol 22:2303–2307
7. Gustafsson LE, Leone AM, Persson MG, Wiklund NP, Moncada S (1991) Endogenous nitric oxide is present in the exhaled air of rabbits, guinea pigs and humans. Biochem Biophys Res Commun 181:852–857
8. Leone AM, Gustafsson LE, Francis PL, Persson MG, Wiklund NP, Moncada S (1994) Nitric oxide is present in exhaled breath in humans: Direct GcMs confirmation. Biochem Biophys Res Commun 201:883–887
9. Persson MG, Midtvedt T, Leone AM, Gustafsson LE (1994) Ca^{2+}- and Ca^{2+}-independent exhaled nitric oxide, presence in germ-free animals, and inhibition by arginine analogues. Eur J Pharmacol 264:13–20
10. Persson MG, Wiklund NP, Gustafsson LE (1993) Endogenous nitric oxide in single exhalations and the change during exercise. Am Rev Respir Dis 148:1210–1214
11. Moncada S (1992) The 1991 Ulf von Euler lecture. The L-arginine: nitric oxide pathway. Acta Physiol Scand 145:201–227

12. Schmidt HHHW, Gagne GD, Nakane M, Pollock JS, Miller MF, Murad F (1992) Mapping of neural nitric oxide synthase in the rat suggests frequent co-localization with NADPH diaphorase but not with soluble guanylyl cyclase, and novel paraneural functions for nitrinergic signal transduction. J Histochem Cytochem 40:1439–1456
13. Fischer AP, Mundel P, Mayer B, Preissler U, Philippin B, Kummer W (1993) Nitric oxide synthase in guinea pig lower airway innervation. Neurosci Lett 149:157–160
14. Kobzik L, Bredt DS, Lowenstein CJ, et al (1993) Nitric oxide synthase in human and rat lung: Immunocytochemical and histochemical localization. Am J Respir Cell Mol Biol 9:371–377
15. Grozdanovic Z, Bruning G, Baumgarten HG (1994) Nitric oxide – a novel autonomic neurotransmitter. Acta Anat 150:16–24
16. Asano K, Chee CB, Gaston B, et al (1994) Constitutive and inducible nitric oxide synthase gene expression, regulation, and activity in human lung epithelial cells. Proc Natl Acad Sci USA 91:10089–10093
17. Robbins RA, Barnes PJ, Springall DR, et al (1994) Expression of inducible nitric oxide in human lung epithelial cells. Biochem Biophys Res Commun 203:209–218
18. Schedin U, Norman M, Gustafsson LE, Herin P, Frostell C (1995) High endogenous nitric oxide in the upper airways of healthy new-born infants. Br J Anaesth (In press)
19. Frostell C, Fratacci MD, Wain JC, Jones R, Zapol WM (1991) Inhaled nitric oxide. A selective pulmonary vasodilator reversing hypoxic pulmonary vasoconstriction. Circulation 83:2038–2047
20. Frostell CG, Blomqvist H, Hedenstierna G, Lundberg J, Zapol WM (1993) Inhaled nitric oxide selectively reverses human hypoxic pulmonary vasoconstriction without causing systemic vasodilation. Anesthesiology 78:427–435
21. Persson MG, Kalze'n H, Gustafsson LE (1994) Oxygen or low concentrations of nitric oxide reverse pulmonary vasoconstriction induced by nitric oxide synthesis inhibition in rabbits. Acta Physiol Scand 150:405–411
22. Sprague RS, Thiemermann C, Vane JR (1992) Endogenous endothelium-derived relaxing factor opposes hypoxic pulmonary vasoconstriction and supports blood flow to hypoxic alveoli in anesthetized rabbits. Proc Natl Acad Sci USA 89:8711–8715
23. Youngson C, Nurse C, Yeger H, Cutz E (1993) Oxygen sensing in airway chemoreceptors. Nature 365:153–155
24. Kourembanas S, McQuillan LP, Leung GK, Faller DV (1993) Nitric oxide regulates the expression of vasoconstrictors and growth factors by vascular endothelium under both normoxia and hypoxia. J Clin Invest 92:99–104
25. Persson MG, Lönnqvist PA, Gustafsson LE (1995) Positive end-expiratory pressure ventilation elicits increases in endogenously formed nitric oxide as detected in air exhaled by rabbits. Anesthesiology 82 (In press)
26. Kinsella JP, McQueston JA, Rosenberg AA, Abman SH (1992) Hemodynamic effects of exogenous nitric oxide in ovine transitional pulmonary circulation. Am J Physiol 263:H875–H880
27. Stewart TE, Valenza F, Ribeiro SP, et al (1995) Increased NO in exhaled gas as an early marker of lung inflammation in a model of sepsis. Crit Care Med (Suppl 23): A170 (Abst)
28. Li CG, Rand MJ (1991) Evidence that part of the NANC relaxant response of guinea-pig trachea to electrical field stimulation is mediated by NO. Br J Pharmacol 102:91–94
29. Belvisi MG, Stretton CD, Yacoub M, Barnes PJ (1992) Nitric oxide is the endogenous neurotransmitter of bronchodilator nerves in humans. Eur J Pharmacol 210:221–222
30. Dupuy PM, Shore SA, Drazen JM, Frostell C, Hill WA, Zapol WM (1992) Bronchodilator action of inhaled nitric oxide in guinea pigs. J Clin Invest 90:421–428
31. Högman M, Frostell C, Arnberg H, Hedenstierna G (1993) Inhalation of nitric oxide modulates methacholine-induced bronchoconstriction in the rabbit. Eur Resp J 6: 177–180
32. Persson MG, Gustafsson LE (1993) Allergen-induced airway obstruction in guinea-pigs is associated with changes in nitric oxide levels in exhaled air. Acta Physiol Scand 149:461–466

33. Persson MG, Friberg SG, Hedqvist P, Gustafsson LE (1993) Endogenous nitric oxide counteracts antigen-induced bronchoconstriction. Eur J Pharmacol 249:R7–R8
34. Nijkamp FP, Van der Linde HJ, Folkerts G (1993) Nitric oxide synthesis inhibitors induce airway hyperresponsiveness in the guinea pig in vivo and in vitro. Role of the epithelium. Am Rev Respir Dis 148:727–734
35. Persson MG, Zetterström O, Agrenius V, Ihre E, Gustafsson LE (1994) Single-breath nitric oxide measurements in asthmatic patients and smokers. Lancet 343:146–147
36. Alving K, Weitzberg E, Lundberg JM (1993) Increased amount of nitric oxide in exhaled air of asthmatics. Eur Respir J 6:1368–1370
37. Kharitonov SA, Yates D, Robbins RA, Logan-Sinclair R, Shinebourne EA, Barnes PJ (1994) Increased nitric oxide in exhaled air of asthmatic patients. Lancet 343:133–135
38. Persson MG, Friberg SG, Gustafsson LE, Hedqvist P (1995) Endogenous nitric oxide promotes patent airways and counteracts allergen-induced bronchoconstriction. Br J Pharmacol (In press)
39. Högman M, Frostell CG, Hedenström H, Hedenstierna G (1993) Inhalation of nitric oxide modulates adult human bronchial tone. Am Rev Respir Dis 148:1474–1478
40. Lundberg JON, Weitzberg E, Lundberg JM, Alving K (1994) Intragastric nitric oxide production in humans: Measurements in expelled air. Gut 35:1543–1546
41. Gerlach H, Rossaint R, Pappert D, Knorr M, Falke KJ (1994) Autoinhalation of nitric oxide after endogenous synthesis in nasopharynx. Lancet 343:518–519
42. Lundberg JON, Weitzberg E, Nordvall SL, Kuylenstierna R, Lundberg JM (1994) Primarily nasal origin of exhaled nitric oxide and absence in Kartagener's syndrome. Eur Resp J 7:1501–1504
43. Schedin U, Frostell C, Persson MG, Jakobsson J, Andersson G, Gustafsson LE (1995) Contribution from upper and lower airways to exhaled endogenous nitric oxide in humans. Acta Anaesthesiol Scand (In press)
44. Persson MG, Agvald P, Gustafsson LE (1994) Detection of nitric oxide in exhaled air during administration of nitroglycerin in vivo. Br J Pharmacol 111:825–828
45. Cederqvist B, Persson MG, Gustafsson LE (1994) Direct demonstration of NO formation in vivo from organic nitrites and nitrates, and correlation to effects on blood pressure and to in vitro effects. Biochem Pharmacol 47:1047–1053
46. Husain M, Adrie C, Ichinose F, Kavosi M, Zapol WM (1994) Exhaled nitric oxide as a marker for organic nitrate tolerance. Circulation 89:2498–2502
47. Persson MG, Gustafsson LE (1992) Ethanol can inhibit nitric oxide formation. Eur J Pharmacol 224:99–100
48. Persson MG, Cederqvist B, Wiklund CU, Gustafsson LE (1994) Ethanol causes decrements in airway excretion of endogenous nitric oxide in humans. Eur J Pharmacol 270:273–278
49. Bauer JA, Wald JA, Doran S, Soda D (1994) Endogenous nitric oxide in expired air: Effects of acute exercise in humans. Life Sci 55:1903–1909
50. Iwamoto J, Pendergast DR, Suzuki H, Krasney JA (1994) Effect of graded exercise on nitric oxide in expired air in humans. Respir Physiol 97:333–345
51. Cremona G, Higenbottam T, Borland C, Mist B (1994) Mixed expired nitric oxide in primary pulmonary hypertension in relation to lung diffusion. QJM 87:547–551
52. Jain B, Rubinstein I, Robbins RA, Leise KL, Sisson JH (1993) Nitric oxide modulates ciliary activity in airway epithelium. Am Rev Respir Dis 147 (Suppl):A937 (Abst)
53. Jain B, Rubinstein I, Robbins RA, Leise KL, Sisson JH (1993) Modulation of airway epithelial cell ciliary beat frequency by nitric oxide. Biochem Biophys Res Commun 191:83–88

Pulmonary Vascular Regulation by Endogenous Nitric Oxide

M. Leeman

Introduction

Endothelial cells contribute to the control of vascular tone by releasing autocrine and paracrine substances that mediate contraction (endothelins, endothelium-derived contracting factor) or relaxation (nitric oxide, prostacyclin, endothelium-derived hyperpolarizing factor) of the vascular smooth muscle [1–3].

In response to several stimuli or mediators, such as shear stress, acetylcholine and bradykinin, nitric oxide (NO) is synthesized from L-arginine within the endothelial cell by a calcium/calmodulin-dependent constitutive NO synthase (cNOS). NO diffuses into the underlying smooth muscle cells where it activates the soluble guanylate cyclase to generate cGMP, the second messenger of vascular relaxation. In the systemic circulation, small amounts of endothelium-derived NO seems continuously released for minute adjustments of vascular tone. Several cell types, such as macrophages and vascular smooth muscle cells, can express an *inducible* form of NOS (iNOS) in response to endotoxin or cytokines. iNOS releases large amount of NO that plays a major role in the host-defense response and that could contribute to the systemic hypotension in circulatory shock [2, 3].

Acetylcholine-induced relaxation of systemic isolated arteries does not occur when the endothelium is stripped from the vessel [4]. Acetylcholine (as well as bradykinin) is considered as an endothelium-dependent dilator, whereas exogenous nitrates, which directly activate guanylate cyclase by releasing NO from their molecule, are considered as endothelium-independent dilators [2–4].

The effects of endogenous NO release can be studied by examining the responses to inhibitors of NOS (such as L-arginine analogs) or action (such as methylene blue, a soluble guanylate cyclase inhibitor, or hemoglobin, a NO scavenger). On the other hand, endothelial dysfunction reflecting altered L-arginine-NO pathway activity (as occurs in systemic arterial hypertension) can be demonstrated by a decreased cGMP content in the vascular wall or by a reduced response to endothelium-dependent vasodilators, contrasting with a preserved response to endothelium-independent vasodilators.

Is NO an Endothelium-Dependent Pulmonary Vasodilator?

Greenberg et al. [5] repeated the experiments of Furchgott and Zawadzki [4] with human pulmonary arteries obtained from resected lungs. As observed in systemic arteries [4], endothelium removal abolished the relaxation to acetylcholine of precontracted pulmonary artery rings, whereas the relaxation in response to vasoactive intestinal peptide was not affected [5].

When administered in spontaneously breathing newborn lambs with pulmonary hypertension induced by hypoxia, by the thromboxane A2 analog U-46619 or by N^G-nitro-L-arginine (L-NNA), exogenous L-arginine, the NO precursor, produced potent pulmonary vasodilation [6]. This effect was inhibited by methylene blue and potentiated by a cGMP phosphodiesterase inhibitor [6].

NO inhibitors, such as L-NNA [7], methylene blue [8, 9] and N^G-nitro-L-arginine methyl ester (L-NAME) [10] blunted the endothelium-dependent relaxation of acetylcholine and/or bradykinin in lambs [6–8] and cats [9, 10]. Of note, after NOS inhibition, the vasodilator response to sodium nitroprusside was enhanced, suggesting that reduced stimulation by endogenous NO could increase the sensitivity of guanylate cyclase to exogenous NO [10].

Very recently, using a modified chemiluminescence assay, NO_2/NO_3 (the degradation products of endogenous NO) have been identified in the effluent of isolated rat lungs, especially when exposed to hypoxia [11].

In conclusion, these data indicate that endothelium-dependent relaxation occurs in the pulmonary circulation and that it is mediated by the release of NO.

Healthy Pulmonary Circulation

Normoxia

It is tempting to suggest that a continuous release of NO is responsible for the low basal pulmonary vascular tone. This issue remains however controversial. Several inhibitors of NOS have been shown to enhance pulmonary vascular tone (Table 1) [7–21]. For example, in intact lambs, L-NNA dose-dependently increased pulmonary artery pressure (with no change in cardiac output), an effect prevented by L-arginine but not by D-arginine [7].

Recently, Stamler et al. [20] invasively studied the hemodynamic effects of intravenous N^G-monomethyl-L-arginine (L-NMMA) in 11 healthy volunteers. This NOS inhibitor decreased cardiac index from 3.6 to 2.6 $L \cdot min^{-1} \cdot m^{-2}$, while mean pulmonary artery pressure and pulmonary artery occluded pressure remained unchanged at 13 and 9 mm Hg, respectively, leading to a significant increase in calculated pulmonary vascular resistance [20]. However, changes in pulmonary vascular resistance at variable cardiac output levels are of uncertain interpretation, because the pulmonary vascular pressure-flow relationship is not strictly linear and does not always

Table 1. NO inhibitors increase basal pulmonary vascular tone

Authors	Ref.	Model	Inhibitors
Fineman et al.	[7]	Intact lambs	L-NNA
Fineman et al.	[8]	Intact lambs	MB
Hyman et al.	[9]	Closed-chest cats	MB
McMahon et al.	[10]	Cat lungs	L-NAME
Isaacson et al.	[11]	Rat lungs	L-NAME
Archer et al.	[12]	Rat PA	L-NMMA
Archer et al.	[13]	Rat PA	MB
Rodman et al.	[14]	Rat PA	MB, Hb
Wiklund et al.	[15]	Anesthetized rabbits	L-NAME
Perella et al.	[16]	Anesthetized dogs	L-NMMA
Barnard et al.	[17]	Rat lungs	L-NAME
Hampl et al.	[18]	Rat lungs, intact rats	L-NAME
Koizumi et al.	[19]	Intact sheep	L-NNA
Stamler et al.	[20]	Healthy volunteers	L-NMMA
Celermajer et al.	[21]	Normal children	L-NMMA

PA: pulmonary arteries; L-NNA: N^G-nitro-L-arginine; L-NAME: N^G-nitro-L-arginine methyl ester; L-NMMA: N^G-monomethyl-L-arginine; MB: methylene blue; Hb: hemoglobin

cross the origin [22, 23]. Moreover, changes in pulmonary blood flow can affect NO release [24]. Thus, whether L-NMMA increases pulmonary vascular tone in normal men is unsettled.

In more controlled experimental condition NO inhibitors did not modify basal pulmonary vascular tone (Table 2) [12, 17, 25–31]. Nishiwaki et al. [29] reported that L-NNA (at a dosage that abolished the bradykinin-induced pulmonary vasodilation) had no effect on the pulmonary vascular pressure-flow relationship in chronically instrumented dogs breathing room air [29]. We [31] recently examined the pulmonary vascular effects of compounds inhibiting different steps of the L-arginine-NO pathway, L-NNA and methylene blue, in intact anesthetized and ventilated dogs. Cardiac output was kept constant (by opening a femoral arteriovenous bypass or by inflating an inferior vena cava balloon catheter) to avoid passive flow-dependent changes in pulmonary vascular pressures [22, 23] and to avoid flow-dependent changes in NO release [24]. Inspiratory fraction of O_2 (FiO_2) varied between 1 to 0.1, allowing to study the effects of endogenous NO in hyperoxia, normoxia and hypoxia in the same dogs. Neither inhibitor significantly influenced the pulmonary driving pressure during hyperoxia and normoxia (Fig. 1), thus suggesting that endogenous NO does not contribute to the low basal pulmonary vascular tone [31]. In accordance, when examined in the lungs from normoxic rats using immunohistochemistry and NADPH diaphorase staining techniques, the expression of NOS was detected in the endothelium of large pulmonary arteries but not in the endothelium of small pulmonary resistance vessels [32].

Table 2. NO inhibitors do not influence basal pulmonary vascular tone

Authors	Ref.	Model	Inhibitors
Archer et al.	[12]	Rat lungs	L-NMMA
Barnard et al.	[17]	Dog lungs	L-NAME
Brashers et al.	[25]	Rat lungs	Several
Mazmanian et al.	[26]	Rat lungs	MB
Hasunuma et al.	[27]	Rat lungs	L-NMMA, Hb
Liu et al.	[28]	Rat lungs	L-NMMA
Nishiwaki et al.	[29]	Conscious dogs	L-NNA
Ogata et al.	[30]	Porcine PA	L-NMMA, Hb
Leeman et al.	[31]	Anesthetized dogs	L-NNA, MB

Abbreviations as in Table 1

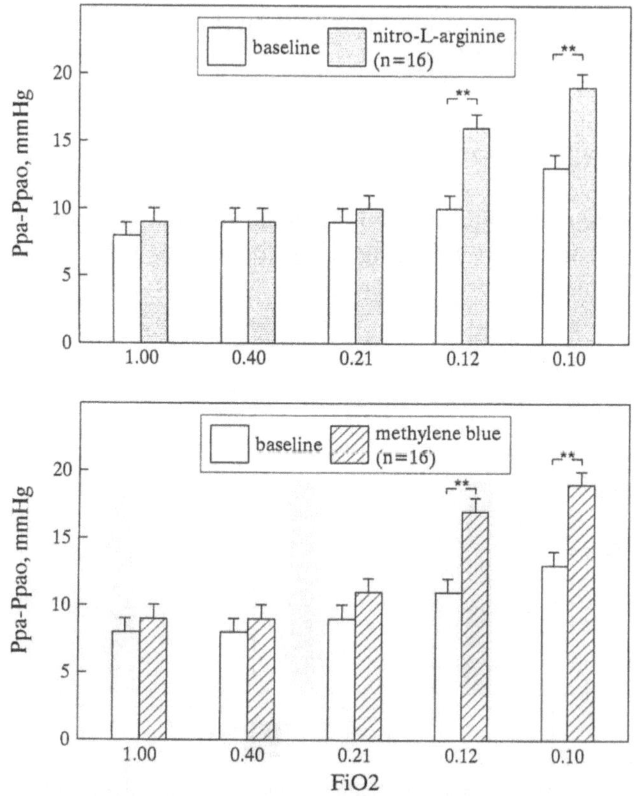

Fig. 1. Pulmonary artery pressure (Ppa) minus occluded Ppa (Ppao) gradient measured at constant cardiac output before and after N^G-nitro-L-arginine (n = 16 dogs) or methylene blue (n = 16 dogs). The (Ppa-Ppao) gradient increased during hypoxia. NO inhibition had no effect on hyperoxic or normoxic (Ppa-Ppao) but further increased it in hypoxia (** $p<0.01$) (Adapted from [30])

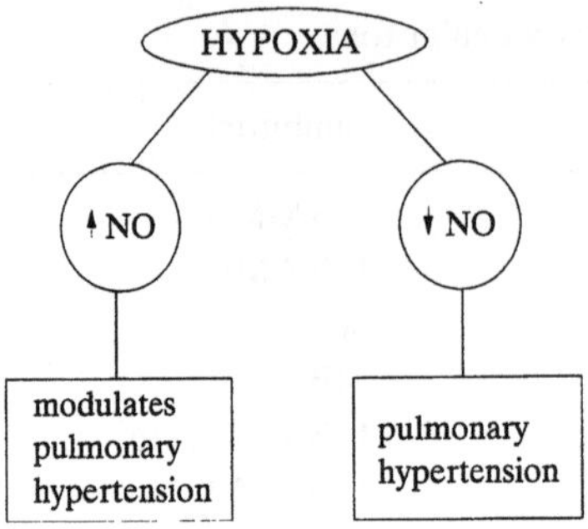

Fig. 2. Possible effects of hypoxia on the L-arginine-NO pathway

In summary, there is no definitive answer on the role of NO in maintaining low basal pulmonary vascular resistance. Part of the inconsistent results yielded in the literature could be species-related. Barnard et al. [17] reported that L-NAME decreased flow in constant-pressure blood-perfused rat lungs (indicating vasoconstriction) but not in dog lungs [17]. In this study, the cyclooxygenase inhibitor indomethacin decreased blood flow in dog lungs but not in rat lungs.

Acute Hypoxia

The contribution of NO in acute hypoxic pulmonary vascular tone has been reported variably (Fig. 2). Exposure of rat [14] or rabbit [33] pulmonary

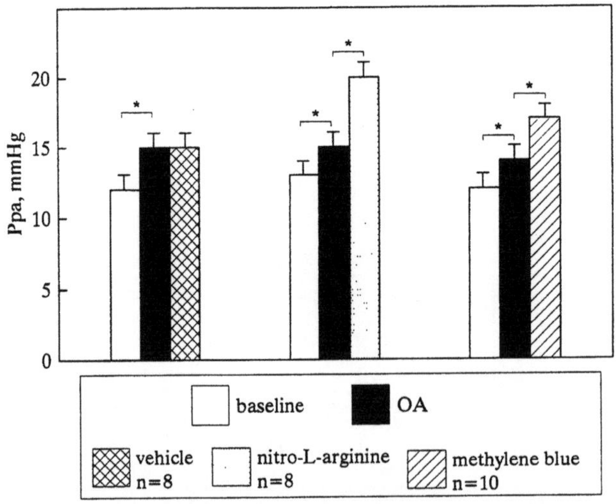

Fig. 3. Pulmonary artery pressure (Ppa) measured at constant cardiac output before and after oleic acid lung injury, and again after vehicle (n = 8 dogs), N^G-nitro-L-arginine (n = 8 dogs) or methylene blue (n = 10 dogs). Oleic acid increased Ppa, which was further enhanced after NO inhibition (* $p < 0.05$) (Adapted from [50])

artery rings to hypoxia decreased their cGMP content and their endothelium-dependent (but not their endothelium-independent) relaxation [14, 33], suggesting that hypoxia impairs NO synthesis or activity, and that hypoxic vasoconstriction results from a reduced NO dilating influence.

On the other hand, numerous studies rather suggest that hypoxia *stimulates* NO release, which then modulates hypoxic vasoconstriction. Thus, inhibitors of NOS have been shown to potentiate the pulmonary vasopressor response to hypoxia in pulmonary artery rings, isolated lungs and intact animals [12, 13, 15, 16, 25–28, 30, 31]. In an intact dog model with controlled cardiac output (as described above), we observed that L-NNA and methylene blue both enhanced pulmonary vascular tone during hypoxia (Fig. 1) [31]. Very recently, an increased accumulation of NO and its metabolite nitrite has been measured in the superfusate of cultured bovine endothelial cells under hypoxic conditions, thus independently of other stimuli such as shear stress or flow [34]. In this model, hypoxia also potentiated the bradykinin-induced NO release and transiently increased cytosolic calcium concentration, an event known to initiate NOS [34].

Chronic Hypoxia

In chronic hypoxic pulmonary hypertension, as in acute hypoxia, NOS could be either reduced (and thus contribute to the elevation in pulmonary artery pressure) or increased (to moderate pulmonary hypertension) (Fig. 2). Experimental data support both possibilities. A reduced response to endothelium-dependent vasodilators has been observed in isolated lungs of rats exposed to chronic hypoxia, whereas the response to sodium nitroprusside was preserved [35]. In this experimental preparation, pretreatment with L-arginine restored endothelium-dependent relaxation, suggesting that chronic hypoxia impairs NOS [36]. In accordance, the endothelium-dependent relaxation to acetylcholine and adenosine diphosphate (ADP) was reduced in pulmonary artery rings obtained from lungs resected in patients with chronic obstructive lung disease, compared to rings from patients without such disease, whereas the relaxation to sodium nitroprusside was similar [37].

Other studies report a blunted ability of both acetylcholine *and* nitroprusside to relax pulmonary arteries and to increase their cGMP content in chronically hypoxic rats [38–40]. A possible mechanism is a desensitization of soluble guanylate cyclase, that could result from a sustained NO release during the initial period of hypoxic exposure. Thus, impairment of the L-arginine-NO pathway during chronic hypoxia could result from altered NOS or guanylate cyclase function.

On the other hand, some studies indicate that chronic hypoxic pulmonary hypertension does not result from a NO deficient state [11, 18, 41, 42]. Chronic L-NAME treatment in rats did not mimic hypoxic pulmonary hypertension: L-NAME, unlike chronic hypoxia, did not increase pulmonary artery pressure nor induce hypertensive pulmonary vascular remodeling [18]. Supplementation with L-arginine had no effect on hypoxic pulmonary hypertension [18]. Lungs from chronically hypoxic rats released more NO_2/NO_3 (degradation products of endogenous NO) than control rats, had greater vasoconstrictor response to L-NAME, and potentiated endothelium-dependent vasodilation to substance P [11]. Thus, increase in both basal and agonist-stimulated pulmonary NO activity was observed in this model [11].

Again, some of the discrepancies could be model-dependent. Acetylcholine vasodilation was impaired *in vitro* in isolated lobar pulmonary arteries but was enhanced *in vivo* in resistance pulmonary arteries in neonatal calves with pulmonary hypertension induced by simulated high altitude in hypobaric chamber for 14 days [41].

Most recently, the expression and localization of NOS were examined in lungs from chronically hypoxic and normoxic rats [32]. NOS was detected in the endothelium of large pulmonary arteries of normoxic and hypoxic rats. NOS was not expressed in the endothelium of normoxic small pulmonary resistance vessels, but prominently appeared in the endothelium of these vessels after chronic hypoxia, which also induced *de novo* NOS expression in the smooth muscle cells of small and large pulmonary arteries [32]. These data could explain some discrepancies between studies on pulmonary artery rings, taken from large conduit arteries upstream to the main site of vascular resistance, and hemodynamic studies on perfused lungs and intact animals [43]. They also allow the speculation that chronic hypoxia downregulates cNOS (that is stimulated by agonists such as acetylcholine), but upregulates endothelial and smooth muscle iNOS (that does not respond to agonist stimulation, but can be inhibited by L-arginine analogs) [32].

Injured Pulmonary Circulation

In lung injury (as previously emphasized for hypoxia), we could hypothesize either a reduced NO release secondary to endothelial damage, resulting in pulmonary hypertension, or an increased NO release, modulating pulmonary hypertension.

Endothelium-Dependent Relaxation in Lung Injury

Experimental as well as clinical acute lung injury is associated with damage of the pulmonary endothelium. To look for endothelial dysfunction, the pulmonary vascular response to acetylcholine has been investigated in several models of lung injury or pulmonary disease. Blunted as well as preserved relaxations in response to acetylcholine have been reported (Table 3) [13, 44–52]. When examined, the endothelium-dependent relaxation to nitroprusside was retained [45, 48, 50, 51]. In sheep, the endotoxin-induced increase in pulmonary artery pressure *in vivo* correlated with the *in vitro* deficit in endothelium-dependent relaxation of the pulmonary vessels [52].

Table 3. Pulmonary endothelium-dependent relaxation in lung injury

Authors	Ref.	Model	Injury	Relaxation to acetylcholine
Feddersen et al.	[44]	Rat lung	ANTU Hyperoxia	Preserved Preserved
Coflesky et al.	[45]	Rat PA	Hyperoxia	Decreased
Ito et al.	[46]	Rat PA	Monocrotaline	Decreased
Tracey et al.	[47]	Guinea pig PA	O_2 radicals	Decreased
Archer et al.	[13]	Rat lung	Monocrotaline	Preserved
Chen et al.	[48]	Rabbit lung	O_2 radicals	Decreased
Crawley et al.	[49]	Rat PA	Bleomycin	Preserved
Liu et al.	[50]	Rat lung	TNF-α	Preserved
Celermajer et al.	[51]	Children	Heart and pulmonary vascular disease	Decreased
Spath et al.	[52]	Awake sheep	Endotoxin	Decreased

ANTU: α-naphtyl thiourea; PA: pulmonary artery; TNF: tumor necrosis factor

It must be pointed out that only a few intact endothelial cells are needed to obtain acetylcholine-induced relaxation [2–4], so that endothelium-dependent dilation could be preserved despite severe lung injury, if a sufficient number of endothelial cells escaped from damage.

NO Release in Lung Injury

Few data have been published about a possible release of NO in acute lung injury [53–56]. In isolated guinea pig lungs perfused at constant flow, Berisha et al. [53] suggested the pathogenic role of NO release in paraquat lung injury. Paraquat increased cGMP levels in lung perfusate, and nitrite and L-citrulline (the coproduct of NOS) production in lung tissue. All signs of lung injury (including the increase in pulmonary artery pressure) were dose-dependently attenuated or prevented by L-NNA or L-NAME pretreatment. Protection conferred by these NOS inhibitors was reversed by L- (but not by D-) arginine [53].

We constructed pulmonary artery pressure-flow relationships before and after NO inhibition in intact anesthetized dogs with oleic acid lung injury [54]. Oleic acid increased pulmonary artery pressure, which was further increased by L-NNA or methylene blue at all levels of cardiac output, suggesting that endogenous NO was released during oleic acid lung injury and moderated pulmonary hypertension [54]. In accordance, pretreatment with L-NAME increased pulmonary artery pressure in isolated rabbit lungs injured by the production of superoxide free radicals from the reaction of purine with xanthine oxidase [55].

Does NO Release Alter Pulmonary Vasoreactivity in Lung Injury?

Blunted response to pulmonary vasoconstrictor stimuli, including hypoxia, has been reported in several models of lung injury such as induced by sepsis [57] or oleic acid [58]. It is attractive to suggest that the release of NO is responsible for this loss of vascular reactivity. However, L-arginine analogs did not restore pulmonary response to various vasoconstrictor agonists in chronic pneumonia [59], nor to hypoxia after sepsis [57] or administration of tumor necrosis factor (TNF-α) [60].

In canine oleic acid lung injury, L-NNA did not increase arterial PO_2 nor reduce intrapulmonary shunt (estimated using a constant infusion of the low-solubility inert gas sulfur hexafluoride), although it increased pulmonary vascular tone, indicating no improvement in the ventilation/perfusion matching, as should be expected if hypoxic pulmonary vasoconstriction had been restored [54]. Of note, methylene blue improved pulmonary gas exchange in the oleic acid model [54], so that a possible role of NO in the regulation of gas exchange during lung injury is not excluded and should be further investigated.

Conclusions

1. NO is released in the pulmonary circulation and accounts for endothelium-dependent pulmonary vasodilation.
2. Whether continuous release of endogenous NO contributes to the low basal pulmonary vascular tone remains unsettled.
3. Acute as well as chronic hypoxia could inhibit NO synthesis or action accounting for hypoxic pulmonary vasoconstriction, or could enhance NO release attenuating hypoxic pulmonary vasoconstriction.
4. Endogenous NO is released during lung injury. In intact dogs, it limits the increase in pulmonary artery pressure, but whether it impairs the hypoxic regulation of gas exchange remains uncertain.

References

1. Vane JR, Änggard KE, Botting RM (1990) Regulatory functions of the vascular endothelium. N Engl J Med 323:27–36
2. Moncada S, Higgs A (1991) The L-arginine-nitric oxide pathway. N Engl J Med 329: 2002–2011
3. Lowenstein CJ, Dinerman JL, Snyder SH (1994) Nitric oxide: A physiologic messenger. Ann Intern Med 120:227–237
4. Furchgott FR, Zawadzki JV (1980) The obligatory role of endothelial cells in the relaxation of arterial smooth muscle by acetylcholine. Nature 299:373–376
5. Greenberg B, Rhoden K, Barnes BJ (1987) Endothelium-dependent relaxation of human pulmonary arteries. Am J Physiol 252:H434–H438
6. Fineman JR, Chang R, Soifer SF (1991) L-arginine, a precursor of EDRF in vitro, produces pulmonary vasodilation in lambs. Am J Physiol 261:H1563–H1569

7. Fineman JR, Heymann MA, Soifer SJ (1991) N^G-nitro-L-arginine attenuates endothelium-dependent pulmonary vasodilation in lambs. Am J Physiol 260:H1299–H1306
8. Fineman JR, Crowley MR, Heymann MA, Soifer SJ (1991) In vivo attenuation of endothelium-dependent pulmonary vasodilation by methylene blue. J Appl Physiol 71:735–741
9. Hyman AL, Kadowitz RJ, Lippton HL (1989) Methylene blue selectively inhibits pulmonary vasodilator responses in cats. J Appl Physiol 66:1513–1517
10. McMahon TJ, Hood JS, Bellan JA, Kadowitz RJ (1991) N^G-nitro-L-arginine methyl ester selectively inhibits pulmonary vasodilator responses to acetylcholine and bradykinin. J Appl Physiol 71:2026–2031
11. Isaacson TC, Hampl V, Weir EK, Nelson DP, Archer SL (1994) Increased endothelium-derived NO in hypertensive pulmonary circulation of chronically hypoxic rats. J Appl Physiol 76:933–940
12. Archer SL, Tolins JP, Raij L, Weir EK (1989) Hypoxic pulmonary vasoconstriction is enhanced by inhibition of the synthesis of an endothelium-derived relaxing factor. Biochem Biophys Res Commun 164:1198–1205
13. Archer SL, Rist K, Nelson DP, DeMaster EG, Cowan N, Weir EK (1990) Comparison of the hemodynamic effects of nitric oxide and endothelium-dependent vasodilators in intact lungs. J Appl Physiol 68:735–747
14. Rodman DM, Yamaguchi T, Hasunuma K, O'Brien RF, McMurtry IF (1990) Effects of hypoxia on endothelium-dependent relaxation of rat pulmonary artery. Am J Physiol 258:L207–L214
15. Wiklund NP, Persson MG, Gustafsson LE, Moncada S, Hedqvist P (1990) Modulatory role of endogenous nitric oxide in pulmonary circulation in vivo. Eur J Pharmacol 185:123–124
16. Perella MA, Edell ES, Krowka MJ, Cortese DA, Burnett JC Jr (1992) Endothelium-derived relaxing factor in pulmonary and renal circulations during hypoxia. Am J Physiol 263:R45–R50
17. Barnard JW, Wilson PS, Moore TM, Thompson WJ, Taylor AE (1993) Effect of nitric oxide and cyclooxygenase products on vascular resistance in dog and rat lungs. J Appl Physiol 74:2940–2948
18. Hampl V, Archer SL, Nelson DP, Weir EK (1993) Chronic EDRF inhibition and hypoxia: Effects on pulmonary circulation and systemic blood pressure. J Appl Physiol 75:1748–1757
19. Koizumi T, Gupta R, Banerjee M, Newman JH (1994) Changes in pulmonary vascular tone during exercise: Effects of nitric oxide synthase inhibition, L-arginine infusion and nitric oxide inhalation. J Clin Invest 94:2275–2282
20. Stamler JS, Loh E, Roddy MA, Currie KE, Creager MA (1994) Nitric oxide regulates basal systemic and pulmonary vascular resistance in healthy humans. Circulation 89: 2035–2040
21. Celermajer DS, Dollery C, Burch M, Deanfield JE (1994) Role of endothelium in the maintenance of low pulmonary vascular tone in normal children. Circulation 89: 2041–2044
22. McGregor M, Sniderman A (1985) On pulmonary vascular resistance : The need for more precise definition. Am J Cardiol 55:217–221
23. Leeman M, Lejeune P, Closset J, Vachiéry JL, Mélot C, Naeije R (1990) Nature of pulmonary hypertension in canine oleic acid pulmonary edema. J Appl Physiol 69: 293–298
24. Hakim TS (1994) Flow-induced release of EDRF in the pulmonary circulation: Site of release and action. Am J Physiol 267:H363–H369
25. Brashers VL, Peach MJ, Rose CE (1988) Augmentation of hypoxic pulmonary vasoconstriction in the isolated perfused rat lung by in vitro antagonists of endothelium-dependent relaxation. J Clin Invest 82:1495–1502
26. Mazmanian GM, Baudet B, Brink C, Cerrina J, Kirkiacharian S, Weiss M (1989) Methylene blue potentiates vascular reactivity in isolated rat lung. J Appl Physiol 66: 1040–1045

27. Hasunuma K, Yamaguchi T, Rodman DM, O'Brien RF, McMurtry IF (1991) Effects of EDRF and EDHF on vasoreactivity of perfused rat lungs. Am J Physiol 260: L97–L104
28. Liu S, Crawley DE, Barnes PJ, Evans TW (1991) Endothelium-derived relaxing factor inhibits hypoxic pulmonary vasoconstriction in rats. Am Rev Respir Dis 143:32–37
29. Nishiwaki K, Nyhan DP, Rock P, et al (1992) N^G-nitro-L-arginine and pulmonary vascular pressure-flow relationship in conscious dogs. Am J Physiol 262:H1131–H1137
30. Ogata M, Ohe M, Katayose D, Takishima T (1992) Modulatory role of EDRF in hypoxic contraction of isolated porcine pulmonary arteries. Am J Physiol 262: H691–H697
31. Leeman M, Zegers de Beyl V, Delcroix M, Naeije R (1994) Effects of endogenous nitric oxide on pulmonary vascular tone in intact dogs. Am J Physiol 266: H2343–H2347
32. Xue C, Rengasami A, Le Cras TD, Koberna PA, Dailey GC, Johns RA (1994) Distribution of NOS in normoxic vs. hypoxic rat lungs: Upregulation of NOS by chronic hypoxia. Am J Physiol 267:L667–L678
33. Johns RA, Linden JM, Peach MJ (1989) Endothelium-dependent relaxation and cyclic GMP accumulation in rabbit pulmonary artery are selectively impaired by moderate hypoxia. Circ Res 65:1508–1515
34. Hampl V, Cornfield DN, Cowan NJ, Archer SL (1995) Hypoxia potentiates NO synthesis and transiently increases cytosolic calcium levels in pulmonary endothelial cells. Eur Respir J 8:515–522
35. Adnot S, Raffestin B, Eddahibi S, Braquet P, Chabrier PE (1991) Loss of endothelium-dependent relaxant activity in the pulmonary circulation of rats exposed to chronic hypoxia. J Clin Invest 87:155–162
36. Eddahibi S, Adnot S, Carville C, Blouquit Y, Raffestin B (1992) L-arginine restores endothelium-dependent relaxation in pulmonary circulation of chronically hypoxic rats. Am J Physiol 263:L194–L200
37. Dinh-Xuan AT, Higenbottam TW, Clelland CA, et al (1991) Impairment of endothelium-dependent pulmonary artery relaxation in chronic obstructive lung disease. N Engl J Med 324:1539–1547
38. Crawley DE, Zhao L, Giembycz MA, et al (1992) Chronic hypoxia impairs soluble guanylate cyclase-mediated pulmonary arterial relaxation in the rat. Am J Physiol 263:L325–L332
39. Zhao L, Crawley DE, Hughes JMB, Evans TW, Winter RJD (1993) Endothelium-derived relaxing factor activity in rat lung during hypoxic pulmonary vascular remodeling. J Appl Physiol 74:1061–1065
40. Maruyama J, Maruyama K (1994) Impaired nitric oxide-dependent responses and their recovery in hypertensive pulmonary arteries of rats. Am J Physiol 266:H2476–H2488
41. Orton EC, Reeves JT, Stenmark KR (1988) Pulmonary vasodilation with stucturally altered pulmonary vessels and pulmonary hypertension. J Appl Physiol 65:2459–2467
42. Russ RD, Walker BR (1993) Maintained endothelium-dependent pulmonary vasodilation following chronic hypoxia in the rat. J Appl Physiol 74:339–344
43. Leeman M, Naeije R (1995) Nitric oxide and hypoxic pulmonary hypertension. Eur Respir J 8:513–514
44. Feddersen CO, McMurtry IF, Henson P, Voelkel NF (1986) Acetylcholine-induced pulmonary vasodilation in lung vascular injury. Am J Respir Dis 133:197–204
45. Coflesky JT, Evans JN (1988) Pharmacologic properties of isolated proximal pulmonary arteries after seven-day exposure to in vivo hyperoxia. Am J Respir Dis 138:945–951
46. Ito K, Nakashima T, Murakami K, Murakami T (1988) Altered function of pulmonary endothelium following monocrotaline-induced vascular injury in rats. Br J Pharmacol 94:1175–1183
47. Tracey WR, Hamilton JT, Craig ID, Paterson NAM (1989) Effect of endothelial injury on the responses of isolated guinea pig venules to reduced oxygen tension. Am J Respir Dis 140:68–74

48. Chen X, Gillis CN (1991) Effects of free radicals on pulmonary vascular response to acetylcholine. J Appl Physiol 71:821–825
49. Crawley DE, McCormack DG, Evans TW (1992) Bleomycin-induced acute lung injury in the rat does not influence pulmonary vascular responsiveness in vitro. Crit Care Med 20:641–644
50. Liu SF, Dewar A, Crawley DE, Barnes PJ, Evans TW (1992) Effect of tumor necrosis factor on hypoxic pulmonary vasoconstriction. J Appl Physiol 72:1044–1049
51. Celermajer DS, Cullen S, Deanfield JE (1993) Impairement of endothelium-dependent pulmonary artery relaxation in children with congenital heart disease and abnormal pulmonary hemodynamics. Circulation 87:440–446
52. Spath JA, Sloane PJ, Gee MH, Albertine KH (1994) Loss of endothelium-dependent vasodilation in the pulmonary vessels of sheep after prolonged endotoxin. J Appl Physiol 76:361–369
53. Berisha HI, Pakbaz H, Absood A, Said SI (1994) Nitric oxide as a mediator of oxidant lung injury due to paraquat. Proc Natl Acad Sci USA 91:7445–7449
54. Leeman M, Zegers de Beyl D, Gilbert E, Mélot C, Naeije R (1993) Is nitric oxide released in oleic acid lung injury? J Appl Physiol 74:650–654
55. Kavanagh BP, Mouchawar A, Goldsmith J, Pearl RG (1994) Effects of inhaled NO and inhibition of endogenous NO synthesis in oxidant-induced acute lung injury. J Appl Physiol 76:1324–1329
56. Zelenkov P, McLoughlin T, Johns RA (1993) Endotoxin enhances hypoxic constriction ot rat aorta and pulmonary artery through induction of EDRF/NO synthase. Am J Physiol 265:L346–L354
57. Fox GA, Paterson NAM, McCormack DG (1994) Effect of inhibition of NO synthase on vascular reactivity in a rat model of hyperdynamic sepsis. Am J Physiol 267: H1377–H1382
58. Leeman M, Delcroix M, Vachiéry JL, Mélot C, Naeije R (1992) Blunted hypoxic vasoconstriction in oleic acid lung injury: Effect of cyclooxygenase inhibitors. J Appl Physiol 72:251–258
59. Yaghi A, Paterson NAM, McCormack DG (1993) Nitric oxide does not mediate the attenuated pulmonary vascular reactivity of chronic pneumonia. Am J Physiol 265: H943–H948
60. Johnson D, Hurst T, Wilson T, et al (1993) N^G-monomethyl-L-arginine does not restore loss of hypoxic pulmonary vasoconstriction induced by TNF-α. J Appl Physiol 75:618–625

Nitric Oxide Inhalation in RDS and ARDS

Inhaled Nitric Oxide: A Review

W. M. Zapol and W. E. Hurford

Introduction

Severe hypoxemia and concurrent pulmonary hypertension can complicate the treatment of acute respiratory failure in neonates and adults. Attempts to treat the pulmonary hypertension with pulmonary vasodilators have been limited by systemic hypotension and worsening of arterial hypoxemia secondary to altered ventilation/perfusion matching [1–3]. Recently, inhaled nitric oxide (NO) was noted to selectively vasodilate the pulmonary circulation [4–6]. Significant systemic hypotension did not occur because NO is inactivated by rapid binding to hemoglobin in the blood. Inhaling NO often improves the systemic arterial oxygen tension in patients with severe acute lung injury (ALI) and pulmonary hypertension [7, 8]. Inhaled NO is receiving intense and widespread examination throughout the world for treating severe hypoxemia and pulmonary hypertension.

Nitric Oxide

NO is a gaseous endogenous vasodilator and has been identified as an endothelium-derived relaxing factor (EDRF) [9–12]. It is an ideal local messenger because of its small size, lipophilic nature, and brief duration of action [13]. NO is synthesized from the terminal guanidino nitrogen of L-arginine and diffuses rapidly into subjacent cells [14, 15]. In vascular smooth muscle, NO binds to the heme iron complex of soluble guanylate cyclase. The resulting nitrosyl-heme activates guanylate cyclase, stimulating the production of cyclic guanosine 3′,5′-monophosphate (cGMP) and the subsequent relaxation of vascular smooth muscle [14, 16, 17]. When NO diffuses into the intravascular space, it is rapidly inactivated by avid binding to hemoglobin [18]. Commonly used chemical nitroso-vasodilators, such as nitroglycerin and nitroprusside, act by releasing NO [19].

NO plays diverse physiologic roles besides smooth muscle relaxation, including neurotransmission, immunoregulation, and the inhibition of platelet aggregation. These functions have been reviewed extensively [17, 20, 21]. NO synthase (NOS) catalyzes the 5 electron oxidation of L-arginine to NO and L-citrulline. Several distinct isoforms of the NOS enzyme, homologous to cytochrome P-450, have been isolated [22–25]. These isoforms differ on the basis of cofactor requirements and structure. Two Ca^{2+}-dependent enzymes are constitutively expressed. Constitutive endothelial NOS (eNOS) is present in vascular endothelium. Neuronal or brain NOS is found in a wide variety of cells such as neurons within the cerebellum, olfactory bulb, hippocampus and ganglion cells within the adrenal medulla [26]. An inducible Ca^{2+}-independent isoform (iNOS) is present in lung, hepatic Kupffer cells, macrophages, and many other tissues [23]. Endotoxin stimulates and glucocorticoids inhibit iNOS production [27]. Inhibitors of NOS, such as N^G-monomethyl-L-arginine (L-NMMA), decrease the production of endogenous NO by the intact lung and enhance hypoxic pulmonary vasoconstriction of isolated pulmonary artery rings [28]. Similarly, chemical inhibitors of guanylate cyclase activation such as methylene blue inhibit the action of NO, whereas specific cGMP phosphodiesterase inhibitors (type 5) such as Zaprinast can enhance or potentiate many of the biologic actions of NO [13, 17, 29, 30].

Endothelium-dependent relaxation of pulmonary arteries occurs in response to a variety of physical and pharmacologic stimuli [31, 32]. Endogenous NO has been measured in the exhaled air of rabbits, guinea pigs, and humans [33–35]. In patients and volunteers, relatively high levels of NO (0.07 to 0. 13 parts per million (ppm)) appear to be produced in the nasopharynx, most likely by bacteria [34, 35]. The NO produced in the nasopharynx is subsequently inhaled and approximately 50 to 70% of inhaled NO is absorbed. When the tracheas of patients were intubated, bypassing the nasopharynx, NO levels were greatly reduced in inhaled and exhaled gas. It is possible that the inhaled endogenous NO produced in the nasopharynx might have a role in the regulation of ventilation/perfusion matching.

In normal lungs, basal release of eNOS contributes to the maintenance of a low resting pulmonary vascular tone [36, 37]. The administration of acetylcholine or the addition of exogenous NO has little or no effect to reduce the basal pulmonary vascular resistance [4, 38–40]. In patients with pulmonary hypertension, on the other hand, acetylcholine infusion or NO inhalation often reduces the pulmonary vascular resistance [4, 38, 39]. It is possible that in some states of acute pulmonary hypertension, such as ARDS, the endogenous production of NO is impaired [41, 42]. This might augment vasoconstriction and foster platelet aggregation. Evidence supporting this hypothesis is indirect at this time. Dinh Xuan and coworkers [43] reported that patients with cor pulmonale secondary to congenital heart disease had an impaired pulmonary vasodilator response to acetylcholine. The same group [41] also studied 18 patients with chronic obstructive pulmonary disease (COPD) and reported that impairment of acetylcholine-induced relaxation was correlated with lower arterial oxygen tensions and greater intimal thickening of pulmonary blood vessels. Such patients may have an intact response to exogenously administered NO even though their response to intravenous acetylcholine is impaired [44].

Zapol and coworkers [5, 6] hypothesized that inhaled NO should diffuse into the pulmonary vasculature of ventilated lung regions and cause relaxation of pulmonary vascular smooth muscle, thereby decreasing pulmonary hypertension. Since NO is inhaled, it should be distributed predominantly to well-ventilated alveoli and not delivered to collapsed or fluid-filled areas of the lung. In the presence of an increased vasomotor tone, local vasodilation of well-ventilated lung regions should cause a "steal" or diversion of pulmonary artery blood flow towards well-ventilated alveoli, improving the matching of ventilation to perfusion ($\dot{V}/\dot{Q}$) and improving arterial oxygenation during ALI. Such an effect would be in marked contrast to the effects of intravenously administered conventional vasodilators (such as nitroprusside, nitroglycerin, or prostacyclin). These intravenous agents also decrease pulmonary artery pressure (PAP), but by non-selectively dilating the pulmonary vasculature, they augment blood flow to non-ventilated areas, thereby increasing the right-to-left shunting of venous blood and reducing the PaO_2. Also unlike available intravenous vasodilators, inhaled NO, because it is avidly bound to hemoglobin and rapidly inactivated, should not produce systemic vasodilation.

Laboratory and Clinical Studies of Inhaled NO

Acute Pulmonary Hypertension

Inhaled NO reduces the pulmonary vascular resistance (PVR) and pulmonary hypertension without altering the systemic vascular resistance (SVR) [5, 6]. In studies of normal awake lambs without pulmonary hypertension, inhaling 80 ppm NO produced no hemodynamic effects. PAP, cardiac output, systemic arterial pressure and SVR were unchanged. When PAP was acutely increased, either by hypoxia, infusing the thromboxane endoperoxide analog U46619, or producing the heparin-protamine reaction, the pulmonary hypertension was rapidly reversed by inhalation of 40 to 80 ppm NO [5, 6]. Pulmonary vasodilation occurred within 3 min, lasted throughout the duration of NO inhalation, and disappeared within 3 min after the discontinuation of NO (Fig. 1). Importantly, SVR was unchanged. The pulmonary vasodilator effect occurred at low levels of inhaled NO (i.e. 5 ppm). Potent vasodilation (65% of the maximal effect) occurred at 20 ppm inhaled NO. This concentration is less than the NIOSH standard for 8 h working exposures (25 ppm). Pulmonary vasodilation was maintained during continuous inhalation of 80 ppm NO for 1 h, no tolerance was observed. That inhaled NO reverses hypoxic pulmonary vasoconstriction has been confirmed by others [7, 45, 46]. Pison and coworkers [7] studied the matching of ventilation to perfusion using the multiple inert gas elimination technique (MIGET) in mechanically-ventilated healthy sheep. They reported that inhaling 20 ppm NO redistributed blood flow towards better ventilated alveoli and reversed the acute pulmonary hypertension caused by breathing a hypoxic gas mixture.

Frostell et al. [39] sequently studied the effects of breathing 40 ppm NO for 10 min in 9 awake, healthy human volunteers exposed to hypoxic breathing

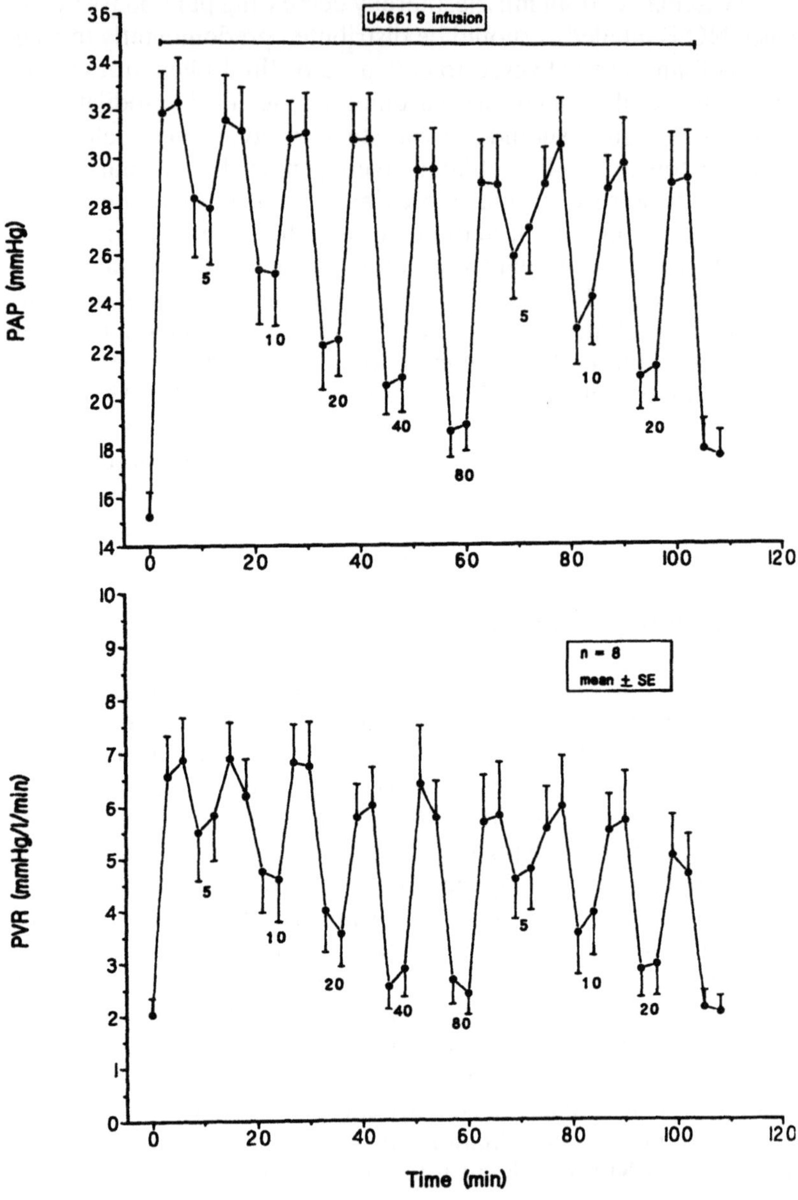

Fig. 1. Plots of mean PAP and PVR during a continuous infusion of U46619. Lambs breathed various levels of NO (5–80 ppm) at FiO_2 0.6 for 6 min, then breathed a gas mixture at FiO_2 0.6 for 6 min without NO (n = 8, mean ± SEM). (From [5] with permission)

conditions. During air breathing, inhaling 40 ppm NO had no pulmonary or systemic vasodilatory effect. When the subjects breathed 12% oxygen, causing mild acute pulmonary hypertension, 40 ppm NO completely reversed the hypoxia-induced increase of PAP and PVR. Systemic blood pressure and vascular resistance were unchanged and methemoglobin levels remained below 1%.

The effects of inhaled NO on endotoxin-induced pulmonary hypertension are more complex. Inhaled NO at a concentration of 10 ppm selectively decreased the acute pulmonary hypertension occurring at least 30 min after the intravenous administration of *E. coli* endotoxin in anesthetized pigs [47]. Arterial oxygenation and pH were also improved during NO inhalation. The early increase of PAP (within 30 min after endotoxin administration), however, was unaffected by this low NO dose. Thromboxane is believed to cause the early increase of PAP observed following endotoxin infusion. In models of acute pulmonary hypertension characterized by an increased plasma thromboxane level, high inhaled NO levels effectively block the vasoconstriction. In a study by Fratacci and coworkers [6], the acute and transient pulmonary hypertension caused by the heparin-protamine reaction in the awake sheep was blocked by inhaling 180 ppm NO [6]. The reasons for differing dose-response relationships in differing species and in different models of acute pulmonary hypertension are not yet understood.

Inhaled NO probably mediates pulmonary vasodilation during lung injury by increasing cGMP levels within the lung's vascular smooth muscle. This increase may be reflected by increased plasma cGMP concentrations. Rovira and coworkers [48] studied a model of ALI induced by bilateral lung lavage in anesthetized lambs. When endogenous NO production was inhibited by infusing N^G-nitro-L-arginine methyl ester (L-NAME), a consistent increase of aortic, as compared with pulmonary arterial, plasma cGMP concentration could be measured within 5 min of breathing 60 ppm NO. Increased aortic plasma cGMP levels were associated with selective pulmonary vasodilation, a reduced venous admixture, and an increased PaO_2. Levels of plasma cGMP returned to baseline within 10 min of discontinuing NO breathing.

NO Inhalation in ARDS

Acute pulmonary hypertension consistently occurs in severe ARDS. In survivors, PVR progressively decreases over time. Non-survivors tend to have a persistently increased PVR. The increased PAP is independent of changes of cardiac output and persists after the correction of systemic hypoxemia [49].

The pulmonary vascular changes in ARDS are produced by a complex combination of primary lung injury (i.e. aspiration, trauma, infection), the consequences of the pulmonary inflammatory response to injury (hypoxia, acidosis, release of cytokines, components of the complement system and arachidonic acid pathway, as well as inhibitors of fibrinolysis), and the iatrogenic complications of intensive care therapy (oxygen toxicity and barotrauma). In severe ARDS, thromboembolic occlusion of the pulmonary vasculature is common [50].

Rossaint and coworkers [8] compared the effects of NO inhalation (18 and 36 ppm) to intravenously infused prostacyclin in 9 patients with ARDS. NO selectively reduced mean pulmonary artery pressure from 37±3 to 30±2 mm Hg (mean±SE). Oxygenation improves due to a decreased level of venous admixture ($\dot{Q}_{VA}/\dot{Q}_t$). During NO breathing, the PaO_2/FiO_2 ratio increased from 152±15 to 199±23 mm Hg (Fig. 2). While the intravenous

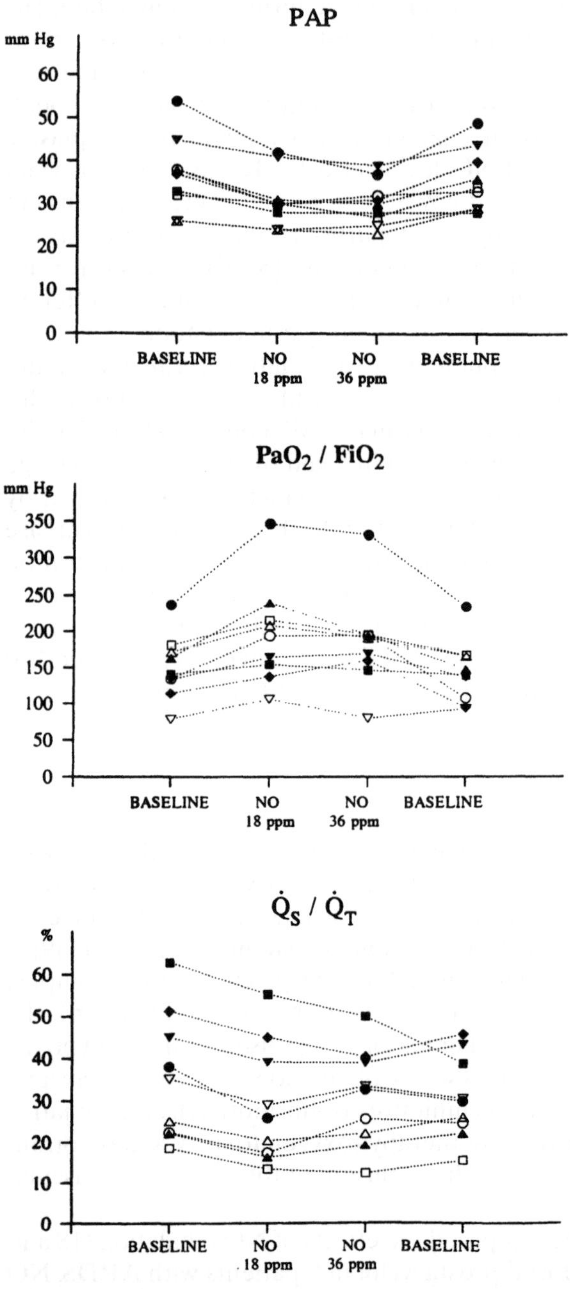

Fig. 2. Mean PAP, arterial oxygenation efficiency (PaO_2/FiO_2), and intrapulmonary shunting ($\dot{Q}_S/\dot{Q}_T$) in 9 patients with ARDS during the inhalation of NO. (From [8] with permission)

infusion of prostacyclin also reduced PAP, mean arterial pressure and PaO_2 decreased as $\dot{Q}_{VA}/\dot{Q}_t$ increased. Subsequent reports documented that inhalation of lower concentrations of NO (<20 ppm) effectively reduced PAP and improved PaO_2 [51–56]. Even very small inhaled concentrations (less than 1 ppm NO) may effectively increase PaO_2 in some patients [55, 57, 58]. Interestingly, improvement of oxygenation may occur at inhaled NO levels (100 ppb or less) which do not cause measurable changes of PVR [5, 58]. Right ventricular ejection fraction may increase in some patients responding to inhaled NO, suggesting that the observed decreases of PAP may be hemodynamically important in relieving the burden placed upon the right ventricle [53, 54].

A marked variation has been reported of the hemodynamic and respiratory effects of NO inhalation, both among patients and within the same patient at different times during illness [51, 59, 60]. It is possible that preexisting pulmonary disease as well as the concomitant administration of

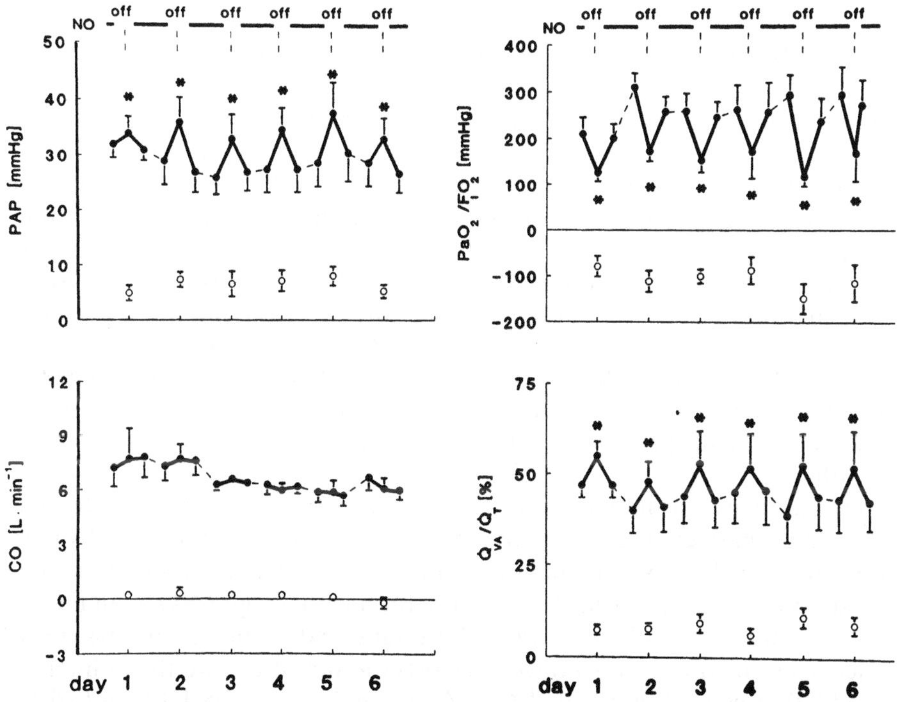

Fig. 3. Hemodynamic function and gas exchange before, during and after brief interruptions (arrows) of NO inhalation (bars) during the first 6 days of treatment in 7 patients with ARDS. Values are means ± SE (solid symbols); also shown (open symbols) are the means ± SE of the individual differences between the values for the effect of treatment and the means of the values determined before and after interrupting NO therapy. The standard errors for the treatment effects were small, indicating that the effects of withdrawal of NO were clear and precisely estimated. * Significant difference from the mean of the values determined before and after interrupting NO therapy. (From [8] with permission)

other vasoactive drugs may contribute to the observed variability. In general, the baseline level of PVR appears to predict the degree of pulmonary vasoconstriction reversible by NO inhalation. Those with the greatest degree of acute pulmonary hypertension appear to respond most to NO inhalation [51, 60]. In addition, the concomitant intravenous infusion of a novel vasoconstricting drug, almitrine, which increases the degree of hypoxic vasoconstriction in the lung, has been reported to enhance the beneficial effect of inhaled NO on PaO_2 [61]. Almitrine, however, also increased mean PAP and cardiac output.

Tachyphylaxis has not been observed even when NO inhalation was continued for up to 53 days [8]. PAP and PaO_2 quickly return to baseline values after discontinuation of NO (Fig. 3). Occasionally, sudden discontinuation of inhaled NO can produce problematic pulmonary vasoconstriction and possibly bronchoconstriction [51, 62, 63]. The reason for this is unclear. Possibly, the addition of exogenous NO may inhibit NOS activity [64, 65] or increase tissue cGMP phosphodiesterase activity.

Inhaled NO in Persistent Pulmonary Hypertension of the Newborn

At birth, there is a sustained decrease of PVR and an increase of pulmonary blood flow, in part due to increasing oxygen tensions. If this does not occur, persistent pulmonary hypertension of the newborn (PPHN) may result. This syndrome is characterized by an increased PVR, increased right-to-left shunting across the ductus arteriosus and foramen ovale, and severe hypoxemia. Extracorporeal membrane oxygenation (ECMO) is often used to support these infants, because conventional vasodilator therapy is limited by severe systemic hypotension. Hypotension may reduce PaO_2 further by increasing right-to-left shunting. It has been hypothesized that endogenous production of NO by the pulmonary vasculature might be decreased in PPHN. If so, then inhaled NO might provide an effective therapy for these severely ill infants [66–70].

Several investigators have studied the effects of inhaled NO on the neonatal pulmonary circulation. In hypoxic newborn lambs studied by Roberts and coworkers [71], NO inhalation decreased PAP and increased pulmonary blood flow without reducing SVR. Ventilation with 80 ppm NO at an FiO_2 of 0.21 caused a three-fold increase of both lung and preductal plasma cGMP concentrations. Severe respiratory acidosis did not attenuate the pulmonary vasodilation caused by inhaled NO. Zayek and coworkers [70] studied the effects of inhaled NO (6 to 100 ppm) in a model of PPHN created by prenatal ligation of the ductus arteriosus in lambs. They reported that inhaled NO caused dose-dependent decreases of PAP and PVR. Decreased right-to-left shunting of blood flow through the foramen ovale during NO inhalation resulted in an increased PaO_2 (from 43 ± 16 to 185 ± 72 mm Hg, mean $\pm$ SEM, while breathing 100 ppm NO) and arterial oxygen saturation (SaO_2) (from 74 ± 8 to $96 \pm 2\%$) and a decreased $PaCO_2$. Systemic blood pressure was unaffected by breathing NO [70]. Using this model, Zayek and coworkers subsequently documented an increased survival rate with NO therapy [72]. After prenatal ligation of the ductus arteriosus and delivery 13 days later, newborn lambs were ventilated at FiO_2 of 0.92 and randomly allocated to breathe 80 ppm NO for 23 h or receive no additional treatment. All 7 control lambs died before the end of the study, whereas 5 of 6 lambs treated with NO survived. Arterial oxygenation was increased in the NO-breathing lambs compared with controls. Interestingly, blood methemoglobin concentration increased to 3% after 4 h of NO breathing, but did not increase further over the 23 h of the study. DeMarco and coworkers [73] also reported that inhaled NO reversed acute pulmonary artery hypertension induced by hypoxia or U46619 in the newborn lamb. This group also reported slightly increased methemoglobin levels after breathing 80 ppm NO for 3 h. In a piglet model of Group B *Streptococcal* sepsis, a common cause of neonatal respiratory failure that was studied by Berger and coworkers [46], inhalation of 150 ppm NO for 30 min reversed the acute pulmonary hypertension caused by the infusion of Group B *Streptococcus*. The inhalation of NO did not alter ventilation/perfusion abnormalities caused by Group B *Streptococcal* sepsis or affect systemic hemodynamics in this study.

In clinical studies by Roberts and coworkers [68], critically ill infants with PPHN rapidly increased preductal oxygen saturation in response to NO inhalation at concentrations up to 80 ppm. In 5 of 6 infants studied, postductual oxygen saturation also increased. In one newborn, the resulting improvement in PaO_2 persisted after the discontinuation of NO, eliminating the need for ECMO.

Kinsella and coworkers [66] reported the effects of inhaling 10–20 ppm NO in 9 infants with severe PPHN. All 9 showed a rapid improvement in oxygenation without systemic hypotension. Clinical improvement continued during treatment with 6 ppm NO for 24 h in 6 of the infants. Subsequently, this group has reported that 13 of 15 patients with PPHN who were candidates for support with ECMO were successfully treated with inhaled NO. ECMO therapy was avoided in these patients [69, 74]. Successful treatment of a 1180-g premature newborn with pulmonary hypertension caused by group B *Streptococcal* sepsis has also been reported by this group [75]. NO inhalation in babies with PPHN currently is being studied in several open [76] and blinded and randomized multicenter trials.

Inhaled NO in Other Pulmonary Hypertensive States

Chronic Pulmonary Hypertension: Clinical studies of brief periods of NO inhalation have been performed in patients with chronic pulmonary hypertension. In 18 patients with either chronic pulmonary hypertension or cardiac disease, Pepke-Zaba and coworkers [4] reported that inhalation of 40 ppm NO in air decreased PVR by 5–68% without affecting SVR. Similar hemodynamic effects have been reported in response to inhaled NO in

6 patients with pulmonary hypertension following mitral valve replacement for mitral stenosis [77], in patients with COPD complicated by pulmonary hypertension [44, 78], and in a 5-month old child with idiopathic pulmonary hypertension which was unresponsive to intravenous vasodilators [79]. Breathing 15 ppm NO decreased PAP by nearly 20% and PVR nearly by 30% in 14 hypoxic patients with COPD [80]. The decrease of PAP was correlated with the level of baseline pulmonary artery hypertension.

Congenital and Acquired Heart Disease: Inhaled NO has been evaluated as a potential therapy for patients with pulmonary hypertension due to cardiac disease. Roberts and coworkers [81] studied 10 children from 3 months to 6.5 years of age with chronic pulmonary hypertension due to congenital heart defects. Inhaling 20 to 80 ppm NO at FiO_2 0.9 for 10 min promptly decreased PAP and PVR. Compared to breathing 0 ppm NO at FiO_2 0.9, NO decreased the mean pulmonary artery pressure from 48 ± 19 to 37 ± 11 mm Hg (mean ± SD). Under the same conditions, NO decreased the PVR index from 536 ± 376 to 308 ± 260 $dyne \cdot sec \cdot cm^{-5} \cdot m^{-2}$, while systemic arterial pressure and resistance remained unchanged. Inhaling 80 ppm NO at FiO_2 0.9 increased pulmonary blood flow in each of the 6 patients with intracardiac shunts. Similar results have been reported when inhaled NO was used to decrease acute pulmonary hypertension in the perioperative period [82–85].

In adult patients, Rich and coworkers [60] have studied the effects of inhaling 20 ppm NO in 20 patients undergoing various types of cardiac surgery requiring cardiopulmonary bypass and 5 patients requiring a ventricular assist device. NO inhalation decreased PAP from 36 ± 3 (mean ± SEM) to 29 ± 2 mm Hg, and 32 ± 2 to 27 ± 1 mm Hg, before and after cardiopulmonary bypass, respectively, and from 68 ± 12 to 55 ± 9 mm Hg in patients requiring ventricular assist devices. The decrease of PVR during NO breathing was proportional to the patients' baseline PVR. Once again, systemic hemodynamics were unaffected during NO breathing.

Bronchodilation and Asthma: Since inhaled NO can diffuse through pulmonary tissue to relax upstream pulmonary artery smooth muscle cells, it seems reasonable that inhaled NO might also dilate constricted airways. However, airways have a thick epithelium and are covered by a mucus coating. Diffusion of NO, a highly lipid soluble substance with a low aqueous solubility, might be impeded by the mucus barrier [78]. Dupuy and coworkers [79] examined the effect of inhaled NO on airway mechanics in anesthetized guinea pigs. In methacholine-bronchoconstricted animals, inhalation of 5 to 300 ppm NO produced a rapid and dose-related reduction of airway resistance and increased dynamic compliance. An inhaled concentration of 15 ppm NO reduced airway resistance by 50%. The bronchodilating effects of NO were additive with the effects of inhaled terbutaline and were stable over 1 h of NO breathing. Additional studies demonstrated that inhaled NO in guinea pigs can reverse bronchoconstriction due to an intravenous infusion of leukotriene D_4, histamine or neurokinin A [81].

Somewhat different results were obtained in studies of anesthetized mechanically-ventilated New Zealand White rabbits [41]. As in the study by Dupuy and coworkers [79], inhalation of 80 ppm NO blocked methacholine-induced increases of airway resistance, which primarily measures central airway tone. Pulmonary compliance, which reflects the behavior of smaller airways, and arterial oxygenation were unaffected by NO inhaled at this concentration. Apparently, peripheral airway tone in this species may be less sensitive to inhaled NO. The effect of inhaled NO on airway tone in humans remains to be well defined [41]. Initial studies suggest that inhaled NO may have a bronchodilator effect in some patients with asthma [82].

Asthmatics also appear to have increased concentrations of NO in exhaled air [35, 86–88]. The increased NO appears to originate in the lower airways [35]. The increased NO levels may be due to an increased iNOS activity of neutrophils and macrophages. Consistent with this theory, inhaled steroid therapy appears to decrease exhaled NO levels in asthmatics [89].

Toxicity of NO

The interest generated by reports of the successful therapeutic use of inhaled NO is tempered by concerns over its toxicity. NO is a common pollutant. It is produced by lightning and the burning of fossil fuels, and exists in the atmosphere at levels near 10 ppb. Its major atmospheric breakdown pathway at these low levels is via combination with ozone [90]. NO is present in cigarette smoke and is routinely inhaled for short periods in concentrations of 400 to 1000 ppm by millions of people [91]. Although the US Occupational Safety and Health Administration has set an NO exposure limit of 25 ppm when breathed for 8 h/day in the workplace, long-term studies of NO toxicology have not been conducted [92].

NO may form several cytotoxic products including nitrogen dioxide (NO_2) [93] and peroxynitrite ($OONO^-$). The rate of oxidation of NO to NO_2 depends upon the oxygen concentration and the square of the NO concentration (Fig. 4) [94]. Nitrogen dioxide is converted in aqueous solutions to nitric and nitrous acids. Occupational safety and health standards limit the NO_2 exposure of workers to 5 ppm [92]. Peroxynitrite is formed in aqueous solutions when NO reacts rapidly with $O_2^{\cdot -}$ [95]. This species is also highly cytotoxic. In addition, NO forms complexes with transitional metal complexes, including those in metalloproteins such as hemoglobin. In tissues, nitrosation of iron-containing enzymes and iron-sulfur proteins of target cells may be responsible for the cytotoxic action of NO generated by activated macrophages [96, 97].

In the circulation, NO combines extremely rapidly with hemoglobin to form nitrosyl Fe(II)-hemoglobin and then methemoglobin [18]. Endogenous NO production probably accounts for the methemoglobin levels found in the blood of non-smokers [98]. Methemoglobinemia may occur if its production is increased or its reduction is diminished [99]. Clinically significant met-

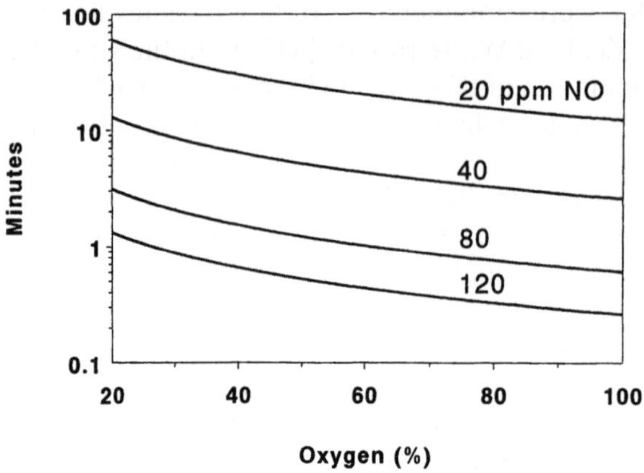

Fig. 4. Time in minutes to yield 5 ppm NO with differing mixtures of NO and oxygen. (Adapted from [119])

hemoglobinemia is rare following low level NO exposure. For example, mice exposed to 10 ppm NO for 6 months had a methemoglobin concentration of 0 to 0.3%, identical to unexposed animals [100]. Also, von Nieding and coworkers [101] found that methemoglobin levels did not rise above 0.7% in 48 normal volunteers breathing 40 ppm NO for up to 15 min. Furthermore, Frostell and coworkers [5] recently reported that breathing 80 ppm NO for 3 h did not increase blood methemoglobin levels in awake spontaneously ventilating lambs. In a clinical study, Wessel and coworkers [102] reported methemoglobin concentrations in excess of 5% in 4 of 123 children who inhaled NO. Certain patients with decreased methemoglobin reductase activity may develop methemoglobinemia in the face of an increased rate of hemoglobin oxidation. The activity of methemoglobin reductase may be decreased as a result of a hereditary deficiency and is normally low in newborn infants [99].

S-nitrosothiols, mostly S-nitrosoproteins, are also formed in plasma. S-nitrosothiols have been identified in human airway lining fluid and in the plasma of normal subjects [103] and patients inhaling NO mixtures [104]. The formation of relatively stable iron-nitrosyl complexes and S-nitrosothiols may provide ways of tailoring the duration of action and transport properties of NO and a means of detoxifying NO in tissues [103, 105]. High levels of intracellular NO have been reported to cause DNA damage and mutations in human cell preparations [106]. Damage to erythrocyte membranes may occur following long-term low-level exposure to NO. Oda and coworkers [100] reported enlarged spleens and increased bilirubin levels, suggestive of increased erythrocyte turnover, in mice following 6 months of exposure to 10 ppm NO. NO also inhibits platelet adhesion to endothelial cells and reverses platelet aggregation *in vitro* [107]. Prolongation of the bleeding time during experimental NO inhalation has been reported [108, 109], but the clinical significance of this effect is uncertain.

High inhaled concentrations of NO have been reported to cause acute pulmonary injury, methemoglobinemia, decreased platelet adhesion, asphyxia and death [93, 110, 111]. Greenbaum et al. [112] exposed anesthetized dogs to 5000–20000 ppm NO. Death ensued secondary to methemoglobinemia, acidemia, and alveolar edema.

Subsequent controlled studies demonstrated that much of the direct pulmonary toxicity of inhaled NO is due to the NO_2 contained in the gas mixtures [93, 113].

Technical Considerations

Several safety issues should be addressed in experimental and clinical studies of inhaled NO [57, 102, 114]:

1. Inhalation circuits designed to deliver NO must insure the accurate continuous delivery of NO while minimizing levels of nitrogen dioxide (NO_2).
2. Calibrated, commercially-purchased stock tanks of NO in nitrogen should be used. The NO concentration of the undiluted gas should be limited to 1000 ppm to decrease the effect of cylinder leakage or inadvertent overdosage.
3. Because the toxicity of NO remains unclear, the smallest effective dose should be sought [115, 116].
4. Sudden discontinuation of inhaled NO may produce severe arterial desaturation and pulmonary hypertension in some patients [51, 63]. Breathing circuits capable of delivering inhaled NO should therefore be available to allow manual ventilation of the patient during tracheal suctioning, transport, etc.
5. Concentrations within a breathing circuit will vary with the NO and oxygen concentrations used and the residence time of NO within the lungs and breathing circuit [115]. Because conversion of NO to NO_2 is offset by minimizing the residence time of NO in the inhaled gas mixture, mixing of a stock gas of NO in nitrogen with the carrier gases (usually oxygen and air) should take place immediately before inhalation. Several studies have used large gas collection bags to store and deliver NO mixed with oxygen-containing gases. Such NO-O_2 mixtures produce steadily increasing levels of NO_2 with concomitantly decreasing levels of NO [116]. This method of administration is unsuitable for clinical application because it does not allow prolonged breathing of constant levels of NO with minimal levels of NO_2. Dry soda lime may be useful in removing NO_2 from the breathing circuit [117].
6. To minimize environmental contamination, exhaust gases from the breathing circuit should be scavenged from the patient's environment.
7. Continuous or intermittent monitoring of NO and NO_2 gas concentrations can be performed easily with commercially available chemiluminescence [118] or electrochemical analyzers [119].
8. In addition to monitoring inhaled and ambient NO and NO_2 concentrations, blood methemoglobin concentrations should be measured regularly in every patient receiving NO.

Conclusion

At present, NO inhalation remains an experimental therapy. Prospective and randomized trials are examining the clinical effectiveness and safety of this novel therapy. Several important areas remain to be addressed: What level and duration of NO exposure is safe for the normal or acutely injured lung? What is the effect of inhaled NO on pulmonary structure and function? What are the best methods to deliver NO to the critically ill patient? Will NO inhalation therapy measurably change outcome? The study of NO biology has already provided us with many new insights into pulmonary vascular function and control [120]. Potentially, the use of inhaled NO may benefit many patients with acute and chronic lung disease.

References

1. Radermacher P, Santak B, Falke KJ (1989) Comparison of prostaglandin E1 and nitroglycerin in patients with ARDS. Prog Clin Biol Res 301:267–270
2. Radermacher P, Santak B, Wust HJ, et al (1990) Prostacyclin for the treatment of pulmonary hypertension in the adult respiratory distress syndrome: Effects on pulmonary capillary pressure and ventilation-perfusion distributions. Anesthesiology 72: 238–244
3. Radermacher P, Santak B, Wust HJ, et al (1990) Prostacyclin and right ventricular function in patients with pulmonary hypertension associated with ARDS. Intensive Care Med 16:227–232
4. Pepke-Zaba J, Higenbottam TW, Dinh-Xuan AT, et al (1991) Inhaled nitric oxide as a cause of selective pulmonary vasodilatation in pulmonary hypertension. Lancet 338:1173–1174
5. Frostell C, Fratacci MD, Wain JC, et al (1991) Inhaled nitric oxide: A selective pulmonary vasodilator reversing hypoxic pulmonary vasoconstriction. Circulation 83: 2038–2047
6. Fratacci MD, Frostell C, Chen TY, et al (1991) Inhaled nitric oxide: A selective pulmonary vasodilator of heparin-protamine vasoconstriction in sheep. Anesthesiology 75:990–999
7. Pison U, Lopez FA, Heidelmeyer CF, et al (1993) Inhaled nitric oxide reverses hypoxic pulmonary vasoconstriction without impairing gas exchange. J Appl Physiol 74:1287–1292
8. Rossaint R, Falke KF, Lopez F, et al (1993) Inhaled nitric oxide for the Adult Respiratory Distress Syndrome. N Engl J Med 328:399–405
9. Ignarro LJ, Buga GM, Wood KS, et al (1987) Endothelium-derived relaxing factor produced and released from artery and vein is nitric oxide. Proc Nat Acad Sci USA 84:9265–9269
10. Palmer RMJ, Ferrige AG, Moncada S (1987) Nitric oxide release accounts for the biological activity of endothelium-derived relaxation factor. Nature 327:524–526
11. Ignarro LJ, Buga GM, Byrns RE, et al (1988) Endothelium-derived relaxing factor and nitric oxide possess identical pharmacologic properties as relaxants of bovine arterial and venous smooth muscle. J Pharm Exp Ther 246:218–226
12. Archer SL, Rist K, Nelson DP, et al (1990) Comparison of the hemodynamic effects of nitric oxide and endothelium-dependent vasodilators in intact lungs. J Appl Physiol 68:735–747
13. Ignarro LJ (1991) Signal transduction mechanisms involving nitric oxide. Biochem Pharmacol 41:485–490

14. Palmer RMJ, Ashton DS, Moncada S (1988) Vascular endothelial cells synthesize nitric oxide from L-arginine. Nature 333:664–666
15. Moncada S, Palmer RMJ, Higgs EA (1989) Biosynthesis of nitric oxide for L-arginine. A pathway for the regulation of cell function and communication. Biochem Pharmacol 38:1709–1715
16. Furchgott RF, Vanhoutte PM (1989) Endothelium-derived relaxing and contracting factors. FASEB J 3:2007–2018
17. Ignarro LJ (1989) Biological actions and properties of endothelium-derived nitric oxide formed and released from artery and vein. Circ Res 65:1–21
18. Rimar S, Gillis CN (1993) Selective pulmonary vasodilation by inhaled nitric oxide is due to hemoglobin inactivation. Circulation 88:2884–2887
19. Gruetter CA, Gruetter DY, Lyon JE, et al (1981) Relationship between cyclic guanosine 3′,5′-monophosphate formation and relaxation of coronary artery arterial smooth muscle by glyceryl trinitrate, nitroprusside, nitrite and nitric oxide. J Pharmacol Exp Ther 219:181–186
20. Ignarro LJ (1989) Endothelium-derived nitric oxide: Actions and properties. FASEB J 3:31–36
21. Snyder SH, Bredt DS (1992) Biologic roles of nitric oxide. Scientific American 266: 68–77
22. Pollock JS, Forstermann U, Mitchell JA, et al (1991) Purification and characterization of particulate endothelium-derived relaxing factor synthase from cultured and native bovine aortic endothelial cells. Proc Natl Acad Sci USA 88:10480–10484
23. Janssens SP, Shimouchi A, Quertermous T, et al (1992) Cloning and expression of a cDNA encoding human endothelium-derived relaxing factor/nitric oxide synthase. J Biol Chem 267:14519–14522
24. Bredt DS, Hwang PM, Glatt CE, et al (1991) Cloned and expressed nitric oxide synthase structurally resembles cytochrome P-450 reductase. Nature 351:714–718
25. Xie QW, Cho HJ, Calaycay J, et al (1992) Cloning and characterization of inducible nitric oxide synthase from mouse macrophages. Science 256:225–228
26. Huang PL, Dawson TM, Bredt DS, Snyder SH, Fishman MC (1993) Targeted disruption of the neuronal nitric oxide synthase gene. Cell 75:1273–1286
27. Knowles RG, Salter M , Brooks SL, et al (1990) Anti-inflammatory glucocorticoids inhibit the induction by endotoxin of nitric oxide synthase in the lung, liver and aorta of the rat. Biochem Biophys Res Commun 172:1042–1048
28. Rees DD, Palmer RMJ, Schulz R, et al (1990) Characterization of three inhibitors of endothelium nitric oxide synthase in vitro and in vivo. Br J Pharmacol 101:746–752
29. Ignarro LJ, Kadowitz PJ (1985) The pharmacological and physiological role of cyclic GMP in vascular smooth muscle relaxation. Annu Rev Pharmacol Toxicol 25: 171–191
30. Martin W, Furchgott RF, Villani GM, et al (1986) Phosphodiesterase inhibitors induce endothelium-dependent relaxation of rat and rabbit aorta by potentiating the effects of spontaneously released endothelium-derived relaxing factor. J Pharmacol Exp Ther 237:539–547
31. Chand N, Altura BM (1981) Acetylcholine and bradykinin relax intrapulmonary arteries by acting on endothlial cells: Role in lung vascular disease. Science 213: 1376–1379
32. Dinh Xuan AT, Higenbottam TW, Clelland C, et al (1990) Acetylcholine and adenosine disphosphate cause endothelium-dependent relaxation of isolated human pulmonary arteries. Eur Respir J 3:633–638
33. Gustafsson LE, Leone AM, Persson MG, et al (1991) Endogenous nitric oxide is present in the exhaled air of rabbits, guinea pigs and humans. Biochem Biophys Res Commun 181:852–857
34. Gerlach H, Roissant R, Pappert D (1994) Autoinhalation of nitric oxide after endogenous synthesis in nasopharynx. Lancet 343:518–519
35. Alving K, Weitzberg E, Lundberg JM (1993) Increased amount of nitric oxide in exhaled air of asthmatics. Eur Respir J 6:1368–1370

36. Celermajer DS, Dollery C, Burch M, Deanfield JE (1994) Role of endothelium in the maintenance of low pulmonary vascular tone in normal children. Circulation 89: 2041–2044
37. Stamler JS, Loh E, Roddy MA, Currie KE, Creager MA (1994) Nitric oxide regulates basal systemic and pulmonary vascular resistance in healthy humans. Circulation 89: 2035–2040
38. Fritts AW, Harris P, Clauss H, et al (1958) The effect of acetylcholine on the human pulmonary circulation under normal and hypoxic condition. J Clin Invest 37: 99–108
39. Frostell CG, Blomqvist H, Hedenstierna G, et al (1993) Inhaled nitric oxide selectively reverses human hypoxic pulmonary vasoconstriction without causing systemic vasodilation. Anesthesiology 78: 427–435
40. Högman M, Frostell C, Arnberg H, et al (1993) Inhalation of nitric oxide modulates methacholine-induced bronchoconstriction in the rabbit. Eur Respir J 6: 177–180
41. Cremona G, Dinh Xuan AT, Higenbottam TW (1991) Endothelium-derived relaxing factor and the Pulmonary circulation. Lung 169: 185–202
42. Dinh Xuan AT, Higenbottam TW, Clelland C, et al (1991) Impairment of endothelium-dependent pulmonary artery relaxation in chronic obstructive lung disease. N Engl J Med 324: 1539–1547
43. Dinh Xuan AT, Higenbottam TW, Clelland C, et al (1990) Impairment of pulmonary endothelium-dependent relaxation in patients with Eisenmenger's syndrome. Br J Pharmacol 99: 9–10
44. Adatia I, Thompson J, Landzberg M, Wessel DL (1993) Inhaled nitric oxide in chronic obstructive lung disease. Lancet 341: 307–308
45. Tönz M, von-Segesser LK, Turina M (1993) Selective pulmonary vasodilatation with inhaled nitric oxide. J Thorac Cardiovasc Surg 105: 760–762
46. Berger JI, Gibson RL, Redding GJ, et al (1993) Effect of inhaled nitric oxide during Group B Streptococcal sepsis in piglets. Am Rev Respir Dis 147: 1080–1086
47. Weitzberg E, Rudehill A, Lundberg JM (1993) Nitric oxide inhalation attenuates pulmonary hypertension and improves gas exchange in endotoxin shock. Eur J Pharmacol 233: 85–94
48. Rovira I, Chen TY, Winkler M (1994) Effects of inhaled nitric oxide on pulmonary hemodynamics and gas exchange in an ovine model of ARDS. J Appl Physiol 76: 345–355
49. Zapol WM, Snider MT (1977) Pulmonary hypertension in severe acute respiratory failure. N Engl J Med 296: 476–480
50. Zapol WM, Jones R (1987) Vascular components of ARDS: Clinical pulmonary hemodynamics and morphology. Am Rev Respir Dis 136: 471–474
51. Bigatello LM, Hurford WE, Kacmarek RM (1994) Prolonged inhalation of low concentrations of nitric oxide in patients with severe adult respiratory distress syndrome. Anesthesiology 80: 761–770
52. Payen D, Gatecel C, Guinard N (1993) Inhalation of low dose of nitric oxide (NO) and IV L-arg in ARDS: Effect on pulmonary hemodynamic, gas exchange and NO metabolites. Am Rev Respir Dis 147: A720 (Abst)
53. Grover R, Smithies M, Bihari D (1993) A dose profile of the physiological effects of inhaled nitric oxide in acute lung injury. Am Rev Respir Dis 147: A350 (Abst)
54. Wysocki M, Vignon P, Roupie E, et al (1993) Improvement in right ventricular function with inhaled nitric oxide in patients with the adult respiratory distress syndrome (ARDS) and permissive hypercapnia. Am Rev Respir Dis 147: A350 (Abst)
55. Puybasset L, Rouby JJ, Mourgeon E, et al (1994) Inhaled nitric oxide in acute respiratory failure: Dose-response curves. Intensive Care Med 20: 319–327
56. Puybasset L, Stewart TE, Rouby JJ, et al (1994) Inhaled nitric oxide reverses the increase in pulmonary vascular resistance induced by permissive hypercapnia in patients with acute respiratory distress syndrome. Anesthesiology 80: 1254–1267
57. Zapol WM, Falke KJ, Rossaint R (1993) Inhaled nitric oxide for the adult respiratory distress syndrome. N Engl J Med 329: 206 (Letter)

58. Gerlach H, Rossaint R, Pappert D, Falke KJ (1993) Time-course and dose-response of nitric oxide inhalation for systemic oxygenation and pulmonary hypertension in patients with adult respiratory distress syndrome. Eur J Clin Invest 23:449–502
59. Ricou B, Suter PM (1993) Variable effects of nitric oxide (NO) in ARDS patients. Am Rev Respir Dis 147:A350 (Abst)
60. Rich GF, Murphy GD, Roos CM, Johns RA (1993) Inhaled nitric oxide: Selective pulmonary vasodilation in cardiac surgical patients. Anesthesiology 78:1028–1035
61. Payen DM, Gatecel C, Plaisance P (1993) Almitrine effect on nitric oxide inhalation in adult respiratory distress syndrome. Lancet 341:1664 (Letter)
62. Dupuy PM, Shore SA, Drazen JM, et al (1992) Bronchodilator action of inhaled nitric oxide in guinea pigs. J Clin Invest 90:421–428
63. Grover R, Murdoch I, Smithies M, et al (1992) Nitric oxide during hand ventilation in patient with acute respiratory failure. Lancet 340:1038–1039
64. Rengasamy A, Johns RA (1993) Regulation of nitric oxide synthase by nitric oxide. Molecular Pharmacology 44:124–128
65. Griscavage JM, Rogers NE, Sherman MP, Ignarro LJ (1993) Inducible nitric oxide synthase from a rat alveolar macrophage cell line is inhibited by nitric oxide. J Immunology 151:6329–6337
66. Kinsella JP, Neish SR, Shaffer E, et al (1992) Low-dose inhalational nitric oxide in persistent pulmonary hypertension of the newborn. Lancet 340:819–820
67. Kinsella JP, McQueston JA, Rosenberg AA, Abman SH (1992) Hemodynamic effects of exogenous nitric oxide in ovine transitional pulmonary circulation. Am J Physiol 263:H875–H880
68. Roberts JD Jr, Polaner DM, Lang P, et al (1992) Inhaled nitric oxide in persistent pulmonary hypertension of the newborn. Lancet 340:818–819
69. Kinsella JP, Abman SH (1993) Inhalational nitric oxide therapy for persistent pulmonary hypertension of the newborn. Pediatrics 91:997–998
70. Zayek M, Cleveland D, Morin FC, et al (1993) Treatment of persistent pulmonary hypertension in the newborn lamb by inhaled nitric oxide. J Pediatr 122:743–750
71. Roberts JD Jr, Chen TY, Kawai N (1993) Inhaled nitric oxide reverses pulmonary vasoconstriction in the hypoxic and acidotic newborn lamb. Circ Res 72:246–254
72. Zayek M, Wild L, Roberts JD, Morin FC III (1993) Effect of nitric oxide on the survival rate and incidence of lung injury in newborn lambs with persistent pulmonary hypertension. J Pediatr 123:947–952
73. DeMarco V, Skimming J, Ellis TM, Cassin S (1994) Nitric oxide inhalation: Effects on the ovine neonatal pulmonary and systemic circulations. Chest 105:91S–92S
74. Kinsella JP, Abman SH (1994) Efficacy of inhalational nitric oxide therapy in the clinical management of persistent pulmonary hypertension of the newborn. Chest 105:92S–94S
75. Abman SH, Kinsella JP, Schaffer MS, Wilkening RB (1993) Inhaled nitric oxide in the management of a premature newborn with severe respiratory distress and pulmonary hypertension. Pediatrics 92:606–609
76. Mercier JC, Dehan M (1994) Inhalational nitric oxide (NO) in persistent pulmonary hypertension of the newborn (PPHN). Results of the French multicenter open study. Am J Respir Crit Care Med 149:A545 (Abst)
77. Girard C, Lehot JJ, Pannetier JC, et al (1992) Inhaled nitric oxide after mitral valve replacement in patients with chronic pulmonary artery hypertension. Anesthesiology 77:880–883
78. Adnot S, Kouyoumdjian C, Defouilloy C, et al (1993) Hemodynamic and gas exchange responses to infusion of acetylcholine and inhalation of nitric oxide in patients with chronic obstructive lung disease and pulmonary hypertension. Am Rev Respir Dis 148:310–316
79. Kinsella JP, Toews WH, Henry D, et al (1993) Selective and sustained pulmonary vasodilation with inhalational nitric oxide therapy in a child with idiopathic pulmonary hypertension. J Pediatr 122:803–806

80. Moinard J, Manier G, Pillet O, Castaing Y (1994) Effect of inhaled nitric oxide on hemodynamics and $\dot{V}_A/\dot{Q}$ inequalities in patients with chronic obstructive pulmonary disease. Am J Respir Crit Care Med 149:1482–1487
81. Roberts JD Jr, Lang P, Bigatello LM, et al (1993) Inhaled nitric oxide in congenital heart disease. Circulation 87:447–453
82. Haydar A, Malhere T, Mauriat P, et al (1992) Inhaled nitric oxide for postoperative pulmonary hypertension in patients with congenital heart defects. Lancet 340:1545
83. Sellden H, Winberg P, Gustafsson LE, et al (1993) Inhalation of nitric oxide reduced pulmonary hypertension after cardiac surgery in a 3.2-kg infant. Anesthesiology 78: 577–580
84. Miller OI, James J, Elliott MJ (1993) Intraoperative use of inhaled low-dose nitric oxide. J Thorac Cardiovasc Surg 105:550–551
85. Berner M, Beghetti M, Ricou B, et al (1993) Relief of severe pulmonary hypertension after closure of a large ventricular septal defect using low dose inhaled nitric oxide. Intensive Care Med 19:75–77
86. Garnier P, Strambu I, Dessanges JF, Dall'ava J, Lockhart A, Dinh-Xuan AT (1994) Increased level of nitric oxide in the exhaled air from asthmatic patients. Am J Respir Crit Care Med 149:A198 (Abst)
87. Kharitonov SA, Yates D, Robbins RA, Logan-Sinclair R, Shinebourne EA, Barnes PJ (1994) Endogenous nitric oxide is increased in the exhaled air of asthmatic patients. Am J Respir Crit Care Med 149:A198 (Abst)
88. Massaro AF, Gaston B, Fanta C, Stamler J, Drazen JM (1994) Expired nitric oxide concentration in normal subjects and patients with asthma. Am J Respir Crit Care Med 149:A198 (Abst)
89. Kharitonov SA, Yates D, Robbins RA, Thomas PS, Barnes PJ (1994) Corticosteroids decrease exhaled nitric oxide. Am J Respir Crit Care Med 149:A201 (Abst)
90. Levine JS (1989) Photochemistry of biogenic gases. In: Rambler MB, Margulis L, Fester R (eds) Global Ecology, Academic Press, pp 51–74
91. Norman V, Keith C (1965) Nitrogen oxides in tobacco smoke. Nature 205:915–916
92. NIOSH recommendations for occupational safety and health standards (1988) MMWR Morb Mortal Wkly Rep 37 (Suppl S-7):21
93. Stavert DM, Lehnert BE (1990) Nitric oxide and nitrogen dioxide as inducers of acute pulmonary injury when inhaled at relatively high concentrations for brief periods. Inhal Toxicol 2:53–67
94. Glasson WA, Tuesday CS (1963) The atmospheric thermal oxidation of nitric oxide. J Am Chem Soc 85:2901–2904
95. Beckman JS, Beckman TW, Chen J, et al (1990) Apparent hydroxyl radical production by peroxynitrite: Implications for endothelial injury from nitric oxide and superoxide. Proc Natl Acad Sci USA 87:1620–1624
96. Hibbs JB, Taintor RR, Vavrin Z, Rachlin EM (1988) Nitric oxide: A cytotoxic activated macrophage effector molecule. Biochem Biophys Res Commun 157:87–94
97. Stuehr DJ, Nathan CF (1989) Nitric oxide: A macrophage product responsible for cytostasis and respiratory inhibition in tumor target cells. J Exp Med 169:1543–1555
98. Borland CDR, Harmes K, Cracknell N, et al (1985) Methaemoglobin levels in smokers and non-smokers. Arch Environ Health 40:330–333
99. Beutler E (1977) Methemoglobinemia and sulfhemoglobinemia. In: Williams WJ, Beutler E, Erslev AJ, Rundles RW (eds) Hematology 2nd edition. McGraw-Hill, New York, pp 491–494
100. Oda H, Nogami H, Kusumoto S, et al (1976) Long-term exposure to nitric oxide in mice. J Jpn Soc Air Pollut 11:150–160
101. von Nieding G, Wagner H, Kockeler H (1975) Investigation of the acute effects of nitrogen monoxide on lung function in man. Staub-Reinhalt Luft 35:175–178
102. Wessel DL, Adatia I, Thompson JE, Hickey PR (1994) Delivery and monitoring of inhaled nitric oxide in patients with pulmonary hypertension. Crit Care Med 22: 930–938

103. Stamler JS, Jaraki O, Osborne J, et al (1992) Nitric oxide circulates in mammalian plasma primarily as an S-nitroso adduct of serum albumin. Proc Natl Acad Sci USA 89:7674–7677
104. Payen D (1993) Inhaled NO in severe ARDS. Presented at the International Conference on Recent Advances in the Treatment of the Adult Respiratory Distress Syndrome, Tutzing, Germany, July 5–8
105. Stamler JS, Singel DJ, Loscalzo J (1993) Biochemistry of nitric oxide and its redox-activated forms. Science 258:1989–1902
106. Nguyen T, Brunson D, Crespi CL, et al (1992) DNA damage and mutation in human cells exposed to nitric oxide in vitro. Proc Natl Acad Sci USA 89:3030–3034
107. Radomski MW, Palmer RMJ, Moncada S (1987) Endogenous nitric oxide inhibits human platelet adhesion to vascular endothelium. Lancet 2:1057–1058
108. Högman M, Frostell C, Arnberg H, Hedenstierna G (1993) Bleeding time prolongation and NO inhalation. Lancet 341:1664–1665
109. Högman M, Frostell C, Arnberg H, Sandhagen B, Hedenstierna G (1994) Prolonged bleeding time during nitric oxide inhalation in the rabbit. Acta Physiol Scand 151: 125–129
110. Clutton-Brock J (1967) Two cases of poisoning by contamination of nitrous oxide with higher oxides of nitrogen during anaesthesia. Br J Anaesth 39:388–392
111. Austin AT (1967) The chemistry of higher oxides of nitrogen as related to the manufacture, storage and administration of nitrous oxide. Br J Anaesth 39:345–350
112. Greenbaum R, Bay J, Hargreaves MD, et al (1967) Effects of higher oxides of nitrogen on the anesthetized dog. Br J Anaesth 39:393–404
113. Hugod C (1979) Effect of exposure to 43 ppm nitric oxide and 3.6 ppm nitrogen dioxide on rabbit lung. Int Arch Occup Environ Health 42:159–167
114. Frostell C (1994) Nitric oxide inhalation – future drug or an invitation to disaster. Paediatric Anaesthesia 4:147–150
115. Foubert L, Fleming B, Latimer R, et al (1992) Safety guidelines for use of nitric oxide. Lancet 339:1615–1616
116. Bouchet M, Renaudin MH, Raveau C, et al (1993) Safety requirement for use of inhaled nitric oxide in neonates. Lancet 341:968–969
117. Oda H, Kusumoto S, Nakajima T (1975) Nitrosyl-hemoglobin formation in the blood of animals exposed to nitric oxide. Arch Environ Health 30:453–456
118. Fontijin A, Sabadell AJ, Ronco RJ (1970) Homogeneous chemiluminescent measurement of nitric oxide and ozone. Anal Chem 42:575–579
119. Petros AJ, Cox PB, Bohn D (1992) Simple method for monitoring concentration of inhaled nitric oxide. Lancet 340:1167 (Lett)
120. Zapol WM, Rimar S, Gillis N, et al (1994) Nitric oxide and the lung. Am J Resp Crit Care Med 149:1375–1380

Monitoring and Safety of Inhaled Nitric Oxide

C. G. Frostell

Introduction

Since 1991, inhalation of nitric oxide (NO) has started to attract attention as a possible therapeutic option in states of pulmonary hypertension of varied etiology [1–4]. In addition, it has become clear that a majority of hypoxemic patients demonstrate improved arterial oxygenation when exposed to minute doses of NO [1, 4, 5]. Reviews of potential indications for clinical use of inhaled NO have been presented elsewhere [6, 7].

Clinical administration of inhaled NO is complicated by the fact that its use as a therapeutic agent is not yet approved anywhere in the world. Registration of NO as a therapeutic agent will have to await definite proof from multicenter clinical trials.

In 1995, therapeutic use of inhaled NO on compassionate (life-saving) grounds will probably be accepted in some parts of the European Union as well as elsewhere in the world. NO official norm presently exists defining possible indications as well as minimum requirements for delivery devices and monitoring. This represents a potentially dangerous situation, which has to be realized by anyone contemplating clinical use of inhaled NO [8]. A further complication of this situation is the existence of two qualities of NO in 1994. There is medical grade NO manufactured under GMP ('good manufacturing practice') rules, used under strict supervision within clinical trials for compassionate reasons in areas where such use is allowed for by the relevant medical product agency. Then there is also an industrial grade NO used for calibration and other purposes, which in many areas can still be delivered to hospitals outside the regulations set up for medicinal gases. On the other hand, therapeutic use of such industrial grade NO is illegal in some parts of the European Union. This chapter will limit itself to use of *medical grade NO*, in situations where the hospital has established approval from the relevant medical product agency, ethical committee as well as hospital pharmacy.

Acceptable clinical safety and good monitoring standards of inhaled NO requires that we know:

1) the clinically effective dose of inhaled NO;
2) what side effects inhaled NO and other oxides of nitrogen may lead to;

3) and how to measure NO and other oxides of nitrogen in inspired gas in a clinical setting.

At present, we interpret 'effective dose' as the delivery of a known and stable NO concentration, resulting in a desired clinical benefit, with a minimized risk for toxic side effects. This chapter will therefore cover:

1) what actions of inhaled NO may confer 'clinical benefit'?
2) which are the toxic side effects of NO and other oxides of nitrogen and how can we minimize them?
3) how to safely deliver NO adequately for a prolonged length of time?
4) how to adequately measure the level of NO and NO_2?
5) what parameters should be monitored during therapy with NO?
6) when should an ICU be equipped for NO delivery?

What Actions of Inhaled NO may Confer 'Clinical Benefit'?

Inhaled NO may improve oxygen exchange and reduce mean pulmonary artery pressure (PAP) in several clinical situations characterized by some degree of lung vascular constriction and/or hypoxemia with or without artificial ventilation [1, 9–11].

We may desire maximum reduction in PAP, which would probably necessitate a high dose of NO [11, 12]. On the other hand, we may prefer a maximum improvement in oxygen exchange in hypoxemia from acute lung injury. The identification of an optimum dose is imperative, as increasing impairment of oxygen exchange can be seen at high doses [12–14]. Alternatively, a reduction in pulmonary microvascular pressure [15] may be desired, necessitating another dose-response curve. NO studies yet exist, establishing which actions of inhaled NO should be given high priority. Indeed, while one patient may benefit from optimized gas exchange, the next patient may derive more clinical benefits from a reduced microvascular pressure!

Which are the Toxic Side Effects of NO and Other Oxides of Nitrogen and How can we Minimize Them?

Methemoglobin Formation

Inhaled NO is to a major extent taken up by the lung. NO entering blood perfusing the lung is inactivated by being bound to hemoglobin in red cells, the endproducts being methemoglobin (MetHb) and nitrate. MetHb is reduced by MetHb reductase back to ferrous hemoglobin. Nitrate is eliminated in the urine [16]. At an inhaled dose below 100 ppm in human adults, MetHb formation is not a big problem [17, 18]. However, at NO doses of over 500 ppm, all tested adult subjects reacted with methemoglobinemia within 1 h [19]. Fullterm and premature neonates as well as small children may display reduced MetHb reductase activity [20], and have reacted with MetHb accumulation during therapy with inhaled NO doses in the range 20–100 ppm [2, 21]. Wessel et al. [2] described clinical experiences from treating 123 patients with inhaled NO, and found 4/123 to develop $>5\%$ methemoglobinemia. All responded well to a dose reduction of NO and recovered without sequelae.

Formation of Nitrogen Dioxide (NO_2)

NO oxidizes spontaneously in an O_2 containing gas mixture to nitrogen dioxide (NO_2), with the rate of NO_2 formation proportional to the O_2 concentration and the square of the NO concentration [22]. NO may in this way cause airway inflammation and damage [22–25]. The acute pulmonary toxicity of NO_2 is often stated to be several times greater than NO. Consequently, the National Institute of Occupational Safety and Health (USA) recommends a TWA (time weighted average) exposure of 5 ppm for NO_2 but 25 ppm for NO [26]. Public Safety agencies from other countries have adopted similar or lower levels. NO_2 levels of less than 0.5 ppm may enhance human airway hyperreactivity [27]. Toxic pulmonary effects of breathing NO_2 at or below 5 ppm are reported. These include altered surfactant chemistry and metabolism [28], epithelial hyperplasia of terminal bronchiole in rats [29], as well as diffuse inflammation and hyperreactivity. At higher inhaled doses (>1000 ppm NO) prolonged inhalation causes pulmonary edema and deaths have been reported [30, 31]. A soda-lime scrubber on the inspiratory limb can reduce the amount of NO_2 in inspired gas [32, 33]. It has to be remembered that also other higher and toxic oxides of nitrogen are formed when NO and oxygen are mixed [23].

Interaction with Coagulation

A prolonged bleeding time has been reported in rabbits [34]. A prolongation of bleeding time was measured also in humans breathing 30 ppm NO for 15 min [35], with the effect absent when the same volunteers were later retested inhaling only 10 ppm NO (unpublished observation).

NO reacting with sulphur in protein may cause S-nitrosylation, forming S-nitrosothiol groups (R–S–NO) [36]. Most proteins, including those involved in coagulation, would consequently be possible targets for nitrosylation when exposed to inhaled NO. Such nitrosylation could in turn alter hemostasis, surfactant function and other activities.

Mutagenicity

NO has mutagenic potential in a bacterial test model [37], and also induce mutations and chromosomal aberrations in rat primary lung cells [38].

Increased antioxidant activity [39] may in the future be a strategy to reduce this risk, together with a low-dose strategy.

Withdrawal Reactions

Rapid rebound reactions (intensified pulmonary vasoconstriction, hypoxemia) after sudden withdrawal of inhaled NO therapy have been described [40, 41]. This may in part be due to interference with NO synthase activity, exogenous NO suppressing endogenous NO formation [42]. Patients displaying such 'NO-dependency' might benefit from careful weaning down to successively lower doses of NO [43], maybe even in the end at a fraction of 1 ppm.

How to Safely Deliver NO Adequately for a Prolonged Length of Time?

At present, only few commercially available delivery devices exist, as most manufactured such devices are used in clinical trials and not available outside these trials. Regulations within the European Union will soon make it impossible to employ home assembled medical equipment that has not passed proper inspection and certification, for the purpose of administering medical drugs. In the meantime, many delivery devices for inhaled NO have been home assembled in Europe and elsewhere. Several of these do not include on-line monitoring at all, or monitor without alarms on the measured level of NO and NO_2.

A large number of experimental exposures of volunteers and patients to inhaled NO for short periods of time (< 1 h) have been made utilizing a non-rebreathing circuit including a reservoir bag [18, 35, 44]. The subjects typically wore a nose-clip and breathed either via a mouthpiece or a facemask. Winberg et al. [45] studied awake children during heart catheterization. They constructed a plexiglass hood that was placed over the head of the infant, into which inspired gas containing NO could be added. Wessel et al. [2] described 5 different systems for the delivery and monitoring of NO and NO_2 during spontaneous breathing, controlled mechanical ventilation, and during anesthesia.

During mechanical ventilation, NO can be added either before the ventilator at a fresh gas inlet, or following the ventilator in the inspiratory limb. Several variants including adding *NO in N_2 to the inspiratory limb* have been described [1, 46–49]. Soda lime scrubbing of NO_2 was not always used in such systems. Ventilator systems are either of a constant-flow type or inspiratory flow triggered. A gas mixing system, adding NO in N_2 to a low pressure gas inlet *before the ventilator*, was described by Stenqvist et al. [32] using mass flow regulators. An earlier home assembled version of this system provided flow-meter controlled gas mixtures of NO in N_2, O_2 and air to a port [50] on a Siemens Servo 900C ventilator. A soda lime canister should be used on all pre-ventilator gas mixing systems. This precaution will minimize the high levels of NO_2 formed due to the prolonged residence time of NO in the elevated O_2 concentration of the inspiratory circuit. Toxic levels of NO_2 is formed in <10 sec at a high oxygen concentration containing >80 ppm NO [51].

The new Siemens Servo 300 (Siemens, Solna, Sweden) ventilator uses another NO mixing method, in which flow regulators blend air, oxygen and NO in N_2 to the final selected concentrations without utilizing a traditional mixing bag. This is yet another example of the prototype delivery devices employed for research but not yet commercially available.

Several delivery systems have been described where NO in N_2 is mixed continuously or intermittently on the inspiratory limb [1, 46–49]. Such systems should as extra safety incorporate an automatic switch-off of the NO/N_2 flow, in the event the main gas delivery (ventilator) is stopped.

A properly functioning soda lime absorber may reduce 35 ppm NO_2 to <1 ppm NO_2 (that is >95% effect) if the ratio NO/NO_2 is close to 1. However, in FiO_2 0.90, high flow of 10 L/min, 100 ppm NO and a ratio of already NO/NO_2 of 40, the scrubber can reduce the NO_2 level only 10–20%. Obviously, the efficiency of an absorber is determined by the pre-absorber ratio of NO to NO_2 and the FiO_2. For the reasons mentioned above, we advocate the use of an absorber, which will push the ratio NO/NO_2 as high as possible, if something fails with the delivery system before the patient.

Back-up devices: Any ICU doing NO therapy should have a manual back-up device next to the patient. If total failure occurs, it would be possible to at least handventilate the patient with NO and pure oxygen until the system can be restarted.

Environmental aspects: Significant leaks of NO into room air is a potential health hazard also for hospital staff, mainly because of NO_2 formation in a poorly ventilated area. We have never worked clinically with higher stock cylinder concentrations than 1000 ppm NO, partly for this reason. Exhaled gas from a patient treated with inhaled NO should if possible be scavenged through the hospital gas evacuation system.

How to measure the Level of NO and NO_2 in Inspired Gas?

At the outset of this section, it must be stated that in a clinical situation we usually attempt to measure a small amount of NO_2 in an oxygen rich gas mixture with a high level of NO. The gas mixture undergoes cyclic pressure changes and may have high humidity or consist of dry gas.

Despite these difficulties we insist that NO and NO_2 levels are measured continuously for days and weeks, for reasons of safety. In addition, most staff around the patient will have limited knowledge of the pitfalls of NO and NO_2 monitoring, have little time to spare for the measurement of these variables, and need to form decisions rapidly on derived numbers if an alarm is activated.

Chemiluminescence (CI)

Present CI monitors were not developed with the intention to work optimally in a high FiO_2 environment with a gas mixture in which typically the NO content is high and the NO_2 level is desired to be a tiny fraction of that number. These machines measure NO and all NOxes , and then subtracts the two to get NO_2. It is typically stated by the manufacturer that this can be performed with an error in NO_2 $<10\%$. Ponder the meaning of that statement in a situation where NO and $NOx \gg NO_2$. Obviously then, the error can be much higher, maybe larger than the calculated number for NO_2 itself! This is probably partly the reason why sometimes we find even negative NO_2 values in gas mixtures with high NO concentrations. However during calibration procedures, we may work with gas mixtures containing only NO_2 in nitrogen or air. In these states we consider the CI to be the 'golden standard' for other techniques.

The literature data on NO_2 levels during experimental and clinical exposure to NO has therefore to be read with considerable caution, keeping the pitfalls just described in mind.

With properly calibrated equipment, we have never been able to produce a NO/NO_2 ratio >50 at $FiO_2=0.90$ and $NO=100$ ppm, as measured with EC cells. That is, even when employing high gas flows rates of >10 L/min and a functioning soda lime absorber, we typically found 2.2–3.0 ppm NO_2.

This is at variance with results recently published by investigators using CI technique [21, 52]. And with CI, even negative NO_2 values may sometimes be obtained [53], for reasons yet to be carefully described. Also others have reported problems with CI measurements of NO and NO_2 in high FiO_2 (personal communication in a recent paper [52]).

Electrochemical Cells (EC)

EC technique for the measurement of NO and NO_2 levels are rapidly spreading as they are not so costly and comparably easy to operate and calibrate. It is our experience that several commercially available such cells monitor NO levels with acceptable accuracy. NO_2 levels, on the other hand, have been much more difficult to measure adequately for several reasons. We have problems such as absorbing of NO_2 to tubing materials, pressure and flow sensitivity of a sensor head, zero drift, long response time, difficulty in handling of calibration gas mixtures, unstable NO_2 levels in oxygenrich and NO containing gas; just to name the more common pitfalls.

When assembling monitoring units outside factories, a host of factors therefore have to be closely considered:
1) how can my monitor tolerate dry gas? (sometimes not)
2) how can I humidify the gas to monitor?
3) is my monitor pressure sensitive? (usually yes)

4) what is the response time of the monitor and how must that affect my monitoring strategy? (usually >10 sec, and tidal variations will be lost)
5) how can I assure gas flow over the sensor? (a pump?)
6) how often is recalibration necessary? (often more than monthly)
7) does the monitor have a zerodrift? (usually yes)
8) where in my delivery system must I monitor?

If inspiratory concentration is to be measured, it makes little sense to monitor mixed gases (Y-piece, trachea) or exhaled gas. However, some researchers anyhow have advocated these approaches since they then have less problems with humidifying gas and/or pressure swings. On the other hand, too low NO/NO_2 values might be seen, as NO and NO_2 are absorbed in the human lung [54]. In addition with a response time > the time for one breathing cycle, only mean values between inspiratory high and expiratory low concentrations will be recorded. This phenomena affects reported dose-response curves and as an example the inspiratory NO doses reported by Puybasset et al. [13] were most probably 30–50% higher than stated in their paper.

UV and IR Technique

Some investigators have recently used such techniques [32] but not reported an improvement upon other more commonly employed methods such as EC or CI based measurements.

Calibration of NO_2 monitors, as we see it, requires access to N_2, variable FiO_2 0.21–0.90, as well as NO_2 in N_2. We first measure NO_2 in a gas mixture with high FiO_2 containing >80 ppm NO which spontaneously should form 1–3 ppm NO_2 in spite of a flow rate through the measuring system of 10 L/min. We then add a few ppm NO_2 in N_2 to that mixture, keeping FiO_2 constant. The addition of NO_2 should be detected as an accurately measured increase in NO_2, which then can be reduced by passing the gas mixture through a soda lime absorber. If this can be reproduced repeatedly, we can be safe that the monitoring system can act as a true detector for what level of NO_2 the patient is exposed to.

What Other Parameters should be monitored during Therapy with NO?

Inspired oxygen fraction (FiO_2)

Maybe the accumulated experience with inhaled NO will show us that the important aspects to monitor during this therapy is different than the NO level itself? One obvious and maybe slightly overlooked danger with inhaled NO is that it is delivered in the suffocating gas nitrogen in order to remain stable, and then is to be mixed into inspired gas to a patient who needs an oxygen enriched gas. Against this background, we consider it vital to put high demands on the delivery device for inhaled NO, and in addition to always measure the resulting FiO_2 distal to the mixing point in such delivery devices. This routine should rule out any delivery strategy based upon the delivery of NO to the Y-piece or intratracheal administration.

Pulmonary Artery Catheter/Monitoring of Central Hemodynamics

In the first clinical exposures of patients to inhaled NO, pulmonary artery catheters were used to monitor effects of NO on central hemodynamics [1]. Although this made good sense from a research perspective, we do not consider such monitoring necessary over the whole treatment period. In a started randomized study over effect of inhaled NO on outcome in acute lung injury, pulmonary artery catheterization is required at therapy start. The major reason for this procedure is to exclude patients with left heart failure from therapy with inhaled NO. In this context, it is interesting to note a side effect of inhaled NO described by Bocchi et al. [55], reporting the development of pulmonary edema in 3 patients with left heart failure exposed to NO during mechanical ventilation. Complete data on pulmonary capillary wedge pressure (PCWP) on 2 of 3 patients were given in the paper. PCWP was >24 mm Hg at the outset in all 3 patients, and rose >5 mm Hg during exposure to NO in the 2 patients for whom data on this aspect are given. At present, we should confirm the absence of left heart failure and degree of pulmonary hypertension before starting inhaled NO as therapy. When such a catheterization is contraindicated, an echocardiography could help to establish this information.

Especially neonates and premature babies are rarely catheterized before/during therapy with inhaled NO. In that situation, it must remain state of the art to rule out major cardiac malformations with echocardiography, establish information on the degree of pulmonary hypertension and extrapulmonary shunting as well as monitor the result of administration of NO to the patient.

Arterial Pressure, Pulse Oximetry, and Backup Delivery Device

In ongoing clinical studies, maximum and minimum alarms on the level of NO in inspired gas has been requested by some authorities, as this therapy is studied for potential life-saving properties. It was therefore considered necessary to at all times be able to identify sudden accidental withdrawal of inhaled NO. Such withdrawal has been described to sometimes cause cardiopulmonary collapse as stated above. Using electrochemical cells for NO and NO_2 monitoring, the lowest level possible to identify is 1–2 ppm. During weaning of patients with 'NO dependency', doses as low as 0.2 ppm may be used for 1–3 days. It is therefore not possible to install a lower limit alarm on presently existing EC cells that would indicate withdrawal of <2 ppm NO.

A meaningful alarm cannot be only optional or easily inactivated for more than short period (few minutes) for medicolegal reasons. Another strategy has consequently to be implemented as follows: A patient who is 1) dependent on inhaled NO; 2) administered a low dose which cannot be measured adequately; and 3) by accident, has acute withdrawal so that NO may rapidly react with falling arterial pressure and desaturation. The continuous monitoring of pulse oximetry and intraarterial pressure will consequently reliably warn for sudden NO withdrawal on a patient that has become 'NO-dependent'. The staff is in such a situation instructed to increase FiO_2 and NO to measurable levels (>2 ppm). If this action for some reasons fails, they will switch the patient to therapy with a manually operated backup device that delivers a preset gas mixture with an $FiO_2 > 0.95$ and 5–10 ppm NO until the error with the delivery equipment can be identified. It follows the backup device needs to be available of the bedside at all times, and fully operational within 3 min.

Methemoglobin

In ongoing clinical European, multicenter studies on the effects of inhaled NO, methemoglobin levels in adults are measured by blood sampling at least daily, and in neonates at least 4 times daily. Methemoglobinemia may not be detected with ordinary pulse oximetry devices [56].

Coagulation Parameters

As stated above, exposure to 30 ppm NO was reported to cause prolongation of bleeding time in rabbits and adult volunteers [35]. The effect on healthy adult volunteers has been difficult to reproduce in an enlarged study on 30 and 80 ppm NO inhaled for 15 min each (unpublished result). Inhaled NO has been used with clinical benefit during extracorporeal membrane oxygenation in children (Kachel and PA Lönnqvist, personal communications) and extracorporeal carbon dioxide removal in adults [1] with concomitant IV anticoagulation therapy. To the knowledge of this author, in the literature so far, no major problems with increased bleeding tendency has been reported with this combination. At present, concomitant intracranial hemorrhage is therefore recognized as a contraindication to NO inhalation in 2 European multicenter studies, but no repeated blood sampling evaluating altered hemostasis is required.

Long-term Follow-up

Chromosomal aberrations and mutations have been ascribed to inhaled NO at doses below 100 ppm [37, 38]. However, some of these observations may be partly due to effects of higher oxides of nitrogen always present during exposure to NO in an oxygen containing atmosphere. Another unknown aspect of NO inhalation is how this therapy may influence pulmonary morphology and the healing process after an acute illness. Consequently, it must at present be considered necessary to enroll all patients exposed to therapeutic inhaled NO for a long-term follow-up program over many years in order to identify any such delayed side effects. Minimum data collected in such a file should include type of delivery and monitor systems used, indication for use of NO therapy, clinical outcome, maximum doses and delivery time of NO and NO_2, total duration of therapy with NO, concomitant diseases, as well as adverse events during therapy. Next of kin and patient must be informed of necessity for a follow-up over many years, and the possibility to trace treated patients must be secured.

When should an ICU be Equipped for NO Delivery?

All medicinal gases are regarded as drugs by authorities in the European Union and North America. This means that therapeutic inhalation of NO must pass drug registration and qualify for such a lable through evidence of efficacy and safety. Thus only multicenter, randomized ('phase 3') trials providing evidence of efficacy and safety characteristics, will legalize clinical use of inhaled NO. It will take several years to generate this knowledge. Meanwhile, all clinical application of inhaled NO should take place on compassionate (life-saving) grounds only, or when significant morbidity/permanent disability is risked.

Several investigators maintain that medical grade NO should only be made available in regions where doctors are able to establish a consensus group of interested clinicians and researchers advising others on management of NO therapy in a pre-registration situation. Such a consensus group could take the initiative and interim responsibility of giving temporary guidelines for a region/country on:

1) compassionate use indications and how to verify their presence;
2) minimum requirements on delivery devices and monitoring routines including training of staff; and
3) a structure for long-term (years) follow-up of exposed patients.

If the clinicians then equipped themselves with ethical committee approvals, and (when required) medical product agencies approval, it should be possible to use medical grade NO. A few such consensus groups or 'working parties' have been established. As an example, ESA (European Society of Anaesthesiology) appointed a working group for establishing guidelines for inhaled NO, at the meeting in Brussels (February 1994).

Conclusion

In my opinion, in 1995 it is still good policy to await data from ongoing randomized, controlled multicenter studies in adults and infants before spreading this potential therapy further outside a clinical trial program. Interested clinicians could support the trials by arranging for transport of suitable patients to participating centers. If therapy with inhaled NO is considered outside clinical trials, it ought to be based on strict national/regional consensus guidelines for compassionate use.

Acknowledgements: This work was in part made possible by a grant from the Swedish Medical Research Council (9073). Helpful comments on monitoring of NO and NO_2 were supplied by colleagues R. Lindwall and L. Gustafsson, Stockholm.

References

1. Rossaint R, Falke KJ, Lopez F, Slama K, Pison U, Zapol WM (1993) Inhaled nitric oxide in adult respiratory distress syndrome. N Engl J Med 328:399–405
2. Wessel DL, Adatia I, Thompson JE, Hickey PR (1994) Delivery and monitoring of inhaled nitric oxide in patients with pulmonary hypertension. Crit Care Med 22: 930–938
3. Kinsella JP, Neish SR, Ivy DD, Shaffer E, Abman SH (1993) Clinical responses to prolonged treatment of persistent pulmonary hypertension of the newborn with low doses of inhaled nitric oxide. J Pediatr 123:103–108
4. The French study group of NO (1994) Pulmonary disease-related responses to inhaled nitric oxide (NO) in severely hypoxemic newborns. Intensive Care Med 20:619 (Abst)
5. Finer NN, Etches PC, Kamstra B, Tierney AJ, Peliowski A, Ryan CA (1994) Inhaled nitric oxide in infants referred for extracorporeal membrane oxygenation: Dose response. J Pediatr 124:302–308
6. Rich GF Johns RA (1994) Nitric oxide and the pulmonary circulation. In: Lake CL, Barash PG Sperry RJ (eds) Advances in Anesthesia. Mosby-Year Book, Inc, St.Louis, pp 1–25
7. Hurford WE, Zapol WM (1994) Nitric oxide inhalation in the intensive care unit. Curr Opin Anesth 7:153–160
8. Frostell CG (1994) Nitric oxide inhalation – future drug or invitation to disaster? Paed Anaesth 4:147–150
9. Pepke-Zaba J, Higenbottam TW, Dinh-Xuan AT, Stone D, Wallwork J (1991) Inhaled nitric oxide as a cause of selective pulmonary vasodilation in pulmonary hypertension. Lancet 338:1173–1174
10. Girard C, Lehot JJ, Pannetier JC, Filley S, Ffrench P, Estanove S (1992) Inhaled nitric oxide after mitral valve replacement in patients with chronic pulmonary artery hypertension. Anesthesiology 77:880–883
11. Roberts JD, Lang P, Bigatello LM, Vlahakes GJ, Zapol WM (1993) Inhaled nitric oxide in congential heart disease. Circulation 87:447–453
12. Gerlach H, Rossaint R, Pappert D, Falke KJ (1993) Time-course and dose-response of nitric oxide inhalation for systemic oxygenation and pulmonary hypertension in patients with adult respiratory distress syndrome. Eur J Clin Invest 23:499–502
13. Puybasset L, Rouby JJ, Mourgeon E, et al (1994) Inhaled nitric oxide in acute respiratory failure: Dose-response curves. Intensive Care Med 20:319–327

14. Lundin S, Westfelt U, Stenqvist O, et al (1994) Dose-response of nitric oxide inhalation in early acute lung injury. Br J Anaesth 72 (Suppl): A106 (Abst)
15. Benzing A, Geiger K (1994) Inhaled nitric oxide lowers pulmonary capillary pressure and changes longitudinal distribution of pulmonary vascular resistance in patients with acute lung injury. Acta Anaesthesiol Scand 38:640–645
16. Wennmalm Å, Benthin G, Edlund A, et al (1993) Metabolism and excretion of nitric oxide in humans. Circ Res 73:1121–1127
17. Yoshida K, Kasama K (1987) Biotransformation of nitric oxide. Environ Health Perspect 73:201–206
18. Frostell C, Blomqvist H, Hedenstierna G, Lundberg J, Zapol WM (1993) Inhaled nitric oxide selectively reverses human hypoxic pulmonary vasoconstriction without causing systemic vasodilation. Anesthesiology 78:427–435
19. Young JD, Dyar O, Xiong L, Howell S (1994) Methaemoglobin production in normal adults inhaling low concentrations of nitric oxide. Intensive Care Med 20:581–584
20. Nilsson A, Engberg G, Henneberg S, Danielson K, De Verdier CH (1990) Inverse relationship between age-dependent erythrocyte activity of methaemoglobin reductase and prilocaine-induced methaemoglobinaemia during infancy. Br J Anaesth 64:72–76
21. Lönnqvist PA, Winberg P, Lundell B, Sellden H, Olsson GL (1994) Inhaled nitric oxide in neonates and children with pulmonary hypertension. Acta Paediatr Scand 83: 1132–1136
22. Austin AT (1967) The chemistry of the higher oxides of nitrogen as related to the manufacture, storage and administration of nitrous oxide. Br J Anaesth 39:345–350
23. Gaston B, Drazen JM, Loscalzo J, Stamler JS (1994) The biology of nitrogen oxides in the airways. Am J Respir Crit Care Med 149:538–551
24. Stavert DM, Lehnert BE (1990) Nitric oxide and nitrogen dioxide as inducers of acute pulmonary injury when inhaled at relatively high concentrations for brief periods. Inhal Toxicol 2:53–67
25. Frampton MW, Morrow PE, Cox C, Gibb FR, Speers DM, Utell MJ (1991) Effects of nitrogen dioxide exposure on pulmonary function and airway reactivity in normal humans. Am Rev Respir Dis 143:522–527
26. Centers for Disease Control (1988) Recommendations for Occupational Safety and Health Standard. MMWR 37:S-7 (21)
27. Bylin G, Hedenstierna G, Lindvall T, Sundin B (1988) Ambient nitrogen dioxide concentrations increase bronchial responsiveness in subjects with mild asthma. Eur Respir J 1:606–612
28. Müller B, Schäfer H, Barth P, von Wichert P (1994) Lung surfactant components in bronchoalveolar lavage after inhalation of NO_2 as markers of altered surfactant metabolism. Lung 172:61–72
29. Evans MJ, Stephens RJ, Cabral LJ, Freeman G (1972) Cell renewal in the lungs of rats exposed to low levels of NO_2. Arch Environ Health 24:180–188
30. Greenbaum R, Bay J, Hargreaves MD, et al (1967) Effects of higher oxides of nitrogen on the anesthetized dog. Br J Anaesth 39:393–404
31. Clutton-Brock J (1967) Two cases of poisoning by contamination of nitrous oxide with the higher oxides of nitrogen during anaesthesia. Br J Anaesth 39:388–392
32. Stenqvist O, Kjelltoft B, Lundin S (1993) Evaluation of a new system for ventilatory administration of nitric oxide. Acta Anaesthesiol Scand 37:687–691
33. Ishibe Y, Sato T, Hayashi T, Kato N, Hata T (1995) Effective absorption of nitrogen dioxide with soda lime. Br J Anaesth 74:107–114
34. Högman M, Frostell C, Arnberg H, Sandhagen B, Hedenstierna G (1994) Prolonged bleeding time during nitric oxide inhalation in the rabbit. Acta Physiol Scand 151: 125–129
35. Högman M, Frostell C, Arnberg H, Hedenstierna G (1993) Bleeding time prolongation and NO inhalation. Lancet 341:1664–1665
36. Stamler JS, Simon DI, Osborne JA, et al (1992) S-Nitrosylation of proteins with nitric oxide: Synthesis and characterization of biologically active compounds. Proc Natl Acad Sci 89:444–448

37. Arroyo PL, Hatch-Pigott V, Mower HF, Cooney RV (1992) Mutagenicity of nitric oxide and its inhibition by antioxidants. Mutation Research 281:193–202
38. Isomura K, Chikahira M, Teranishi K, Hamada K (1984) Induction of mutations and chromosome aberrations in lung cells following in vivo exposure of rats to NO. Mutation Research 136:119–125
39. Walker MW, Kinter MT, Roberts RJ, Spitz DR (1995) Nitric oxide-induced cytotoxicity: Involvement of cellular resistance to oxidative stress and the role of glutathione in protection. Pediatr Res 37:41–49
40. Berner M, Beghetti M, Ricou B, Rouge JC, Pretre R, Friedli B (1993) Relief of severe pulmonary hypertension after closure of a large ventricular septal defect using low dose inhaled nitric oxide. Intensive Care Med 19:75–77
41. Girard C, Durand PG, Vedrinne C, et al (1993) Case conference: Inhaled nitric oxide for right ventricular failure after heart transplantation. J Cardiothorac Vasc Anesth 7:481–485
42. Assreuy J, Cunha FQ, Liew FY, Moncada S (1993) Feedback inhibition of nitric oxide synthase activity by nitric oxide. Br J Pharmacol 108:833–837
43. Petros AJ (1994) Down-regulation of endogenous nitric oxide production after prolonged administration. Lancet 344:191
44. Högman M, Frostell CG, Hedenstrom H, Hedenstierna G (1993) Inhalation of nitric oxide modulates adult human bronchial tone. Am Rev Respir Dis 148:1474–1478
45. Winberg P, Lundell B, Gustafsson LE (1994) Effect of inhaled nitric oxide on raised pulmonary vascular resistance in children with congenital heart disease. Br Heart J 71:282–286
46. Roberts JD, Chen TY, Kawai N, et al (1993) Inhaled nitric oxide reverses pulmonary vasoconstriction in the hypoxic and acidotic newborn lamb. Circ Res 72:246–254
47. Pison U, Lopez FA, Heidelmeyer CF, Rossaint R, Falke KJ (1993) Inhaled nitric oxide reverses hypoxic pulmonary vasoconstriction without impairing gas exchange. J Appl Physiol 74:1287–1292
48. Benzing A, Beyer U, Kiefer P, Geiger K (1993) Inhaliertes Stickstoffmonoxid. Anaesthetist 42:175–178
49. Young JD (1994) An universal nitric oxide delivery system. Br J Anaesth 73:700–702
50. Blomqvist H, Wickerts CJ, Andreen M, Ullberg U, Örtqvist Å, Frostell C (1993) Enhanced pneumonia resolution by inhalation of nitric oxide? Acta Anaesthesiol Scand 37:110–114
51. Foubert L, Fleming B, Latimer R, et al (1993) Safety guidelines for use of nitric oxide. Lancet 339:1615–1616
52. Karamanoukian HL, Glick PL, Zayek M, et al (1994) Inhaled nitric oxide in congenital hypoplasia of the lungs due to diaphragmatic hernia or oligohydramnios. Pediatrics 94:715–718
53. Miller CC (1994) Chemiluminescence analysis and nitrogen dioxide measurement. Lancet 343:300–301
54. Wagner HM (1970) Absorption von NO und NO_2 in MIK- und MAK-Konzentrationen bei der Inhalation. Staub Reinhalt Luft 30:380–381
55. Bocchi EA, Bacal F, Auler JOC Jr, Carmone MJ de C, Bellotti G, Pileggi F (1994) Inhaled nitric oxide leading to pulmonary edema in stable severe heart failure. Am J Cardiol 74:70–72
56. Rieder HU, Frei FJ, Zbinden AM, Thomson DA (1989) Pulse oximetry in methaemoglobinaemia. Anaesthesia 44:326–327

Persistent Pulmonary Hypertension of the Newborn

J. C. Mercier, A. T. Dinh-Xuan, and S. H. Abman

Introduction

Persistent pulmonary hypertension of the newborn (PPHN) is a complex disorder associated with a wide array of cardiopulmonary diseases. PPHN is characterized by marked pulmonary hypertension and altered vasoreactivity, leading to right-to-left shunting of blood across the patent ductus arteriosus and foramen ovale [1–3]. Critical hypoxemia results from extrapulmonary shunting, and is often poorly responsive to high inspired oxygen concentrations and pharmacologic vasodilation [4]. However, the PPHN syndrome is often associated with severe parenchymal lung disease (e.g. lung hypoplasia, meconium aspiration pneumonitis, bacterial pneumonia, and surfactant deficiency) which causes intrapulmonary shunting and further complicates the clinical course and response to treatment. Moreover, disturbances in cardiac performance, hypovolemia, and decreased systemic vascular resistance may compromise the tenuous balance between the systemic and pulmonary circulation in PPHN. Thus, effective therapy for neonatal hypoxemic respiratory failure complicated by extrapulmonary veno-arterial admixture requires vigilant attention to all aspects of these cardiopulmonary interactions (Fig. 1).

Treatment of PPHN has been limited so far by the lack of vasodilator agents with selective pulmonary vascular effects, and ineffective therapies for severe parenchymal lung disease and systemic hemodynamic collapse (e.g. sepsis syndrome). However, recent insights into fetal and transitional pulmonary vasoregulation, newer approaches to management of severe parenchymal lung disease in newborn infants, and the potential role of inhaled nitric oxide (NO) as a selective pulmonary vasodilator may markedly modify our current therapeutic approaches. In this chapter, we will review:

1) the recent developments in vascular biology concerning the endothelial-derived products NO and endothelin; and
2) the potential applications and limitations of inhalational NO therapy in the management of PPHN.

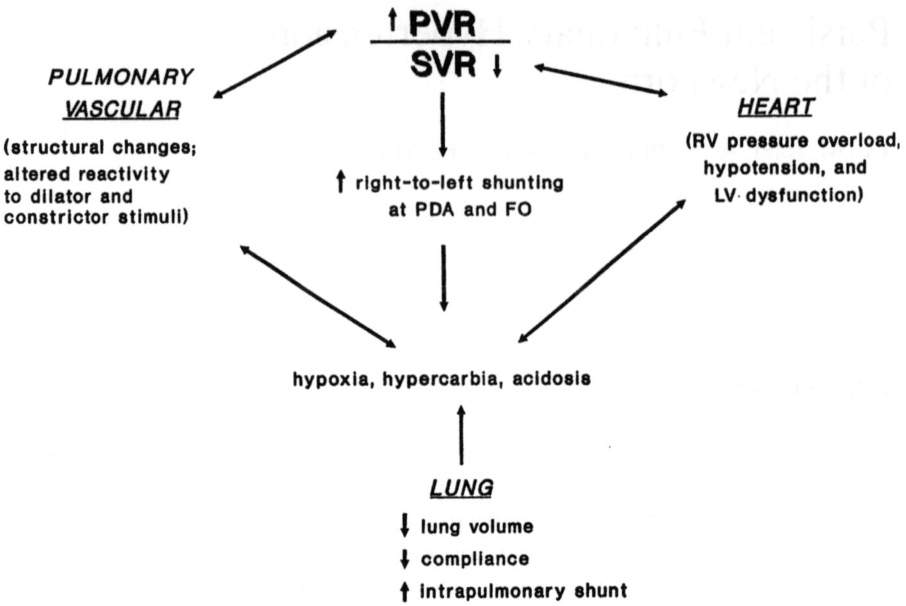

Fig. 1. Schematic of cardiopulmonary interactions in PPHN

Fetal and Perinatal Pulmonary Circulation

The fetal pulmonary circulation is characterized by high resistance and low flow, accepting less than 10% of the combined ventricular output [5]. This unique physiologic state ensures enough pulmonary blood flow to provide for lung growth and development [6], yet allows blood to shunt right to left at the levels of the ductus arteriosus and foramen ovale to perfuse the placenta, the fetal organ of gas exchange. Mechanisms that contribute to the maintenance of high resting tone in the fetal lung are not completely understood but include mechanical factors such as the lack of rhythmic distention, the absence of an air-liquid interface, the presence of a low resting alveolar and arteriolar oxygen tension, and a different 'balance', when compared to the postnatal lung, between vasodilating and vasoconstricting mediators [7–10].

A second feature of the fetal pulmonary circulation, in addition to high resting tone, is its capacity for autoregulation. Upon exposure to several dilating stimuli including shear stress, oxygen, and several pharmacologic agents which acts on endothelium, the fetal pulmonary vasculature exhibits a time-limited response [11–14]. Initial vasodilation is followed by a return of vascular resistance to its resting state despite continued exposure to the vasodilating stimulus (Fig. 2). This time-limited vasodilatory response limits excessive pulmonary blood flow, which is detrimental to the developing fetus, and allows for preservation of placental perfusion. In contrast, agents which act directly on the vascular smooth muscle cell, such as atrial natri-

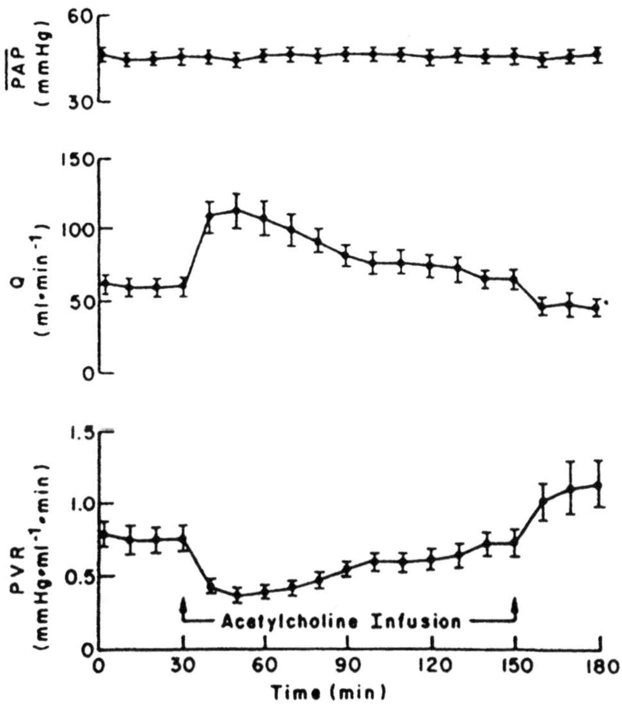

Fig. 2. The fetal pulmonary circulation is characterized by its capacity for autoregulation. Upon exposure to several pharmacologic agents which act on endothelium including acetylcholine infusion at 1.5 $\mu g \cdot min^{-1}$, the fetal pulmonary vasculature exhibits a time-limited vasodilation. PAP = pulmonary arterial pressure, Q = left pulmonary artery blood flow, PVR = pulmonary vascular resistance, mean of 9 fetal lambs. (From [12] with permission)

uretic peptide and 8-bromo-cGMP, cause sustained vasodilation in the fetal lung [15]. This suggests that factors responsible for the autoregulatory response include the inability of the endothelial cell to sustain the release of locally acting vasodilatory mediators, an 'uncoupling'of these mediators from the adjacent smooth muscle cell, or alternatively, enhanced release of vasoconstricting mediators by the endothelial cell to counterbalance vasodilation.

At birth, as the lungs assume the function of gas exchange, pulmonary blood flow increases 8 to 10-fold, and pulmonary vascular resistance (PVR) rapidly drops (Fig. 3) [7–16]. The fall in PVR reflects both structural adaptation of the pulmonary bed to lung inflation and the effects of locally acting vasoactive mediators released at birth [17]. Central to this transition from the high resistance bed *in utero* to the low resistance circuit *ex utero* is the endothelial cell which acts as a paracrine modulator of vascular resistance by releasing various vasoactive products [18]. Therefore, the intimate relationships between the endothelial cell and smooth muscle cell and the mechanisms which elicit vasodilation are essential in understanding the transitional circulation.

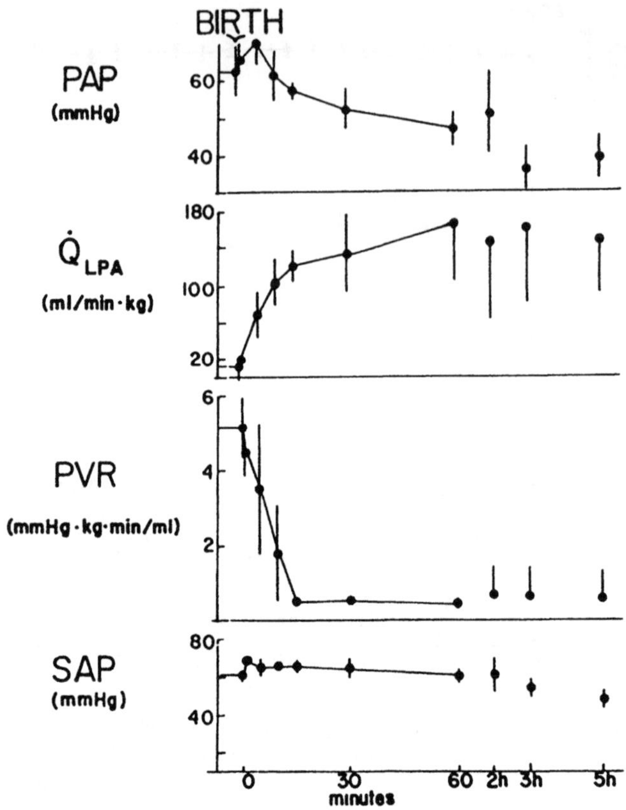

Fig. 3. At birth, pulmonary blood flow increases 8 to 10-fold, and pulmonary vascular resistance (PVR) rapidly drops. PAP = pulmonary arterial pressure, Q_{LPA} = left pulmonary arterial blood flow, SAP = systemic arterial pressure. (From [31] with permission)

Endothelial Cell's Role in Regulating Vascular Tone

The endothetial cell produces and releases several mediators which act on the subjacent smooth muscle cell to modulate local vascular tone. Several stimuli influence the release of these mediators, and complex mechanisms determine the overall balance between endothelial-derived vasoconstricting and vasodilating factors. For example, the onset of ventilation at birth directly lowers PVR, presumably by recruitment and a change in the geometry of small pulmonary vessels [17]. Ventilation also stimulates the release of the endothelial-derived vasodilators, prostacyclin and NO [19, 20]. This chapter will focus on three of these endothelial-derived mediators, prostaglandins (PG), NO, and endothelin.

The interactions of these mediators with the smooth muscle cell is schematically summarized in Fig. 4. An endothelial-dependent agonist relies on the endothelial cell's production of NO to elicit a vasodilatory response. An endothelial-independent dilator acts directly on the smooth muscle cell,

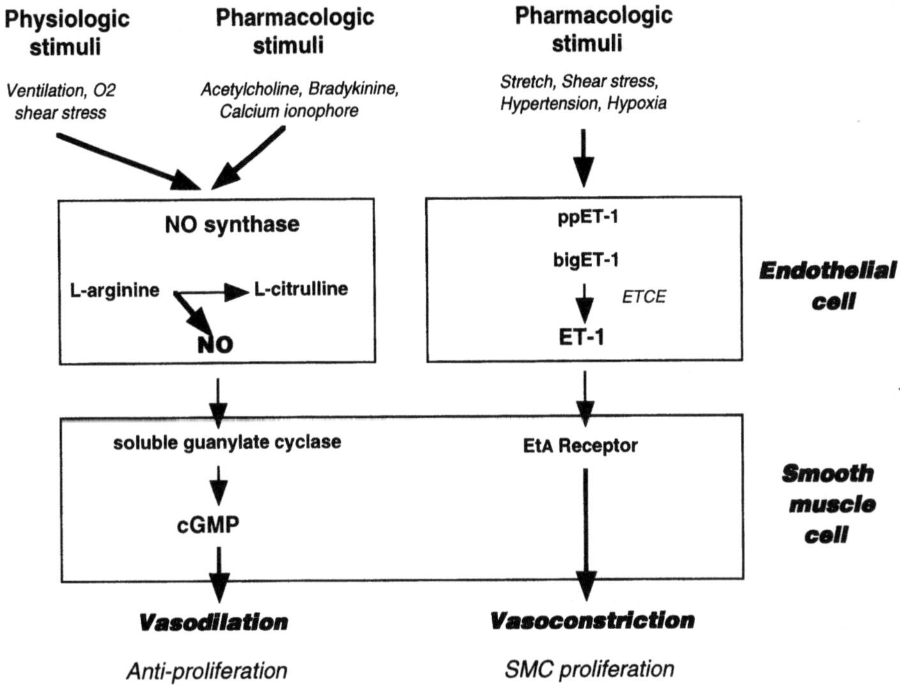

Fig. 4. Schematic illustrating the effects of physiologic and pharmacologic stimuli on pulmonary vascular endothelial cell production of NO and ET. NO is formed from L-arginine and molecular oxygen by an endothelial constitutive (type III) NO synthase. NO diffuses to subjacent vascular smooth muscle, stimulating cGMP production and causing vasodilation. Severe hypertension and hypoxemia lead to endothelial ET-1 production from preproendothelin-1 and Big-ET-1. ET-1 acts on the smooth vascular muscle ET_A receptor; causing vasoconstriction as well as smooth muscle cells (SMC) proliferation

either through the NO-cGMP system as in the case of exogenous NO, sodium nitroprusside or 8-bromo-cGMP, or via other mechanisms.

Eicosanoids: Eicosanoids are locally produced, vasoactive lipid metabolites of arachidonic acid. Within the endothelial cell, arachidonic acid can be metabolized through the lipoxygenase pathway to form the leukotrienes, or through the cyclooxygenase pathway to form different prostaglandins. In the fetal and transitional circulations, the cyclooxygenase pathway predominates with release of PGI_2 or prostacyclin, the most potent prostanoid mediator formed [21].

Leukotrienes: Early work suggested that the leukotrienes, powerful vasoconstrictors when infused into the fetal pulmonary circulation, played an important role in maintaining high tone within the fetal pulmonary vascular bed. It was further suggested that leukotriene production decreased at birth contributing to the drop in PVR during postnatal transition. These hypotheses were based on two findings: 1) the administration of non-specific leukotriene antagonists caused fetal pulmonary vasodilation and blocked the pressor response to exogenous administration of leukotrienes [22]; and 2) leukotrienes were detected in tracheal aspirates from human newborns with PPHN but not in intubated newborns without pulmonary hypertension [23]. However, the non-specificity of the leukotriene blockers used made speculations regarding leukotriene importance in the fetal circulation questionable [24]. Furthermore, leukotriene levels in lung fluid in newborns with PPHN remain detectable even after resolution of pulmonary hypertension [25]. Cassin et al. [26] measured leukotriene content in fetal lamb tracheal fluid and demonstrated very low leukotriene activity. To further clarify the role of leukotrienes in fetal and transitional pulmonary vascular regulation, Abman and Stenmark [27] measured leukotriene concentrations in fetal and postnatal lambs both in lung liquid and lung tissues. Leukotriene content in fetal lung liquid was found extremely low, in comparison with adult ewes. Fetal lung tissue levels of leukotrienes were also very low. In addition, no significant change in lung leukotriene levels was observed during the transition in normal or hypertensive ovine fetuses. Thus, direct measurement of leukotriene levels in the fetal lung fails to support leukotriene activity as a major determinant of fetal or transitional pulmonary vascular tone.

Prostaglandins: The PG comprise a family of chemically-related compounds formed via the cyclooxygenase pathway of arachidonic acid metabolism. Although several of these compounds display a vasoactive effect when infused into the fetal pulmonary circulation, prostacyclin appears to be the major prostanoid metabolite within the fetal lung.

Prostacyclin is produced by fetal pulmonary blood vessels *in vitro* [21, 28], and when infused into the fetal lung *in vivo* causes very potent vasodilation [29]. Several studies have demonstrated an increase in prostacyclin production during the normal transition, suggesting that prostacyclin synthesis is stimulated by birth-related events [30, 31]. In accordance with this hypothesis, acute cyclooxygenase inhibition with indomethacin or meclofenamate attenuates the drop in PVR at birth in term animals [32]. However, this attenuation is modest, and postnatal adaptation is not significantly changed suggesting that although increased production of prostacyclin contributes to the pulmonary vascular changes during the transition, prostacyclin is not essential for postnatal adaptation. Increased prostacyclin synthesis at birth was suggested as primarily stimulated by the onset of breathing [31]. To further determine which type of event at birth accounted for increased prostacyclin production, Velvis et al. [33] performed a series of experiments independently evaluating the effects of static distention, rhythmic distention, and oxygenation on prostacyclin production in near-term animals. These investigators demonstrated that cyclooxygenase inhibition abolished the drop in PVR associated with rhythmic distention of the lungs after birth but had no effect on PVR during static distention or during rhythmic distention *with* oxygenation. Thus, under normal circumstances, i.e. ventilation with oxygenation, prostacyclin is not essential for normal postnatal transition.

PG of the D- and E-series manifest vasodilatory activity when infused into the fetal lung [34, 35]. PGE_1 is also important in maintaining patency of the ductus arteriosus. A switch in the production of local PG stimulated by a change in the oxygen environment at birth is likely to postnatally close the ductus arteriosus. PG of the F-series induce vasoconstriction in the fetal pulmonary vasculature [36], but these PG are probably not important.

Finally, *thromboxane* (TX) is a PG mediator with potent vasoconstricting and platelet aggregation properties. This prostanoid is increased in experimental models of PPHN associated with group B streptococcal sepsis, and cyclooxygenase inhibition attenuates the marked pulmonary hypertension which accompanies experimental neonatal group B streptococcal sepsis [37, 38]. It is thus feasible that TX, or other PG with vasopressor activity, may contribute to failed postnatal adaptation.

Nitric Oxide: In the early 1980s, the obligatory role of endothelial cells in the relaxation of arterial smooth muscle by several dilating stimuli including acetylcholine was recognized, and the production and release of an endothelial-derived relaxing factor (EDRF) was hypothesized [18]. This labile compound was identified as NO in 1987 [39, 40]. Since then, much has been learned about the mechanisms of NO release and activity, and its importance in several vascular beds, including the perinatal circulation, has been established.

Catalyzed by a constitutive enzyme endothelial NO synthase (cNOS), NO and L-citrulline are produced from L-arginine and molecular oxygen [41, 42]. Due to its high lipophily, NO rapidly diffuses to the subjacent smooth muscle cell where it 'activates' soluble guanylate cyclase resulting in an increased smooth muscle cell concentration of cyclic guanosine-mono-phosphate (cGMP) [43]. cGMP then initiates a cascade of intracellular events resulting in smooth muscle relaxation [44]. cGMP is rapidly hydrolyzed and inactivated within the smooth muscle cell by phosphodiesterase enzymes [45, 46]. Thus, continuous release of NO and production of cGMP are important in maintaining low basal tone. This basal production of NO increases in response to endothelial-dependent vasodilators.

Using molecular and immunostaining techniques, Halbower et al. [47] have found endothelial NOS protein in the ovine fetal pulmonary vasculature as early as 0.29 term. In addition, intrapulmonary infusion of L-nitro-arginine (L-NA), an inhibitor of NOS and thus of NO production, results in an increase in pulmonary vascular tone as early as 0.77 term [48]. Thus, endogenous NO release is physiologically important in modulating *basal* fetal pulmonary vascular tone, early in gestation.

In contrast, a diminished ability of the pulmonary vascular endothelial cell to increase production of NO has been found in *in vitro* isolated rings [49] and in *in vivo* whole animal studies [50] in response to endothelial-dependent dilators, early in gestation. For instance, the endothelial-dependent agonists acetylcholine and oxygen exhibit progressive vasodilator potency between 115 and 140 days gestation (term is 150 days) [12, 51]. Unlike endothelial-dependent dilators, the early fetus responds normally to endothelial-independent agonists such as inhaled NO or the direct NO donor, sodium nitroprusside [15]. These findings suggest that the developing smooth muscle cell's ability to respond to NO precedes the immature endothelium's ability to increase and sustain the *stimulated* production and release of NO.

At birth, ventilation with rhythmic distention of the lungs, and increased oxygenation and shear stress contribute to the normal postnatal decline in PVR. The vasodilatory response to these birth-related stimuli is markedly attenuated by pretreatment with L-NA suggesting that increased release of endogenous NO is necessary for a smooth transition of the pulmonary circulation at birth [52–55]. To confirm these findings, the effects of exogenous NO have been studied in the ovine transitional circulation [56]. Without changing fetal PaO_2, inhaled NO causes potent, selective pulmonary vasodilation equivalent to that achieved with the combination of ventilation plus 100% oxygen. It is thus tempting to speculate that endothelial cell dysfunction and inability to sustain the release of NO in response to birth-related stimuli might contribute to failure of postnatal adaptation and result in the clinical state PPHN. This hypothesis was supported by findings in a hypertensive fetal model [57]. Late gestation, ovine fetuses made hypertensive by partial ductal compression or ligation develop histologic changes of pulmonary hypertension including right ventricular hypertrophy and pulmonary vascular remodeling [58]. Upon delivery, the hypertensive animals demonstrate much lower increase in pulmonary blood flow and failure to decrease PVR than the controls (non-hypertensive) [59], as well as hypoxic episodes contemporary of suprasystemic pulmonary arterial pressure very similar to the ones observed in newborns with severe PPHN [60]. Furthermore, they exhibit preferential impairment of endothelial-dependent vasodilation, while responsiveness to exogenous NO remains relatively intact [57].

Although exogenous NO causes vasodilation in the hypertensive ovine fetus, the degree of vasodilation achieved is less than in controls. This difference in maximal vasodilation between hypertensives and controls might be explained on the basis of structural changes within the pulmonary vascular bed. However, the NO response might be pharmacologically increased by the use of cGMP-phosphodiesterase inhibitors, as zaprinast or dipyridamole [61, 62].

Much remains to be learned about the NO/cGMP system. Recent work has suggested that NO may have a direct effect on the smooth muscle cell [64, 65], independent of guanylate cyclase activation, and further elucidation of NO's interaction with the smooth muscle cell is an active area of research. Other areas include the mechanisms of NO expression and regulation, the intracellular interactions between NO, cGMP, cAMP, and the phosphodiesterase enzymes, as well as the mechanisms of cGMP-induced vasodilation.

Endothelin: Recognition of the endothelial cell's important role in regulating vascular tone, and the identification of EDRF, prompted the search for counterbalancing endothelial-derived vasoconstrictor substances. In 1988, a novel potent vasoconstrictor peptide produced by vascular endothelial cells was isolated and named endothelin [65]. Further work established that this agent was one of a family of three structurally similar peptides, Endothelin-1, -2, and -3 (ET-1, ET-2, and ET-3), each produced by a separate gene [66]. Investigations in the perinatal circulation have focused primarily on ET-1.

Within the endothelial cell, ET-1 is formed during a 2 step reaction. Preproendothelin-1, a 212-amino acid precursor, is cleaved by an endopeptidase to form proendothelin-1 or Big ET-1, a 38-amino acid peptide. Big ET-1 is then cleaved by a putative endothelin converting enzyme to the biologically active ET-1, a 21-amino acid peptide. ET-1 is primarily formed and released at the abluminal surface of the endothelial cell. ET-1 then binds to specific receptors to elicit its physiologic effect.

Two different receptors (ET_A and ET_B) for endothelin have been characterized (Fig. 5) [67, 68]. The ET_A receptor displays much greater binding affinity for ET-1 and ET-2, and is generally present on the smooth muscle cell in fetal lung. Binding of ET-1 to this receptor results in vasoconstriction through a complex series of intracellular events involving activation of phospholipase C, production of inositol triphosphate, and release of intracellular stores of calcium [69]. In the fetal and newborn lung, the ET_B receptor is located on the endothelial cell. ET_B receptor binding results in vasodilation which may be mediated by endogenous NO [70, 71], prostacyclin release [72], or via ATP-gated potassium channels [73].

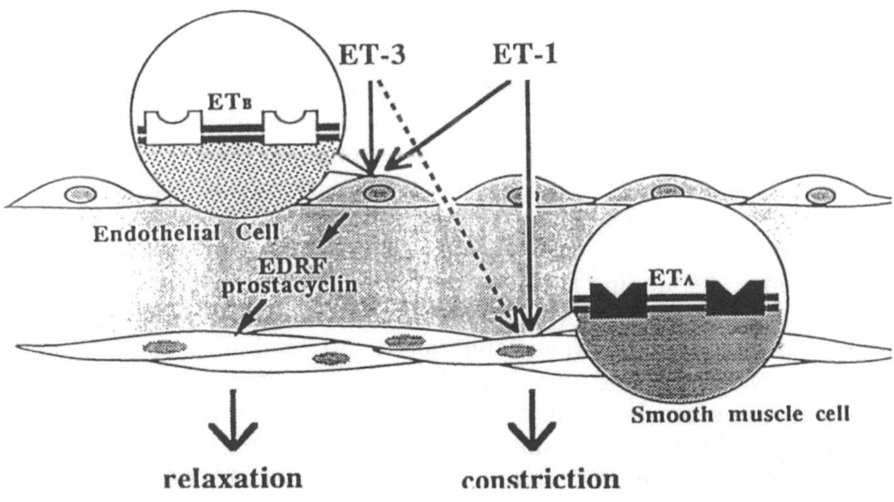

Fig. 5. Vasodilation and vasoconstriction induced by exogenously administered endothelin (ET-1) or ET-3. ET_A is localized on the smooth vascular muscle, inducing vasoconstriction. ET_B receptor is localized on the endothelial cell, releasing EDRF or prostacyclin. Apparent response induced by ET depends on the balance of activities between ET_A and ET_B. (From [69] with permission)

The hemodynamic response to ET-1 in the fetus depends on the dose and route of administration, the vascular bed examined, and resting vascular tone. Pulmonary infusion of ET-1 in the fetus and newborn results in a biphasic response with initial, potent vasodilation followed by vasoconstriction [74]. Prolonged or repeated infusions of ET-1 cause progressively increased vasoconstriction. Systemic administration of ET-1 in the fetus and newborn results in a sustained vasoconstriction without preceding vasodilation [74, 75]. The mechanisms responsible for this difference between the fetal pulmonary and systemic response to exogenous ET-1 are unknown but probably involve differential expression of receptor subtypes within these vascular beds. The hemodynamic response to Big ET-1 in the fetal lung differs from the response to ET-1 [76, 77]. Big ET-1 causes vasoconstriction without vasodilation. Since Big ET-1 requires processing by the endothelial cell before release as ET-1, the response to the exogenous administration of Big ET-1 may be more representative of the physiologic response to endogenous ET-1. It also suggests that ET-1 is preferentially released abluminally by the endothelial cell where it is more likely to bind to smooth muscle cell ET_A receptors and cause vasoconstriction.

To further clarify the physiologic role of ET_A and ET_B receptor activity within the fetal lung, the hemodynamic effects of BQ 123, a selective ET_A receptor antagonist, and Sarafotoxin S6c, a selective ET_B receptor agonist were studied [78]. Blockade of the ET_A receptor with BQ 123 caused a dose-dependent increase in pulmonary blood flow and decrease in PVR; stimulation of the ET_B receptor with Sarafotoxin S6c also resulted in vasodilation. Both of these vasodilatory responses were blocked by L-NA suggesting that selective ET_A receptor blockade within the fetal lung allows for preferential binding of ET-1 to the ET_B receptor resulting in vasodilation through the NO-cGMP pathway [78, 79]. Consistent with this hypothesis is the finding that CGS 27830, a blocker of both ET_A and ET_B receptors has no effect on basal pulmonary vascular tone. These findings are consistent with a role for ET-1 in the maintenance of high PVR in the fetus [69, 78, 80].

Cassin et al. [81] have shown that the pulmonary vascular response to exogenous ET-1 is tone-dependent. Administration of ET-1 to the fetus results in vasodilation whereas after ventilation with 100% oxygen, ET-1 causes vasoconstriction. This finding led to the interesting speculation that a spike release of ET-1 at birth might be one factor which increases NO production during the transition. However, abnormally sustained release of ET-1 might 'exhaust' NO stores and stimulate ET_A receptors causing vasoconstriction and failure of postnatal pulmonary vascular adaptation. Endothelin is known to stimulate fibroblast and smooth muscle cell proliferation *in vitro* [82], and it is speculated that increased production and release of ET-1 *in utero* might cause chronic hypertension with pulmonary vascular remodeling. This abnormal production and release of ET-1 might account for several of the characteristic abnormalities noted in PPHN. This hypothesis is supported by the fact that elevated immunoreactive ET-1 levels are present in newborns with PPHN [83].

In summary, ET-1 is physiologically active in the fetal and transitional pulmonary circulations. ET-1 may play an important role in modulating fetal and transitional vascular tone. Abnormal production and/or release of ET-1, or alternatively a change in the normal balance between the two receptors, ET_A eliciting vasoconstriction and ET_B inducing vasodilation may contribute to abnormal postnatal adaptation of the pulmonary vascular bed. However, much remains to be learned about the mechanisms controlling ET-1 expression by the endothelial cells.

Clinical Trials of Inhaled NO in Infants with PPHN

Recent studies have suggested that inhaled NO causes marked improvement in oxygenation in many newborn infants with PPHN. Roberts et al. [84] reported that brief (30 min) inhalation of NO at 80 ppm improved oxygenation in patients with PPHN, but this response was sustained in only one patient after NO was discontinued. Kinsella et al. [85] demonstrated rapid improvement in oxygenation in 6 neonates with severe PPHN with inhaled NO, using doses of 20 ppm for 4 h, then decreasing the dose to 6 ppm with sustained improvement in oxygenation for the duration of treatment. In a subsequent study [86], this low-dose NO strategy was used in an additional 9 infants with severe PPHN, including two with congenital diaphragmatic hernia. The management incorporated the use of high frequency oscillatory ventilation to optimize lung inflation during NO therapy. All neonates had resolution of the underlying PPHN, but two infants with overwhelming sepsis subsequently required treatment with extracorporeal membrane oxygenation (ECMO) for hemodynamic support. Thus, in these pilot studies, 13 of 15 infants with severe PPHN recovered without the need for ECMO [87].

Two recent studies has corroborated the efficacy of low dose NO ($\leq$20 ppm) in causing acute improvement in oxygenation in patients with severe PPHN. Finer et al. [88] randomly evaluated the improvement in oxygenation with doses of NO ranging from 5–80 ppm, and found that the acute improvement in oxygenation in PPHN was similar at all doses of NO studied. They also found that oxygenation increased in patients who did not have extrapulmonary shunting by echocardiography, suggesting that NO could improve oxygenation in some cases of neonatal hypoxemic respiratory failure without severe PPHN. We organized the French pediatric study group on inhaled NO, and undertook an open trial in 100 neonates including 25 preterms and 75 terms with severely hypoxemic respiratory failure [89]. Effective doses ranged from 5 to 20 ppm in the responders, whereas NO levels were unsuccessfully increased in the non-responders. Sustained improvement in the oxygenation was observed in most of the neonates with persistent fetal circulation (PFC) or respiratory distress syndrome (RDS) with sepsis, provided there was no associated shock, and to a lower extent in those with meconium aspiration syndrome (MAS). In contrast, very few infants with congenital diaphragmatic hernia (CDH) responded to inhaled NO

Table 1. Analysis of prognostic factors in 100 severely hypoxemic newborns treated with inhaled NO in the French open trial [89]

Variable	[n]	Survival without ECMO [%]	p value
Gestational age			
<29 weeks	9	22	
<34 weeks	16	56	<0.01
≥34 weeks	51	78	
Diagnosis			
RDS/ARDS	28	64	
PFC	11	85	NS
Sepsis	23	64	
MAS	14	57	
CDH	24	12	<0.01*
OI after 30-min NO			
<20	46	83	
20–40	23	56	<0.01
>40	7	29	

ARDS: (acute) respiratory distress syndrome; PFC: persistent fetal circulation; MAS: meconium aspiration syndrome; CDH: congenital diaphragmatic hernia; OI: oxygenation index. * CDH *vs* all other diagnoses

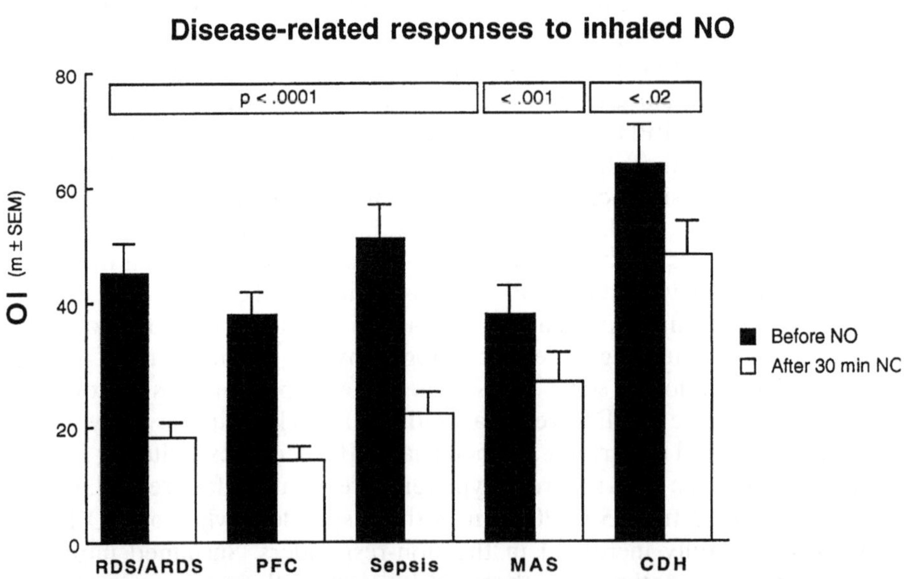

Fig. 6. Response to inhaled NO was disease-specific in term neonates treated in the French open trial [89]. Oxygenation index (OI) were significantly reduced (paired t-test) after only 30-min on NO in RDS/ARDS (n = 9), PFC (n = 11), sepsis (n = 17), and even MAS (n = 14). In contrast, OI remained above 40 in most infants with CDH (n = 24) more or less associated with severe pulmonary hypoplasia

(Fig. 5). Survival to NO mainly depended upon the gestational age, the associated diagnosis, and the magnitude of the improvement in the oxygenation (Table 1). In a subset of these hypoxemic infants in whom sequential echos could have been completed, half of them exhibiting massive right-to-left ductal shunting were found to dramatically respond to NO with increases both in pulmonary perfusion and in oxygenation along with reversal of ductal shunting, whereas the other half having no clear evidence of extrapulmonary shunting more slowly responded to NO along with unsignificant increase in pulmonary blood flow rather suggesting improvement in ventilation/perfusion matching [90]. More striking was the fact that inhaled NO, in combination with exogenous surfactant, high frequency oscillatory ventilation (HFOV), and possibly better general care, have dramatically decreased, as in some US centers, the numbers of ECMO in France over the last two years.

Yet, a few infants do not well respond to inhaled NO, and the reasons of this failure remain speculative. However, in cases complicated by severe lung disease, inhaled NO alone will unlikely be effective without optimal pulmonary management and adequate alveolar recruitment.

Management strategies in PPHN

Increasing clinical experience with inhaled NO in PPHN has led to better understanding of the potential role of this potent and selective pulmonary vasodilator. In infants with *idiopathic* PPHN, the so-called 'persistent fetal circulation', there is severe pulmonary hypertension without apparent parenchymal lung disease. Thus, the pulmonary management should aim at minimizing overzealous mechanical ventilation and avoiding lung overinflation. Selective pulmonary vasodilation with inhaled NO indeed causes rapid and marked improvement in oxygenation in these infants. However, neonatal respiratory failure is quite often associated with other severe lung diseases such as surfactant deficiency, infectious pneumonia, meconium aspiration, lung hypoplasia ..., and many pathophysiologic mechanisms can cause pulmonary hypertension, and PPHN. Parenchymal consolidation which in turn induces focal hypoxemic vasoconstriction is likely to be the main factor responsible for hypoxemia due to ventilation-perfusion mismatching. Disturbances in cardiac performance as well as decreased systemic arterial resistance can result in further clinical deterioration [91].

Thus, meticulous analysis of the relative roles of the pulmonary parenchymal, pulmonary vascular, and cardiac components contributing to hypoxemic respiratory failure is essential to assess the effects of new therapies in PPHN. It is well examplified in the meconium aspiration syndrome in which the pulmonary parenchymal disease and severe reactive pulmonary hypertension may be quite variable. For example, the chest radiograph may initially show markedly heterogeneous atelectasis associated with aspiration of particulate meconium, but subsequent progression of diffuse parenchymal disease perhaps related to "secondary" surfactant deficiency or surfactant inactivation may complicate the clinical course [92, 93]. Elsewhere, there may be predominantly large airway diseases associated with aspiration of particulate meconium, leading to "check-valve" type obstruction, focal overinflation, and air leak. Moreover, decreased cardiac performance and reduced left ventricular output can decrease systemic vascular resistance (SVR), exacerbating the right-to-left shunting of blood at the ductal level. Another disease in which severe hypoxemia may be caused by multiple factors is congenital diaphragmatic hernia. In infants with CDH, increased PVR may be related to the small cross sectional area of pulmonary vessels causing a *fixed* component of high PVR, but altered pulmonary vasoreactivity may lead to *reactive* pulmonary hypertension. A diminished left ventricular mass has been found associated with poor prognosis in CDH [94, 95], and this may also contribute to pulmonary hypertension and decreased left ventricular output. Moreover, lung immaturity [96], underinflation, and susceptibility to lung injury during mechanical ventilation further complicate the management of patients with CDH given the major role of parenchymal lung disease in many cases of PPHN, pharmacologic pulmonary vasodilation alone might not be expected to cause sustained clinical improvement.

The effects of inhaled NO may be suboptimal when lung volume is decreased in association with pulmonary parenchymal disease [97]. Atelectasis and air space disease (pneumonia, pulmonary edema) will decrease effective delivery of this inhalational agent to its site of action in terminal lung units. In PPHN associated with *heterogeneous* (patchy) parenchymal lung disease, inhaled NO may be effective in optimizing ventilation-perfusion matching by preferentially causing vasodilation in lung units which are well ventilated. The effects of inhaled NO on ventilation-perfusion matching appear to be optimal at lower doses ($\leqslant$20 ppm) [98, 99]. In cases complicated by *homogeneous* (i.e. diffuse on the chest X-ray) parenchymal lung disease and underinflation, pulmonary hypertension may be exacerbated by the adverse mechanical effects of underinflation on PVR, resulting in *functional* hypoplasia. In this setting, effective treatment of the underlying lung disease is essential, and sometimes sufficient, to cause resolution of the accompanying pulmonary hypertension.

The fetus is characterized by both structural and functional pulmonary immaturity, including surfactant deficiency. Pulmonary immaturity and hyaline membrane disease (HMD) lead to respiratory failure after premature delivery, and exogenous surfactant therapy has been demonstrated both to improve blood gas exchange and survival [100]. However, exogenous surfactant therapy results in suboptimal responses in many human newborns thought to have HMD [101], suggesting that other problems of prematurity (e.g. structural lung immaturity or altered vascular reactivity) or complications of mechanical ventilation contribute to respiratory failure. Furthermore, progressive deterioration after surfactant therapy occurs in few premature human neonates with HMD, leading to severe pulmonary hypertension with right-to-left shunting across the ductus arteriosus and increased mortality [102–106].

It has been recently reported that endogenous NO production modulates basal pulmonary vascular tone as early as 115 days gestation in the ovine fetus, and contributes to the increase in pulmonary blood flow after extremely premature delivery [48]. In addition, inhaled NO caused marked increases in pulmonary blood flow after NO synthase inhibition, demonstrating marked responsiveness of the vascular smooth muscle cell to NO very early in gestation. After delivery of these extremely premature lambs, altered pulmonary vasoregulation led to increased PVR and decreased pulmonary blood flow during mechanical ventilation with high oxygen concentrations, and worsening gas exchange despite a good initial response to exogenous surfactant. The pulmonary hypertension associated with severe HMD in the extremely premature lamb is responsive to exogenous NO, and early treatment with low-dose NO attenuates the decline in gas exchange and pulmonary perfusion during prolonged mechanical ventilation [107].

It is possible that the early use of low-dose inhaled NO could play a role in the management of the premature infants with severe respiratory failure unresponsive to exogenous surfactant therapy [89, 108], by improving ventilation-perfusion matching and reducing PVR. However, little is known of the potential pulmonary and systemic toxicities which could occur in the premature infant exposed to inhalational NO, and studies designed to carefully assess such toxicities are essential before routine clinical application.

There is some evidence that surfactant deficiency contributes to decreased lung compliance and atelectasis in some patients with PPHN [109–111]. Recent studies have suggested that exogenous surfactant therapy can cause sustained clinical improvement in term infants with pneumonia and meconium aspiration syndrome [112], in infants with CDH [113], and reduce the duration of ECMO [114]. Another mode of recruting non-ventilated lung areas while minimizing baro/volotrauma is to decrease air-liquid surface tension area using perfluorocarbon derivatives i.e.total or partial liquid ventilation (PAGE). This has been successfully tested in animals, and used in a few infants with refractory hypoxemic respiratory failure [115].

For many years, alkalosis-induced mechanical hyperventilation induced with the hope of decreasing pulmonary arterial pressure and improving oxygenation has been the cornerstone of the management in infants with PPHN [116–119]. However, concerns have been raised over the potential complications associated with baro/volotrauma, and sustained hypocapneic alkalosis may have adverse neurologic effects [120, 121]. Wung et al. [122] described a rather conservative approach to the management of PPHN which is characterized by individualized ventilator strategies based on the underlying disease state, avoiding hypocapneic alkalosis, minimizing the use of paralysis, and pulmonary vasodilation using tolazoline. A corroborative study using a similar strategy reported survival in 5 of 6 infants with PPHN [123]. However, controlled trials have not been conducted.

Thus, alternative approaches to mechanical ventilation of the newborn infant with PPHN have attracted considerable interest. HFOV allows for effective alveolar ventilation and recruitment, while reducing the potential parenchymal and airway injury associated with large phasic pressure and lung volume changes [124]. High end-inspiratory lung volume, as it occurs during conventional ventilation, has been implicated in the development of pulmonary edema and lung injury [125]. However, the *strategy* employed during HFOV is critically important to optimize lung inflation and minimize the risk of adverse cardiopulmonary interactions [126, 127]. Without a strategy which does not adequately recruit and sustain lung volume, HFOV was found somehow ineffective [128]. In contrast, with a strategy designed to recruit atelectatic lung and sustain lung volume, HFOV was more efficacious than conventional mechanical ventilation in severe neonatal hypoxemic respiratory failure [129]. Because of marked attenuation of airway pressure across the endotracheal tube and airways during HFOV, the proximal airway pressure measurements do not accurately reflect the pressure transmitted to the lung [130]. Therefore, conventional indices of severity of illness (e.g. oxygenation index, ventilation index) may not be applicable during HFOV. An effective strategy during HFOV often requires stepwise increases in mean airway pressure by 3–5 cm H_2O increments, sometimes up to 30–35 cm H_2O, until improvement in oxygenation occurs and adequate lung inflation (with 7–8 ribs on the chest X-ray) is obtained. Since airway pressure is poorly transmitted to the pleural space in the underinflated and poorly compliant lung, this is usually well tolerated. However, as lung inflation and compliance improve, ventilator support must be rapidly reduced to avoid lung injury. Because of the diverse nature of diseases complicating PPHN, there is no single approach to mechanical ventilation. Ventilator adjustments must be made with attention to radiographic changes of the accompanying lung disease and to hemodynamic status.

Serial echocardiographic measurements are also very useful during NO treatment to determine the effects of inhaled NO therapy on *extrapulmonary* shunting. In conditions associated with suboptimal or transient responses but resulting in resolution of the extrapulmonary shunting, the primary pathophysiologic disturbance is likely caused by severe *intrapulmonary* shunting, and effective alveolar recruitment will be essential.

In some infants, the effects of inhaled NO may be time-dependent, with gradual improvement in oxygenation over several hours. The rate of response to inhaled NO is likely to depend upon multiple factors, including the nature of the underlying lung disease, the presence of circulating vasoconstrictor agents, cardiac performance, and structural changes in the pulmonary vasculature. Potential mechanisms for suboptimal NO responses and loss of NO responsiveness require, however, further study. Excessive vasoconstrictor mediators could potentially abate the vasodilatory effects of inhaled NO. Moreover, both functional changes in the pulmonary vascular smooth muscle cell (e.g. decreased soluble guanylate cyclase activity) and structural alterations in the pulmonary vascular tree may alter responsiveness to exogenous NO. It is thus possible that the combination of other agents with inhaled NO, such as the phospodiesterase inhibitor zaprinast or dipyridamole, could augment pulmonary vascular cGMP concentrations and allow for more effective vasodilation at lower doses of NO [62, 63, 131]. Serial echocardiographic determinations of cardiac performance may yield predictors for refractory cardiac dysfunction, and serve as an indicator for the need for the cardiac support provided by ECMO.

Toxicity

Although "non-toxic" (i.e. $\leq$20 ppm) concentrations of inhaled NO have been mainly used in the clinical studies described above, concerns remain regarding potential toxicities, including methemoglobinemia, lung injury due to NO_2, peroxynitrite and hydroxyl radical formation [132, 133]. Methemoglobin reductase activity has been found decreased in preterm and term neonates, and frequent assessments of methemoglobin levels are mandatory. There is so far little evidence for NO lung toxicity at low concentrations in adult animals [134, 135]. Inhaled NO may have indeed a protective lung effect by attenuating the increases in capillary permeability associated with oxidant-induced acute lung injury [136]. However, antioxidant host defenses are decreased in the immature lung, and there are potential risks of surfactant inactivation [137]. Chronic exposure to high concentrations of NO may cause downregulation of endogenous NO synthesis [138], and it is now well recognized that slow weaning of NO is advisable in newborns who have been treated by inhaled NO. However, the potential effects of NO on inhibition of DNA synthesis and deamination of DNA [139, 140] are of particular concern.

Another potentially adverse effect of NO inhalation is the prolongation in bleeding time at doses of 30–300 ppm that has been observed in rabbits [141]. Endogenous NO inhibits platelet adhesion to the vascular endothelium [142], and this effect may be related to the formation of stable S-nitrosoproteins [143]. However, no change in bleeding time occurred in a few premature infants treated with inhaled NO [144].

As with most therapeutic agents, the complications associated with NO are likely dose-dependent, and NO should be used in the lowest effective dose. Controlled clinical trials are ongoing to clarify the potential toxicities of this new therapy, including NO-induced target cell toxicity, pulmonary inflammatory cell function, and pathologic consequences of chronic exposure [145].

Conclusion

Our understanding of fetal and transitional pulmonary circulation has greatly improved over the recent years. Regulation of pulmonary vascular tone is, among several other neurohumoral agents, insured by two finely balanced endothelial-derived systems, i.e. prostacyclin and maybe thromboxane, and more importantly NO and endothelin. Although the pathophysiologic events which culminate in PPHN remain poorly understood, PPHN likely results from endothelial cell dysfunction with an alteration in the normal 'balance' of these endothelial-derived mediators. The latest speculations are: 1) spike release of ET-1 at birth might be one factor which increases NO production through its ET_B receptor, and initiates pulmonary vasodilation during normal transition; 2) abnormally sustained release of ET-1 might 'exhaust' NO stores and stimulate ET_A receptors causing vasoconstriction, and failure of postnatal pulmonary vascular adaptation; and 3) while more prolonged increased production and release of ET-1 might cause chronic hypertension with pulmonary vascular remodeling, i.e. PPHN. Manifestations of PPHN often involve dysfunctional pulmonary vasoregulation with suprasystemic PVR causing extrapulmonary shunting, pulmonary parenchymal disease causing intrapulmonary shunting, and systemic hemodynamic deterioration. Successful management of severe PPHN, therefore, depends upon the application of appropriate strategies to manage the various cardiopulmonary interactions which characterize this syndrome. Inhaled NO should obtain a marked improvement in oxygenation when optimal lung inflation has been achieved and systemic blood volume and vascular resistance have been maintained. Thus, meticulous attention should be given both to adequate lung recruitment using frequent chest X-rays, and to ductal and pulmonary blood flows using serial echocardiographic studies, while vigilant attention should be given to their changes in the clinical course. This combined approach appears to have indeed dramatically decreased the number of required ECMO, while successful application of inhaled NO in PPHN seems to have significantly reduced costs of hospitalization and length of hospital stay by approximately 50 and 40%, respectively (SH Abman, personal communication 1994). Nevertheless, only ongoing randomized controlled-studies will definitely demonstrate the potential value and limitations of this very promising therapy and prove its hopefully limited potential for toxicity.

References

1. Levin D, Heymann M, Kitterman J, Gregory G, Phibbs R, Rudolph A (1976) Persistent pulmonary hypertension of the newborn infant. J Pediatr 88:626–630
2. Gersony WM, Duc GV, Sinclair JC (1969) "PFC" syndrome (persistence of the fetal circulation). Circulation 40 (Suppl 3):87a (Abst)
3. Gersony WM (1984) Neonatal pulmonary hypertension: Pathophysiology, classification, and etiology. Clin Perinatol 11:517–524
4. Heymann J, Adams A, Gardner T (1984) Persistent pulmonary hypertension of the newborn: Trends, incidence, diagnosis, and management. Am J Dis Child 138: 592–595
5. Teitel DF, lwamoto HS, Rudolph AM (1987) Effects of birth-related events on central blood flow patterns. Pediatr Res 22:557–566
6. Wallen LD, Perry SF, Alston JT, Maloney JE (1990) Morphometric study of the role of pulmonary arterial flow in fetal lung growth in sheep. Pediatr Res 27:122–127
7. Dawes GS, Mott JC, Widdicombe JG, Wyatt DG, Wyatt DG (1953) Changes in the lungs of the newborn lamb. J Physiol (Lond) 121:141–162
8. Cassin S, Dawes GS, Ross BB (1964) Pulmonary blood flow and vascular resistance in immature foetal lambs. J Physiol (Lond) 171:80–89

9. Enhorning G, Adams FH, Norman A (1966) Effects of lung expansion on the fetal lamb circulation. Acta Pediatr Scand 55:441–451
10. Walker AM, Ritchie BC, Adamson TM, Maloney JE (1988) Effect of changing lung volume on the pulmonary circulation of fetal lambs. J Appl Physiol 64:61–67
11. Abman SH, Accurso FJ, Ward RM, Wilkening RB (1986) Adaptation of fetal pulmonary blood flow to local infusions of tolazoline. Pediatr Res 20:1131–1135
12. Accurso FJ, Alpert B, Wilkening RB, Petersen RG, Meschia G (1986) Time-dependent response of fetal pulmonary blood flow to an increase in fetal oxygen tension. Respir Physiol 63:43–52
13. Accurso FJ, Wilkening RB (1988) Temporal response of the fetal pulmonary circulation to pharmacologic vasodilators. Proc Soc Exp Biol Med 187:89–98
14. Abman SH, Accurso FJ (1989) Acute effects of partial compression of the ductus arteriosus in the fetal pulmonary circulation. Am J Physiol 257:H626–H634
15. Abman SH, Accurso FJ (1991) Sustained fetal pulmonary vasodilation during prolonged infusion of atrial natriuretic factor and 8-bromo-GMP. Am J Physiol 266: H183–H192
16. Cassin S, Dawes GS, Mott JC, Strang LB (1964) The vascular resistance of the foetal and newly ventilated lung of the lamb. J Physiol (Lond) 171:61–79
17. Haworth SG (1988) Pulmonary vascular remodeling in neonatal pulmonary hypertension: State of the art. Chest 93:133S–138S
18. Furchgott RF, Zawadski JV (1980) The obligatory role of endothelial cells in the relaxation of arterial smooth muscle by acetylcholine. Nature 288:373–376
19. Velvis H, Moore R, Heymann MA (1991) Prostaglandin inhibition prevents the fall in pulmonary vascular resistance as a result of rhythmic distention of the lungs in fetal lambs. Pediatr Res 30:62–68
20. Cornfield DN, Chatfield BA, McQueston JA, McMurtry IF, Abman SH (1992) Effects of birth-related stimuli on L-arginine-dependent pulmonary vasodilation in ovine fetus. Am J Physiol 262:H1474–H1481
21. Terragno NA, Terragno A (1979) Prostaglandin metabolism in fetal and maternal vasculature. Fed Proc 38:75–77
22. Soifer SJ, Loitz RD, Roman C, Heymann MA (1985) Leukotriene and organ antagonists increase pulmonary blood flow in fetal lambs. Am J Physiol 249:H570–H576
23. Stenmark KR, James SL, Voelkel NF, Toews WH, Reeves JT, Murphy RC (1983) Leukotriene C_4 and D_4 in neonates with hypoxemia and pulmonary hypertension. N Engl J Med 309:77–80
24. Gause GE, Baker R, Cassin S (1988) Specificity of RPL 57231 for leukotriene D_4 receptors in fetal pulmonary circulation. Am J Physiol 254:H120–H125
25. Kulik TJ, Lock JE (1987) Leukotrienes and the immature circulation. Am Rev Respir Dis 136:220–222
26. Cassin S, Stenmark KR, Gause G, Zapp LM, Kuck H, Westcott JY (1990) Leukotrienes and prostaglandins in fetal lung liquid. J Appl Physiol 68:2214–2222
27. Abman SH, Stenmark KR (1992) Changes in lung eicosanoid content during normal and abnormal transition in perinatal lambs. Am J Physiol 262:L214–L222
28. Remuzzi G, Misiani R, Muratore D, et al (1979) Prostacyclin and human fetal circulation. Prostaglandins 18:341–348
29. Cassin S, Winikor I, Tod ML, et al (1981)The effects of prostacyclin on the fetal pulmonary circulation. Pediatr Pharmacol 1:197–203
30. Leffler CW, Hessler JR, Terragno NA (1980) Ventilation-induced release of prostaglandin-like material from fetal lungs. Am J Physiol 238:H282–H286
31. Leffler CW, Hessler JR, Green RS (1984) The onset of breathing at birth stimulates pulmonary vascular prostacyclin synthesis. Pediatr Res 18:938–942
32. Davidson D (1988) Pulmonary hemodynamics at birth: Effects of acute cyclooxygenase inhibition in lambs. J Appl Physiol 64:1676–1682
33. Velvis H, Moore P, Heymann MA (1991) Prostaglandin inhibition prevents the fall in pulmonary vascular resistance as a result of rhythmic distention of the lungs in fetal lambs. Pediatr Res 30:62–68

34. Cassin S, Tyler T, Leffler C, Wallis R (1979) Pulmonary and systemic vascular responses of perinatal goats to prostaglandin E_1 and E_2. Am J Physiol 236: H828–H832
35. Cassin S, Tod ML, Phillips J, Frisinger J, Jordan J, Gibbs C (1981) Effects of prostaglandin D_2 on the perinatal circulation. Am J Physiol 240:H755–H760
36. Kadowitz PJ, Joiner PD, Hyman AL (1974) Influence of prostaglandins E_1 and F_2alpha on pulmonary vascular resistance in the sheep. Proc Soc Exp Biol Med 145: 1257–1262
37. Rojas J, Larson LE, Ogletree ML, Brigham KL, Stahlman MT (1983) Effects of cyclooxygenase inhibition on the response to Group B streptococcal toxin in sheep. Pediatr Res 17:107–110
38. Runkle B, Goldberg RN, Streiffeld MM, et al (1984) Cardiovascular changes in Group B streptococcal sepsis in the piglet: Response to indomethacin and relationship to prostacyclin and thromboxane A_2. Pediatr Res 18:874–878
39. Palmer RMJ, Ferrige AG, Moncada S (1987) Nitric oxide release accounts for the biological activity of endothelial derived relaxing factor. Nature 327:524–526
40. Ignarro LJ, Buga GM, Wood KS, Byrns RE, Chaudhuri G (1987) Endothelium-derived relaxing factor produced and released from artery and vein is nitric oxide. Proc Natl Acad Sci USA 84:9265–9269
41. Forstermann U, Schmidt HHHW, Pollock JS, et al (1991) Isoforms of nitric oxide synthase: Characterization and purification from different cell types. Biochem Pharmacol 42:1849–1857
42. Busse R, Mulsch A, Fleming I, Hecker M (1993) Mechanisms of nitric oxide release from the vascular endothelium. Circulation 87 (Suppl. V):18–25
43. Moncada S, Palmer RMJ, Higgs EA (1991) Nitric oxide: Physiology, Pathophysiology, and Pharmacology. Pharmacol Rev 43:109–142
44. Somlyo AP, Somlyo AV (1994) Signal transduction and regulation in smooth muscle. Nature 372:231–236
45. Lugnier C, Schoeffter P, LeBec A, Strouthou E, Stoclet JC (1986) Selective inhibition of cyclic nucleotide phosphodiesterases of human, bovine, and rat aorta. Biochem Pharmacol 35:1743–1751
46. Thomas MK, Francis SH, Corbin JD (1990) Characterization of a purified bovine lung cGMP-binding cGMP phosphodiesterase. J Biol Chem 265:14964–14970
47. Halbower AC, Tuder RM, Franklin WA, Pollock JS, Föstermann U, Abman SH (1994) Maturation-related changes in endothelial nitric oxide synthase immunolocalization in developing ovine lung. Am J Physiol 267:L585–L591
48. Kinsella JP, Ivy DD, Abman SH (1994) Ontogeny of NO activity and response to inhaled NO in the developing ovine pulmonary circulation. Am J Physiol 267: H1955–H1961
49. Abman SH, Chatfield BA, Rodman DM, Hall SL, McMurtry IF (1991) Maturational changes in endothelial-derived relaxating factor activity in ovine pulmonary arteries in vitro. Am J Physiol 260:L280–L285
50. Morin FC III, Egan EA, Ferguson W, Lundgren CEG (1988) Development of pulmonary vascular response to oxygen. Am J Physiol 254:H542–H546
51. Abman SH, Chatfield BA, Hall SL, McMurtry IF (1990) Role of endothelium-derived relaxing factor during transition of pulmonary circulation at birth. Am J Physiol 259: H1921–H1927
52. Fineman JR, Heymann MA, Soifer SJ (1991) N^{ω}-nitro-L-arginine attenuates endothelium-dependent pulmonary vasodilation in lambs. Am J Physiol 260: H1299–H1306
53. Cornfield DN, Chatfield BA, McQueston JA, et al (1992) Effects of birth-related stimuli on L-arginine-dependent pulmonary vasodilation in ovine fetus. Am J Physiol 262:H1474–H1481
54. Moore P, Velvis H, Fineman JR, Soifer SJ, Heymann MA (1992) EDRF inhibition attenuates the increase in pulmonary blood flow due to oxygen ventilation in fetal lambs. J Appl Physiol 73:2151–2157

55. Fineman JR, Wong J, Morin FC III, Wild LM, Soifer SJ (1994) Chronic nitric oxide in utero produces persistent pulmonary hypertension in newborn lambs. J Clin Invest 93:2675–2683
56. Kinsella JP, McQueston JA, Rosenberg M, Abman SH (1992) Hemodynamic effects of exogenous nitric oxide in ovine transitional pulmonary circulation. Am J Physiol 263:H875–H880
57. McQueston JA, Kinsella JP, Ivy DD, McMurtry IF, Abman SH (1995) Chronic pulmonary hypertension *in utero* impairs endothelium-dependent vasodilation. Am J Physiol 268:H288–H294
58. Morin FC III (1989) Ligating the ductus arteriosus before birth remodels the pulmonary vasculature of the lamb. Pediatr Res 25:251–257
59. Abman SH, Shanley PF, Accurso FJ (1989) Failure of postnatal adaptation of the pulmonary circulation after chronic intrauterine pulmonary hypertension in fetal lambs. J Clin Invest 83:1849–1858
60. Morin FC III (1989) Ligating the ductus arteriosus before birth causes persistent pulmonary hypertension in the newborn lamb. Pediatr Res 25:245–250
61. Steinhorn RH, Thusu KG, Russell JA, Gugino SF, Paradisis M, Morin FC (1994) The cGMP specific phosphodiesterase inhibitor M&B 22948 enhances the effect of nitric oxide in newborn lambs with persistent pulmonary hypertension. Pediatr Res 35:89A (Abst)
62. Ziegler JW, Ivy DD, Kinsella JP, Clarke WR, Abman SH (1994) Dipyridamole, a cGMP phospodiesterase inhibitor augments inhaled nitric oxide-induced pulmonary vasodilation in the ovine transitional circulation. Pediatr Res 35:90A (Abst)
63. Archer SL, Huang JMC, Hampl V, Nelson DP, Shultz PJ, Weir EK (1994) Nitric oxide and cGMP cause vasorelaxation by activation of a charybdotoxin-sensitive K channel by cGMP dependent protein kinase. Proc Nat Acad Sci USA 91:7583–7587
64. Bolotina VM, Najibi S, Palacino JJ, Pagano PJ, Cohen RA (1994) Nitric oxide directly activates calcium-dependent potassium channels in vascular smooth muscle. Nature 368:850–853
65. Yanagisawa M, Kurihara H, Kimura S, et al (1988) A novel potent vasoconstrictor peptide produced by vascular endothelial cells. Nature (Lond) 332:411–415
66. Inoue A, Yanagisawa M, Kimura S, et al (1989) The human endothelin family: Structurally and pharmacologically distinct isopeptides predicted by three separated genes. Proc Nat Acad Sci USA 86:2863–2867
67. Arai H, Hari S, Aramori 1, Ohkubo H, Nakanishi S (1990) Cloning and expression of a cDNA encoding endothelin receptor. Nature (Lond) 348:730–732
68. Sakurai T, Yanagisawa M, Takuwa Y, et al (1990) Cloning of a cDNA encoding a non-isopeptide-selective subtype of the endothelin receptor. Nature (Lond) 348: 732–735
69. Masaki T, Kimura S, Yanagisawa M, Goto K (1991) Molecular and cellular mechanism of endothelin regulation. Implications for vascular function. Circulation 84: 1457–1468
70. Hirata Y, Emori T, Eguchi S, et al (1993) Endothelin receptor subtype B mediates synthesis of nitric oxide by cultured bovine endothelial cells. J Clin Invest 91: 1367–1373
71. Wong J, Vanderford PA, Fineman JR, Soifer SJ (1993) Endothelin-1 produces pulmonary vasodilation in the intact newbom lamb. Am J Physiol 265:H1318–H1325
72. D'Orleans P, Telemaque S, Claing A, et al (1992) Human big-endothelin-1 and endothelin-1 release prostacyclin via the activation of ET-1 receptors in the rat perfused lung. Br J Pharmacol 103:773–776
73. Wong J, Fineman JR, Heymann MA (1994) The role of endothelin and endothelin receptor subtypes in regulation of fetal pulmonary vascular tone. Pediatr Res 35: 664–670
74. Chatfield BA, McMurtry IF, Hall SL, Abman SH (1991) Hemodynamic effects of endothelin-1 on ovine fetal pulmonary circulation. Am J Physiol 261:R182–R187

75. Bradley LM, Czaja JF, Goldstein RE (1990) Circulatory effects of endothelin in newborn piglets. Am J Physiol 259: H1613–H1617
76. Liben S, Stewart DS, De Marte J, Perreault T (1993) Ontogeny of Big-Endothelin-1 effects in the newborn piglet pulmonary vasculature. Am J Physiol 265: H139–H145
77. Jones OW III, Abman SH (1994) Systemic and pulmonary hemodynamic effects of Big endothelin-1 and phosphoramidon in the ovine fetus. Am J Physiol 266: R929–R935
78. Ivy DD, Kinsella JP, Abman SH (1994) Physiologic characterization of endothelin A and B receptor activity in the ovine fetal pulmonary circulation. J Clin Invest 93: 2141–2148
79. Tod ML, Cassin S (1992) Endothelin-1-induced pulmonary arterial dilation is reduced by N-nitro-L-arginine in fetal lambs. J Appl Physiol 72: 1730–1734
80. Wang Y, Coceani F (1992) Isolated pulmonary resistance vessels from fetal lambs. Contractile behavior and responses to indomethacin and endothelin-1. Circ Res 71: 320–327
81. Cassin S, Kristova V, Davis T, Kadowitz P, Gause G (1991) Tone-dependent responses to endothelin in the isolated perfused fetal sheep pulmonary circulation in situ. J Appl Physiol 70: 1228–1234
82. Janakidevi K, Fisher MA, Del Vecchio PJ, Tiruppathi C, Figge J, Malik AB (1992) Endothelin-1 stimulates DNA synthesis and proliferation of pulmonary artery smooth muscle cells. Am J Physiol 263: C1295–C1301
83. Rosenberg M, Kennaugh J, Koppenhafer SL, Loomis M, Chatfield BA, Abman SH (1993) Elevated immunoreactive endothelin-1 levels in infants with persistent pulmonary hypertension of the newborn. J Pediatr 123: 109–114
84. Roberts JD, Polaner DM, Lang P, Zapol WM (1992) Inhaled nitric oxide in persistent pulmonary hypertension of the newborn. Lancet 340: 818–819
85. Kinsella JP, Neish SR, Shaffer E, Abman SH (1992) Low-dose inhalational nitric oxide in persistent pulmonary hypertension of the newborn. Lancet 340: 819–820
86. Kinsella JP, Neish SR, Ivy DD, Shaffer E, Abman SH (1993) Clinical responses to prolonged treatment of persistent pulmonary hypertension of the newborn with low doses of inhaled nitric oxide. J Pediatr 123: 103–108
87. Kinsella JP, Abman SH (1993) Inhalational nitric oxide therapy for persistent pulmonary hypertension of the newborn. Pediatrics 91: 997–998
88. Finer NN, Etches PC, Kamstra B, Tierney AJ, Peliowski A, Ryan CA (1994) Inhaled nitric oxide in infants referred for extracorporeal membrane oxygenation: Dose response. J Pediatr 124: 302–308
89. Mercier JC, Dehan M, and the Groupe français pédiatrique d'étude du NO (1994) Inhalational nitric oxide in persistent pulmonary hypertension of the newborn. Results of the French multicenter open study. Am J Respir Crit Care Med 149: A545 (Abst)
90. Rozé JC, Storme L, Zupan V, Morville P, Dinh-Xuan AT, Mercier JC (1994) Echocardiographic investigation of inhaled nitric oxide in newborn babies with severe hypoxaemia. Lancet 344: 303–305
91. Kinsella JP, McCurnin DCw Clark RH, Lally K, Null DM (1992) Cardiac performance in ECMO candidates: Echocardiographic predictors for ECMO. J Pediatr Surg 27: 44–47
92. Clark DA, Nieman GF, Thompson AM, Paskanik AM, Rokhar JE, Bredemberg CE (1987) Surfactant displacement by meconium free fatty acids: An alternate explanation for atelectasis in meconium aspiration syndrome. J Pediatr 110: 765–770
93. Moses D, Holm BA, Spitale P, Liu M, Enhorning G (1991) Inhibition of pulmonary surfactant function by meconium. Am J Obstet Gynecol, 164: 477–481
94. Siebert H, Haas J, Beckwith J (1984) Left ventricular hypoplasia in congenital diaphragmatic hernia. J Pediatr Surg 19: 567–571
95. Schwart SM, Vermillion RP, Hirschl RB (1994) Evaluation of left ventricular mass in children with left-sided congenital diaphragmatic hernia. J Pediatr 125: 447–451

96. Glick PL, Stannard VA, Leach CL, et al (1992) Pathophysiology of congenital diaphragmatic hernia II: The fetal lamb CDH model is surfactant deficiency. J Pediatr Surg 27:382–388
97. Antunes MJ, Greenspan JS, Holt WJ, Vallieu DS, Spitzer AR (1994) Assessment of lung function pre-nitric oxide therapy: A predictor of response? Pediatr Res 35:212A (Abst)
98. Rossaint R, Falke KJ, Lopez F, Slama K, Pison U, Zapol WM (1993) Inhaled nitric oxide for the adult respiratory distress syndrome. N Engl J Med 328:399–405
99. Gerlach H, Rossaint R, Pappert D, Falke KJ (1993) Time-course and dose-response of nitric oxide inhalation for systemic oxygenation and pulmonary hypertension in patients with adult respiratory distress syndrome. Eur J Clin Invest 23:499–502
100. Jobe AH (1993) Pulmonary surfactant therapy. N Engl J Med 328:861–868
101. Jobe AH (1992) Surfactant in the perinatal period. Early Human Development 26: 57–62
102. Strang LB, MacLeish MH (1961) Ventilatory failure and right-to-left shunt in newborn with respiratory distress. Pediatrics 28:17–27
103. Stahlman M, Blankenship WJ, Shepard FM, Gray J, Young WC, Malan AF (1972) Circulatory studies in clinical hyaline membrane disease. Biol Neonate 20:300–320
104. Evans NJ, Archer LNJ (1991) Doppler assessment of pulmonary artery pressure and extrapulmonary shunting in the acute phase of hyaline membrane disease. Arch Dis Child 66:6–11
105. Skinner JR, Boys RJ, Hunter S, Hey EN (1992) Pulmonary and systemic arterial pressure in hyaline membrane disease. Arch Dis Child 67:366–373
106. Walther FJ, Benders MJ, Leighton JO (1992) Persistent pulmonary hypertension in premature neonates with severe respiratory distress syndrome. Pediatrics 90:899–904
107. Kinsella JP, Ivy DD, Abman SH (1994) Inhaled nitric oxide lowers pulmonary vascular resistance and improves gas exchange in severe experimental hyaline membrane disease. Pediatr Res 36:402–408
108. Abman SH, Kinsella JP, Schaffer MS, Wilkening RB (1993) Inhaled nitric oxide therapy in a premature newborn with severe respiratory distress and pulmonary hypertension. Pediatrics 92:606–609
109. Hallmann M, Kankaanpaa K (1980) Evidence of surfactant deficiency in persistence of the fetal circulation. Eur J Pediatr 134:129–134
110. James DK, Chiswick JL, Havier A, Williams M, Hallworth J (1985) Nonspecificity of surfactant deficiency in neonatal respiratory disorders. Br Med J 105:1635–1637
111. Lotze A, Whitsett JA, Kammerman LA, Ritter M, Taylor GA, Short BL (1990) Surfactant protein A concentrations in tracheal aspirate fluid from infants requiring extracorporeal membrane oxygenation. J Pediatr 116:435–440
112. Auten RL, Notter RH, Kendig JW, Davis JM, Shapiro DL (1991) Surfactant treatment of full-term newborns with respiratory failure. Pediatrics 87:101–107
113. Glick PL, Leach CL, Besner GE, et al (1992) Pathophysiology of congenital diaphragmatic hernia III: Exogenous surfactant therapy for the high-risk neonate with CDH. J Pediatr Surg 27:866–869
114. Lotze A, Knight FR, Martin GR, et al (1993) Improved pulmonary outcome after exogenous surfactant therapy for respiratory failure in term infants requiring extracorporeal membrane oxygenation. J Pediatr 122:261–268
115. Leach CL, Fuhrman BP, Morin FC, Rath MG (1993) Perfluorocarbon associated gas exchange (partial liquid ventilation) in respiratory distress syndrome. Crit Care Med 21:1270–1278
116. Peckham GJ, Fox WW (1978) Physiologie factors affecting pulmonary artery pressure in infants with persistent pulmonary hypertension. J Pediatr 93:1005–1010
117. Fox WW, Duara S (1983) Clinical management of persistent pulmonary hypertension of the neonate. J Pediatr 103:505–514
118. Drummond WH, Gregory GA, Heymann MA, Phibbs RA (1981) The independent effects of hyperventilation, tolazoline, and dopamine on infants with persistent pulmonary hypertension. J Pediatr 98:603–611

119. Schreiber MD, Heymann MA, Soifer SJ (1986) Increased arterial pH, not decreased $PaCO_2$, attenuates hypoxia induced pulmonary vasoconstriction in newborn lambs. Pediatr Res 20:113–117
120. Bifano EM, Pfannenstiel A (1988) Duration of hyperventilation and outcome in infants with persistent pulmonary hypertension. Pediatrics 81:657–661
121. Hendricks-Munoz KD, Walton JP (1988) Hearing loss in infants with persistent fetal circulation. Pediatrics 81:650–656
122. Wung JT, James LS, Kilchevsky E, James E (1985) Management of infants with severe respiratory failure and persistence of the fetal circulation, without hyperventilation. Pediatrics 76:488–494
123. Dworetz AR, Moya FR, Sabo B, Gross I (1989) Survival of infants with PPHN without ECMO. Pediatrics 84:1–6
124. Clark RH (1994) High-frequency ventilation. J Pediatr 124:661–670
125. Dreyfuss D, Saumon G (1993) Role of tidal volume, FRC, and end-inspiratory volume in the development of pulmonary edema following mechanical ventilation. Am Rev Respir Dis 148:1194–1203
126. Meredith KS, deLemos RA, Coalson JJ, et al (1989) Role of lung injury in the pathogenesis of hyaline membrane disease in premature baboons. J Appl Physiol 66: 2150–2158
127. Kinsella JP, Gerstmann DD, Clark RH, et al (1991) HFOV vs. IMV: Early hemodynamic effects in the premature baboon with HMD. Pediatr Res 29:160–166
128. Froese AB, McCulloch PR, Sugiura M, Vaclavik S, Possmayer F, Moller F (1989) A comparison of ventilation strategies for the use of high-frequency oscillatory ventilation in the treatment of hyaline membrane disease. Acta Anaesthesiol Scand 33: 102–107
129. Clark RH, Yoder BA, Sell MS (1994) Prospective, randomized comparison of high-frequency oscillation and conventional ventilation in candidates for extracorporeal membrane oxygenation. J Pediatr 124:447–454
130. Gerstmann DR, Fouke JM, Winter DC, Taylor AF, deLemos RA (1990) Proximal, tracheal, and alveolar pressures during high-frequency oscillatory ventilation in a normal rabbit model. 28:367–373
131. Braner DA, Fineman JR, Change R, Soifer SJ (1993) M&B 22948, a cGMP phosphodiesterase inhibitor, is a pulmonary vasodilator in lambs. Am J Physiol 264: H252–H258
132. Beckman JS, Beckman TW, Chen J, Marshall PA, Freeman BA (1990) Apparent hydroxyl radical production by peroxynitrite: Implications for endothelial injury from nitric oxide and superoxide. Proc Nat Acad Sci USA 87:1620–1624
133. Gaston B, Drazen JM, Loscalzo J, Stamler JS (1994) The biology of nitrogen dioxides in the airways. Am J Respir Crit Care Med 149:538–551
134. Stavert DM, Lehnert BE (1990) Nitric oxide and nitrogen dioxide as inducers of acute pulmonary injury when inhaled at relatively high concentrations for brief periods. Inhalation Toxicology 2:53–67
135. Hugod C (1979) Effect of exposure to 43 ppm nitric oxide and 3.6 ppm nitrogen dioxide on rabbit lung. Int Arch Occup Environ Health 42:159–167
136. Kavanagh BP, Mouchawar A, Goldsmith J, Pearl RG (1994) Effects of inhaled NO and inhibition of endogenous NO synthesis in oxidant-induced acute lung injury. J Appl Physiol 76:1324–1329
137. Haddad IY, Ischiropoulos H, Holm BA, Beckman JS, Baker JR, Matalon S (1993) Mechanisms of peroxynitrite-induced injury to pulmonary surfactants. Am J Physiol 265:L555–L564
138. Bult H, Meyer GRY, Jordaens FH, Herman AG (1991) Chronic exposure to exogenous nitric oxide may suppress its endogenous release and efficacy. J Cardiovasc Pharmacol 17:S79–S82
139. Lepoivre M, Fieschi F, Coves J, Thelander L, Fontocave M (1991) Inactivation of ribonucleotide reductase by nitric oxide. Biochem Biophys Res Commun 179: 442–448

140. Wink DA, Kasprzak KS, Maragos CM, et al (1991) DNA deaminating ability and genotoxicity of nitric oxide and its progenitors. Science 254:1001–1003
141. Högman M, Frostell C, Arnberg H, Hedenstierna G (1993) Bleeding time prolongation and NO inhalation. Lancet 341:1664–1665
142. Radomski MW, Palmer RMJ, Moncada S (1987) Endogenous nitric oxide inhibits platelet adhesion to vascular endothelium. Lancet 2:1057
143. Stamler JS, Simon DI, Osborne JA, et al (1992) S-Nitrosylation of proteins with NO: Synthesis and characterization of biologically active compounds. Proc Natl Acad Sci USA 89:444–448
144. Ahluwalia J, Kelsall A, Raine J, et al (1994) Nitric oxide improves oxygenation in neonates with respiratory distress syndrome and pulmonary hypertension. Pediatr Res 36:3A (Abst)
145. Freeman B (1994) Free radical chemistry of nitric oxide. Chest 150:79S–84S

Almitrine and Inhaled Nitric Oxide in Acute Respiratory Failure

D. M. Payen, R. Kong, and C. Gatecel

Introduction

The treatment of hypoxia is perhaps one of the most challenging facets of the intensive care of patients with acute lung injury (ALI) such as acute respiratory distress syndrome (ARDS) and severe pneumonia. Arterial oxygenation can be improved by techniques that range from simply increasing the inspired oxygen concentration (FiO_2) to the use of sophisticated extracorporeal lung assist devices, but none is without adverse consequences. High FiO_2 may generate oxidative radicals and have detrimental effects on lung tissue [1]. Positive end-expiratory pressure (PEEP) is frequently used to recruit alveoli during mechanical ventilation, but the level of PEEP is often limited to diminish its effects on systemic [2] and regional [3] hemodynamics and lung tissue [4]. Extracorporeal lung assistance techniques are difficult, expensive and potentially dangerous, and the risk of hemorrhagic complications weighs heavily against their expected benefits [5]. A safe and simple maneuver to improve arterial oxygenation is still required.

In this chapter, we present a novel approach to the treatment of hypoxia in ALI. Our data on arterial oxygenation and pulmonary hemodynamics, obtained when nitric oxide (NO) inhalation is combined with almitrine infusion, serve to illustrate this. First, we discuss the pathophysiology of hypoxia, and focus in particular on the perfusion term of the ventilation/perfusion (VA/Q) ratio.

Pathophysiologic Considerations

Hypoxemia is the most immediate threat to survival in ALI and is often the result of abnormal VA/Q ratios. Considering the perfusion term, the ideal modification to improve PaO_2 would be to reduce blood flow in shunting areas and increase it in ventilated areas.

The division of the lung into venous admixture, dead space and ideal gas exchange compartments provides a conceptually simple model for considering pulmonary gas exchange. However, the quantitative estimation of the distribution of VA/Q ratios, using the multiple inert gas elimination technique (MIGET) developed by Wagner and coworkers [6], has permitted much greater understanding of pulmonary gas exchange. Despite valid theoretical limitations, the value of this method has now been established in a wide variety of experimental and clinical states in animal and man [7].

Hypoxic pulmonary vasoconstriction (HPV) provides a critical link between gas exchange and blood flow distribution characteristics of the lung. By regulating regional vascular conductance (the inverse of resistance) and reducing blood flow to lung regions with low VA/Q ratios [8, 9], HPV contributes to the efficiency of gas exchange. The stimulus and site of HPV has motivated much research and this can be summarized as follows. The hypoxic stimulus seems to be a simple function of both mixed venous [10] and alveolar oxygen tension [11], which trigger HPV via a cellular mechanism or sensor that has yet to be found. All the pulmonary vessels can respond to hypoxia by vasoconstriction [9, 12], but the *in vivo* HPV response occurs predominantly in small arteries with diameters less than 500 μm. These vessels are close to the stimulus, being surrounded by alveolar gas on the outside, and mixed venous blood on the inside. The smooth muscle cells then respond to the local oxygen tension gradient across the vessel wall. The larger vessels have *vasa vasorum* derived from bronchial arteries and their constriction is determined by the systemic arterial oxygen tension [9]. Their response is biphasic: constriction when PaO_2 decreases to 40–50 mm Hg, and dilation with further decreases in PaO_2 [9]. Therefore, in large pulmonary vessels, HPV is not stimulated in the absence of systemic hypoxemia and is less profound than in small vessels.

If many substances and circumstances can alter the HPV response, it is only recently that endogenous released NO has been shown to modify HPV [13, 14]. In isolated perfused lungs, the administration of NO synthesis (NOS) inhibitors during ventilation with a hypoxic gas mixture, augments the increase in pulmonary arterial pressure, suggesting that NO release may attenuate HPV. The influence of endogenous NO on VA/Q matching has been well demonstrated in an anesthetized rabbit model [14]. The lungs were separately ventilated to produce unilateral alveolar hypoxia. Administration of a NO inhibitor resulted in a dose-dependent reduction in the intrapulmonary distribution of the blood flow to hypoxic alveoli and a concomitant increase in PaO_2. This effect was reversed by L-arginine infusion thus implicating the role of NO. The authors suggest that NO release by the lung may attenuate regional HPV, support blood flow to hypoxic lung units, and result in a decrease in PaO_2. Moreover, this *in vivo* study demonstrated that substantial intrapulmonary redistribution of blood flow away from hypoxic alveoli can occur without absence of any change in pulmonary arterial pressure. These and other results strongly suggest that endogenous NO is present in the lung during alveolar hypoxia and acts to oppose HPV.

In ARDS patients, increased VA/Q abnormalities, atelectasis, loss of vascular bed and small vessel constriction from endogenous mediators, in addition to HPV, result in pulmonary hypertension [15]. Some of these elements can be tested as follows: if the shift from maintenance FiO_2 to FiO_2 1.0 leads to a modest increase in PaO_2, a large shunt is the predominant cause of hypoxia; a decrease in pulmonary artery pressure (PAP) during increasing FiO_2 suggests an efficient HPV phenomenon in the rest of the lung; if pulmonary hypertension persists, the part of vasoconstriction, narrowing or loss vascular bed cannot be determined.

The original and crucial feature of NO inhalation therapy is the exclusive delivery of NO to ventilated alveoli so that HPV is not affected in atelectatic areas and oxygenation is better preserved. Several human studies [16–19] have reported results supporting such a concept. Rossaint et al. [20] found an increase in PaO_2 from 152 ± 15 to 199 ± 23 mm Hg in ARDS patients inhaling 18 ppm of NO. Using MIGET, the authors showed a reduction in blood flow to low VA/Q regions and a significant reduction in oxygen shunt. This observation implies that an active vasoconstriction that was present in ventilated lung areas was reduced by NO inhalation, and the diversion of the blood flow away from atelectatic regions reduced the oxygen shunt. Most clinical investigators, however, have observed variable effects during NO inhalation. From the above discussion, it appears that oxygen content is only increased with NO when, in addition to HPV, some constriction is present in small arteries or veins [21]. The major requirement is that NO should reduce PVR in the ventilated lung regions so that shunt flow declines, but at the same time does not reduce the end-capillary oxygen so as to offset the benefit to gas exchange [22].

This background suggests that the ability of NO to improve oxygenation could be enhanced by adding an infused vasoconstrictor [22]. This combination might be particularly worthwhile in conditions where the HPV response is reduced or abolished, such as in pulmonary infection [23], septicemia [24] direct lung trauma, and septic shock [25]. However, the combined therapy is unlikely to be as effective if the small pulmonary vessels are not actively constricted, but fibrotic or irreversibly narrowed, as observed after one or two weeks of ARDS [26].

Ideally, the vasoconstrictor drug should:

1) act preferentially on vessels in poorly or unventilated lung regions to potentiate or restore altered HPV;
2) preserve a gradient of vascular resistance between ventilated regions receiving NO and regions where HPV has reduced perfusion;
3) act mainly on prealveolar vessels to minimize pulmonary venous hypertension that may promote pulmonary edema, and
4) have negligible effects on systemic circulation and oxygenation.

Consequently, we have chosen almitrine bimesylate [27] as numerous experimental and clinical studies have shown that its action involves HPV.

Almitrine Bimesylate and Pulmonary Vasculature

Pharmacology of Almitrine Bimesylate

The respiratory stimulant property of almitrine was first described by Laubie and Diot in 1972 [27]. The drug's effects on lung volumes, blood gases [28], ventilation/perfusion distribution [29, 30] and large pulmonary vessel compliance [31] have been studied (the latter by cineangiography).

Almitrine bimesylate is the salt of a lipid-soluble drug which is poorly ionized at normal physiological pH. It is an agonist of the peripheral arterial chemoreceptors and increases discharge in the carotid sinus nerve [32], but the mechanism of action is not known. Dose-response studies have been performed in several animal species. In man, the acute effects on ventilation, pulmonary hemodynamics and gas exchange have been studied after oral (50 to 200 mg) [29, 33, 34] and intravenous (0.25–0.3 mg/kg bolus, or 4–33 mcg/kg/min infusion) [31, 35, 36] administration. The oral drug is rapidly absorbed in healthy human volunteers. Peak plasma concentrations are found 1.5 to 2.4 h after a 100 mg dose and a wide inter-individual variation is observed that is not related to body weight. Optimal therapeutic plasma concentration appears to be 200 to 300 mcg/mL.

Almitrine is highly (>99%) bound over a wide dose range, predominantly to low density lipoprotein. It does not bind to hemoglobin or carbonic anhydrase. After a bolus intravenous dose, plasma concentrations decrease biexponentially with an initial plasma half-life of 2.5 h, but a longer terminal half-life which ranges from 27 h in rats to 25 days in man. The drug distributes throughout the body and high tissue levels are found in liver, small intestine and stomach. Tissue uptake is fully reversible and the elimination is time-dependent. Tissue levels peak 6 to 24 h after the intravenous dose and decline over the subsequent 96 h with a tissue half-life of around 12 h.

The metabolism of almitrine results in at least 7 metabolites. The two major metabolic pathways are: 1) hydroxylation of the allyl side chains to hydroxy derivatives with subsequent glucuronide conjugation; and 2) deallylation to mono- and bi-deally metabolites. Only the hydroxy metabolites show some almitrine-like activity, though they are less potent and have a shorter duration of action than the parent drug. The metabolites are excreted either in the bile, or directly into the intestines from the gut circulation. Urinary elimination is low because of the high molecular weight. Both the drug and its metabolites are completely eliminated, though this occurs slowly and can take more than 900 h. Almitrine does not interact with commonly prescribed drugs, although cimetidine can reduce its absorption (Clinical expert report from IRIS, Paris, France).

The only side effect deserving special mention is a sensory peripheral neuropathy which usually affects lower limbs [37]. This has been reported in some patients after prolonged oral treatment ranging from 9 to 25 months [38]. Axonal degeneration is seen in histological study of affected nerves. Recovery can take from 3 to more than 12 months after stopping the drug.

The incidence as defined by electrophysiological changes is higher than the clinical syndrome (around 30%) and has diminished considerably with the reduction of the oral dose. It can be reasonably assumed that for the proposed use in ALI, at low intravenous dose and for a short period (a few days), such a complication would be unusual.

Effects of Almitrine on Oxygenation

Experimental studies have been performed to analyze the mechanism by which almitrine may improve PaO_2. This can occur without stimulating ventilation [28, 30, 39] possibly because of a redistribution of blood flow away from poorly ventilated areas.

Isolated perfused lung studies have the advantage that interference from pulmonary reflexes and hormones are eliminated. Similar results have been found in studies using isolated ferret [40] and rat lung [40–43]. These show that almitrine causes constriction of the relaxed vessels of normoxic lung [40]. In the presence of HPV (in response to ventilation with a hypoxic gaz mixture), almitrine has a dual action: in low doses further vasoconstriction occurs, but high doses lead to vasodilation [40, 42, 43]. Similar findings have been found in chronic hypoxic rats [40]. Conflicting results have been published about the magnitude of the effect during hypoxia compared to normoxia. Attempts to identify a substance which may mediate dilator response of almitrine have been unsuccessful. These initial results suggesting a similarity between hypoxic and almitrine-induced pulmonary vasoconstriction have stimulated further studies. Recently, Gottschall et al. [42] have verified that almitrine mimics HPV. If almitrine and hypoxia are similar in their impact on the lung circulation, one would anticipate that almitrine would: 1) produce pulmonary vasoconstriction; 2) potentiate or enhance HPV; 3) enhance HPV induced by other agents; 4) be inhibited by agents inhibiting HPV; 5) act at the precapillary level, implying that the drug is not predominantly a pulmonary vasoconstrictor; and 6) have little or no effect on systemic vessels.

The authors [42] have documented most of the anticipated actions of almitrine. Addition of 10 mcg of almitrine in the reservoir of the blood-perfused rat lung increased the perfusion pressure at constant flow with no effect on microvascular pressure, this accounting for a precapillary vasoconstriction. The increase in pressure with almitrine was dose-dependent. Moreover, almitrine potentiated the subsequent hypoxic pressure response, but had no effect on established hypoxic vasoconstriction. The latter observation was not related to the elevated pulmonary vascular tone before almitrine *per se*, since almitrine further increased pressure after KCl-induced constriction. The inhibition of prostaglandin synthesis by meclofenamate did not modify the almitrine response. Agents known to enhance or inhibit HPV had similar effect on almitrine-induced vasoconstriction. Finally, almitrine had no effect on pulmonary conduit arteries. It can be concluded from these results that almitrine closely mimics hypoxia, but not exactly.

The mechanism by which hypoxia is sensed remains uncertain. It has been suggested that O_2 tension is detected in the vasculature through its effects on energy status, as reflected by the level of high energy phosphates such as ATP [44, 45]. It has been shown in beef heart and in rat liver mitochondria [46], that almitrine may affect energy metabolism such as oxidative phosphorylation. This is the major source of ATP, formed as electrons are transferred from NADPH to O_2 through a series of electron carriers, in aerobic organisms. It is carried out by respiratory assemblies that are located in the inner membrane of mitochondria. Almitrine inhibits ATP synthesis or impairs electron transport downstream from the rotenone blocking site, localized in the NADPH-cytochrome-Q reductase complex. These actions may correspond to the locus of hypoxic sensing [28]. The signal may be generated by ATP reduction in a distinct cellular compartment in the pulmonary vasculature, as recently showed by Archer et al. [45].

Some *in vivo* studies, essentially in dog models, have reported conflicting results on the effects of almitrine on pulmonary vasculature. Hugues et al. [47] showed that an intravenous bolus of almitrine (10 mcg/kg) specifically vasoconstricted pulmonary vessels in normoxic, but not in hypoxic conditions. This vasoconstriction was not abolished by α or β adrenergic receptor blockade, indomethacin, vagotomy, or bilateral clamping of the common carotid arteries. The potential enhancement of HPV by almitrine had also been tested in normal dogs [28]. Using MIGET, almitrine was shown to increase pulmonary vascular resistance by 60% during ventilation with hypoxic mixture, but did not change on 100% O_2. These results have been interpreted as pulmonary vascular effects, since systemic hemodynamics, blood gases and parameters of oxygenation were not altered. The authors suggested that almitrine enhanced hypoxic vasoconstriction in the lung; an effect which may explain the observed PaO_2 increase in hypoxic patients. In a dog model where one lung was ventilated with hyperoxic gas and the other with an hypoxic mixture, Chen et al. [39] tested the impact of almitrine on HPV in the dose range of 0,003 to 3 mcg/kg/min. The significant effects of almitrine occurred only at 3 mcg/kg/min. Almitrine amplified the HPV effect, since the percentage of blood flow to the hypoxic lung was further reduced by the addition of almitrine and there was a concomitant increase in PaO_2. Naeije et al. [48] have recently published important results concerning the effect of almitrine in HPV in dogs. They outlined the contradictory results in the literature: in the healthy dog [28] and human [36], the increase in PVR induced by hypoxia is enhanced by almitrine, whereas in the isolated canine lung model, almitrine constricted hyperoxic pulmonary lung vessels, but not preconstricted hypoxic vessels [47]. These observations suggest that the pulmonary vasoconstricting activity of almitrine may inhibit the function of HPV. Accordingly, in an oleic acid dog model, the increase in pulmonary vascular tone induced by high dose of almitrine worsened gas exchange [49]. Expecting a dose-depending biphasic effect of almitrine, the authors tested lower doses of almitrine (2 and 4 mcg/kg/min) than those in their previous studies. In dogs exhibiting HPV response, 2 mcg/kg/min had no vascular effect, whereas 4 mcg/kg/min inhibited HPV. In dogs without HPV response to hypoxic gas mixture, almitrine at 2 and 4 mcg/kg/min restored HPV, as did intravenous administration of a chemoreceptor agonist doxapram. The authors concluded that almitrine at low doses restore HPV in dogs, probably by a direct effect on the pulmonary vessels.

In parallel with these experimental data, almitrine has been studied in man, particularly in patients with chronic obstructive pulmonary disease (COPD). Hypoxia was reduced as a result, as shown by MIGET, of improved VA/Q mismatching [29]. Some of these studies were in ARDS patients with severe hypoxia that could be explained by shunt and VA/Q mismatching [30]. Reports based on conventional gas exchange studies have shown that, in patients with severe acute respiratory failure secondary to shock or sepsis, intravenous almitrine significantly improved PaO_2 with negligible pulmonary and systemic hemodynamic changes [50]. Reyes et al. [30] have shown in 9 ARDS patients ventilated at constant FiO_2, that PaO_2 increased from 78 ± 15 to 140 ± 49 mm Hg (+79%) during almitrine infusion at approximately 16 mcg/kg/min. They used MIGET to show that almitrine redistributed pulmonary blood flow from shunt areas to lung units with normal VA/Q ratio. Pulmonary pressures increased from 26 ± 5 to 30 ± 5 mm Hg. These transient changes returned to baseline in 30 min, after the drug infusion was stopped.

Almitrine has also been tested during severe hypoxemia caused by shunting, in the clinical setting of unilateral pneumonia [51]. Based on the earlier discussion, one can expect a positive effect of the drug in this situation, since redistribution of pulmonary flow might reduce VA/Q inequalities. Using a high dose of almitrine (17 mcg /kg/min), the authors [23, 24] failed to find any increase in PVR and PaO_2. These surprising negative results may have resulted, as suggested by the dog studies, from the use of a high dose and perhaps also the particular lung disease, since HPV has been shown to be abolished in acute unilateral bacterial pneumonia.

Summary

Almitrine is a large molecular weight lipophilic compound which stimulates the carotid body and improves ventilation/perfusion inequality by a direct action on pulmonary vessels and also through the carotid body. The remarkable similarities between the effects of almitrine and hypoxia suggest a close mechanism of action, possibly at mitochondrial level through oxidative phosphorylation. The observed benefit on PaO_2 has been related to the pulmonary vasoconstrictor effect of the drug. The dose range seems to be crucial since at low doses (< 10 mcg/kg/min), almitrine reinforces or restores HPV, whereas at higher doses the drug has no effect or dilates pulmonary vessels. If HPV response is not common, even in humans (20%), the effect of almitrine on HPV restoration is also extremely attractive to improve the VA/Q inequality. Although the molecular mechanisms remain unknown, the use of such a drug as a continuous intravenous infusion seems to be efficient. In term of risk/benefit ratio, since we propose to use almitrine intravenously, at low dose and for a short period of time, the risk of its most notable adverse effect (sensory peripheral neuropathy) should be minimal. Moreover it would be difficult, given the known multifactorial risks (such as ischemia, neuromuscular blocking agents, nutrition, and compression) for neuropathy in ICU patients, to attribute such a side effect to the drug.

Clinical Results of the NO-Almitrine Combination

The investigation was done according to the recommendations of the Ethical French Law, after obtaining informed consent from patient's closest relatives. Eighteen consecutive patients suffering from severe ALI based on a PaO_2/FiO_2 value <150 mm Hg were prospectively studied. Group I corresponded to the diagnosis of ARDS (Murray Score: 3.1 ± 0.2; $n = 8$) based on classical criteria as a compatible underlying disease, a pulmonary artery occlusion pressure <20 mm Hg during mechanical ventilation and bilateral lung opacities. The etiology of ARDS was direct lung injury in 62%, and an indirect lung injury in 38% of patients. Group II corresponded to severe ALI ($n = 10$) with a heterogeneous lung disease assessed by chest X-ray and thoracic CT scan. The characteristics of these two groups are shown in Table 1.

Mechanical ventilation was achieved by a Servo ventilator Siemens 900 C or 300 (Elema Siemens, Solna, Sweden) using different ventilatory modes. PEEP level averaged 12 ± 2 cm H_2O (ranging from 3 to 21 cm H_2O) in group I and 6 ± 1.1 (ranging from 0 to 10 cm H_2O) in Group II. All the patients were sedated, paralyzed and mechanically ventilated with FiO_2 1.0. When a pharmacologic cardiovascular support was needed, it was maintained throughout the study period.

NO administration

Since NO reacts rapidly with oxygen [21], the delivery system was choosen to limit the time contact between NO and oxygen. NO was administered continuously via a fenestrated silicone catheter (1.3 mm ∅; Vygon, France) positioned within the tracheal tube so that the distal tip of the catheter was 2 cm above the distal end of the tube. The position of tube and catheter was confirmed by fibroscopy and chest X-ray. NO was given by intratracheal continuous flow from a cylinder (225 ppm NO in nitrogen, CFPO, Meudon La Forêt, France) and the NO concentration was calculated as follows:

$$\text{Inhaled NO concentration (ppm)} = 225 \times \text{NO flow rate/minute ventilation}$$

This calculation was validated *in vitro*. A volumetric ventilator (Cesar, CFPO or Siemens Servo 900C) was linked via a standard patient circuit to a

Table 1. Patients clinical characteristics in Group I (ARDS) and Group II (Acute Lung Injury) expressed in mean ± SEM

	Age/ sex	GSI	Murray score	PEEP level (mm Hg)	Cqs (mL/mm Hg)	PaO_2/FiO_2 (mm Hg)	$PaCO_2$ (mm Hg)	Venti-lation (days)	Inhaled NO (days)	Almitrine infusion (days)	Alive (%)
Group I	41.1 ± 6.5	13.4 ± 1.2	3.1 ± 0.2	12.3 ± 2	37.2 ± 6	79 ± 11.3	47.3 ± 2.1	8.2 ± 2.2	5.1 ± 1.1	1.4 ± 0.3	63
Group II	48 ± 7	14.9 ± 1.6		6.1* ± 1.1	38.4 ± 3.8	89.4 ± 7.6	38* ± 1	4.6 ± 1.1	4.9 ± 1.1	3.3 ± 0.9	50

Pts: patients; GSI: gravity score index; PEEP: positive end-expiratory pressure; Cqs: quasi static pulmonary compliance; Ventilation: ventilation duration before the protocol; Inhaled NO: duration of NO treatment; Almitrine infusion: duration of almitrine treatment

* $p < 0.05$ Group I vs Group II

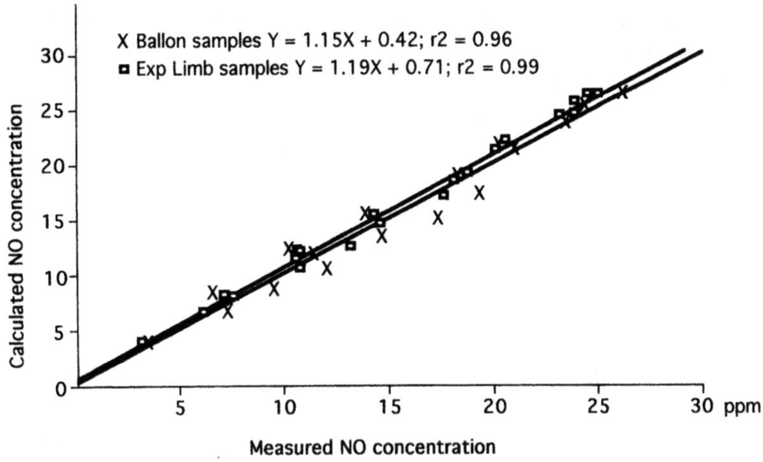

Fig. 1. Correlations between measured NO concentrations and values calculated as follows:

$$\text{Inhaled NO concentration (ppm)} = \frac{225 \times \text{NO flow rate}}{\text{minute ventilation}}$$

NO concentration measurements were performed in the balloon (×) or in the expiratory limbs of the ventilator (□). NO concentrations are expressed in ppm

4000 mL balloon mimicking the usual vital capacity and residence time. The ventilator supplied saturated gas at FiO_2 1.0. The catheter was pushed into the tracheal tube and connected to the NO cylinder equipped with a pressure regulator and a flowmeter (CG 0205, CFPO). Gas samples drawn from the balloon and from the expiratory limb of the circuit were analyzed by chemiluminescence (Topaze 2020, Cosma, Igny, France). We obtained a close correlation (Fig. 1) between the concentrations of measured and calculated NO. The similar correlation with values measured from either the balloon or the expiratory limb suggests a rapid and adequate mixing of both gases. Nitrogen dioxide formation was negligible (lower than 0.3 ppm). Table 2 summarizes the comparative data obtained by varying ventilatory settings such as PEEP

Table 2. Correlations between measured NO concentrations and calculated values at different ventilatory settings

Correlation	PEEP level (cm H_2O)	Inspiratory time (sec)	Gas temperature (°C)
r	0.999	0.996	0.993
slope	1.023	0.9	1.256
intercept (ppm)	0	−1.048	−1.542
n	6	13	13

PEEP application from 0 to 20 cm H_2O, inspiration time from 0.3 to 1.5 sec, gas temperature from 20 to 37°C

(0 to 20 cm H_2O), inspiration time (0.3 to 1.5 sec) and gas temperature (20 to 37°C).

A similar circuit was used *in vivo* to deliver 15 ppm of NO continuously to study patients and the administered NO concentration was sequentially verified on the ventilator expiratory limb by chemiluminescence (Echo Physic, Massy, France). The accuracy of the flowmeter delivering NO was verified by the difference between expiratory and inspiratory minute ventilation measured by the ventilator. This difference corresponded to the value of flow of the added gas.

Measurements

Routine clinical monitoring for these patients consisted of an oxymetric thermodilution pulmonary artery catheter (Abbott Laboratories, Chicago, IL) for all patients (except 4 in Group II), which permitted the monitoring of mean pulmonary arterial pressure (PAP), right atrial pressure (Pra), pulmonary artery occlusion pressure (Ppao) and mean systemic arterial pressure (Psa) with disposable transducers linked to a monitoring system. Pressure values were the average over the respiratory cycle from the registered data on a magnetic recorder, except for Ppao which was measured during the end-expiratory phase. Cardiac output (CO) was measured by thermodilution technique when available ($n = 14$) and each value was the average of three injections of 10 mL of 5% dextrose solution. In 3 patients, cardiac output was measured by a transesophageal Doppler system (ODM 1, Abbott Laboratories, France) [52]. Systemic (SVR) and pulmonary vascular (PVR) resistances were calculated according to the standard formulae as the direct ratio between the driving pressure and CO in IU.

Arterial and mixed venous blood gases were measured with standard blood gas electrodes (ABL 300, Radiometer, Copenhagen, Denmark) and the total hemoglobin concentration, hemoglobin oxygen saturation, and methemoglobin levels were measured by spectrophotometry (OSM 2 Hemoximeter, Radiometer, Copenhagen, Denmark). Qs/Qt was calculated from standard shunt equation using the FiO_2 measured by the ventilator sensor and oxygen contents. Oxygen consumption (VO_2 in mL O_2/min) and oxygen delivery (DO_2 in mL O_2/min) were classically calculated.

Ventilatory parameters including peak inspiratory pressure (Ppi) and quasi-static compliance (Cqs) were continuously monitored by a ventilatory module (Siemens).

To ensure a constant FiO_2 before and during NO administration, nitrogen (N_2) was administered with O_2 "before NO" at the same flow rate as the NO-nitrogen mixture "during NO". The same protocol steps were applied to all patients. Hemodynamic and arterial blood gas parameters were recorded after a 30 min control period (during nitrogen inhalation). NO-nitrogen was then given and after 60 min of inhalation, the same measurements were made. Almitrine (Servier, Suresnes, France) was then infused intravenously and titrated from 8 to 16 mcg/kg/min in group I [30] and from 4 to 8 mcg/kg/min in Group II to achieve the best PaO_2/FiO_2. Measurements were done 60 min after the start or the stabilization of the infusion. Since NO inhalation acts as an "on-off" phenomenon [53], the impact of almitrine alone was verified 60 min after withdrawal of NO inhalation and replacement by N_2 when it was possible. This step of the protocol was not performed in 3 patients from Group I and 3 patients from Group II because of severe hypoxia.

After the above sequence of the protocol, the association of NO and almitrine was continued and FiO_2 was then reduced at a level which ensured a PaO_2 ranging from 75 to 120 mm Hg. When FiO_2 was reduced to 0.4, the level of PEEP was reduced according to the defined ranges of PaO_2. As expected, this was only possible in group II. Finally, almitrine was stopped if PaO_2 remained stable and higher than 200 mm Hg for 4 h at FiO_2 1.0. The response to NO withdrawal was tested daily. NO was discontinued if PaO_2 remained around 100 mm Hg at FiO_2 ≤ 0.50 after stopping NO for 4 h.

Conclusion

Except for the difference in lung disease, by definition, between the two groups, the clinical characteristics were similar for many aspects such as age, severity score, duration of mechanical ventilation prior to the protocol, outcome and lung quasi static compliance. However, for a similar level of PaO_2/FiO_2, the levels of PEEP and $PaCO_2$ were higher in the ARDS group – the latter because of permissive hypercapnia (Table 1).

Hemodynamic parameters (Table 3): In group I, the inhalation of NO at 15 ppm produced a prompt (delay less than 2 min) decrease in pulmonary artery pressure of 3 ± 1.5 mm Hg from baseline ($p < 0.02$) and which returned to N_2 value when NO inhalation was combined with almitrine infusion. NO withdrawal during almitrine infusion led to a significant increase in PAP (from 31.4 ± 1.7 to 37 ± 3.9 mm Hg; $p < 0.02$). PVR decreased by 0.9 ± 0.35 IU ($p < 0.002$) during NO inhalation, while the systemic hemodynamic parameters, particularly CO and PAP, did not change.

In group II, PAP increase concerned only the comparison between NO inhalation without almitrine, and almitrine infusion without NO. PVR during NO plus almitrine (4 ± 0.6 IU) did not differ from nitrogen point (3.7 ± 06 IU), while they were higher than those of NO alone (3.2 ± 0.4 IU; $p < 0.002$). The other parameters did not change significantly.

Gas exchange parameters (Table 4): In group I (ARDS): PaO_2 increased during NO inhalation (from 88 ± 7.1 to 137 ± 12 mm Hg; $p < 0.006$) and increased further when almitrine was infused (238 ± 31 mm Hg; $p < 0.006$). This effect was associated with a parallel increase in arterial oxygen saturation and a decrease in Qs/Qt. It should be mentioned that almitrine alone improved PaO_2 to a similar extent as NO alone. Partial pressure of oxygen in

Table 3. Hemodynamic parameters evolution in group I (ARDS) and group II (Acute Lung Injury)

Group I	N_2	NO	NO+ALM	ALM	anova
HR (bpm)	99.2±9.3	105.8±6.1	111.8±7.2	105.5±5.3	ns
Pra (mm Hg)	11 ±0.6	10.3±0.8	11.9±0.8	11.8±1.4	ns
Ppao (mm Hg)	11.1±0.5	10.7±0.5	12.6±1	15 ±2.7	ns
Ppam (mm Hg)	31.1±1.6	28 ±1.4*	31.4±1.7**	37 ±3.9	p<0.02
Psa (mm Hg)	71.1±4.1	67 ±4.1	74.5±4.6	70.8±5.9	ns
CO (L/min)	7.2±0.8	7.2±0.7	7.9±0.6	7.8±0.5	ns
PVR (IU)	5.2±0.7	4.3±0.6*	4.4±0.5**	4.9±1.3	p<0.002
SVR (IU)	15.8±1.1	15.2±1.3	12.9±1.4	13.4±1.7	ns
Group II					
HR (bpm)	91.9±5.8	90.3±4.2	88.3±4.9	98.5±5.8	ns
Pra (mm Hg)	10 ±1.2	10.3±1.4	9.6±0.5	11.7±0.7	ns
Ppao (mm Hg)	12.7±2.8	12.2±2.5	11.3±1.8	14.2±1.5	ns
Ppam (mm Hg)	25.7±2.9	23.3±2.6	25.3±2.1	28.5±3.1**	p<0.03
Psa (mm Hg)	73.7±2.9	75.3±3.5	69.5±2.4	68.3±4.9	ns
CO (L/min)	6.1±0.7	5.9±0.8	5.9±0.7	5.5±1.1	ns
PVR (IU)	3.7±0.6	3.2±0.4	4 ±0.6**	5.8±0.7**	p<0.002
SVR (IU)	16.8±2.5	17.7±2.9	15.2±2.4	18.3±2.9	ns

N_2: nitrogen control situation; NO: 60 min 15 ppm continuous NO inhalation; ALM: almitrine intravenous infusion; HR: heart rate; Pra: right atrial pressure; Ppao: occlusion arterial pulmonary pressure; Ppam: mean pulmonary arterial pressure; Psa: mean systemic arterial pressure; CO: cardiac output; PVR: pulmonary vascular resistances; SVR: systemic vasculary resistances (mean±SEM)

* $p<0.05$ in comparison with N_2 situation

** $p<0.05$ in comparison with NO situation

mixed venous blood also increased during NO and more importantly during NO combined with almitrine. Arterial oxygen content and oxygen delivery increased significantly during NO+almitrine compared to N_2. Partial pressure of arterial carbon dioxide did not change nor did oxygen consumption and lung mechanics.

In group II (ALI): PaO_2 increased during NO inhalation (from 89±7.6 to 196±23 mm Hg; $p<0.0001$) and further increased when almitrine was added (366±32 mm Hg; $p<0.0001$). This effect was followed by a concomitant increase in arterial oxygen saturation, arterial oxygen content and a decrease in Qs/Qt. As in group I, almitrine alone improved PaO_2 similarly to NO. Partial pressure of oxygen in mixed venous blood increased during NO inhalation and NO+almitrine. Arterial partial pressure of carbon dioxide did not change nor oxygen delivery, oxygen consumption and lung mechanics.

Table 4. Gazometric, ventilatory and metabolic parameters evolution in group 1 (ARDS) and group Il (Acute Lung Injury)

Group I	N_2	NO	NO+ALM	ALM	anova
PaO_2 (mm Hg)	88 ±7.1	137 ±12*	238 ±31**	102 ±15.4*†	p<0.006
SaO_2 (%)	94 ±0.6	98 ±0.7*	99 ±0.4**	97 ±1.2*	p<0.02
$P\bar{v}O_2$ (mm Hg)	37 ±0.7	40 ±1*	43 ±2**	40 ±1.2*†	p<0.002
$S\bar{v}O_2$ (%)	65 ±0.9	69 ±1.4*	74 ±2.5**	68 ±3†	p<0.02
Qs/Qt (%)	37 ±1.8	30 ±1.4*	25 ±1.3**	33 ±1.7*†	p<0.02
$PaCO_2$ (mm Hg)	47 ±2.1	49 ±2.7	47 ±2.6	49 ±3	ns
CaO_2 (mL O_2/min)	12.6±1.6	13.1±1.5	13.6±1.8*	12.7±1.8	p<0.03
DO_2 (mL O_2/min)	864 ±81	907 ±71	1041 ±79**	972 ±51	p<0.004
VO_2 (mL O_2/min)	265 ±23	270 ±23	287 ±22	290 ±22	ns
Ppi (cm H_2O)	33.1±2.2	33 ±2	33.2±1.8	32 ±1.8	ns
QSC (mL/cm H_2O)	40.7±7	40.5±6.4	39.8±6.3	38.8 ±5.9	ns
Group II					
PaO_2 (mm Hg)	89 ±7.6	196 ±23*	366 ±32**	251 ±51*†	p<0.0001
SaO_2 (%)	96 ±1	99 ±1*	100**	99 ±1*	p<0.0003
$P\bar{v}O_2$ (mm Hg)	36 ±1	40 ±1.8*	43 ±2.1*	43 ±3.2	P<0.007
$S\bar{v}O_2$ (%)	70 ±3	77 ±2*	81 ±2*	79 ±4	p<0.001
Qs/Qt (%)	31 ±4	22 ±3*	15 ±3**	17 ±4*	p<0.0002
$PaCO_2$ (mm Hg)	38 ±1	36 ±1.3	36 ±1.5	38 ±2.6	ns
CaO_2 (mL O_2/min)	12.6±0.9	13.4±0.9*	14.1±0.9**	14.4±1.7	p<0.008
DO_2 (mL O_2/min)	729 ±65	765 ±82	822 ±89	725 ±114	ns
VO_2 (mL O_2/min)	218 ±19	215 ±24	226 ±30	195 ±39	ns
Ppi (cm H_2O)	23 ±0.7	24 ±0.7	23 ±0.8	23 ±1.1	ns
QSC (mL/cm H_2O)	39 ±3.8	40 ±3.9	40 ±3.4	39 ±4.5	ns

N_2: nitrogen control situation; NO: 60 min 15 ppm continuous NO inhalation; ALM: almitrine intravenous infusion; PaO_2: partial arterial oxygen pressure; SaO_2: arterial hemoglobin oxygen saturation; $PaCO_2$: partial arterial carbon dioxide pressure; $P\bar{v}O_2$: partial mixed venous oxygen pressure; $S\bar{v}O_2$: mixed venous hemoglobin oxygen saturation; Qs/Qt: calculated intra pulmonary shunt; CaO_2: arterial oxygen content; DO_2: oxygen transport; VO_2: calculated oxygen consumption; Ppi: peak inspiratory pressure; Cqs: quasi static pulmonary compliance (mean ± SEM)

* $p<0.05$ in comparison with N_2 situation
** $p<0.05$ in comparison with NO situation
† $p<0.05$ in comparison between NO + ALM and ALM situations

Prolonged inhalation of NO and almitrine infusion (Fig. 2): According to the protocol, under NO inhalation plus almitrine infusion, it was possible to substantially reduce FiO_2 in both groups. FiO_2 reduction was more pronounced in group II than in group I (−59 versus −31%; $p<0.001$). PEEP reduction was only possible in 5 patients in group II from 6±1 to 4±1 cm H_2O

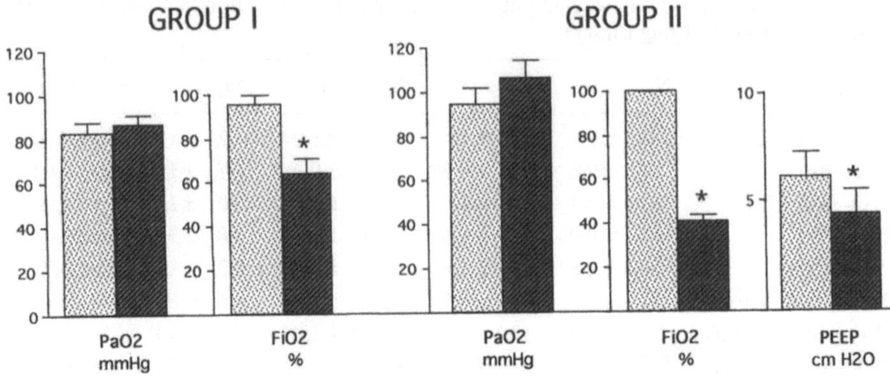

Fig. 2. Impact of inhaled NO and intravenous almitrine association on oxygen and PEEP requirements to maintain a sufficient PaO_2 in group I (ARDS) and Group II (ALI). Dotted bars: control situation. Hached bars: continuous treatment by NO inhalation + almitrine infusion
* $p<0.001$ in comparison between control situation

($p<0.05$). The mean duration of NO inhalation (around 5 days) was similar in both groups. Although not significant, almitrine infusion tended to be longer in group II than in group I (3.3±0.9 versus 1.4±3 days). For the patients who died, the cause of the death was never related to respiratory failure, but to multiple organ failure (n=1), severe sepsis (n=2), brain failure (n=3), cardiogenic shock (n=1), and terminal cancer (n=1).

Pulmonary hemodynamic parameters changes were similar in both groups following the different steps of thc protocol and did not *per se* preclude the gas exchange response. Since CO and PAP values were not different during NO + almitrine compared to control, gas exchange can be interpreted in light of the modifications of VA/Q mismatching.

PAP decreased with NO and increased 3 and 6 mm Hg in ALI and ARDS, respectively, when almitrine was given without NO. In the presence of NO, infusion of almitrine induced a 2 to 3 mm Hg increase in PAP which is statistically significant, but probably of limited clinical relevance as the resulting PAP values are not significantly different from control (N_2) values. Changes in PVR paralleled changes in PAP as expected (i.e. it decreased with NO and increased with almitrine). These changes were of minimal quantitative and qualitative interest, this variable being a weak indicator of the modifications of the vasoactive status of the pulmonary circulation following pharmacologic interventions.

Interestingly, although PaO_2 values of the two groups during the control condition and after NO did not differ, with the combination of NO + almitrine, the improvement in PaO_2 was more prominent in group II ($p<0.001$). Two hypotheses for the superior improvement in PaO_2 in group II following the addition of almitrine to NO can be considered: 1) a more pronounced effect of almitrine on HPV in group II; and 2) the quantitatively smaller percentage of lung remaining with normal VA/Q ratio in ARDS patients (Group I), which would limit the potential for reinforcement of HPV.

A marked reinforcement of the pre-existing HPV in ALI patients compared to ARDS patients could be responsible for the larger improvement in PaO_2. This result is apparently contrary to previous reports where HPV is impaired during unilateral pneumonia [23], possibly involving pulmonary vascular hypocontractility, especially during septic pneumonia. In our ALI patients, the improvement in PaO_2 was observed in 9 out of 10 patients and did not depend upon the presence of bacterial pneumonia. In light of the supposedly important remaining surface of normal VA/Q areas, even a small HPV response (or reinforcement) may divert a large proportion of blood from shunting areas, explaining the large increase in PaO_2 observed. In addition to a better VA/Q matching, the effect of $P\bar{v}O_2$ has to be mentioned [10], since the correlation between the increase in PaO_2 and the increase in $P\bar{v}O_2$ from control to NO + almitrine, had a steeper slope in ALI patients (Fig. 3). This suggests the existence of more extended low VA/Q zones in ALI than in ARDS.

At the opposite, an alternate hypothesis could be that the potentiation of HPV induced by almitrine was less efficient in increasing PaO_2 in ARDS patients. Even if baseline and post-NO PaO_2 levels were similar in ARDS and ALI patients, one might reasonably assume that the extent of the alteration of the lung parenchyma was quantitatively more important in ARDS than in ALI group. Therefore, following almitrine, the pulmonary blood diverted from non- or poorly-ventilated areas to normally ventilated areas would flow in a relatively small fraction of the remaining lung parenchyma available for normal gas exchange, resulting in lower PaO_2 increase than in ALI patients.

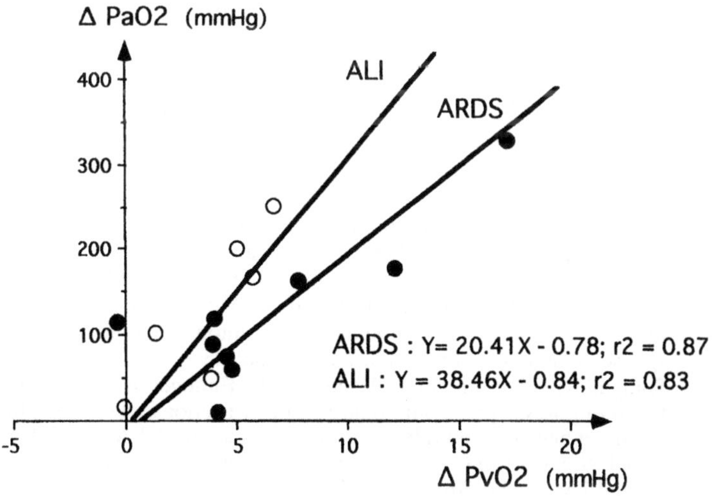

Fig. 3. Correlations between PaO_2 and $P\bar{v}O_2$ modifications during NO inhalation + almitrine infusion, in ARDS (*plain circles*) and ALI (*open circles*) patients

Finally, with the combination of NO and almitrine therapy, FiO_2 could be reduced by 60 and 40% in ARDS and ALI groups, respectively, while maintaining a similar PaO_2.This beneficial effect on gas exchange could potentially lead to important therapeutic and prognostic implications since it should reduce hyperoxic-induced lung injury, and limit the pulmonary baro-volotrauma associated with "aggressive" modes of mechanical ventilation.

Acknowledgements: This work was supported in part by the Grants of University Paris VII 1994, the Délégation à la Recherche Clinique AP-HP 1993, and the Direction de la Recherche et de l'Enseignement Doctoral (D.R.E.D.) 1994.

References

1. Riley D, Kramer M, Kerr J, et al (1987) Damage and repair of lung connective tissue in rats exposed to toxic levels of oxygen. Am Rev Respir Dis 135:441–447
2. Payen D, Brun Buisson C, Carli P, et al (1987) Hemodynamic, gas exchange and hormonal consequences of LBPP during PEEP ventilation. J Appl Physiol 62:61–70
3. Payen D, Farge D, Beloucif S, et al (1987) No involvement of ADH in acute antidiuresis during PEEP ventilation in human. Anesthesiology 66:17–23
4. Rouby J, Lherm T, Martin de Lassale E, et al (1993) Histologic aspects of pulmonary barotrauma in critically ill patients with acute respiratory failure. Intensive Care Med 19:383–389
5. Brunet F, Belghith M, Mira J, et al (1993) Extracorporeal carbon dioxide removal and low-frequency positive-pressure ventilation. Chest 104:889–898
6. Wagner P, Saltzman H, West J (1974) Measurement of continuous distribution of ventilation-perfusion ratios: Theory. J Appl Physiol 36:588–599
7. West J, Wagner P (1991) Ventilation-perfusion relationship.In: Crystal R, West J (eds) The lung: Scientific Foundations. 1. Raven Press, New York, pp 1289–1306
8. Voelkel N (1986) Mechanisms of hypoxic pulmonary vasoconstriction. Am Rev Respir Dis 33:1186–1195
9. Marshall B, Marshall C, Frasch F, Hanson C (1994) Role of hypoxic pulmonary vasoconstriction in pulmonary gas exchange and blood flow distribution. Intensive Care Med 20:291–297
10. Payen D, Carli P, Brun-Buisson C, et al (1985) Lower body positive pressure versus dopamine during PEEP ventilation. J Appl Physiol 58:77–82
11. Marshall C, Marshall B (1983) Site and sensitivity for stimulation of hypoxic pulmonary vasoconstriction. J Appl Physiol 55:711–716
12. Marshall B, Marshall C, Magno M, Lilagan P, Pietra G (1991) Influence of bronchial arterial PO_2 on pulmonary vascular resistance. J Appl Physiol 70:405–415
13. Persson MG, Gustafsson LE, Wiklund NP, Moncada S, Hedqvist P (1990) Endogenous nitric oxide as a probable modulator of pulmonary circulation and hypoxic pressor response in vivo. Acta Physiol Scand 140:449–457
14. Sprague R, Thiemermann C, Vane J (1992) Endogenous endothelium-derived relaxing factor opposes hypoxic pulmonary vasoconstriction and supports blood flow to hypoxic alveoli in anesthetized rabbits. Proc Natl Acad Sci USA 89:8711–8715
15. Zapol W, Snider M (1977) Pulmonary hypertension in severe acute respiratory failure. N Engl J Med 296:476–482
16. Payen D, Gatecel C, Plaisance P (1993) Almitrine effect on nitric oxide inhalation in adult respiratory distress syndrome. Lancet 341:1664 (Letter)

17. Payen D, Gatecel C, Guinard N (1993) Inhalation of low dose of nitric oxide (NO) and IV L-Arg in ARDS patients: Effect on pulmonary hemodynamic, gas exchange and NO metabolites. Am Rev Respir Dis 147: A720 (Abst)
18. Puybasset L, Rouby J, Mourgeon E, et al (1994) Inhaled nitric oxide in acute respiratory failure: Dose-response curves. Intensive Care Med 20:319–327
19. Wysocki M, Delclaux C, Roupie E, et al (1994) Additive effect on gas exchange of inhaled nitric oxide and intravenous almitrine bismesylate in the adult respiratory distress syndrome. Intensive Care Med 20:254–259
20. Rossaint R, Falke K, Lopez F, et al (1993) Inhaled nitric oxide for the adult respiratory distress syndrome. N Engl J Med 328:399–405
21. Zapol W, Rimar S, Gillis N, Marletta M, Bosken C (1994) Nitric oxide and the lung. Am J Respir Crit Care Med 149:1375–1380
22. Marshall B, Hanson C, Frasch F, Marshall C (1994) Role of hypoxic pulmonary vasoconstriction in pulmonary gas exchange and blood flow distribution. Intensive Care Med 20:379–389
23. Graham L, Vasil A, Vasil M, Voelkel N, Stenmark K (1990) Decreased pulmonary vasoreactivity in an animal model of chronic Pseudomonas pneumonia. Am Rev Respir Dis 142:221–229
24. Spaden H, Vincken W (1987) Pulmonary arterial hypertension in sepsis and the adult respiratory distress syndrome. Acta Clin Belg 47:30–40
25. Reeves J, Grover R (1974) Blockade of acute hypoxic pulmonary hypertension by endotoxin. J Appl Physiol 36:328–332
26. Jones R, Reid L, Zapol W, et al (1985) Pulmonary vascular pathology. In: Zapol W, Falke K (eds) Acute respiratory failure. Dekker, New York, pp 23–160
27. Laubie M, Diot F (1972) A pharmacological study of the respiratory stimulant action of S 2620. J Pharmacol 3:363–374
28. Romaldini H, Rodriguez-Roisin R, Wagner P, West J (1983) Enhancement of hypoxic pulmonary vasoconstriction by almitrine in the dog. Am Rev Respir Dis 128:288–293
29. Mélot C, Naeije R, Rotschild T, et al (1983) Improvement in ventilation-perfusion matching by almitrine in COPD. Chest 83:528–533
30. Reyes R, Roca J, Rodriguez-Roisin R, et al (1988) Effect of almitrine on ventilation-perfusion distribution in adult respiratory distress syndrome. Am Rev Respir Dis 137:1062–1067
31. Herve P, Musset D, Simmoneau G, Wagner W Jr, Duroux P (1989) Almitrine decreases the distensibility of the large pulmonary arteries in man. Chest 96:572–577
32. Laubie M, Schmitt H (1980) Long-lasting hyperventilation induced by almitrine: Evidence for a specific effect on carotid and thoracic chemoreceptors. Eur J Pharmacol 61:125–136
33. Georgopoulos D, Walker S, Anthonisen A (1989) Increased chemoreceptor output and ventilatory response to sustained hypoxia. J Appl Physiol 67:1157–1163
34. Castaing Y, Manier G, Guenard H (1986) Improvement in ventilation-perfusion relationships by almitrine in patients with chronic obstructive pulmonary disease during mechanical ventilation. Am Rev Respir Dis 143:910–916
35. Clergue F, Ecoffey C, Derenne J, Viars P (1984) Oxygen drive to breathing during halothane anesthesia: Effects of almitrine bismesylate. Anesthesiology 61:627–628
36. Melot C, Dechamps P, Hallemans R, Decroly P, Mols P (1989) Enhancement of hypoxic pulmonary vasoconstriction by low dose almitrine bimesylate in normal humans. Am Rev Respir Dis 139:111–119
37. Gherardi R, Louarn F, Benvenuti C, et al (1985) Peripheral neuropathy in patients treated with almitrine bimesylate. Lancet 2:1247–1250
38. Voisin C, Howard P, Ansquer J (1987) Almitrine bimesylate. A long-term, placebo-controlled, double blind study in COAD. Vectarion international multicentre study group. Bull Eur Physiopath Respir 23 (suppl 11):169–182
39. Chen L, Miller F, Clarke W, et al (1990) Low-dose almitrine bimesylate enhances hypoxic pulmonary vasoconstriction in closed chest dogs. Anesth Analg 71:475–483

40. Bee D, Gill G, Emery C, et al (1983) Action of almitrine on the pulmonary vasculature in ferrets and rats. Bull Europ Physiopath Respir 19:539–545
41. Falus F, Herget J, Hampl V (1991) Almitrine in low dose potentiates vasoconstrictor responses of isolated rat lungs to moderate hypoxia. Eur Respir J 4:688–693
42. Gottschall EB, Fernyak S, Wuertemberger G, Voelkel N (1992) Almitrine mimics hypoxic vasoconstriction in isolated rat lungs. Am J Physiol 263:H383–H391
43. Russel P, Emery C, Nicholl J, Barer G, Howard P (1992) Interactions between hypoxic and almitrine-induced vasoconstriction in the rat lung. Clin Sci 82:265–272
44. Rounds S, McMurtry I (1981) Inhibitors of oxidative ATP production cause transient vasoconstriction and block subsequent pressor responses in rat lungs. Circ Res 48:393–400
45. Archer S, Huang J, Henry T, Peterson D, Weir E (1993) A redox-based O_2 sensor in rat pulmonary vasculature. Circ Res 73:1100–1112
46. Rigoulet M, Fraisse L, Ouhabi R, et al (1990) Flux-dependent increase in the stoichiometry of charge translocation by mitochondrial ATPase/ATP synthase induced by almitrine. Biochem Biophys Acta 1018:91–97
47. Hughes J, Allison D, Goatcher A, Tripathi A (1986) Influence of alveolar hypoxia on pulmonary vasomotor responses to almitrine in the dog. Clin Sci 70:555–564
48. Naeije R, Lejeune P, Vachiéry J, et al (1990) Restored hypoxic pulmonary vasoconstriction by peripheral chemoreceptor agonists in dogs. Am Rev Respir Dis 142:789–795
49. Leeman M, Lejeune P, Hallemans R, Melot C, Naeije R (1988) Effects of increased pulmonary vascular tone on gas exchange in canine oleic acid pulmonary edema. J Appl Physiol 65:662–668
50. Reyes A, Lopes-Messa J, Alonso P (1987) Almitrine in acute respiratory failure. Effects on pulmonary gas exchange and circulation. Chest 91:388–393
51. Dreyfuss D, Djedaini K, Lanore J, et al (1992) A comparative study of the effects of almitrine bimesylate and lateral position during unilateral bacterial pneumonia with severe hypoxemia. Am Rev Respir Dis 146:295–299
52. Valtier B, Mottin D, Nolland B, Molis T, Payen D (1992) Cardiac output measurement by four methods in intensive care units: Interest of oesophageal Doppler route. Am Rev Respir Dis 145:A781 (Abst)
53. Frostell C, Fratacci MD, Wain J, Jones R, Zapol W (1991) Inhaled nitric oxide, a selective pulmonary vasodilator reversing hypoxic pulmonary vasoconstriction. Circulation 83:2038–2047

Nitric Oxide Inhalation in ARDS

H. Gerlach, R. Rossaint, and K. J. Falke

Introduction

The acute adult respiratory distress syndrome (ARDS), a disorder described by Ashbaugh et al. [1] with a mortality of approximately 60–70%, is characterized by a heterogeneous deterioration of both alveolar ventilation and pulmonary perfusion due to edema and local inflammation, associated with gas space collapse, vasoconstriction and vascular obliteration [2, 3]. The mechanisms of the acute, diffuse pulmonary inflammation cause non-cardiogenic pulmonary edema, pulmonary hypertension, a decrease in lung compliance, and progressive hypoxemia due to an increasing intrapulmonary shunt. Pulmonary hypertension and an aggressive mechanical ventilatory strategy using high inspiratory oxygen concentrations (FiO_2) and high peak inspiratory pressures (PIP) are possible pathogenic mechanisms which contribute to continuing pure results [4]. Pulmonary hypertension induces a rise in the microvascular filtration pressure [5], enhancing the development of interstitial pulmonary edema [6] and right ventricular dysfunction [7, 8]. Reduction of pulmonary vascular resistance (PVR) by systemically infused vasodilating drugs was shown to exert a beneficial effect in ARDS by lowering pulmonary artery pressure (PAP), although the intrapulmonary right-to-left shunt increases by this treatment; because of the global effect on the vasculature, the use of systemic vasodilators is limited [9]. The concomitant dilation of the systemic circulation causes arterial hypertension possibly affecting the blood flow to various organs. In addition, vasodilation in the pulmonary vasculature gives rise to increase blood flow to areas of intrapulmonary shunt, thereby compromising oxygenation even further [9, 10]. This dilation may require further increase in the FiO_2 and airway pressures needed for the maintenance of normal arterial blood gases. High FiO_2 and high PIP have to be regarded as factors which contribute to the progression of the disease [11, 12].

Modern management of ARDS aims to reduce its damaging effects by using a less aggressive ventilatory therapy. This involves pressure-limited mechanical ventilation with positive end-expiratory pressure (PEEP) and permissive hypercapnia, selective ventilation of the lungs, placing patients in the prone or lateral positions, and dehydration by drugs or hemofiltration systems. The pure prognosis of ARDS, however, often demands the use of additional methods like extracorporeal respiratory support in specialized departments or hospitals [13–16].

The vascular endothelium, which was shown to play an active role in host response [17], also contributes to the regulation of vascular resistance [18]. A factor, which is synthesized and expressed by endothelial cells and reduces vasodilation, described as endothelium-derived relaxing factor (EDRF) [19], was demonstrated to be equivalent to nitric oxide (NO), synthesized from L-arginine by the enzyme NO-synthase (NOS) [20–22]. NO binds to the enzyme guanylate cyclase in smooth muscle cells, enhances synthesis of cyclic guanosine monophosphate (cGMP), and thus exerts a relaxing effect by the induction of dephosphorylation of myosin light chain filaments [23]. It is now acknowledged that the nitro-vasodilators, which have been in therapeutic use for decades, react with the thiol compounds in the endothelial cells of the vessels. The spontaneous desintegration of the compounds releases NO which brings about the vasodilation as described above [24]. NO, generated either by intrinsic or by inducible NOS, is only effective for a few seconds after it has been generated; it is than oxidized in an aqueous solution and is transformed into nitric dioxide (NO_2), which is transformed into nitrite (NO_2^-) and nitrate (NO_3^-) by hydrolysis. The lipophilic characteristic of NO allows the mediator to diffuse freely through tissues and surrounding blood vessels. In the blood, NO becomes inactivated within seconds by binding to hemoglobin; the hemoglobin molecule has a 1500 times higher affinity for NO than for carbon monoxide [25]. Nitrosyl-hemoglobin (NO-Hb) is produced and, in the presence of oxygen, oxidized into methemoglobin from which ferrous Hb is rapidly regenerated by methemoglobin reductase in red blood cells with nitrate as a byproduct [24, 26]. Most of the nitrate is excreted by the kidneys. Some of the nitrate is secreted in the saliva and is converted to nitrite by bacteria. This nitrite is partially converted to N_2 in the stomach and excreted as gas. The nitrate in the intestines is partly reduced to NH_3, reabsorbed, and metabolized into urea. Most of the inorganic endproducts of inhaled NO are excreted in the urine within 48 h.

The question arose if inhaled NO might be able to induce selective vasodilation in the pulmonary vessels without influencing the systemic circulation, since binding of NO to hemoglobin inactivates the molecule within seconds. There is little evidence of toxicity of inhaled NO when the concentration is below 50 parts per million (ppm), whereas inhaling high concentrations can be lethal because NO causes severe acute pulmonary edema and methemoglobinemia (see "Toxicology of NO"). Former toxicological studies demonstrated that animals, which had breathed the gas in concentrations up to 40 ppm for 6 days to 6 months, showed no evidence of any toxic effect [26, 27].

First Experiences with NO Inhalation

Low doses of inhaled NO have been shown to selectively dilate the pulmonary vasculature in patients with primary pulmonary hypertension [28, 29].

In parallel, animal studies have confirmed that inhaled NO acts selectively on the pulmonary vessels [30, 31]. Inhaled NO reversed pulmonary hypertension without a reduction in systemic blood pressure in awake, spontaneously breathing lambs which were acutely vasoconstricted either by an infusion of the stable thromboxane endoperoxide analog, U46619, or by breathing a hypoxic gas mixture [30]. Furthermore, it was reported that this selective vasodilatory effect on the pulmonary vasculature occurred without impaired gas exchange [31].

The encouraging results with inhaled NO as a selective pulmonary vasodilator for pulmonary hypertension gave rise to further hypothesis with regard to the typical features of ARDS: as mentioned above, ARDS is characterized by a maldistribution of ventilation and perfusion in the lung in favor of intrapulmonary right-to-left shunt areas. If, as a matter of fact, it is possible to achieve vasodilation by NO inhalation exclusively in the pulmonary vasculature, this phenomenon is actually explicable only by the described attribute of NO to bind to hemoglobin immediately after diffusion into the intravascular space and, hence, to be inactivated before reaching the systemic circulation. Since intrapulmonary shunt areas, however, are defined as vessels without any ventilation, this implies that these areas are not included in NO inhalation. Therefore, NO inhalation in ARDS should exert its vasodilating effect not only selectively in the pulmonary vasculature, but also selectively in ventilated areas of the lung, i.e. non-shunt areas, decreasing the relative portion of intrapulmonary right-to-left shunt areas, and, thus, improving systemic oxygenation. To test this hypothesis, 9 ARDS-patients were treated in our clinic in Berlin [32] with a short-term NO inhalation, and the registered effects on hemodynamics and pulmonary gas exchange were compared with the effects after treatment of the same patients by intravenous infusion of prostacyclin (PGI_2), using a dose of 4 ng/kg/min. NO inhalation was performed by discontinuous insufflation of a primary gas of 400 or 800 ppm NO into the inspiratory limb of the ventilator; the insufflation was regulated by a conventional nebulizer which was triggered by the ventilator, opening the valve only during inspiration. NO/NO_2 concentrations were continuously monitored by a specified NO/NO_2 chemiluminometer, taking samples from the inspiratory limb near to the patient. The results of these studies, published by Rossaint et al. in 1993, revealed that both inhaled NO and infused prostacyclin are able to reduce the pulmonary resistance about 20% [32]. In contrast to prostacyclin, which simultaneously caused a systemic hypertension and a decrease of arterial oxygen saturation, inhaled NO did not induce any change of systemic hemodynamics, but improved arterial oxygenation in a significant way, in the mean from 152 (PaO_2 without NO) to 199 mm Hg (with NO). Measuring the ventilation/perfusion ratio of the patients with a standardized multiple-inert-gas-elimination technique [33], it was confirmed that the portion of intrapulmonary shunt areas was expanded by the infusion of prostacyclin, but, in contrast, reduced by inhalation of 18 or 36 ppm NO due to a redistribution of pulmonary blood gas towards areas with nearly normal ventilation/perfusion ratios. This beneficial effect of NO could be registered also in long-term studies up to 53 days; neither critical side effects such as methemoglobinemia or other consequences like a distinct bronchodilation which had been described in animal experiments after NO inhalation [34], occurred during these tests.

Inhaled NO has also been used in the therapy of persistent pulmonary hypertension of the newborn (PPHN) [35, 36]. Since in these newborns the increased PVR results in right-to-left shunting of blood accross the patent ductus arteriosus and foramen ovale [37], the reduction of pulmonary hypertension by inhaled NO must probably cause a redistribution of blood flow from the patent ductus arteriosus and intracardiac shunt to the lung, thereby reversing the hypoxemia [36]. To achieve these facts, NO was inhaled in concentrations between 6–80 ppm.

NO Inhalation and Cardiac Function

The pathogenic mechanisms of pulmonary hypertension in ARDS induce an increase in the outflow pressure load on the right ventricle. This may decrease right ventricular ejection fraction (RVEF) and increase right ventricular volume [7, 38]. Animal studies suggest that severe acute pulmonary hypertension possibly results in insufficient myocardial blood flow and impaired contractility of the right ventricle, eventually inducing right ventricular pump failure [8]. The use of vasodilators has been advocated to treat pulmonary hypertension and improve right ventricular function in patients with severe ARDS [7, 39]. In a group of hemodynamically stable ARDS patients, inhaled NO additionally increased RVEF, which was, in contrast to intravenously infused PGI_2, not accompanied by an increase in cardiac output. In accordance with the observations of other authors [7, 38–41], in our group of patients, the RVEF was below the normal range of about 45–65%, and right ventricular end-diastolic and end-systolic volumes were above the normal range [41, 42]. The absolute values of RVEF we found were lower than those reported by others [39, 41, 43]. This might partly be explained by higher PAP values in our group of patients. In addition, this observation may be due to the critical situation of the patients who are not only suffering from isolated acute respiratory failure but from severe ARDS combined with multiple organ failure and sepsis. Although none of our patients was in septic shock, the measured RVEF values are in the range reported for patients with sepsis [41, 44, 45]. The concomitant right ventricular end-diastolic and end-systolic dilation observed in our patients and in septic patients may represent a circulatory adaptation utilizing the Frank-Starling mechanism to maintain cardiac output [38, 40].

Inhaled NO, as well as intravenously infused PGI_2, increases the RVEF consistently and thus significantly by about 3%, representing an over 10% improvement; the magnitude of this increase has been considered significant by others [39, 46]. In contrast to other authors [7, 39, 41], we did not find a significant correlation between PAP and RVEF. This may be explained by the fact that we investigated only a small number of patients with a relative uniform degree of pulmonary hypertension. Moreover, the changes in PAP and RVEF induced by both vasodilators were modest, and no interventions to alter PAP over a wider physiological range were performed. Since during NO inhalation, the increase of RVEF was accompanied by decreased right ventricular volumes at a constant cardiac index (CI), these changes reflect improved right ventricular function. However, these data also show that CI could not be influenced by the obtained alterations in right ventricular parameters in this group of hemodynamically stable ARDS patients. The increase in CI observed during the infusion of PGI_2 probably was due to decreased systemic vascular resistance, since the decrease of PAP and increase of RVEF were similar during the administration of either vasodilator. The right ventricular end diastolic volume did not decrease during the infusion of PGI_2, probably due to an augmented venous return [40]. Inhaled NO may cause an increase RVEF and CI in selected patients with severe ARDS as demonstrated for 4 patients by Wysocki et al. [47]. We cannot exclude that higher concentrations of inhaled NO, which have been described to decrease PAP more effectively [48], would have decreased PAP further, possibly resulting in a greater increase of RVEF and also an increase of CI.

It still remains unclear when and for what reasons acute pulmonary artery hypertension limits cardiac output in patients with severe ARDS. However, using NO as a selective gaseous pulmonary vasodilator, it was demonstrated that reduction in PAP secondary to a pharmacological intervention may increase RVEF but not necessarily cardiac output. This suggests that altered right ventricular function in ARDS is not necessarily a limiting factor to cardiac output since both vasodilators, NO and PGI_2, decreased PAP and improved RVEF, it cannot be determined so far which drug should be used to treat pulmonary hypertension in ARDS. Since inhaled NO improved arterial oxygenation without systemic vasodilatation [30–32], NO – also taking into account the toxicology of this vasodilator which has been discussed extensively [30, 32] – may be advantageous in the treatment of ARDS. On the other hand, although PGI_2 increased venous admixture and decreased PaO_2, the augmented CI prevented reduction of the oxygen delivery. Moreover, other investigators have reported an increase in oxygen delivery when infusing larger doses of PGI_2 [39].

Dose-response Characteristics of NO Inhalation in ARDS

First animal experiments demonstrated an obvious dose-dependence for the selective pulmonary vasodilation by NO up to 80 ppm [30]; the first studies with NO inhalation in ARDS patients, however, revealed no difference between the two doses of 18 or 36 ppm NO regarding pulmonary resistance and systemic oxygenation [32]. On the other hand, the question for possible toxic side effects of NO, especially during long-term inhalation, was not answered even if no serious consequences were found so far. To test the dose-response of selective pulmonary vasodilation by NO in patients with severe ARDS, studies were performed in our clinic [48], using NO doses between 10 parts per billion (ppb) and 100 ppm. Surprisingly, it was found that an improvement of systemic oxygenation by NO can be reached with much lower doses of NO than used before (the ED_{50} for increasing oxygenation was approximately 0.1 ppm), and that this improved oxygenation does not necessarily goes parallel with a reduction of a pulmonary resistance ($ED_{50} > 2$ ppm) [48]. In addition, the data revealed that the optimal improvement of systemic oxygenation was found with 10 ppm NO, and that higher doses of NO reversed the beneficial effect; the pulmonary resistance, in contrast, was continuously reduced correlating to the increasing NO doses [48]. These results might be declared by the theory that, with lower NO doses, there is a more or less "strictly selective" vasodilation in ventilated areas of the pulmonary vasculature, reducing intrapulmonary shunt areas and, thus, increasing systemic oxygenation, and that higher NO doses, however, induced diffusion of the lipophilic NO through the lung tissue into areas which are not ventilated, i.e. shunt areas, still resulting in a selective pulmonary vasodilation with further reduction of pulmonary resistance, but reversing the beneficial effect on oxygenation. These studies also revealed that there was an impressive variance between patients with severe ARDS [48]; the optimal NO dose regarding systemic oxygenation varied from 0.5 to 100 ppm; some patients showed no or only moderate effects after induction of NO inhalation ("non-responders").

The finding about the dose-response characteristics regarding PaO_2 and PAP produce a dilemma for those clinicians planning randomized trials for NO inhalation therapy in ARDS. The question arises if they should aim to lower PAP and possibly reduce pulmonary edema formation by using higher NO levels, or if they should only provide for an increase of PaO_2 by using lower inhaled NO levels. At this time, the answer is uncertain and requires further experimental and clinical studies. Determining optimal NO levels will also depend on better understanding of how to establish in each patient, whether and at what high levels NO will augment shunt, which pulmonary vasoconstrictor drugs such as almitrine should be added to NO inhalation, thereby constricting shunting regions, and improving arterial oxygenation. No randomized trials should be done until there are answers to these questions.

We must also bear in mind that it is possible that the great variation in response to inhaled NO will eventually be found to be related to the patient's pulmonary vascular levels of guanylate cyclase, the enzyme which is the putative NO receptor. It is likely that various systemic and pulmonary diseases can alter the expression and activity of this heme-containing enzyme. Levels of plasma cGMP, the product of guanylate cyclase have been shown to be elevated approximately 4-fold after 10 min of NO breathing [49]. Measuring the transpulmonary plasma levels of cGMP may allow investigators to examine *in vivo* the activity of guanylate cyclase in the lungs of ARDS patients. Surely, this enzyme must have great avidity for NO if 5–40 ppm inhaled NO levels can reduce PAP.

Long-term studies with low-dose inhalation of NO in ARDS patients demonstrated that the reduction of PAP is obviously not necessary to improve systemic oxygenation during inhalation of NO, i.e. that, as considered by previous studies [32], the redistribution, not the enhancement of pulmonary perfusion causes the increasing PaO_2. Significant improvement of PaO_2 is already induced by doses of NO (60–230 ppb) which are much lower than those used in former studies [29, 30, 32, 35, 36], exert a dose-response, and are similar to those measured in the free atmosphere. Hence, adding these low doses of NO to the inspiratory gas mixture is not much different to breathing room air relative to the NO content [50].

Long-term inhalation of NO, however, sometimes induces rebound effects for systemic oxygenation and PAP, which requires systemic weaning-off procedures by stepwise reduction of inhalatory NO doses down to 10–70 ppb before the final shut-off. This effect might be due to a feed-back inhibition of the endothelial NOS by the exogenously supplied NO; thus, vasoconstriction in the ventilated areas my re-occur after shut-off of NO. These findings, combined with other studies showing that the absolute level of PAP is a marker for the severity of pulmonary microvascular injury in ARDS [51], and that pulmonary hypertension is associated with impaired NO production [52], confirm the hypothesis that even low doses of NO might reduce PAP in most severe cases of ARDS as presented in former studies [32, 48, 50], since the endogenous NO production by the pulmonary vascular endothelium is considerably impaired under these circumstances. This was underlined by the observation that intravenous application of inhibitors of endogenous NOS was shown to increase systemic and pulmonary artery resistance, but had no effect in ARDS patients, at least not in our clinic.

Toxicology of NO

NO is a potentially toxic molecule and its biochemistry is not fully understood [53]. Under physiological conditions, there exist an area of interrelated redox-forms: NO, nitrosonium cation (NO^+) and nitroxyl-anion (NO^-). The chemistry of NO^+ is characterized by addition and substitution reactions with nucleophiles and aromatic compounds, and this N-nitrosilation has been implicated in the pathogenesis of cancer [53]. NO^- converts rapidly to NO_2 which is a strong oxidant with significant cytotoxic potential, while the toxic radicals, peroxinitrite ($ONOO^-$) and hydroxyl anion (OH^-) are formed by reactions of NO with superoxide [54]. Thus, the exogenous administration of NO is potentially harmful, particularly in patients receiving high dose oxygen therapy, and in those with acute inflammatory disorders (e.g. sepsis) associated with enhanced superoxide formation. Furthermore, NO oxidizes hemoglobin to methemoglobin, and this should be monitored carefully [55]. It is likely that these potential toxic effects will be minimized by using low concentrations of inhaled NO. Initial studies in humans used concentrations of NO in the order of 40–100 ppm, but it is encouraging to know that (at least for ARDS) the dose-response curve for a decrease in PAP yields an ED_{50} in the order of 2–3 ppm and for improved gas exchange from about 100 ppb [48]. The reason why the ED_{50} differs for pulmonary artery pressure and gas exchange will require further evaluation but is consistent with NO causing selective dilation of vessels supplying ventilated alveoli when given in low concentrations and a more widespread dilation and lowering of overall pulmonary respiratory resistance at higher concentrations.

Although nitrous oxide, the production of which gives rise to NO and NO_2, has been in use for over 60 years for narcosis, the anesthetist's attention was only drawn to the toxicology of NO so far in 1967 by a case report in the United Kingdom [56]. A healthy 39-year-old woman died due to massive pulmonary edema after anesthesia during which she inadvertently inhaled NO and NO_2 for 30 min from a contaminated nitrous oxide cylinder. The inquiry which followed confirmed that the nitrous oxide cylinder has been contaminated with NO, and it was later calculated that NO must have been present in a concentration of 10000 ppm [57]. How far this intoxication can be attributed solely to NO remains unclear because NO oxidizes to NO_2 depending on the concentration and the surrounding FiO_2. Thus, a 50% oxidation of 10000 ppm NO takes place within 24 sec at room temperature whereas 10 ppm NO oxidize to NO_2 by 50% only after 7 h [58]. Our own measurements using chemiluminescense [59] revealed that at 25°C room temperature, 20 ppm NO are 50% converted to NO_2 after 50 min when exposed to FiO_2 0.9, and after 120 min with FiO_2 0.5. NO_2, which is a strong oxidizing gas, initiates lipid peroxidation resulting in cell injury and cell death [60]. Acute effects inducing pulmonary edema have been considered during controlled exposure to NO_2 in concentrations above 50 ppm [56, 57]. However, even through prolonged exposure to such low concentrations as 2.3 ppm, NO_2 may cause delayed effects as it reduces glutathione peroxidase activity which offers the antioxidative defense by 14% [61]. The same concentration also decreased the alveolar permeability [61], measured as the 99m-Tc-diethylentriamin-pentaacidic-acid clearance rate, which is considered to be a non-invasive assessment of the integrity of the alveolar epithelium, by 22%. These findings stress the importance of avoiding or at least minimizing NO_2 exposure.

NO (and also NO_2 in small quantities), is produced during various combustion processes as an oxide of nitrogen. For example, NO develops on the glowing ends of cigarettes, and smokers inhale it in concentrations of 600 to 10000 ppm [62, 63]. Studies from our group showed that the environmental air NO concentration is usually between 2 and 80 ppb. In animal experiments [57] and in humans [56], NO concentrations way above 5000 ppm seem to rapidly lead to methemoglobinemia and a toxic pulmonary edema. However, if NO is inhaled in concentrations of less than 50 ppm, it appears to have no acute toxic effects. When 10 ppm NO is inhaled, there is only a very low NO-Hb concentration in the blood (0. 13% of the total Hb) and a low methemoglobin level (0.2% of the total Hb) [26]. Rabbits have inhaled 43 ppm NO and 3.6 ppm NO_2 for 6 days without pulmonary edema (assessed by gravimetry and light electron microscopy) [27]. Mice have been exposed to 10 ppm NO for 10 months without a rise in methemoglobin but with a slight enlargement of the spleen and a small rise in bilirubin [26]. Therefore, the US Occupational Safety and Health Administration has set the average NO exposure level at 25 ppm at places of work [64]. As a result of our own work, the treatment of patients with severe ARDS using NO inhalation has been undertaken with concentrations below 50 ppm; a significant increase of methemoglobin during NO inhalation is only seen by the first onset of inhalation from a mean of 0.7 to a mean of 1.4% of the total Hb. Over the following long-term inhalation period, no further increase and no toxic side effects could be determined so far. However, to minimize any possible problems of inhaled NO, we recommend the use of the lowest effective concentration of NO, evaluated by standardized and repeated dose-response analysis [48].

Lowering Inhaled NO Doses: Back to Endogenous Concentrations?

With an increased number of studies examining the role of endogenous NO in disease states, it will be important to determine whether the dose-response curve to inhaled NO differs between diseases with reduced endogenous NO-mediated pulmonary vasodilation (e.g. chronic obstructive lung disease) [52] and those where production of NO is normal or increased [65]. It would also be interesting to determine whether the NO present in cigarette smoke or the currently permitted level of NO contaminant in cylinders of nitrous oxide dilates pulmonary vessels; there is already some preliminary evidence to suggest that inhalation of cigarette smoke is associated with a fall in PVR [66].

The finding that, in some ARDS patients, therapeutic doses of NO are not much different from environmental NO concentrations, might imply toxicity of NO to be a secondary problem; the complexity of the biological mediator EDRF/NO, however, should not be underestimated. NO is not only an exogenous, inhalable gas, but also a quickly synthesized, endogenous mediator, which is highly lipophilic and can diffuse through body tissues in a free and uninhibited manner. Therefore, it was not surprising when, in 1991, Gustafsson et al. [67] described that they were able to measure NO in the exhaled air of animals and humans which inhaled pure, NO-free air. They interpreted their findings as a hint for exhalation of NO from the alveolar space of the lung after diffusion from the capillary endothelium.

Meanwhile, studies from our clinic revealed that NO is actually synthesized in the upper airways and *autoinhaled* by healthy persons [68]; smokers were demonstrated to produce less NO in their airways than non-smokers. Breath-holding can lead to nasal concentrations of NO up to 1 ppm – former shown to exert reproducible beneficial effects on systemic oxygenation in patients with severe ARDS [48, 50]! If patients are intubated and ventilated during normal operations, the nasal and prelaryngeal NO concentrations increase, probably by cumulation, whereas the intratracheal inspiratory and expiratory NO concentrations decrease; these findings imply that the intubation interrupts the normal NO autoinhalation. Finally, long-term ventilated patients, e.g. with severe ARDS, were found to have no or only low concentrations of NO in their upper airways [68]. At the moment, the presented findings only allow to speculate about the physiological relevance: disruption of natural NO autoinhalation by artificial ventilation might be a reason for ventilation/perfusion maldistributions in the lung. Hence, low-dose NO inhalation in ARDS patients might be considered as NO replacement, since the inspiratory NO concentrations in volunteers are similar to those that are effective in ARDS [48, 50, 68]. In addition, it was shown that acute hypoxia like in ARDS may reduce the release of NO and/or limit the effect of NO by impairing the function of the soluble guanylate cyclase, thus inducing an absolute or relative deficiency of endogenous NO [69].

NO Inhalation: Therapy or Experiment?

The first experimental treatments of patients with severe ARDS or pulmonary hypertension for a short-term or a long-term period up to 53 days demonstrated that the beneficial effect of NO on PAP, intrapulmonary shunt and systemic oxygenation was reproducible, i.e. no tachyphylaxis in responding patients occurred. Side-effects like toxic lung edema or increase of extravascular lung water were not diagnosed, and, as mentioned above, no patient showed a dangerous increase of methemoglobin. Results from other studies suppose that NO inhalation has a positive effect on the release of an interstitial lung edema during ARDS due to a reduction of the pulmonary capillary pressure [70]. During our daily tests of the effects by a short on-off measurement, however, it was noticed that at least in some patients a rebound phenomenon occurred: after set-off of NO after long-term inhalation, some patients had an increase of PAP over the baseline; the PaO_2 also decreased dramatically, but after a few hours, it increased again to reach a constant value, which was somewhere between the values of NO inhalation and immediately after set-off of inhalation [50]. This phenomenon probably is declarable by a negative feed-back inhibition, because *in vitro* studies demonstrated that both constitutive and inducible NOS are inhibited by high exogenous NO concentrations [71, 72].

Another side effect of prolonged NO inhalation is a change in bleeding time: Hogman et al. [73] demonstrated that an inhalation of 30 ppm NO increased the bleeding time in rabbits from 51 ± 5 to 72 ± 7 sec. A similar effect on bleeding time was also noticed in healthy volunteers [74]. Future data will show if these findings have to be considered as beneficial or negative effects; however, in patients suffering from bleeding problems, NO inhalation should be performed with care.

So far, there are no prospective, randomized studies of NO inhalation on the overall outcome of the patients. It will not be possible to perform these studies in the next 2 to 3 years; therefore we tested our own data retrospectively on the effect of NO inhalation regarding mortality. Matched pairs were formed with respect to etiology and grade of ARDS, as well as on other organ failures, comparing patients with or without NO inhalation. The first data of this analysis revealed that both the patients with or without NO inhalation (a total of 26 pairs) have a survival rate of 69%. Hence, no beneficial effect of NO inhalation in ARDS patients on the overall outcome is proven so far! Controlled studies in the future will show whether inhalation of NO in ARDS and pulmonary hypertension is a new therapeutic concept; until results are presented, NO inhalation must be considered experimental!

Conclusion and Future Aspects

Inhaled NO presents many challenges as we try to understand this novel signal transduction system, and we will undoubtly gain a great deal of knowledge from *in vitro* as well as laboratory animal studies. The rapidity with which this advance in basic science has gone from bench to bedside is truly breathtaking. Inhalation of NO appears to be an exciting advance in the treatment of ALI and pulmonary hypertension. Low doses of inhaled NO have been shown to redistribute blood flow from areas with intrapulmonary shunt to lung regions with normal ventilation/perfusion ratios in adult patients suffering from acute respiratory failure. In this way, inhaled NO reduces PAP and increases arterial oxygenation. Inhaled NO selectively dilates vessels and ventilates lung areas and does not cause systemic vasodilation due to the inhalation strategy and a rapid inactivation of NO in blood by its binding to hemoglobin. Studies revealed that low dose NO inhalation might be considered as a replacement for the autoinhalation of endogenous NO, synthesized in the upper airways, which is interrupted in intubated patients.

The major advantage over alternative therapies, e.g. intravenous prostacyclin or nitrates, is the lack of systemic effects, however this may be due to the route of administration. Infants, children, and adult patients all will benefit more as we discover ways to deliver this powerful gas which promises saver, more selective treatment of pulmonary hypertension and hypoxemia. It is possible that also the inhalation of prostacyclin or short-acting nitrovasodilators would be equally selective for the vessels in ventilated pulmonary regions. Potential toxic effects of NO should be limited by the use of low concentrations, and, it will be important to develop standardized methods for delivery and handling of the gas. Recently, a European working party on the use of inhaled NO in the clinical setting has been established under the auspices of the European Society of Anaesthesiologists. This group should establish how to administer NO (it corrods metals and interacts with plastics, and there are problems with exposure to high concentrations of oxygen), how to monitor therapy (expensive and cumbersome chemiluminescence, or cheaper and less sensitive methods like as "sensor sticks" [75]), and what are the practical clinical and scientific implications for the further use of NO.

Issues such as tolerance, dependence and rebound vasospasm need to be evaluated. The description of the essential role of endothelium-derived NO in mediating vascular relaxation and stimulating soluble guanylate cyclase in many other tissues has thrown into many areas of research.

The reaction of the individual patient is not predictable; trying different NO doses seems to be indicated. On the other hand, how can we guess if the best systemic oxygenation or the most effective reduction of pulmonary resistance or something in between is the optimal treatment for the patient? So far, no studies have investigated the effect of NO inhalation on mortality in ARDS, and many questions regarding dose-response relationships, the toxicity of inhaled NO, effects on platelet aggregation and on macrophage activity, and possible interactions of the inhaled NO with endogenous NO production are still to be answered. At the moment, final statements about the physiological relevance of NO cannot be presented; the rapid development of NO science let us expect new aspects every day. One important point for the future are the possible effects of NO inhalation on immunogical and neurological functions, which must be investigated before NO inhalation can be accepted as a standard therapy for severe ARDS or other forms of pulmonary hypertension. Finally, we have to insist on the unmistakable demand for a cautious elevation of patients, sufficient NO/NO_2-measuring systems including ranges of less than 1 ppm, and for the development and usage of modern standardized and safe NO application systems. After all, NO could be a cheap, effective and selective pulmonary vasodilator and, if randomized, controlled studies show improvement in morbidity and mortality, its widespread use would be justified.

References

1. Ashbaugh DG, Bigelow DB, Petty TL, Levine BE (1967) Acute respiratory distress in adults. Lancet 2:319–323
2. Zapol WM, Snider MT (1977) Pulmonary hypertension in severe acute respiratory failure. N Engl J Med 296:476–480
3. Rinaldo JE, Rogers RM (1982) Adult respiratory distress syndrome. Changing concepts of lung injury and repair. N Engl J Med 306:900–909
4. European ARDS Collaborative Working Group (1988) Adult respiratory distress syndrome (ARDS): Clinical predictors, prognostic factors and outcome. Intensive Care Med 14 (Suppl. 1): A300 (Abst)
5. Erdmann AJ, Vaughan TR Jr, Brigham KL, et al (1975) Effect of increased vascular pressure on lung fluid balance in unanesthetized sheep. Circ Res 37:271–284
6. Gottlieb SS, Wood LDH, Hansen DE, Long GR (1987) The effect of nitroprusside on pulmonary edema, oxygen exchange, and blood flow in hydrochloric acid aspiration. Anesthesiology 67:203–210
7. Sibbald WJ, Driedger AA, Myers ML, Short Al, Wells GA (1983) Biventricular function in the adult respiratory distress syndrome. Chest 84:126–134
8. Vlahakes GJ, Turley K, Hoffmann JI (1981) The pathophysiology of failure in acute right ventricular hypertension: Hemodynamic and biochemical correlations. Circulation 63:87–95
9. Radermacher P, Santak B, Becker H, Falke KJ (1989) Prostaglandin E_1 and nitroglycerin reduce pulmonary capillary pressure but worsen ventilation-perfusion distri-

butions in patients with adult respiratory distress syndrome. Anesthesiology 70: 601–606
10. Zapol WM, Snider MT, Rie MA, Frikker M, Quimm DA (1985) Pulmonary circulation during adult respiratory distress syndrome. In: Zapol WM, Falke KJ (eds) Acute Respiratory Failure. Marcel Dekker, New York, pp 241–273
11. Haschek WM, Reiser KM, Klein-Szanto AJ, et al (1983) Potentiation of butylated hydroxytoluene-induced acute lung damage by oxygen. Cell kinetics and collagen metabolism. Am Rev Respir Dis 127:28–34
12. Kolobow T, Moretti MP, Fumagalli R, et al (1987) Severe impairment in lung function induced by high peak airway pressure during mechanical ventilation. An experimental study. Am Rev Respir Dis 135:312–315
13. Zapol WM, Snider MT, Hill JD, et al (1979) Extracorporeal membrane oxygenation in severe acute respiratory failure. A randomized prospective study. JAMA 242: 2193–2196
14. Rossaint R, Slama K, Lewandowski K, et al (1992) Extracorporeal lung assist with heparin-coated systems. Int J Artif Organs 15:29–34
15. Tharratt RS, Allen RP, Albertson TE (1988) Pressure-controlled inverse ratio ventilation in severe adult respiratory failure. Chest 94:755–762
16. Suchyta MR, Clemmer TP, Orme JFJ, Morris AH, Elliott CG (1991) Increased survival of ARDS patients with severe hypoxemia (ECMO criteria). Chest 99:951–955
17. Gerlach H, Esposito C, Stern DM (1990) Modulation of endothelial hemostatic properties: An active role in the host response. Annu Rev Med 41:15–24
18. Lie M, Sejersted OM, Kiil F (1970) Local regulation of vascular cross section during changes in femoral arterial blood flow in dogs. Circ Res 27:727–737
19. Furchgott RF, Zawadzki JV (1980) The obligatory role of endothelial cells in the relaxation of arterial smooth muscle by acetylcholine. Nature 288:373–376
20. Ignarro LJ, Buga GM, Wood KS, Byrns RE, Chaudhuri G (1987) Endothelium-derived relaxing factor produced and released from artery and vein is nitric oxide. Proc Natl Acad Sci USA 84:9265–9269
21. Palmer RM, Ferrige AG, Moncada S (1987) Nitric oxide release accounts for the biological activity of endothelium-derived relaxing factor. Nature 327:524–526
22. Palmer RM, Rees DD, Ashton DS, Moncada S (1988) L-arginine is the physiological precursor for the formation of nitric oxide in endothelium-dependent relaxation. Biochem Biophys Res Commun 153:1251–1256
23. Brenner BM, Troy JL, Ballermann BJ (1989) Endothelium-dependentvascular responses. J Clin Invest 84:1373–1378
24. Ignarro LJ, Lippton H, Edwards JC, et al (1981) Mechanism of vascular smooth muscle relaxation by organic nitrates, nitrites, nitroprusside and nitric oxide: Evidence for the involvement of S-nitrosothiols as active intermediates. J Pharmacol Exp Ther 218: 739–749
25. Gibson QH, Roughton FJW (1957) The kinetics of equilibria of the reactions of nitric oxide with sheep hemoglobin. J Physiol 136:507–526
26. Oda H, Nogami H, Kusumoto S, Nakajima T, Kurata A, Imai K (1976) Long-term exposure to nitric oxide in mice. J Jpn Soc Air Pollut 11:150–160
27. Hugod C (1979) Effect of exposure to 43 ppm nitric oxide and 3.6 ppm nitrogen dioxide on rabbit lung. A light and electron microscopic study. Int Arch Occup Environ Health 42:159–167
28. Higenbottam T, Pepke-Zaba J, Scott J, Woolman P, Coutts C, Wallwork J (1988)Inhaled "endothelium-derived relaxing factor" (EDRF) in primary hypertension. Am Rev Respir Dis 137:107A (Abst)
29. Pepke-Zaba J, Higenbottam TW, Dinh-Xuan AT, Stone D, Wallwork J (1991) Inhaled nitric oxide as a cause of selective pulmonary vasodilatation in pulmonary hypertension. Lancet 338:1173–1174
30. Frostell C, Fratacci MD, Wain JC, Jones R, Zapol WM (1991) Inhaled nitric oxide. A selective pulmonary vasodilator reversing hypoxic pulmonary vasoconstriction. Circulation 83:2038–2047

31. Pison U, Lopez FA, Heidelmeyer CF, Rossaint R, Falke K (1993) Inhaled nitric oxide selectively reverses hypoxic pulmonary vasoconstriction without impairing pulmonary gas exchange. J Appl Physiol 74:7287–7292
32. Rossaint R, Falke KJ, Lopez F, Slama K, Pison U, Zapol WM (1993) Inhaled nitric oxide in adult respiratory distress syndrome. N Engl J Med 328:399–405
33. Wagner PD, Saltzman HA, West JB (1974) Measurement of continuous distributions of ventilation-perfusion ratios: Theory. J Appl Physiol 36:588–599
34. Dupuy PM, Shore SA, Drazen JM, Frostell C, Hill WA, Zapol WM (1992) Bronchodilator action of inhaled nitric oxide in guinea pigs. J Clin Invest 90:421–428
35. Roberts JD, Polander DM, Lang P, Zapol WM (1992) Inhaled nitric oxide in persistent pulmonary hypertension of the newborn. Lancet 340:818–819
36. Kinsella JP, Neish SR, Shaffer E, Abman SH (1992) Low-dose inhalational nitric oxide in persistent pulmonary hypertension of the newborn. Lancet 340:819–820
37. Levin DL, Heymann MA, Kitterman JA, et al (1976) Persistent pulmonary hypertension of the newborn infant. J Pediatr 89:626–630
38. Dhainaut JF, Brunet F (1990) Right ventricular performance in adult respiratory distress syndrome. Eur Respir J 3(Suppl 11):490–495
39. Radermacher P, Santak B, Wüst HJ, Tarnow J, Falke KJ (1990) Prostacyclin and right ventricular function in patients with pulmonary hypertension associated with ARDS. Intensive Care Med 16:227–232
40. Sibbald WJ, Driedger AA (1983) Right ventricular function in acute disease states: Pathophysiologic considerations. Crit Care Med 11:339–345
41. Brunet F, Dhainaut JF, Devaux JY, Huyghebaert MF, Villenant D, Monsallier JF (1988) Right ventricular performance in patients with acute respiratory failure. Intensive Care Med 14:474–477
42. Eddy AC, Rice CL (1989) The right ventricle: An emerging concern in the multiply injured patient. J Crit Care 4:58–66
43. Sibbald WJ, Driedger AA, Cunningham DG, Cheung H (1986) Right and left ventricular performance in acute hypoxemic respiratory failure. Crit Care Med 14: 858–857
44. Dhainaut JF, Lanore JJ, De Gournay JM, et al (1988) Right ventricular dysfunction in patients with septic shock. Intensive Care Med 14:488–491
45. Vincent JL, Reuse C, Frank N, Contempre B, Kahn RJ (1989) Right ventricular dysfunction in septic shock: Assessment by measurements of right ventricular ejection fraction using the thermodilution technique. Acta Anaesthesiol Scand 33:34–38
46. Vincent JL, Reuse C, Kahn RJ (1988) Effect on right ventricular function of a change from dopamine to dobutamine in critically ill patients. Crit Care Med 16:659–662
47. Wysocki M, Vignon P, Roupie E, et al (1993) Improvement in right ventricular function with inhaled nitric oxide in patients with the adult respiratory distress syndrome (ARDS) and permissive hypercapnia. Am Rev Respir Dis 147:A350 (Abst)
48. Gerlach H, Rossaint R, Pappert D, Falke KJ (1993) Time-course and dose-response of nitric oxide inhalation for systemic oxygenation and pulmonary hypertension in patients with adult respiratory distress syndrome. Eur J Clin Invest 23:499–502
49. Roberts JD, Chen TY, Kawai N, et al (1993) Inhaled nitric oxide reverses pulmonary vasoconstriction in the hypoxic and acidotic newborn. Circ Res 72:246–254
50. Gerlach H, Pappert D, Lewandowski K, Rossaint R, Falke KJ (1993) Long-term inhalation with evaluated low doses of nitric oxide for selective improvement of oxygenation in patients with adult respiratory distress syndrome. Intensive Care Med 19: 443–449
51. Villar J, Blazquez MA, Lubilio S, Quintana J, Manzano JL (1989) Pulmonary hypertension in acute respiratory failure. Crit Care Med 17:523–526
52. Dinh-Xuan AT, Higenbottam TW, Clelland CA, et al (1991) Impairment of endothelium-dependent pulmonary-artery relaxation in chronic obstructive lung disease. N Engl J Med 324:1539–1547
53. Stamler JS, Singel DJ, Loscalzo J (1992) Biochemistry of nitric oxide and its redox-activated forms. Science 258:1898–1902

54. Radi R, Beckman JS, Bush KM, Freeman BA (1991) Peroxynitrite oxidation of sulfhydryls. J Biol Chem 266:4244–4250
55. Keilin D, Hartree EF (1937) Reaction of nitric oxide with haemoglobin and methaemoglobin. Nature 139:548–551
56. Clutton-Brock J (1967) Two cases of poisoning by contamination of nitrous oxide with higher oxides of nitrogen during anaesthesia. Br J Anaesth 39:388–392
57. Greenbaum R, Bay J, Hargreaves MD, et al (1967) Effects of higher oxides of nitrogen on the anaesthetized dog. Br J Anaesth 39:393–404
58. Austin AT (1967) The chemistry of the higher oxides of nitrogen as related to the manufacture, storage and administration of nitrous oxide. Br J Anaesth 39:345–350
59. Fontijn A, Sabadell AJ, Ronco RJ (1970) Homogenous chemiluminescent measurement of nitric oxide with ozone. Anal Chem 42:575–579
60. Thomas HV, Mueller PK, Lyman RL (1968) Lipid peroxidation of lung lipids in rats exposed to nitrogen dioxide. Science 159:532–534
61. Rasmussen TR, Kjaergaard SK, Tarp U, Pedersen OF (1992) Delayed effects of NO_2 exposure on alveolar permeability and glutathione peroxidase in healthy humans. Am Rev Respir Dis 146:654–659
62. Norman V, Keith CH (1965) Nitrogen oxides in tobacco smoke. Nature 205:915–916
63. Higenbottam T, Borland C (1987) NO yields of contemporary UK, US and French cigarettes. Int J Epidemiol 16:31–34
64. Centers for Disease Control (1988) Recommendations for occupational safety and health standard. MMWN 37 (Suppl):5–7
65. Vallance P, Moncada S (1993) The role of endogenous nitric oxide in septic shock. New Horizons 1:77–86
66. Skwarski KM, Gorecka D, Sliwinski P, Hogg J, MacNee W (1993) The effects of cigarette smoking on pulmonary hemodynamics. Chest 103:1166–1172
67. Gustafsson LE, Leone AM, Persson MG (1991) Endogenous nitric oxide is present in the exhaled air of rabbits, guinea pigs and humans. Biochem Biophys Res Commun 181:852–857
68. Gerlach H, Rossaint R, Pappert D, Knorr M, Falke KF (1994) Autoinhalation of nitric oxide after endogenous synthesis in nasopharynx. Lancet 343:518–519
69. Cremona G, Dinh-Xuan AT, Higenbottam TW (1991) Endothelium-derived relaxing factor and the pulmonary circulation. Lung 169:185–202
70. Benzing A, Geiger K (1994) Inhaled nitric oxide lowers pulmonary capillary pressure and changes longitudinal distribution of pulmonary vascular resistance in patients with acute lung injury. Acta Anaesthesiol Scand 38:640–645
71. Rogers NE, Ignarro LJ (1992) Constitutive nitric oxide synthase from cerebellum is reversibly inhibited by nitric oxide formed from L-arginine. Biochem Biophys Res Commun 189:242–249
72. Assreuy J, Cunha FQ, Liew FY, Moncada S (1993) Feedback inhibition of nitric oxide synthase activity by nitric oxide. Br J Pharmacol 108:833–837
73. Hogman M, Frostell C, Arnberg H, Sandhagen B, Hedenstierna G (1994) Prolonged bleeding time during nitric oxide inhalation in the rabbit. Acta Physiol Scand 151: 125–129
74. Hogman M, Frostell C, Arnberg H, Hedenstierna G (1993) Bleeding time prolongation and NO inhalation . Lancet 341:1664–1665
75. Petros AJ, Cox PB, Bohn D (1992) Simple method for monitoring inhaled nitric oxide. Lancet 340:1167

Inhaled Nitric Oxide: The Light and Shadow of a Therapeutic Breakthrough

A. T. Dinh-Xuan, F. Brunet, and J. F. Dhainaut

Introduction

More than a decade ago, the existence of an extremely labile and potent endogenous vasodilator synthesized by the endothelium, termed endothelium-derived relaxing factor (EDRF), was established by Furchgott and Zawadzki [1]. The nature of EDRF since remained elusive until experimental results from several laboratories identified it as the free radical nitric oxide (NO) [2, 3]. Because it has a single unpaired electron, NO is also considered to be a free radical [4]. Molecular targets of NO are diverse, including heme proteins, non-heme iron-sulfur enzymes, DNA, and reactive oxygen species such as superoxide anion [5, 6]. Depending on its nature, the target molecule can either be activated (e.g. the heme-protein soluble guanylate cyclase) or inhibited (e.g. the non-heme iron protein ferritin) as a result of reacting with NO. In vascular smooth muscle, the molecular target of NO is the soluble enzyme guanylate cyclase [7]. Stimulation of the latter by NO increases the level of the second messenger cyclic guanosine monophosphate (cGMP) within vascular smooth muscle, thereby causing vasorelaxation [8]. The biochemistry of NO synthesis is remarquably simple, as to the nature of its precursors. The nitrogen atom of NO is derived from the N-guanidino terminal of the amino acid, L-arginine, whereas the oxygen atom is provided by molecular oxygen [7]. NO is synthesized from these two precursors by a newly discovered family of enzymes, the NO synthases (NOS). The complementary DNA for various isoforms of the NOS family have been recently cloned, and their amino acid primary structure sequenced [9]. There are two major subgroups of NOS isoforms, the constitutive and the inducible one. Endothelial NOS are predominantly constitutive, and most certainly play a key role in the modulation of systemic [10] and pulmonary [11] vascular tone. One of the inducible NOS isoforms probably mediates the cytotoxic activity of activated macrophages against numerous pathogens and intracellular microorganisms [12]. Another inducible NOS isoform, which accounts for pathological synthesis of large amounts of NO, is thought to be the major cause of the refractory hypotension seen in human septic shock [13]. The activity of these different NOS isoforms is highly regulated, requiring the presence of specific cofactors, including the calcium/calmodulin complex and reduced nicotinamide adenine dinucleotide phosphate (NADPH) [7]. The synthesis of NO is stereospecifically inhibited by various L-arginine analogs, which act as competitive inhibitors of both the constitutive and inducible NOS isoforms [7].

Physiological Role of NO in the Human Pulmonary Circulation

As in other mammalian species, pulmonary endothelium-dependent relaxation mediated by NO is also present in humans [14]. *In vivo* studies suggest that endogenous NO may have physiological importance in the modulation of human pulmonary circulation [11]. Indeed, endogenous NO is present in the exhaled air from spontaneously breathing healthy subjects [15, 16]. Furthermore, significant increases in NO output are observed in conditions which might alter pulmonary and bronchial vasoreactivity, such as exertion and hypoxia [16]. Data from studies using isolated vascular rings or perfused lungs also strongly suggest that endothelium-derived NO probably modulates pulmonary vasoreactivity in normoxia [17], and during acute alveolar hypoxia [18]. Although some controversies still exist as to its rate of production during chronic hypoxia [19, 20], endogenous NO probably has a pivotal role in the modulation of pulmonary vascular tone in health and pulmonary vascular disease [11, 21]. Indeed, most authors agree that NO is probably the major paracrine mediator that can rapidly adapt its rate of synthesis and/or release in response to acute increases in pulmonary vascular tone. Hence, a possible physiological role for NO would be to protect the pulmonary vasculature against disproportionate vasoconstriction that might result from various chemical and physical stimuli [11, 21].

Inhaled NO

Pulmonary hypertension may complicate various chronic and acute pulmonary disorders, including ARDS. The treatment of pulmonary hypertension, irrespective of its etiology, has always represented a formidable challenge for physicians, the main difficulties stemming from the fact that no pulmonary vasodilator is actually devoid of systemic side-effects. The "ideal" pulmonary vasodilator must therefore be selective, having activity restricted to the pulmonary circulation. Knowing that NO is rapidly inactivated by hemoglobin [22], it has been hypothesized that NO, given by inhalation, could be such a selective pulmonary vasodilator [23]. After entering the airways, inhaled NO primarily reaches the pulmonary vascular smooth muscle through diffusion from alveolar spaces, thereby causing pulmonary vasodilatation (Fig. 1). Once it reaches the luminal site of pulmonary endothelial cells, NO is rapidly bound to – and therefore inactivated by – circulating hemoglobin (Fig. 1). Thus, no downstream vasodilatory effect is likely to occur with inhaled NO.

However, the recent suggestion that NO could be carried away by serum albumin, which subsequently releases NO downstream, makes systemic hypotension a theoretical possibility [24]. However, to the best of our knowledge, such an effect has not been reported in humans.

Inhaled NO has been successfully applied in animals to reverse hypoxic and thromboxane-induced pulmonary vasoconstriction [25]. In humans, the short-term inhalation of NO also produces significant and selective pulmonary vasodilatory effects in newborn infants with persistent pulmonary hypertension [26–28] or in children with severe hypoxemic respiratory failure [29], and in adults with pulmonary hypertension, either primary [30], or secondary to chronic obstructive pulmonary disease (COPD) [31] and ARDS [32–49] (Table 1). Furthermore, when given by inhalation, NO has a unique activity to improve gas exchange, hence arterial oxygenation in both adults with ARDS (Table 1) and newborn babies with refractory hypoxemia [26–28]. Thus, unlike infused prostacyclin, which dilates pulmonary vessels of ventilated as well as non-ventilated lung units (Fig. 2A), inhaled NO preferentially induces vasodilatation in the former (Fig. 2B). As a result, ventilation/perfusion mismatching is reduced as shown by a decrease in intrapulmonary shunting (Q_S/Q_T); and gas exchange is improved as shown by an increase in PaO_2/FiO_2 (Fig. 2B). Such beneficial effects of inhaled NO have been convincingly further documented in a recent study using echocardiography in newborn infants with severe hypoxemia [28]. In these babies with respiratory distress syndrome, hypoxemia may result from right-to-left extrapulmonary shunt or intrapulmonary shunt [50]. Extrapulmonary shunt through ductus arteriosus or foramen ovale, or both, is the result of increased pulmonary vascular resistance (PVR) [50]. Inhaled NO reduces intrapulmonary shunt by improving ventilation/perfusion matching [51] and reduces extrapulmonary shunt by decreasing PVR. Some newborn babies without significant extrapulmonary shunting may benefit from inhaled NO by improvement of ventilation/perfusion matching rather than by a decrease in PVR [28]. Nevertheless, careful Doppler echocardiographic assessment could help to predict the likelihood of beneficial effects of inhaled NO in neonatal hypoxemia. NO being more effective in newborn infants with extrapulmonary shunting than in those without [28].

An elegant way to improve arterial oxygenation is to combine the use of inhaled NO with infused almitrine [47–49], a known pulmonary vasoconstrictor, which has already been shown to improve ventilation/perfusion matching in patients with COPD [52]. The resulting pulmonary vasoconstriction, which predominates in the shunting vascular bed induced by almitrine will further reduce Q_S/Q_T, hence improving arterial oxygenation (Fig. 2C). It is possible that the inhaled route of administration, as much as NO itself, is instrumental in obtaining such beneficial effects on arterial oxygenation, as improvements of a similar magnitude are observed in ARDS using aerosolized prostacyclin [53]. Together with the marked increase in arterial oxygenation, a slight but significant decrease in partial pressure of arterial carbon dioxide ($PaCO_2$) is seen with inhaled NO in patients with ARDS [42]. It

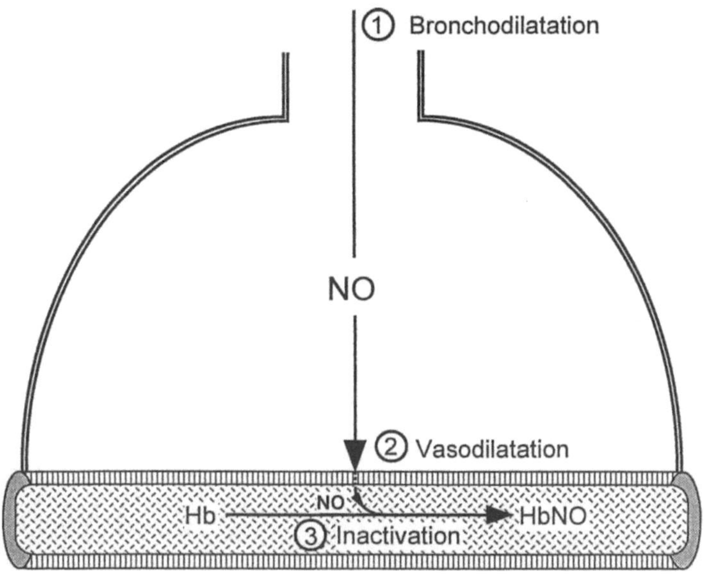

Fig. 1. Sites of action of inhaled NO.
When given by inhalation, NO first dilates the respiratory tract ①, then relaxes the pulmonary vasculature ②. NO diffusing from alveolar spaces to the blood stream is rapidly bound to hemoglobin (Hb). The product HbNO is devoid of vasodilating effects ③, hence the absence of systemic hypotension

is likely that the bronchodilatory effects of NO [54] may account for the decrease in $PaCO_2$ by opening up parts of the lungs which were previously poorly ventilated. This contention has been recently supported by the demonstration that in patients with stable COPD, alveolar ventilation may increase with inhaled NO in some occasions [55].

Effects of NO on Right Ventricular Function

Inhaled NO also improves right ventricular function in patients with ARDS [40–42] (Table 1). Indeed, pulmonary hypertension may, in some occasions, cause right ventricular (RV) dysfunction [56, 57], which could be a factor limiting survival in ARDS [58]. Reducing the pulmonary hypertension by infusing vasodilators decreases RV loading conditions and therefore improves RV function [59]. However, pulmonary hypertension reaching a level which is severe enough to compromise RV function seldom occurs in ARDS. Nevertheless, in the occasional patient in whom pulmonary hypertension and resulting RV failure are a clinically important problem, it would be impor-

Table 1. Clinical studies with inhaled NO in ARDS

Authors (years)	Ref.	n	NO (ppm)	Duration	PaO_2	mPAP	RVEF	Q_S/Q_T
Inhaled NO								
Rossaint et al. (1993)	[32]	10	5 to 36	40 min to 53 days	↑	↓	?	↓
Gerlach et al. (1993)	[33]	3	0.06 to 0.23	9 to 13 days	↑	±	?	?
Gerlach et al. (1993)	[34]	12	0.01 to 100	15 min	↑	↓	?	?
Payen et al. (1993)	[35]	15	12 to 18	1 h	↑	↓	?	↓
Bigatello et al. (1994)	[36]	7	2 to 20	2 to 27 days	↑	↓	?	↓
Puybasset et al. (1994)	[37]	6	0.1 to 5	15 to 20 min	↑	↓	?	↓
Puybasset et al. (1994)	[38]	11	2	30 min	↑	↓	?	↓
Levy et al. (1994)	[39]	13	5	?	↑	↓	?	↓
Grover et al. (1993)	[40]	10	5 to 40	30 min	↑	↓	?	↓
Wysocki et al. (1993)	[41]	4	10 to 20	1 h	↑	↓	↑	?
Fierobe et al. (1994)	[42]	13	5	15 min	↑	↓	↑	↓
Ricou & Suter (1993)	[43]	5	15 to 45	?	±	±	?	?
Mira et al. (1994)	[44]	6	15	30	±	±	?	?
Monchi et al. (1994)	[45]	11	15	30	±	±	?	?
Roupie et al. (1994)	[46]	17	10	?	±	±	?	±
Inhaled NO + almitrine i.v.								
Payen et al. (1993)	[47]	2	18	1 h	↑↑	↓	?	↓↓
Wysocki et al. (1994)	[48]	17	5 to 10	30 min	↑↑	↓	?	?
Page et al. (1994)	[49]	7	20	45 min	↑↑	±	?	?

n: number of patients; PaO_2: partial pressure of arterial oxygen; mPAP: mean pulmonary artery pressure; RVEF: right ventricular ejection fraction; Q_S/Q_T: intrapulmonary shunting tant to attempt to improve RV function by reducing the RV afterload. The effect of inhaled NO on RV function was studied in a series of 13 consecutive patients with ARDS [42]. RV function was assessed by thermodilution technique using a pulmonary artery catheter equipped with a rapid response thermistor before and 15 min after initiation of inhalation of NO (5 ppm). We found, with NO inhalation, that mean pulmonary artery pressure fell and concomitantly RV ejection fraction (RVEF) increased. This increase could not be solely explained by the presumed decrease in this component of RV afterload. Analysis of individual patient data reveals that substantial changes in RV end-systolic and end-diastolic volumes occurred in response to inhaled NO, owing to selective changes in RVEF with a constant cardiac output and heart rate. At baseline, despite pulmonary artery hypertension, the RV output of our patients with ARDS was maintained within normal range due to the use of the Frank-Starling mechanism, e.g. RV dilation. These findings are consistent with several reports in patients with ARDS [56, 57]. Only, if severe contractile depression of the RV coexists, the ability to use the Frank-Starling mechanism to maintain RV pump function seems potentially limited, as previously demonstrated in severe bacterial sepsis [57], viral myocarditis [57], or RV contusion [56]. With addition of inhaled NO, an unchanged stroke volume was associated with an increase in RVEF related to a greater decrease in RV end-systolic volume than the concurrent decrease in RV end-diastolic volume. In the absence of change in intrathoracic pressure, decrease in both pulmonary artery pressure and PVR suggests that RV afterload may decrease with NO inhalation. It seems reasonable to submit that the decrease in RV afterload is the main mechanism of the increase in RVEF with NO inhalation. As a consequence of decreased RV dilation, a decrease in pulmonary artery occluded pressure suggests that NO inhalation also improves left ventricular filling.

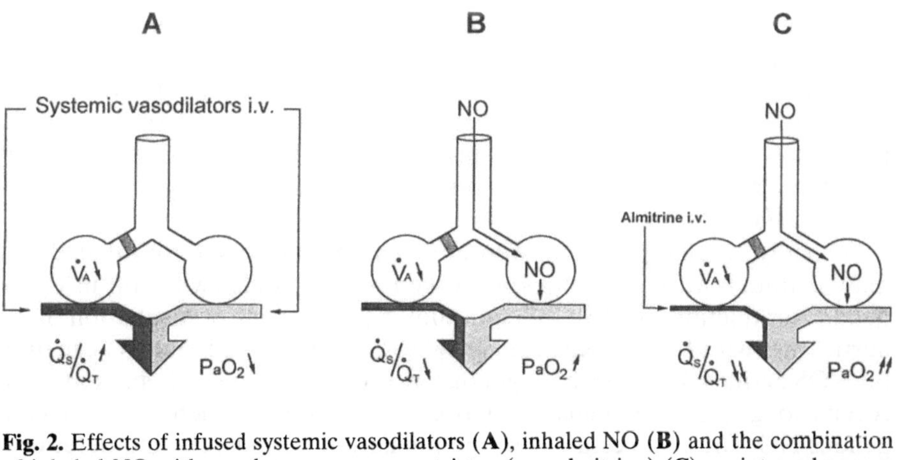

Fig. 2. Effects of infused systemic vasodilators (**A**), inhaled NO (**B**) and the combination of inhaled NO with a pulmonary vasoconstrictor (e.g. almitrine) (**C**) on intrapulmonary shunting (Q_S/Q_T) and partial pressure of arterial oxygen (PaO_2)

The Remaining Unsolved Questions

To date, three main difficulties remain and need to be addressed as to the use of inhaled NO in ARDS. Firstly, despite initial optimistic reports [32], it is now apparent that not all patients are likely to benefit from inhaled NO [43–46], even with doses higher than 40 ppm [44]. It is important to consider such a possibility, as this could imply that inhaled NO should only be used in selected subsets of ARDS patients in whom this gas would be the most likely to have beneficial effects. The effects of inhaled NO were assessed in 6 patients whose ARDS severity scores were between 3 and 3.5, and simplified acute physiologic scores between 6 and 23 (mean score 15 ± 2) at entry [44]. Hemodynamic variables were measured daily, at baseline and 30 min after starting NO inhalation. Individual responses to NO were deemed positive when more than 20% increase in PaO_2/FiO_2 ratios from baseline, or more than 10% decrease in mean PAP from baseline, or both, were seen. None of the 6 patients initially responded to NO (Fig. 3). Amongst these, 2 patients rapidly died 5 and 19 days after onset of the syndrome, respectively. The subsequent responses of the 4 remaining patients are shown in Fig. 3. However, 3 of these 4 patients subsequently died whilst on NO inhalation 48, 28 and 12 days after onset of the syndrome, respectively. Although compelling evidence suggests that inhaled NO could represent a major breakthrough in ARDS treatment, our observation highlights the fact that NO might sometimes fail to improve arterial oxygenation and to reduce pulmonary hypertension in some ARDS patients. The reason of this apparent lack of effect of inhaled NO is not well understood. The dominant abnormality in ARDS remains the presence of shunting area defined by the existence of perfused lung regions which are not ventilated. Positive end-expiratory pressure (PEEP) reduces shunt, but also creates areas of extremely high ventilation/perfusion ratio (>10). This local overexpansion of healthy lung units, probably by causing mechanical compression on the pulmonary vasculature, may limit the vasodilatory effect of inhaled NO. Alternatively, lack of responsiveness of the target enzyme – i.e. the soluble guanylate cyclase of pulmonary vascular smooth muscle – to respond to NO. However, direct evidence for such an abnormality remains to be established. As some ARDS patients who initially did not respond to inhaled NO could later become responsive, we recommend daily testing of NO, even in those who did not respond, in order to start long-term NO inhalation in patients who will subsequently favorably respond to NO.

Secondly, weaning patients who are treated with long-term inhaled NO may be fraught with difficulties, as exogenous NO could switch off its endogenous production [60–63]. Indeed, studying the effects of incubation of rat aortic rings with NO gas, Kiff et al. [64] have demonstrated that NO can inhibit NOS activity, hence its endogenous production. As a result, patients who receive long-term NO inhalation may eventually become dependent on exogenous NO. Thus, on cessation of the inhalation period, rebound deterioration of arterial oxygenation and reappearance of pulmonary hypertension

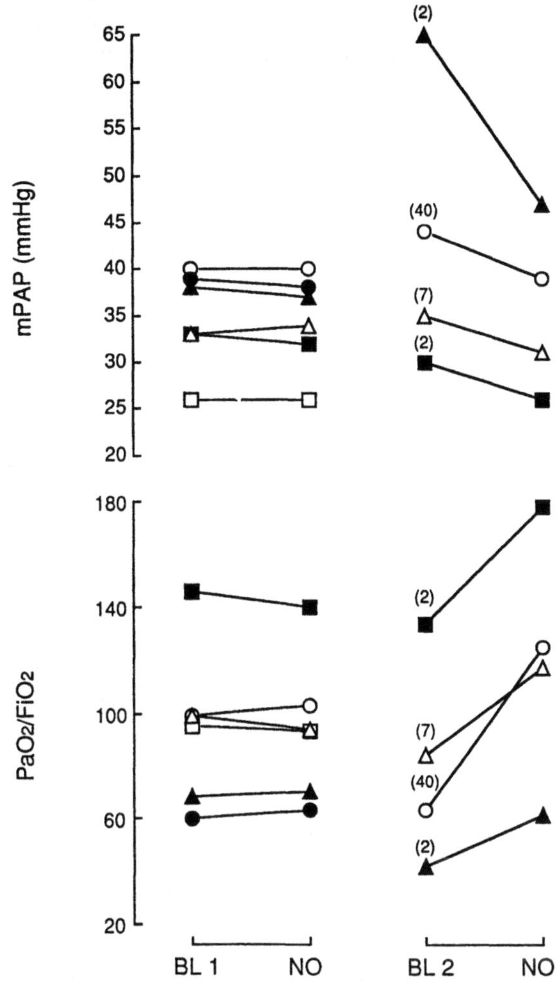

Fig. 3. Individual changes in mean pulmonary artery pressure (mPAP) and PaO_2/FiO_2 ratio after inhalation of NO on the first and the second tests, baseline 1 (BL 1) and baseline 2 (BL 2), respectively. Numbers in brackets represent numbers of days separating BL 2 from BL 1. (Adapted from [44])

may sometimes occur [33]. A progressive and stepwise weaning-off procedure is therefore recommended to avoid such rebound phenomenon [65].

Finally, potential harmful effects of this oxygen-derived free radical remains to be defined. Under physiological conditions, there are at least 3 redox forms of NO which, either by gaining or loosing an electron, are converted to nitroxyl anion (NO^-) or nitrosonium (NO^+), respectively [4]. A better understanding of the biochemistry of nitrogen oxides (NOx), which include NO, is mandatory in order to predict, and therefore avoid, the toxicity induced by these molecules. Indeed, experimental data suggest that both endogenous and exogenous NOx react readily with oxygen, superoxide, water, nucleotides, metalloproteins, thiols, amines and lipids [4]. Biochemical endproducts of these reactions may account for an array of toxic effects, including impaired mitochondrial respiration, lipid peroxidation, and mutagenesis [4]. In the meantime, the most practical way of reducing NO toxicity is reducing the concentration of NO administered to patients. In this perspective, a recent study is encouraging in reporting significant beneficial affects of NO on gas exchange with doses ranging from 60 to 230 parts per billion [33], i.e. several hundred times less than doses initially used in both ARDS [32] and primary pulmonary hypertension [30].

Conclusion

Inhaled NO undoubtedly has made a recent, and dramatic breakthrough in the way in which patients with ARDS are managed; at least when considering its specific and beneficial effects on pulmonary hemodynamics and gas exchange. However, as with other therapeutic advances, inhaled NO still awaits to undergo large, multicenter, controlled trials before we can determine whether the future management of ARDS will rely on the use of a gas which was not long ago regarded by many physicians as an atmosphere pollutant [66]. Furthermore, new insights are needed to address the as yet unsolved questions, among which the last, but not least, relates to acute and chronic toxic effects of this highly reactive oxygen-derived free radical.

References

1. Furchgott RF, Zawadzki JV (1980) The obligatory role of endothelial cells in the relaxation of arterial smooth muscle by acetylcholine. Nature 288:373–376
2. Ignarro LJ, Buga GM, Wood KS, Byrns RE, Chaudhuri G (1987) Endothelium-derived relaxing factor produced and released from artery and vein is nitric oxide. Proc Natl Acad Sci USA 84:9265–9269
3. Palmer RMJ, Ferrige AG, Moncada S (1987) Nitric oxide release accounts for the biological activity of endothelium-derived relaxing factor. Nature 327:524–526
4. Stamler JS, Singel DJ, Loscalzo J (1992) Biochemistry of nitric oxide and its redox-activated forms. Science 258:1898–1902
5. Nathan CF (1992) Nitric oxide as a secretory product of mammalian cells. FASEB J 6: 3051–3064
6. Henry Y, Lepoivre M, Drapier JC, Ducrocq C, Boucher JL, Guissani A (1993) EPR characterization of molecular targets for NO in mammalian cells and organelles. FASEB J 7:1124–1134
7. Moncada S, Palmer RMJ, Higgs EA (1991) Nitric oxide: Physiology, pathophysiology, and pharmacology. Pharmacol Rev 43:109–142
8. Murad F (1986) Cyclic guanosine monophosphate as a mediator of vasodilation. J Clin Invest 78:1–5
9. Förstermann U, Schmidt HH, Pollock JS, et al (1991) Isoforms of nitric oxide synthase: Characterization and purification from different cell types. Biochem Pharmacol 10: 1849–1857
10. Moncada S, Higgs EA (1993) The L-arginine-nitric oxide pathway. N Engl J Med 329: 2002–2012

11. Dinh-Xuan AT (1992) Endothelial modulation of pulmonary vascular tone. Eur Respir J 5:757–762
12. Nathan CF, Hibbs JB Jr (1991) Role of nitric oxide synthesis in macrophage antimicrobial activity. Curr Opin Immunol 3:65–70
13. Petros A, Bennett D, Vallance P (1991) Effect of nitric oxide synthase inhibitors on hypotension in patients with septic shock. Lancet 338:1557–1558
14. Dinh-Xuan AT, Higenbottam TW, Clelland CA, Pepke-Zaba J, Wells FC, Wallwork J (1990) Acetylcholine and adenosine diphosphate cause endothelium-dependent relaxation of isolated human pulmonary arteries. Eur Respir J 3:633–638
15. Gustafsson LE, Leone AM, Persson MG, Wiklund NP, Moncada S (1991) Endogenous nitric oxide is present in the exhaled air of rabbits, guinea pigs and humans. Biochem Biophys Res Commun 181:852–857
16. Garnier P, Strâmbu I, Dessanges JF, et al. (1994) Endogenous nitric oxide in expired air increases on exercise and isocapnic hyperventilation in normal subjects. Am J Respir Crit Care Med 149:A778 (Abst)
17. Dinh-Xuan AT, Higenbottam TW, Clelland CA, et al. (1991) Impairment of endothelium-dependent pulmonary-artery relaxation in chronic obstructive lung disease. N Engl J Med 324:1539–1547
18. Liu SF, Crawley DE, Barnes PJ, Evans TW (1991) Endothelium-derived relaxing factor inhibits hypoxic pulmonary vasoconstriction in rats. Am Rev Respir Dis 143:32–37
19. Adnot S, Raffestin B, Eddhahibi S, Braquet P, Chabrier PE (1991) Loss of endothelium-dependent relaxant activity in the pulmonary circulation of rats exposed to chronic hypoxia. J Clin Invest 87:155–162
20. Isaacson TC, Hampl V, Weir EK, Nelson DP, Archer SL (1994) Increased endothelium-derived NO in hypertensive pulmonary circulation of chronically hypoxic rats. J Appl Physiol 76:933–940
21. Dinh-Xuan AT (1993) Disorders of endothelium-dependent relaxation in pulmonary disease. Circulation 87 (Suppl V):V81–V87
22. Martin W, Smith JA, White DG (1986) The mechanisms by which haemoglobin inhibits the relaxation of rabbit aorta induced by nitrovasodilators, nitric oxide or bovine retractor penis inhibitory factor. Br J Pharmacol 89:563–571
23. Higenbottam TW, Pepke-Zaba J, Scott JP, Woolman P, Coutts C, Wallwork J (1988) Inhaled endothelium-derived relaxing factor in primary pulmonary hypertension. Am Rev Respir Dis 137 (Suppl):107 (Abst)
24. Stamler JS, Jaraki O, Osborne J, et al (1992) Nitric oxide circulates in mammalian plasma primarily as an S-nitroso adduct of serum albumin. Proc Natl Acad Sci USA 89:7674–7677
25. Frostell CG, Fratacci MD, Wain JC Jr, Jones R, Zapol WM (1991) Inhaled nitric oxide: A selective pulmonary vasodilator reversing hypoxic pulmonary vasoconstriction. Circulation 83:2038–2047
26. Roberts JD Jr, Polaner DM, Lang P, Zapol WM (1992) Inhaled nitric oxide in persistent pulmonary hypertension of the newborn. Lancet 340:818–819
27. Kinsella JP, Neish SR, Shaffer E, Abman SH (1992) Low-dose inhalational nitric oxide in persistent pulmonaly hypertension of the newborn. Lancet 340:819–820
28. Rozé JC, Storme L, Zupan V, Morville P, Dinh-Xuan AT, Mercier JC (1994) Echocardiographic investigation of inhaled nitric oxide in newborn babies with severe hypoxaemia. Lancet 344:303–305
29. Abman SH, Griebel JL, Parker DK, Schmidt JM, Swanton D, Kinsella JP (1994) Acute effects of inhaled nitric oxide in children with severe hypoxemic respiratory failure. J Pediatr 124:881–888
30. Pepke-Zaba J, Higenbottam TW, Dinh-Xuan AT, Stone D, Wallwork J (1991) Inhaled nitric oxide as a cause of selective pulmonary vasodilatation in pulmonary hypertension. Lancet 338:1173–1174
31. Adnot S, Kouyoumdjian C, Defouilloy C, et al. (1993) Hemodynamic and gas exchange responses to infusion of acetylcholine and inhalation of nitric oxide in patients with chronic obstructive lung disease and pulmonary hypertension. Am Rev Respir Dis 148:310–316

32. Rossaint R, Falke K J, López FA, Slama K, Pison U, Zapol WM (1993) Inhaled nitric oxide for the adult respiratory distress syndrome. N Engl J Med 328:399–405
33. Gerlach H, Pappert D, Lewandowski K, Rossaint R, Falke KJ (1993) Long-term inhalation with evaluated low doses of nitric oxide for selective improvement of oxygenation in patients with adult respiratory distress syndrome. Intensive Care Med 19: 443–449
34. Gerlach H, Rossaint R, Pappert D, Falke KJ (1993) Time-course and dose-response of nitric oxide inhalation for systemic oxygenation and pulmonary hypertension in patients with adult respiratory distress syndrome. Eur J Clin Invest 23:499–502
35. Payen DM, Gatecel C, Guinard N (1993) Inhalation of low dose of nitric oxide and IV L-arginine in ARDS: Effect on pulmonary hemodynamic, gas exchange and NO metabolites. Am Rev Respir Dis 147 (suppl): A720 (Abst)
36. Bigatello LM, Hurford WE, Kacmarek RM, Roberts JD Jr, Zapol WM (1994) Prolonged inhalation of low concentrations of nitric oxide in patients with severe adult respiratory distress syndrome. Anesthesiology 80:761–770
37. Puybasset L, Rouby JJ, Mourgeon E, et al (1994) Inhaled nitric oxide in acute respiratory failure: Dose-response curves. Intensive Care Med 20:319–327
38. Puybasset L, Rouby JJ, Mourgeon E, et al (1994) Inhaled nitric oxide reverses the increase in pulmonary vascular resistance induced by permissive hypercapnia in patients with acute respiratory distress syndrome. Anesthesiology 80:1254–1267
39. Levy B, Bollaert PE, Bauer P, et al (1994) Early use of inhaled nitric oxide associated with therapeutic optimization in severe adult respiratory distress syndrome. Am J Respir Crit Care Med 149 (suppl): A425 (Abst)
40. Grover R, Smithies M, Bihari D (1993) A dose profile of the physiological effects of inhaled nitric oxode in acute lung injury. Am Rev Respir Dis 147 (suppl): A350 (Abst)
41. Wysocki M, Vignon P, Roupie E, et al (1993) Improvement in right ventricular function with inhaled nitric oxide in patients with the adult respiratory distress syndrome and permissive hypercapnia. Am Rev Respir Dis 147 (suppl): A350 (Abst)
42. Fierobe L, Brunet F, Dhainaut JF, et al (1995) Effect of inhaled nitric oxide on right ventricular function in adult respiratory distress syndrome. Am J Respir Crit Care Med 151 (In press)
43. Ricou B, Suter PM (1993) Variable effects of nitric oxide in ARDS patients. Am Rev Respir Dis 147 (suppl): A350 (Abst)
44. Mira JP, Monchi M, Brunet F, Fierobe L, Dhainaut JF, Dinh-Xuan AT (1994) Lack of efficacy of inhaled nitric oxide in ARDS. Intensive Care Med 20:532 (Letter)
45. Monchi M, Fierobe L, Mira JP, et al (1994) Predictive factors of inhaled nitric oxide inefficiency in ARDS patients. Am J Respir Crit Care Med 149 (suppl): A423 (Abst)
46. Roupie E, Wysocki M, Langeron O, et al (1994) Effects of inhaled nitric oxide on oxygenation and hemodynamic parameters in the acute respiratory distress syndrome: Responders versus non-responders. Am J Respir Crit Care Med 149 (suppl): A424 (Abst)
47. Payen DM, Gatecel C, Plaisance P (1993) Almitrine effect on nitric oxide inhalation in adult respiratory distress syndrome. Lancet 341:1664 (Letter)
48. Wysocki M, Delclaux C, Roupie E, et al (1994) Additive effects on gas exchange of inhaled nitric oxide and intravenous almitrine bimesylate in the adult respiratory distress syndrome. Intensive Care Med 20:254–259
49. Page Y, Bertrand M, Page D, et al (1994). Effects of inhaled nitric oxide in ARDS patients are variable and may be enhanced by almitrine. Am J Respir Crit Care Med 149 (suppl): A426 (Abst)
50. Walthers FJ, Benders M, Leighton JO (1992) Persistent pulmonary hypertension in premature neonates with severe respiratory distress syndrome. Pediatrics 90:899–904
51. Pison U, López FA, Heidelmeyer CF, Rossaint R, Falke KJ (1993) Inhaled nitric oxide reverses hypoxic pulmonary vasoconstriction without impairing gas exchange. J Appl Physiol 74:1287–1292
52. Mélot C, Naeije R, Rothschild T, Mertens P, Mols P, Hallemans R (1983) Improvement in ventilation-perfusion matching by almitrine in COPD. Chest 83:528–533

53. Walmrath D, Schneider T, Pilch J, Grimminger F, Seeger W (1993) Aerosolised prostacyclin in adult respiratory distress syndrome. Lancet 342:961–962
54. Högman M, Frostell CG, Hedenström H, Hedenstierna G (1993) Inhalation of nitric oxide modulates adult human bronchial tone. Am Rev Respir Dis 148:1474–1478
55. Moinard J, Manier G, Pillet O, Castaing Y (1994) Effect of inhaled nitric oxide on hemodynamics and V_A/Q inequalities in patients with chronic obstructive pulmonary disease. Am J Respir Crit Care Med 149:1482–1487
56. Sibbald WJ, Driedger AA, Myers ML, Short AIK, Wells GA (1983) Biventricular function in the adult respiratory distress syndrome. Chest 84:126–134
57. Brunet F, Dhainaut JF, Devaux JY, Huyghebaert MF, Villemant D, Monsallier JF (1988) Right ventricular performance in patients with acute respiratory failure. Intensive Care Med 14:474–477
58. Zapol WM, Snider MT (1977) Pulmonary hypertension in severe acute respiratory failure. N Engl J Med 296:476–480
59. Radermacher P, Huet Y, Pluskwa F, et al (1988) Comparison of ketanserin and sodium nitroprusside in patients with severe ARDS. Anesthesioloy 68:152–157
60. Bult H, de Meyer GRY, Jordaens F, Herman AG (1991) Chronic exposure to exogenous nitric oxide may suppress its endogenous release and efficacy. J Cardiovasc Pharmacol 17 (Suppl 3):S79–S82
61. Rogers NE, Ignarro LJ (1992) Constitutive nitric oxide synthase from cerebellum is reversibly inhibited by nitric oxide formed from L-arginine. Biochem Biophys Res Commun 189:242–249
62. Buga GM, Griscavage JM, Rogers NE, Ignarro LJ (1993) Negative feedback regulation of endothelial cell function by nitric oxide. Circ Res 73:808–812
63. Assreuy J, Cunha FQ, Liew FY, Moncada S (1993) Feedback Inhibition of nitric oxide synthase activity by nitric oxide. Br J Pharmacol 108:833–837
64. Kiff RJ, Moss DW, Moncada S (1994) Effect of nitric oxide gas on the generation of nitric oxide by isolated blood vessels: Implications for inhalation therapy. Br J Pharmacol 113:496–498
65. Zapol WM, Rimar S, Gillis N, Marletta M, Bosken CH (1994) Nitric oxide and the lung. Am J Respir Crit Care Med 149:1375–1380
66. Quinn AC, Vallance P (1993) Inhaled nitric oxide – from pollutant to patent. Eur J Clin Invest 23:445–447

Subject Index

Springer-Verlag and the Environment

We at Springer-Verlag firmly believe that an international science publisher has a special obligation to the environment, and our corporate policies consistently reflect this conviction.

We also expect our business partners – paper mills, printers, packaging manufacturers, etc. – to commit themselves to using environmentally friendly materials and production processes.

The paper in this book is made from low- or no-chlorine pulp and is acid free, in conformance with international standards for paper permanency.

MIX
Papier aus verantwortungsvollen Quellen
Paper from responsible sources
FSC® C105338

If you have any concerns about our products,
you can contact us on
ProductSafety@springernature.com

In case Publisher is established outside the EU,
the EU authorized representative is:
Springer Nature Customer Service Center GmbH
Europaplatz 3, 69115 Heidelberg, Germany

Printed by Libri Plureos GmbH
in Hamburg, Germany